Assessment of Climate Change over the Indian Region

R. Krishnan · J. Sanjay ·
Chellappan Gnanaseelan ·
Milind Mujumdar · Ashwini Kulkarni ·
Supriyo Chakraborty
Editors

Assessment of Climate Change over the Indian Region

A Report of the
Ministry of Earth Sciences (MoES),
Government of India

Editors
R. Krishnan
Centre for Climate Change Research
Indian Institute of Tropical Meteorology
(IITM-MoES)
Pune, India

J. Sanjay
Centre for Climate Change Research
Indian Institute of Tropical Meteorology
(IITM-MoES)
Pune, India

Chellappan Gnanaseelan
Short Term Climate Variability and Prediction
Indian Institute of Tropical Meteorology
(IITM-MoES)
Pune, India

Milind Mujumdar
Centre for Climate Change Research
Indian Institute of Tropical Meteorology
(IITM-MoES)
Pune, India

Ashwini Kulkarni
Short Term Climate Variability and Prediction
Indian Institute of Tropical Meteorology
(IITM-MoES)
Pune, India

Supriyo Chakraborty
Centre for Climate Change Research
Indian Institute of Tropical Meteorology
(IITM-MoES)
Pune, India

ISBN 978-981-15-4326-5   ISBN 978-981-15-4327-2   (eBook)
https://doi.org/10.1007/978-981-15-4327-2

This Springer imprint is published by the registered company Springer Nature Singapore Pte Ltd.
The registered company address is: 152 Beach Road, #21-01/04 Gateway East, Singapore 189721, Singapore

Foreword by M. Rajeevan

Earth's climate is changing, primarily as a result of human activities. We are adding green-house gases to the atmosphere that is driving the climate to a warmer state. The warming is evident in the long-term observations from the top of the atmosphere to the depths of the oceans. Climate change presents growing challenges to human health and safety and quality of life and economy of the country. While climate change is global, changes in climate are not expected to be uniform across the planet.

Greenhouse gas emissions from human activities will continue to affect Earth's climate for decades and even centuries. Future climate change is expected to further disrupt many areas of life. The impacts are already evident and are expected to become increasingly disruptive across the globe. The severity of future climate change impacts will depend largely on actions taken to reduce greenhouse gas emissions and to adapt to the changes that will occur.

For policy makers, it is important to have a clear comprehensive view on the possible future climate change projections. Climate models are often used to project how our world will change under future scenario. Climate models have proven remarkably accurate in simulating the climate change we have experienced to date. Global climate models project a continuation of human-induced climate change during the twenty-first century and beyond. Climate change projections, however, may have large uncertainties. The largest uncertainty in climate change projections is the level of greenhouse gas emissions in future.

The need for a comprehensive assessment report on climate change was felt for a long time. This is the first-ever climate change assessment report for India. The report consists of 12 chapters describing the observed changes and future projections of precipitation, temperature, monsoon, drought, sea level, tropical cyclones, and extreme weather events, etc. This report will be very useful for policy makers, researchers, social scientists, economists, and students. I congratulate the editors and contributors for bringing out this valuable national climate change assessment report.

M. Rajeevan
Secretary
Ministry of Earth Sciences
New Delhi, India

Foreword by J. Srinivasan

In the twenty-first century, the impact of human-induced climate change will pose a great challenge to humanity. The increase in global mean temperature can be directly linked to the increase in greenhouse gases like carbon dioxide and methane. The causes of local climate change is, however, much more complex. The local climate change is influenced not only by the increase in the greenhouse gases but also by the increase in air pollution and the local changes in land-use pattern. In order to understand local climate change, we need more observations and a detailed analysis of the factors that lead to local changes in climate. India is a vast country with many climate zones, and hence, local climate change and their causes can be quite complex. India is concerned about the impact of climate change on the vulnerable population in both urban and rural areas.

The present book is the first attempt to document climate changes in different parts of India. Chapter 1 discusses the global climate change, regional climate change, Indian monsoon variability, and the development of the first earth system model in India. Chapter 2 focuses on the changes in mean temperature and the extremes. The evolution of these parameters in the twenty-first century is explored in detail. The variability of precipitation is of great concern in India in view of its impact of agriculture. Chapter 3 examines the past changes in precipitation, the impact of climate change on precipitation, and the projections of changes in precipitation in future. Chapter 4 presents the variations in greenhouse emissions and concentrations in the twentieth century and their projected change in the twenty-first century. The increase in air pollution in India has been a cause for major concern. Chapter 5 provides a detailed account of the increase in aerosols in India and their radiative impact. The frequency of droughts and floods is discussed in Chap. 6. The changes in synoptic scale events and extreme storms are covered in Chaps. 7 and 8. India has a long coastline, and hence, the impact of sea-level rise is a matter of concern to policy makers. The impact of ocean warming and glacier melting on regional sea-level rise is highlighted in Chap. 9. Chapter 10 deals with the impact of warming of the Indian Ocean, while Chap. 11 looks at climate change in the Himalayas. Chapter 12 discusses the benefits of mitigating climate change through the reduction in emission of greenhouse gases and aerosols.

This book will be a valuable source material since it contains important data and hence will become a reference material in future.

J. Srinivasan
Divecha Centre for Climate Change
Indian Institute of Science
Bengaluru, India

Preface

Human activities since the nineteenth century have contributed to substantial increases in the atmospheric concentrations of heat-trapping greenhouse gases (GHG), such as carbon dioxide (CO_2), methane (CH_4), nitrous oxide (N_2O), and fluorinated gases. Carbon dioxide is the main long-lived GHG in the atmosphere related to human activities. Burning of fossil fuels, deforestation, and land use changes, among other human (anthropogenic) activities, have led to a rapid increase of atmospheric CO_2 levels from 280 parts per million during 1850 to more than 416 parts per million in February 2020. The series of assessment reports of the United Nations Intergovernmental Panel on Climate Change (IPCC) provides unequivocal evidence for the role of anthropogenic forcing in driving the observed warming of the Earth's surface by about 1°C during the last 150 years. Consequences of this warming have already manifested in several other global-scale changes such as melting glaciers, rising sea levels, changing precipitation patterns, and an increasing tendency of weather and climate extremes. These changes are projected to continue through the twenty-first century, as the GHG concentrations continue to rise.

Robust attributions and projections of regional-scale changes to anthropogenic forcing are inherently more complex than global-scale changes because of the strong internal variability at local scales. For example, the IPCC Fifth Assessment Report (AR5) reported large inter-model spread in the climate change response of the Indian monsoon precipitation, Indian Ocean regional sea-level rise, Himalayan snow cover, and other aspects of the regional climate system. In this context, this book presents a comprehensive assessment of climate change over the Indian region and its links to global climate change. This assessment report is based on peer-reviewed scientific publications, analyses of long-term observed climate records, paleoclimate reconstructions, reanalysis datasets and climate model projections from the Coupled Model Intercomparison Project (CMIP) and the COordinated Regional climate Downscaling EXperiment (CORDEX). This book is the first ever climate change report for India from the Ministry of Earth Sciences, Government of India, and its preparation was led and coordinated by the Centre for Climate Change Research (CCCR) at the Indian Institute of Tropical Meteorology (IITM), Pune.

The aim of this assessment report is to describe the physical science basis of regional climate change over the Indian subcontinent and adjoining areas. The first chapter briefly introduces global climate change, sets the regional context, and synthesizes the key points from the subsequent chapters. Apart from GHGs, emissions of anthropogenic aerosols over the Northern Hemisphere have substantially increased during the last few decades, and their impacts on the regional climate are also assessed. Chapters 2–11 of the report assess changes in several aspects of regional climate and their drivers, viz., temperature, precipitation, GHGs, atmospheric aerosols and trace gases, droughts and floods, synoptic systems, tropical cyclones and extreme storms, Indian Ocean warming and sea-level rise, and the Himalayan cryosphere. While impacts and policy lie beyond the scope of this report, Chapter 12 closes this report with a brief outline of the potential implications of climate change for the country's natural ecosystems, water resources, agriculture, infrastructure, environment, and public health, along with some policy-relevant messages towards realizing India's sustainable development goals by mitigating these risks.

This report also documents various aspects of the natural variability of the global and regional climate system, teleconnection mechanisms, and coupled feedback processes of the atmosphere–ocean–land–cryosphere system. A brief discussion on key knowledge gaps is included in Chapters 2–11. A salient feature of this report is the inclusion of introductory results based on the CMIP Phase 6 (CMIP6) projections of the IITM Earth System Model (IITM-ESM)—the first climate model from India, developed at the CCCR-IITM, that is contributing to the Sixth IPCC Assessment Report (i.e, IPCC AR6) to be released in 2021.

We hope that the material included in this regional climate change assessment report will benefit students, researchers, scientists and policy makers, and help in advancing public awareness of India's changing climate, and to inform adaptation and mitigation strategies.

Pune, India   R. Krishnan
June 2020   J. Sanjay
                                                        Chellappan Gnanaseelan
                                                             Milind Mujumdar
                                                             Ashwini Kulkarni
                                                          Supriyo Chakraborty

Acknowledgements

Expert Reviewers

M. Rajeevan, Ministry of Earth Sciences, Government of India, New Delhi, India
Mathew Collins, Centre for Engineering, Mathematics and Physical Sciences, University of Exeter, Exeter, UK
Dev Niyogi, Purdue University, West Lafayette, IN and University of Texas at Austin, Austin, TX, USA
J. Srinivasan, Indian Institute of Science, Bengaluru, India
Ravi Nanjundiah, Indian Institute of Tropical Meteorology (IITM-MoES), Pune, India
Donald Wuebbles, University of Illinois, Urbana, IL, USA
Raghu Murtugudde, University of Maryland, College Park, MD, USA

Science Communication Experts

TROP-ICSU Team (Rahul Chopra, Aparna Joshi, Anita Nagarajan, Megha Nivsarkar): TROP ICSU—Climate Education Project of the International Science Council at the Indian Institute of Science Education and Research (IISER), Pune, India
Rajeev Mehajan, Scientist 'G'/Advisor, Science and Engineering Research Board, Department of Science and Technology, Government of India, New Delhi, India
Abha Tewari, Independent Researcher (Formerly with Indian Air Force, National Health Systems Resource Centre, and Ministry of Environment Forest and Climate Change).

Design and Copy Editing

Sandip Ingle, Indian Institute of Tropical Meteorology (IITM-MoES), Pune, India
Abhay Singh Rajput, Indian Institute of Tropical Meteorology (IITM-MoES), Pune, India

Graphics Design

T. P. Sabin, Indian Institute of Tropical Meteorology (IITM-MoES), Pune, India
Jyoti Jadhav, Indian Institute of Tropical Meteorology (IITM-MoES), Pune, India
Mahesh Ramadoss, Indian Institute of Tropical Meteorology (IITM-MoES), Pune, India

Library and Logistics

Shompa Das, Indian Institute of Tropical Meteorology (IITM-MoES), Pune, India
Rohini Ovhal, Indian Institute of Tropical Meteorology (IITM-MoES), Pune, India
Parthasarathi Mukhopadhyay, Indian Institute of Tropical Meteorology (IITM-MoES), Pune, India
Keshav Barne, Indian Institute of Tropical Meteorology (IITM-MoES), Pune, India
Rajiv Khapale, Indian Institute of Tropical Meteorology (IITM-MoES), Pune, India

Yogita Kad, Indian Institute of Tropical Meteorology (IITM-MoES), Pune, India
Yogesh Pawar, Indian Institute of Tropical Meteorology (IITM-MoES), Pune, India
Shafi Sayyed, Indian Institute of Tropical Meteorology (IITM-MoES), Pune, India
Sandip Kulkarni, Indian Institute of Tropical Meteorology (IITM-MoES), Pune, India
Ajit Prasad, Indian Institute of Tropical Meteorology (IITM-MoES), Pune, India

Executive Summary

Observed Changes in Global Climate

The global average temperature[1] has risen by around 1°C since pre-industrial times. This magnitude and rate of warming cannot be explained by natural variations alone and must necessarily take into account changes due to human activities. Emissions of greenhouse gases (GHGs), aerosols and changes in land use and land cover (LULC) during the industrial period have substantially altered the atmospheric composition, and consequently the planetary energy balance, and are thus primarily responsible for the present-day climate change. Warming since the 1950s has already contributed to a significant increase in weather and climate extremes globally (e.g., heat waves, droughts, heavy precipitation, and severe cyclones), changes in precipitation and wind patterns (including shifts in the global monsoon systems), warming and acidification of the global oceans, melting of sea ice and glaciers, rising sea levels, and changes in marine and terrestrial ecosystems.

Projected Changes in Global Climate

Global climate models project a continuation of human-induced climate change during the twenty-first century and beyond. If the current GHG emission rates are sustained, the global average temperature is likely to rise by nearly 5°C, and possibly more, by the end of the twenty-first century. Even if all the commitments (called the "Nationally Determined Contributions") made under the 2015 Paris agreement are met, it is projected that global warming will exceed 3°C by the end of the century. However, temperature rise will not be uniform across the planet; some parts of the world will experience greater warming than the global average. Such large changes in temperature will greatly accelerate other changes that are already underway in the climate system, such as the changing patterns of rainfall and increasing temperature extremes.

Climate Change in India: Observed and Projected Changes

Temperature Rise Over India

India's average temperature has risen by around 0.7°C during 1901–2018. This rise in temperature is largely on account of GHG-induced warming, partially offset by forcing due to anthropogenic aerosols and changes in LULC.

---

[1]Unless otherwise specified, "temperature" refers to the sea surface temperature (SST) for oceanic areas and near surface air temperature over land areas.

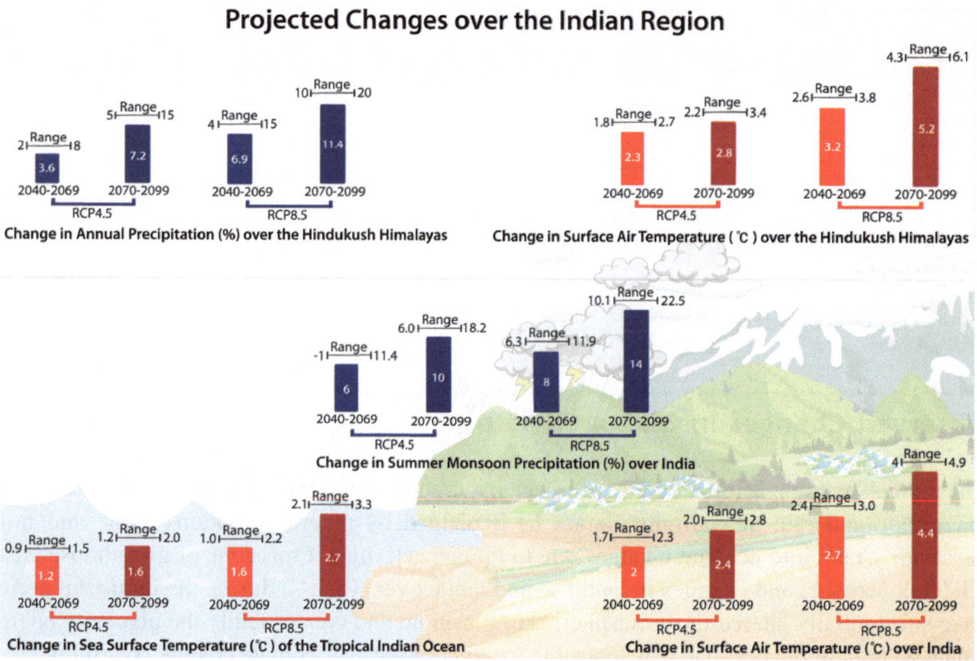

Projected Changes over the Indian Region

Fig. 1 Best estimate and range in climate model projections of future changes in 1. Surface air temperature over India (°C; bottom right panel), 2. Sea surface temperature of the tropical Indian Ocean (°C; bottom left panel), 3. Surface air temperature over the Hindu Kush Himalayas (°C; top right panel), 4. Summer monsoon precipitation over India (% change; centre panel), 5. Annual precipitation over the Hindu Kush Himalayas (% change; top left panel). All the changes are computed relative to their climatological average over the 30-year period 1976–2005. Projected changes are reported for the middle and end of the 21st century under the RCP4.5 and RCP8.5 scenarios (defined in Box 1). Details regarding the models and computations are discussed in the respective chapters

By the end of the twenty-first century[2], average temperature over India is projected to rise by approximately 4.4°C relative to the recent past (1976–2005 average[3]) (Fig. 1), under the RCP8.5 scenario (see Box 1).

Box 1: Description of future forcing scenarios

Projections by climate models of the Coupled Model Intercomparison Project Phase 5 (CMIP5) are based on multiple standardized forcing scenarios called Representative Concentration Pathways (RCPs). Each scenario is a time series of emissions and concentrations of the full suite of GHGs, aerosols, and chemically active gases, as well as LULC changes through the twenty-first century, characterized by the resulting Radiative Forcing* in the year 2100 (IPCC 2013). The two most commonly analyzed scenarios in this report are "RCP4.5" (an intermediate stabilization pathway that results in a Radiative Forcing of 4.5 W/m² in 2100) and "RCP8.5" (a high concentration pathway resulting in a Radiative Forcing of 8.5 W/m² in 2100).

*A measure of an imbalance in the Earth's energy budget owing to natural (e.g., volcanic eruptions) or human-induced (e.g., GHG from fossil fuel combustion) changes.

---

[2]Projections referring to the "middle of the 21st century" pertain to the climatological average over the period 2040–2069 and "end of the 21st century" to the climatological average over the period 2070–2099.

[3]Unless otherwise noted, projected changes are reported w.r.t. this baseline period throughout the report.

In the recent 30-year period (1986–2015), temperatures of the warmest day and the coldest night of the year have risen by about 0.63°C and 0.4°C, respectively. By the end of the twenty-first century, these temperatures are projected to rise by approximately 4.7°C and 5.5°C, respectively, relative to the corresponding temperatures in the recent past (1976–2005 average), under the RCP8.5 scenario.

By the end of the twenty-first century, the frequencies of occurrence of warm days and warm nights[4] are projected to increase by 55% and 70%, respectively, relative to the reference period 1976-2005, under the RCP8.5 scenario.

The frequency of summer (April–June) heat waves over India is projected to be 3 to 4 times higher by the end of the twenty-first century under the RCP8.5 scenario, as compared to the 1976–2005 baseline period. The average duration of heat wave events is also projected to approximately double, but with a substantial spread among models.

In response to the combined rise in surface temperature and humidity, amplification of heat stress is expected across India, particularly over the Indo-Gangetic and Indus river basins.

Indian Ocean Warming

Sea surface temperature (SST) of the tropical Indian Ocean has risen by 1°C on average during 1951–2015, markedly higher than the global average SST warming of 0.7°C, over the same period. Ocean heat content in the upper 700 m (OHC700) of the tropical Indian Ocean has also exhibited an increasing trend over the past six decades (1955–2015), with the past two decades (1998–2015) having witnessed a notably abrupt rise.

During the twenty-first century, SST (Fig. 1) and ocean heat content in the tropical Indian Ocean are projected to continue to rise.

Changes in Rainfall

The summer monsoon precipitation (June to September) over India has declined by around 6% from 1951 to 2015, with notable decreases over the Indo-Gangetic Plains and the Western Ghats. There is an emerging consensus, based on multiple datasets and climate model simulations, that the radiative effects of anthropogenic aerosol forcing over the Northern Hemisphere have considerably offset the expected precipitation increase from GHG warming and contributed to the observed decline in summer monsoon precipitation.

There has been a shift in the recent period toward more frequent dry spells (27% higher during 1981–2011 relative to 1951–1980) and more intense wet spells during the summer monsoon season. The frequency of localized heavy precipitation occurrences has increased worldwide in response to increased atmospheric moisture content. Over central India, the frequency of daily precipitation extremes with rainfall intensities exceeding 150 mm per day increased by about 75% during 1950–2015.

With continued global warming and anticipated reductions in anthropogenic aerosol emissions in the future, CMIP5 models project an increase in the mean (Fig. 1) and variability of monsoon precipitation by the end of the twenty-first century, together with substantial increases in daily precipitation extremes.

Droughts

The overall decrease of seasonal summer monsoon rainfall during the last 6–7 decades has led to an increased propensity for droughts over India. Both the frequency and spatial extent of droughts have increased significantly during 1951–2016. In particular, areas over central India,

---

[4]Warm days (nights) correspond to cases when the maximum (minimum) temperature exceeds the 90th percentile.

southwest coast, southern peninsula and north-eastern India have experienced more than 2 droughts per decade, on average, during this period. The area affected by drought has also increased by 1.3% per decade over the same period.

Climate model projections indicate a high likelihood of increase in the frequency (>2 events per decade), intensity and area under drought conditions in India by the end of the twenty-first century under the RCP8.5 scenario, resulting from the increased variability of monsoon precipitation and increased water vapour demand in a warmer atmosphere.

Sea Level Rise

Sea levels have risen globally because of the continental ice melt and thermal expansion of ocean water in response to global warming. Sea-level rise in the North Indian Ocean (NIO) occurred at a rate of 1.06–1.75 mm per year during 1874–2004 and has accelerated to 3.3 mm per year in the last two and a half decades (1993–2017), which is comparable to the current rate of global mean sea-level rise.

At the end of the twenty-first century, steric sea level[5] in the NIO is projected to rise by approximately 300 mm relative to the average over 1986–2005 under the RCP4.5 scenario, with the corresponding projection for the global mean rise being approximately 180 mm.

Tropical Cyclones

There has been a significant reduction in the annual frequency of tropical cyclones over the NIO basin since the middle of the twentieth century (1951–2018). In contrast, the frequency of very severe cyclonic storms (VSCSs) during the post-monsoon season has increased significantly (+1 event per decade) during the last two decades (2000–2018). However, a clear signal of anthropogenic warming on these trends has not yet emerged.

Climate models project a rise in the intensity of tropical cyclones in the NIO basin during the twenty-first century.

Changes in the Himalayas

The Hindu Kush Himalayas (HKH) experienced a temperature rise of about 1.3°C during 1951–2014. Several areas of HKH have experienced a declining trend in snowfall and also retreat of glaciers in recent decades. In contrast, the high-elevation Karakoram Himalayas have experienced higher winter snowfall that has shielded the region from glacier shrinkage.

By the end of the twenty-first century, the annual mean surface temperature over HKH is projected to increase by about 5.2°C under the RCP8.5 scenario (Fig. 1). The CMIP5 projections under the RCP8.5 scenario indicate an increase in annual precipitation (Fig. 1), but decrease in snowfall over the HKH region by the end of the twenty-first century, with large spread across models.

Conclusions

Since the middle of the twentieth century, India has witnessed a rise in average temperature; a decrease in monsoon precipitation; a rise in extreme temperature and rainfall events, droughts, and sea levels; and an increase in the intensity of severe cyclones, alongside other changes in the monsoon system. There is compelling scientific evidence that human activities have influenced these changes in regional climate.

Human-induced climate change is expected to continue apace during the twenty-first century. To improve the accuracy of future climate projections, particularly in the context of

---

[5]Steric sea-level variations refer to changes arising from ocean thermal expansion and salinity variations.

regional forecasts, it is essential to develop strategic approaches for improving the knowledge of Earth system processes, and to continue enhancing observation systems and climate models.

Drafting Authors[6]: R. Krishnan and Chirag Dhara

---

[6]The Executive Summary (ES) is drafted based on assessments from the individual chapters.

Contents

About the Editors

R. Krishnan specializes in climate modelling studies on scientific issues relating to the "Dynamics, variability, and predictability of the Asian monsoon, climate change and its impacts on monsoon precipitation, weather and climate extremes, phenomenon of monsoon breaks and droughts". Currently, he is leading the Centre for Climate Change Research (CCCR) at the Indian Institute of Tropical Meteorology, Pune, and is involved in developing in-house capability in Earth System Modeling to address various scientific issues related to climate change and monsoon. He carried out Ph.D. research in Atmospheric Sciences at the Physical Research Laboratory, Ahmedabad, and obtained Ph.D. degree from the University of Pune in 1994. He has published more than 100 scientific articles/papers, advised Ph.D.s (11 awarded, 12 ongoing) and Master (6 M.Sc/M.Tech) dissertations, and offered training lectures in Geophysical Fluid Dynamics & Atmospheric Science. He is Fellow of the Indian Academy of Sciences (IASc), Indian National Science Academy (INSA), and the Indian Meteorological Society (IMS). He is a Member of the Joint Scientific Committee (JSC) of the World Climate Research Programme (WCRP), Coordinating Lead Author (CLA) of the Chapter on Water Cycle Changes in the IPCC WG1 Sixth Assessment Report (AR6), and CLA of the Chapter on Climate Change in the Hindu Kush Himalayan (HKH) Monitoring and Assessment Programme (HIMAP). He also served as a Member of the CLIVAR Monsoon Panel and the CORDEX Science Advisory Team of the WCRP, WMO. He is an Editor for the scientific journals—Earth System Dynamics (EGU Journal), Mausam (IMD Journal), and Journal of Indian Society of Remote Sensing.

J. Sanjay specializes in the area of regional climate change with a focus on the generation of future climate change scenarios for the Indian monsoon region using dynamical downscaling techniques with high resolution regional climate models (RCMs). He is leading the CCCR team for coordinating the data archiving, management, and dissemination activities of the South Asia component of the international coordinated Regional Climate Downscaling Experiment (CORDEX) initiative by the World Climate Research Program (WCRP) of WMO. Prior to joining IITM in 1988, Sanjay completed M.Sc. in Meteorology at the Cochin University of Science and Technology. He carried out Ph.D. research in Atmospheric Sciences at IITM and obtained Ph.D. degree from the University of Pune in 2007. He has 40 publications, including 21 papers in peer-reviewed journals. Sanjay is a Member of the WCRP CORDEX Science Advisory Team (SAT), a Contributing Author (CA) of the IPCC Working Group I (WGI) Sixth Assessment Report Chapter Atlas, and a Lead Author (LA) of the Chapter on Climate Change in the Hindu Kush Himalayan (HKH) Monitoring and Assessment Programme (HIMAP). He was also a scientific knowledge partner on climate science for a segment of the International CARIAA research programme on climate change Adaptation at Scale in Semi-Arid Regions (ASSAR) of India.

Chellappan Gnanaseelan is a Senior Scientist and Project Director of Short Term Climate Variability and Prediction, at the Indian Institute of Tropical Meteorology, Pune. He received his Ph.D. and M.Tech. degrees from the Indian Institute of Technology (IIT), Kharagpur, India.

He also served as visiting fellow at Florida State University. His research interests include climate variability, Indian Ocean dynamics and variability, annual to decadal climate prediction and variability, ocean modelling, air–sea interaction, monsoon variability and teleconnection, and data assimilation. He has authored over 100 research papers in peer-reviewed journals. He has successfully guided 15 Ph.D. and 30 Master's students and presently guiding many doctoral and postdoctoral students and also contributed considerably to human resource development by teaching Master's and Ph.D. students at Pune University, IITM, University of Hyderabad, IIT Bhubaneswar, and IIT Delhi and trainees of India Meteorological Department. He is an Associate Editor of Journal of Earth System Science and Editor of Ocean Digest, the quarterly newsletter of Ocean Society of India. He is also an Adjunct Professor of Pune University.

Dr. Milind Mujumdar is a Senior Scientist at the Centre for Climate Change Research (CCCR), Indian Institute of Tropical Meteorology, Pune. He is currently engaged in the field scale soil moisture monitoring using Cosmos Ray Soil-Moisture Monitoring System (COSMOS) and wireless network of various surface hydro-meteorological sensors to study the soil water dynamics. He has carried out diagnostic and modelling studies to understand the Asian monsoon variability and its response to warming climate. He completed his education up to M.Sc. (Mathematics) from Khandwa (M.P.). He obtained his M.Phil. with focus on "Mathematical Modelling" during 1989 and Ph.D. on studies related to "Climate Modelling" during 2002, from the University of Pune. He has published more than 30 research articles in national and international journals. He is also associated with Universities and Institutes for guiding M.Sc./M.Tech. and Ph.D. students. During his research career, he had scientific visits to the University of Tokyo, Hokkaido University, Nagoya University Japan; University of Hawaii, USA; University of Reading and European Centre for Medium Range Weather Forecast (ECMWF), UK; CSIRO (Melbourne, Australia); University of Cape Town (South Africa).

Ashwini Kulkarni has a Ph.D. in Statistics (1992) from the Pune University. She has 30 years of research experience in the field of Atmospheric Science. Her main research interests include climate variations and teleconnections over South, East, and Southeast Asia; Asian monsoon variability in coupled and regional climate model simulations and projections; applications of statistics in climate research. She has published more than 50 research papers in reviewed scientific international journals. She received the 10th IITM Silver Jubilee award for best research contribution in 1997, and she is "Adjunct Professor" of Department of Atmospheric and Space Science, Savitribai Phule Pune University since 2000. She has been providing guidance to Ph.D., M.Tech./M.Sc. students for their thesis/projects and has a vital contribution into human resource development of IITM, IMD, and SPPU. She is a member of international committees, viz. Empirical Statistical Downscaling Group-Asia and Upper Indus Basin Network. She is also a Senior Scientist at International CLIVAR Monsoon Project Office at IITM. She is a contributory author of Chapter 14 of IPCC AR5 and Reviewer of IPCC special reports.

Supriyo Chakraborty Scientist-F, is presently Head of the Mass Spectrometry Group and the MetFlux India Project that investigates the atmosphere–biosphere exchanges of CO_2 and energy fluxes at various ecosystems across the country. He obtained his B.Sc. (Hons) from the Indian Institute of Technology, Kharagpur, in 1984, and M.Sc. in Exploration Geophysics from the same institute in 1986. He worked as a graduate student at the Physical Research Laboratory, Ahmedabad, and obtained his Ph.D. in 1995 of M.S. University of Baroda. He undertook postdoctoral work at the University of California, Santa Barbara, during 1995-1996 and also at the University of California, San Diego, during 1996-1998. Afterwards, he worked at the Physical Research Laboratory, Ahmedabad, and Birbal Sahni Institute of Paleosciences, Lucknow, and currently at IITM since 2007. He has been working in the fields of monsoon reconstruction using the isotopic analysis of natural archives, stable isotopic characteristics of precipitation and study of moisture dynamical processes, ecosystem GHGs fluxes, and energy transfer processes at various natural ecosystems. He has been recognized as an Adjunct Professor, at the Savitribai Phule Pune University, Pune. He has supervised/co-supervised several students for Ph.D. and M.Sc. dissertation. He has published over 60 research papers in various national and international journals.

Introduction to Climate Change Over the Indian Region

Coordinating Lead Authors

R. Krishnan, Indian Institute of Tropical Meteorology (IITM-MoES), Pune, India
C. Gnanaseelan, Indian Institute of Tropical Meteorology (IITM-MoES), Pune, India
J. Sanjay, Indian Institute of Tropical Meteorology (IITM-MoES), Pune, India
P. Swapna, Indian Institute of Tropical Meteorology (IITM-MoES), Pune, India
Chirag Dhara, Indian Institute of Tropical Meteorology (IITM-MoES), Pune, India,
e-mail: chirag.dhara@tropmet.res.in (corresponding author)

Lead Authors

T. P. Sabin, Indian Institute of Tropical Meteorology (IITM-MoES), Pune, India
Jyoti Jadhav, Indian Institute of Tropical Meteorology (IITM-MoES), Pune, India
N. Sandeep, Indian Institute of Tropical Meteorology (IITM-MoES), Pune, India
Ayantika Dey Choudhury, Indian Institute of Tropical Meteorology (IITM-MoES), Pune, India
Manmeet Singh, Indian Institute of Tropical Meteorology (IITM-MoES), Pune, India
M. Mujumdar, Indian Institute of Tropical Meteorology (IITM-MoES), Pune, India
Anant Parekh, Indian Institute of Tropical Meteorology (IITM-MoES), Pune, India
Abha Tewari, (Formerly with Indian Air Force, National Health Systems Resource Center and Ministry of Environment Forest and Climate Change)
Rajeev Mehajan, Science and Engineering Research Board, Department of Science and Technology, Government of India, New Delhi, India

Contributing Authors

Rahul Chopra, TROP ICSU—Climate Education Project of the International Science Council at the Indian Institute of Science Education and Research (IISER), Pune, India
Aparna Joshi, TROP ICSU—Climate Education Project of the International Science Council at the Indian Institute of Science Education and Research (IISER), Pune, India
Anita Nagarajan, TROP ICSU—Climate Education Project of the International Science Council at the Indian Institute of Science Education and Research (IISER), Pune, India
Megha Nivsarkar, TROP ICSU—Climate Education Project of the International Science Council at the Indian Institute of Science Education and Research (IISER), Pune, India

Review Editors

M. Rajeevan, Ministry of Earth Sciences, Government of India, New Delhi, India
M. Collins, Centre for Engineering, Mathematics and Physical Sciences, University of Exeter, Exeter, UK
Dev Niyogi, Purdue University, West Lafayette, IN and University of Texas at Austin, Austin, TX, USA

Corresponding Author

Chirag Dhara, Indian Institute of Tropical Meteorology (IITM-MoES), Pune, India,
e-mail: chirag.dhara@tropmet.res.in

The original version of this chapter was revised. The caption of figure 1.5 has been changed. The correction to this chapter is available at https://doi.org/10.1007/978-981-15-4327-2_13

© The Author(s) 2020, corrected publication 2021
R. Krishnan et al. (eds.), *Assessment of Climate Change over the Indian Region*,
https://doi.org/10.1007/978-981-15-4327-2_1

1.1 General Introduction

The Earth's climate has varied considerably throughout its history. Periodic and episodic natural changes caused by natural climate forcings such as orbital variations and volcanic eruptions, and amplified by feedback processes intrinsic to the climate system, have induced substantial changes in planetary climate on a range of timescales.

Multiple independent lines of investigation have provided increasingly compelling evidence that human activities have significantly altered the Earth's climate since the industrial revolution (Stocker et al. 2013). A distinctive aspect of the present-day climate change is the rapid pace at which it is proceeding relative to that of natural variations alone; a pace that is unprecedented in the history of modern civilization.

The Earth's energy budget, which is the balance between the energy that the Earth receives from the Sun and the energy that it radiates back into space, is a key factor that determines the Earth's global mean climate. The composition of the atmosphere alters climate by modulating the incoming and outgoing radiative fluxes at the surface. The key drivers of present-day climate change are anthropogenic (human-caused) emissions of greenhouse gases (GHGs), aerosols and changes in land use and land cover (LULC).

GHGs warm the surface by reducing the amount of Earth's terrestrial radiation escaping directly to space. Atmospheric concentrations of the key GHGs—carbon dioxide, methane, and nitrous oxide—are now at higher levels than they have been at any time over the last 800,000 years, according to ice core records. In addition, their mean rates of increase over the past century are, with high confidence, unprecedented in the last 22,000 years (Stocker et al. 2013).

Aerosols are small particles or droplets suspended in the atmosphere produced from both natural and anthropogenic sources. Natural sources include mineral dust from soil erosion, sea salt and volcanic eruptions. Key anthropogenic sources are industrial air pollution, transport, and biomass burning, which produce airborne sulfates, nitrates, ammonium and black carbon, and dust produced by land degradation processes such as desertification. Aerosols tend to cool the surface by scattering or absorption of solar radiation (direct effect), or by enhancing cloud formation (indirect effect). Aerosol pollution, due to human activities, has thus offset a part of the warming caused by anthropogenic GHG emissions (Myhre et al. 2013).

Much of the Earth's land surface has been affected by considerable changes in land use and land cover (LULC) over the past few centuries (and even earlier), mainly because of deforestation and the expansion of agriculture. Deforested areas have a diminished capacity to act as a carbon dioxide sink and, if accompanied by biomass burning, are a direct source of GHGs. Conversion of land from natural vegetation to agriculture or pasturage also alters the terrestrial albedo, contributing to changes in the surface radiative balance.

The net effect of human-induced climate forcing has been an increase in the global average near-surface air temperature[1] by approximately 1 °C since pre-industrial times (Allen et al. 2018). Each of the last three decades has been successively warmer at the Earth's surface than any preceding decade since 1850 (Stocker et al. 2013) and 2001–2018 have been 18 of the 19 warmest years in the observational record. Trends in other important global climate indicators such as rate and patterns of precipitation, temperature and precipitation extremes, atmospheric water vapour concentration, continental ice melt, sea-level rise, ocean heat content, ocean acidification and the frequency of powerful cyclones are consistent with the response expected from a warming planet (Stocker et al. 2013). At the current rate of temperature rise, it is *likely*[2] that global warming will reach 1.5 °C between 2030 and 2052 (high confidence) (Allen et al. 2018) and 3–5 °C by the end of the century relative to pre-industrial times. Even if warming is limited to 1.5 °C in the twenty-first century, certain slowly evolving changes such as ocean thermal expansion would persist well beyond 2100 causing sea levels to continue rising (high confidence).

While climate change is global, changes in climate are not expected to be uniform across the planet. For instance, Arctic temperatures are rising much faster than the global average (Stocker et al. 2013), and rates of sea-level rise vary significantly across the world (Church et al. 2013). Changes in climate at regional scales are not understood as robustly as at the global scale due to insufficient local observational data or understanding of physical phenomena specific to given regions (Flato et al. 2013). Yet, knowledge of present and expected changes in regional climate is critical to people and policymakers to plan for disaster management, risk mitigation and for formulating locally relevant adaptation strategies (Burkett et al. 2014).

The regional climate over the Indian subcontinent involves complex interactions of the atmosphere–ocean–land–cryosphere system on different space and time scales. In addition, there is evidence that anthropogenic activities have influenced the regional climate in recent decades. Impacts associated with human-induced climate change such as increasing heat extremes, changing monsoon patterns and sea-level rise pose serious threats to lives and livelihoods on the subcontinent. This makes it necessary to understand how and why the climate is changing across India and how these changes are expected to evolve in the future. The following

---

[1]The 'near surface air temperature' is defined as the temperature 2 m above the surface over land areas, and as the sea surface temperature (SST) for oceanic areas.
[2]Defined in Box 1.4.

section provides a brief overview of the mean climate of the Indian subcontinent and sets the context for understanding the key aspects of climate change in the region.

1.1.1 Setting the Regional Context

The distinct topographical and geographical features of the Indian subcontinent endow the region with widely varying climatic zones ranging from the arid Thar desert in the north-west, Himalayan tundra in the north, humid areas in the southwest, central and northeastern parts, together with diverse microclimatic areas that spread across the vast subcontinent. A dominant feature of the regional climate is the Indian Summer Monsoon (ISM), which is characterized by pronounced seasonal migrations of the tropical rain belts associated with the Inter-tropical Convergence Zone (ITCZ), along with large-scale seasonal wind reversals (Gadgil 2003; Schneider et al. 2014).

The Himalayas and the Hindu Kush mountains protect the Indian subcontinent from large-scale incursions of cold extra-tropical winds during the winter season. Additionally, the seasonal warming of the Himalayas and the Tibetan Plateau during the boreal summer sets up a north–south thermal contrast relative to the tropical Indian Ocean, which is important for initiating the large-scale summer monsoon circulation. The climatological seasons in India are broadly classified as the winter (December–January–February), pre-monsoon (March–April–May), summer monsoon (June–July–August–September) and the post-monsoon (October–November) seasons. A distinction of India's climate is the exceptionally strong seasonal cycle of winds and precipitation (Turner and Annamalai 2012). The Indian summer monsoon, also known as the South Asian monsoon, is a major component of the global climate (see Box 1.1 for a summary of monsoon processes over the Indian subcontinent). In addition to monsoonal rains, areas in the western Himalayas (WH) also receive substantial precipitation during the winter and early spring months from eastward propagating synoptic-scale weather systems known as the Western Disturbances that originate from the Mediterranean region (e.g. Dimri et al. 2015; Hunt et al. 2018; Krishnan et al. 2019a, b) (Chap. 11). The Indian region is also prone to a wide range of severe weather events and climate extremes, including tropical cyclones, thunderstorms, heat waves, floods, droughts, among others.

Box 1.1: Monsoon Processes over the Indian subcontinent
Large-scale orographic features such as the Himalayas and the Tibetan Plateau (e.g. Boos and Kuang 2010; Turner and Annamalai 2012 and references therein); as well as narrow mountains such as the Western Ghats along the Indian west coast and the Arakan Yoma

mountains along the Myanmar coast (e.g. Xie et al. 2006; Rajendran and Kitoh 2008; Krishnan et al. 2013; Sabin et al. 2013) exert control on the distribution of monsoon precipitation over the Indian subcontinent. With moisture-laden winds from the Arabian Sea, the Bay of Bengal and the Indian Ocean feeding bountiful rains over vast areas in central-north and northeast India, Western Ghats and peninsular India, central-eastern Himalayas; the summer monsoon activity is sustained through feedbacks between the monsoon circulation and the release of latent heat of condensation by moist convective processes (Rao 1976; Krishnamurti and Surgi 1987).

The ISM is home to a variety of precipitation producing systems, which include—monsoon onset vortices, meso-scale systems and orographic precipitation, west–north-west moving synoptic systems (lows and depressions) from the Bay of Bengal and Southeast Asia, slow northward and westward propagating large-scale organized rainbands and mid-tropospheric cyclones, to name a few (Rao 1976). Interactions among multiple scales of motion (i.e. planetary, regional, synoptic, meso and cumulus scales) render significant spatio-temporal heterogeneity in the monsoon rainfall distribution over the region. Warm rain processes during the summer monsoon region are recognized to be dominant over the Western Ghats and other areas in India, as evidenced from the Tropical Rainfall Measurement Mission (TRMM) Precipitation Radar (PR) satellite (e.g. Shige et al. 2017) and aircraft measurements (e.g. Konwar et al. 2014). There has been improved understanding of the three-dimensional structure of latent heating associated with convective and stratiform clouds during the summer monsoon season based on the TRMM satellite observations (e.g. Houze 1997; Houze et al. 2007; Stano et al. 2002; Romatschke and Houze 2011); as well as the monsoonal circulation response to latent heating based on numerical simulation experiments (e.g. Choudhury and Krishnan 2011; Choudhury et al. 2018) in the recent decades.

Several areas in south-eastern peninsular India, including areas covering the states of Tamil Nadu and Andhra Pradesh, receive considerable rainfall during the northeast monsoon (October–December) (Rajeevan et al. 2012). The northeast monsoon develops following the withdrawal of the summer monsoon rainy season when the northern landmass of India and the Asian continent begins to cool off rapidly so that high-pressure builds over northern India. The northeasterly monsoon winds from the northern areas gather moisture from the Bay of Bengal and contribute to precipitation over peninsular India and parts of Sri Lanka (Turner and Annamalai 2012).

1.1.2 Key Scientific Issues

Climate over the Indian subcontinent has varied significantly in the past century in response to natural variations (e.g. Box 1.2 on the variability of the ISM) and anthropogenic forcing (see Box 1.3). In recent times, there has been considerable progress in understanding the influence of anthropogenic climate change over the Indian subcontinent, particularly the regional monsoon.

State-of-the-art climate models project a continuation of anthropogenic global warming and associated climate change during the twenty-first century, the impacts of which have profound implications for India. Yet, there remain substantial knowledge gaps with regard to climate projections, particularly at smaller spatial and temporal scales. For instance, CMIP5 simulations of historical and future changes in the monsoon rainfall exhibit wide variations across the Indian region (Sperber et al. 2013; Turner and Annamalai 2012), posing difficulties for policy making. Likewise, it is necessary to reduce the range among climate models projections of future changes in Indian Ocean warming, regional sea-level rise, tropical cyclone activity, weather and climate extremes, changes in the Himalayan snow cover, etc. It is essential to deepen our understanding of the science of climate change, improve the representation of key processes in climate models (e.g. clouds, aerosol–cloud interactions, vegetation–atmosphere feedbacks, etc.) and also build human capacity to address these challenges.

Efforts in these directions have already begun in India. One such initiative is the development of an Earth System Model (IITM-ESM) at the Centre for Climate Change Research (CCCR) in the Indian Institute of Tropical Meteorology (Swapna et al. 2018). A brief discussion of the IITM-ESM is provided in the following section.

Box 1.2: Indian Summer Monsoon Variability

The ISM also exhibits a rich variety of natural variations on different timescales ranging across sub-seasonal/ intra-seasonal, interannual (year-to-year), multi-decadal and centennial timescales, which are evident from instrumental records and paleoclimate reconstructions (e.g. Turner and Annamalai 2012; Sinha et al. 2015). The sub-seasonal/intra-seasonal variability of the ISM is dominated by active and break monsoon spells (e.g. Rajeevan et al. 2010) and the interannual variability is associated with excess or deficient seasonal monsoon rainfall over India (e.g. Pant and Parthasarathy 1981). The interannual and decadal timescale variations in the ISM rainfall are known to have links with the tropical Pacific, Indian and Atlantic oceans, particularly with climate drivers such as the El Nino/Southern Oscillation (ENSO), Indian Ocean Dipole (IOD), Equatorial Indian Ocean Oscillation (EQUINOO), Pacific Dedacal Oscillation

(PDO), etc. The reader is referred to Chap. 3 for more details on the ISM variability and associated teleconnections.

Box 1.3 Anthropogenic Drivers of Climate Change

Changes in the atmospheric concentration of GHGs, aerosols and LULC are the key anthropogenic drivers of global climate change.

The global atmospheric carbon dioxide concentration has increased from an average of 280 ppm in the pre-industrial period to over 407 ppm in 2018 (https://scripps.ucsd.edu/programs/keelingcurve/), contributing a radiative forcing (RF) of about 2.1 W/m^2 at the top of the atmosphere.

Unlike GHGs, which are well-mixed in the atmosphere, the concentration of anthropogenic aerosols in the atmosphere exhibits large spatio-temporal variability and complex interactions with clouds and snow, giving rise to uncertainties in the estimation of the aerosol RF. The IPCC AR5 estimated the globally averaged total aerosol effective radiative forcing (excluding black carbon on snow and ice) to be in the range -1.9 to -0.1 W/m^2. Over India, the direct aerosol RF is estimated to range from -15 to $+8$ W/m^2 at the top of the atmosphere and -49 to -31 W/m^2 at the surface (Nair et al. 2016) (Chap. 5). The implications are that aerosol RF can be significantly larger than the GHG forcing at regional scales and large gradients in the aerosol RF can significantly perturb the regional climate system.

The IPCC AR5 reported a globally averaged RF due to anthropogenic changes in LULC to be about -0.2 W/m^2 although it was anticipated that this estimate may be revised downwards with emerging research. As with aerosols, there are large spatio-temporal variations in RF due to LULC changes at regional scales.

Despite the uncertainties in the estimation of RF due to anthropogenic aerosol and LULC changes, there is high confidence that they have offset a substantial portion of the effect of GHGs on both temperature and precipitation (IPCC AR5).

1.1.3 IITM-ESM: A Climate Modelling Initiative from India

The Coupled Model Intercomparison Project (CMIP) organized under the auspices of the World Climate Research Programme (WCRP) forms the basis of the climate projections in the IPCC Assessment Reports. The CMIP experiments have evolved over six phases (Meehl et al. 2000;

Fig. 1.1 Schematic of IITM Earth system model (IITM-ESM)

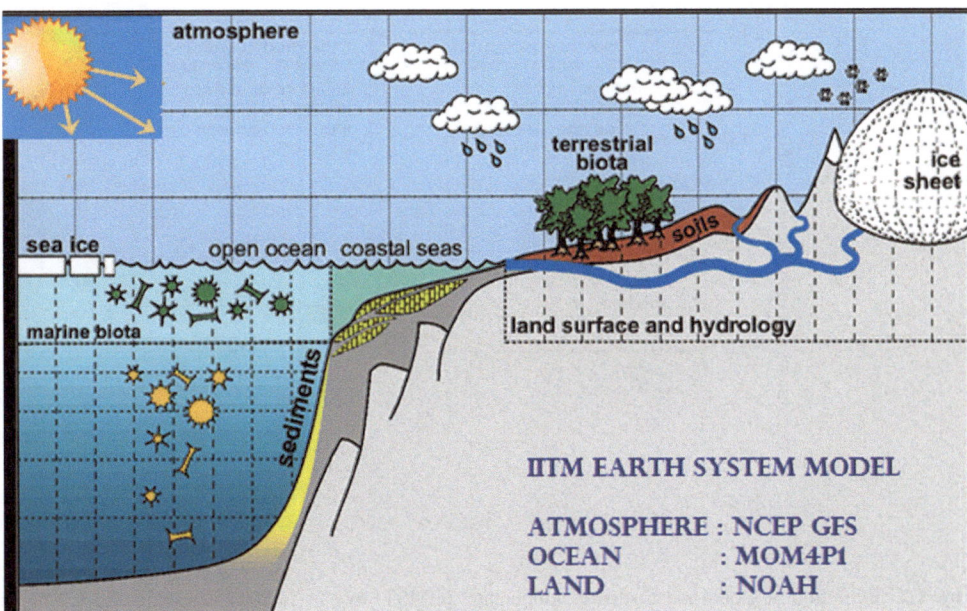

Meehl and Hibbard 2007; Taylor et al. 2012; Eyring et al. 2016) and become a central element of national and international assessments of climate change.

The IITM—Earth System Model (IITM-ESM) has contributed to the CMIP6 and IPCC AR6 assessments, the first time from India. The philosophy behind the development of the IITM-ESM is to create capabilities in global modelling, with special emphasis on the South Asian monsoon, to address the science of climate change, including detection, attribution and future projections of global and regional climate.

The IITM-ESM is configured using an atmospheric general circulation model based on the National Centers for Environmental Prediction (NCEP) Global Forecast System (GFS) with a global spectral triangular truncation of 62 waves (T62, grid size $\sim$200 km) and 64 vertical levels with top model layer extending up to 0.2 hPa, a global ocean component based on the Modular Ocean Model Version 4p1 (MOM4p1) having a resolution of $\sim$100 km with finer resolution ($\sim$35 km) near-equatorial regions with 50 levels in the vertical, a land surface model (Noah LSM) with four layers and a dynamical sea ice model known as the Sea Ice Simulator (SIS). The details about IITM-ESM are described in Swapna et al. (2015). A schematic of the IITM-ESM is shown in Fig. 1.1.

1.2 Global and Regional Climate Change

This section provides a summary of assessments of the observed and projected changes in the global climate and regional climate over India, based on published scientific literature, key findings from the individual chapters of this report, together with analyses of observed and reanalysis datasets, and diagnoses from the CMIP, CORDEX and IITM-ESM model simulations.

1.2.1 Observed Changes in Global Climate

The evidence for a warming world comes from multiple independent climate indicators in the atmosphere and oceans (Hartmann et al. 2013). They include changes in surface, atmospheric and oceanic temperatures, glaciers, snow cover, sea ice, sea-level rise, atmospheric water vapour, extreme events and ocean acidification.

Inferences of past climate from paleo-climate proxies suggest that recent changes in global surface temperature are unusual and natural processes alone cannot explain the rapid rate of warming in the industrial era. Computer-based climate models are unable to replicate the observed warming unless the effect of human-induced changes such as emissions of GHGs and aerosols, and changes in land use and land cover are included (Bindoff et al. 2013).

The Global Mean Surface Temperature (GMST) comprising the global land surface air temperature (LSAT) and sea surface temperature (SST) is a key metric in the climate change policy framework. Historical records of GMST extend back farther than any other global instrumental series making it the key to understanding the patterns and magnitude of natural climate variations and distinguishing them from anthropogenically forced climate change (Fig. 1.2).

The Fifth Assessment Report by the Intergovernmental Panel on Climate Change (IPCC AR5) concluded that it is

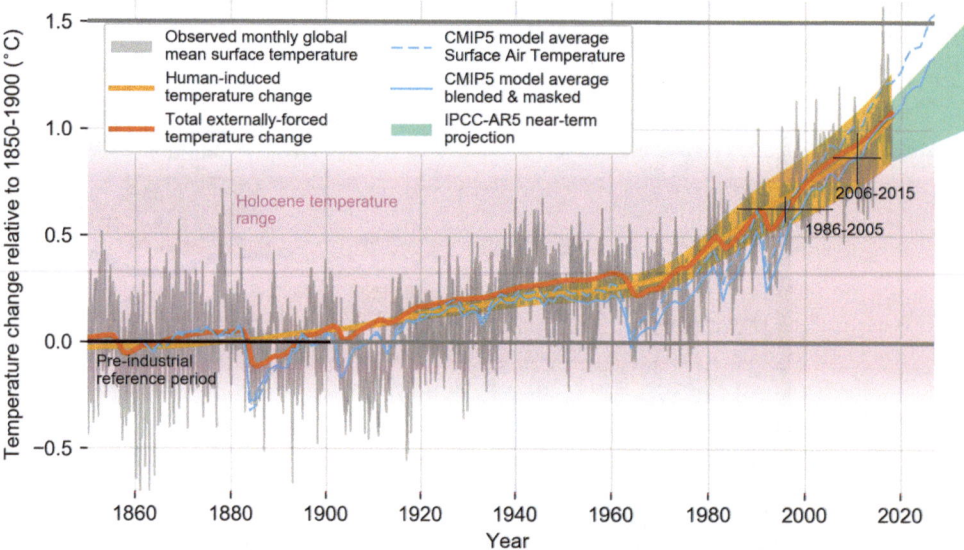

Fig. 1.2 Evolution of global mean surface temperature (GMST) over the period of instrumental observations. Grey-shaded line shows monthly mean GMST in the HadCRUT4, NOAAGlobalTemp, GISTEMP and Cowtan-Way datasets, expressed as departures from 1850 to 1900, with varying grey line thickness indicating inter-dataset range. All observational datasets shown represent GMST as a weighted average of near-surface air temperature over land and sea surface temperature over oceans. Human-induced (yellow) and total (human- and naturally forced, orange) contributions to these GMST changes are shown calculated following Otto et al. (2015) and Haustein et al. (2017). Fractional uncertainty in the level of human-induced warming in 2017 is set equal to ±20% based on multiple lines of evidence. Thin blue lines show the modelled global mean surface air temperature (dashed) and blended surface air and sea surface temperature accounting for observational coverage (solid) from the CMIP5 historical ensemble average extended with RCP8.5 (Defined in the following section) forcing (Cowtan et al. 2015; Richardson et al. 2018). The pink shading indicates a range for temperature fluctuations over the Holocene (Marcott et al. 2013). Light green plume shows the AR5 prediction for average GMST over 2016–2035 (Kirtman et al. 2013). Reproduced from Allen et al. 2018 (Fig. 1.2)

certain that GMST has increased since the late nineteenth century. Each of the past three decades has been successively warmer at the Earth's surface than all previous decades in the observational record, and the first decade of the twenty-first century has been the warmest (Hartmann et al. 2013). Global warming attributed to human activities has now reached approximately 1 °C (*likely* between 0.8 and 1.2 °C) in the global mean above the mean pre-industrial temperature (period 1850–1900) and is increasing at 0.2 °C (*likely* between 0.1 and 0.3 °C) per decade (Allen et al. 2018). IPCC AR5 also summarized that it is *very likely* that the numbers of cold days and nights have decreased and the numbers of warm days and nights have increased globally since about 1950. Regional trends are sufficiently complete over 1901–2012 and show that almost the entire globe, including both land and ocean, has experienced surface warming.

IPCC AR5 assessed that it is virtually certain that globally the troposphere has warmed and the lower stratosphere has cooled since the mid-twentieth century based on multiple independent analyses of measurements from radiosondes and satellite sensors. However, there is low confidence in the rate of temperature change, and its vertical structure, in most areas of the planet.

According to IPCC AR5, anthropogenic forcing has contributed substantially to upper-ocean warming (above 700 m). On a global scale, ocean warming is largest near the surface, with the upper 75 m having warmed by 0.11 [0.09 to 0.13] °C per decade over the period 1971–2010 (Stocker et al. 2013).

Warming is expected to elevate the rate of evaporation and increase the moisture content of the atmospheric. Indeed, the amount of water vapour in the atmosphere, measured as specific humidity, has increased globally over both the land and ocean. IPCC AR5 summarized that it is *very likely* that global near-surface air specific humidity has increased since the 1970s (Hartmann et al. 2013). However, during recent years the near-surface moistening over land has abated (medium confidence). As a result, fairly widespread decreases in relative humidity near the surface are observed over land in recent years.

IPCC AR5 concluded that confidence in precipitation change averaged over global land areas since 1901 is low for years prior to 1951 and medium afterwards. Averaged over the mid-latitude land areas of the northern hemisphere, precipitation has *likely* increased since 1901 (medium confidence before and high confidence after 1951) (Hartmann et al. 2013). Precipitation in tropical land areas has increased (medium confidence) over the decade ending in 2012, reversing the drying trend that occurred from the mid-1970s to mid-1990s. Human influence has also contributed to large-scale changes in precipitation patterns over land (medium confidence; Bindoff et al. 2013). It is *likely* that, since about 1950, the number of heavy precipitation events over land has increased in more regions than it has decreased.

Local changes in temperature affect the cryosphere. The amount of ice contained in glaciers globally has been declining every year for over 20 years. Total ice loss from the Greenland and Antarctic ice sheets during 1992–2011 (inclusive) has been 4260 [3060–5460] Gt, equivalent to 11.7 [8.4–15.1] mm of sea level. However, the rate of change has increased with time and most of this ice has been lost in the second decade of the 20-year period (Vaughan et al. 2013).

The global average sea level rose by 19 cm from 1901 to 2010 (Stocker et al. 2013). The average rate of rise measured by satellites has been 3.2 [2.9–3.5] mm/year since the 1990s up from 1.7 [1.5–1.9] mm/year during the twentieth century, obtained from historical tide gauge records (Hartmann et al. 2013). Thermal expansion and glacier melt because of anthropogenic global warming have been the major drivers of rise in global sea levels over the past century.

Substantial losses in Arctic sea ice have been observed since satellite records began, particularly at the time of the minimum extent, which occurs in September, at the end of the annual melt season. In contrast, there has been an increase in Antarctic sea ice, but with a smaller rate of change than in the Arctic.

Snow cover is sensitive to changes in temperature, particularly during the spring, when the snow starts to melt. Spring snow cover has shrunk across the northern hemisphere since the 1950s. IPCC AR5 concluded that it is *likely* that snowfall events are decreasing in most regions (North America, Europe, Southern and East Asia) where increased winter temperatures have been observed (Hartmann et al. 2013). The total seasonal snowfall is reported to be declining along with increase in maximum and minimum temperatures in the western Himalaya. Confidence is low for changes in snowfall over Antarctica.

Uptake of anthropogenic CO_2 by the ocean increases the hydrogen ion concentration in the ocean water, causing acidification. There is high confidence that the global average pH of the surface ocean has decreased by 0.1 pH units since the beginning of the industrial era, corresponding to an approximately 30% increase in acidity (Stocker et al. 2013).

1.2.2 Projected Changes in Global Climate

This section assesses projected long-term changes in the global climate system during the twenty-first century. These changes are expected to be larger than the internal variability of the climate system and to depend primarily on how anthropogenic emissions change the atmospheric composition in the future. Aerosol emissions are projected to decline in the coming decades, and it is expected that future changes will be dominated by the increasing concentrations of GHGs.

Climate models are used to study the response of the climate system to anthropogenic activity (also referred to as 'external forcing'). Towards studying how the climate will change in the twenty-first century, several standardized scenarios have been developed, each with a specific description of how human-induced changes would affect the planet's energy budget. Differences between scenarios are based on underpinning assumptions about future changes in fossil fuel consumption, land use change, etc., and were developed using integrated assessment models that combined economic, demographic and policy modelling, with simplified physical climate models in order to simulate the global economic impacts of climate change under different mitigation scenarios (Calel and Stainforth 2017).

Earth system models, developed by climate modelling groups worldwide, perform climate change simulations for these forcing scenarios, whose standardization facilitates easy intercomparison between the results of these studies.

The scenarios used by models participating in the Coupled Model Intercomparison Project Phase 5 (CMIP5) that contributed to the IPCC AR5 were termed the Representative Concentration Pathways (RCPs) and covered the period from 2006 to 2100 (van Vuuren et al. 2011). The four RCPs that were defined were the RCP2.6, representing a low emissions pathway resulting in radiative forcing (RF) of roughly 2.6 W/m^2 at the end of the twenty-first century, RCP4.5 and RCP6 representing intermediate emission pathways resulting in an RF of 4.5 W/m^2 and 6 W/m^2, respectively and the high emissions scenario RCP8.5 representing a pathway with continued growth in GHG emissions leading to an RF of roughly 8.5 Wm^{-2} at the end of the twenty-first century. Various chapters of this report mainly use the RCP pathways to study future changes in the climate system.

The AR5 assessment concluded that GMST will continue to rise over the twenty-first century with increasing GHGs. The increase in GMST for 2081–2100, relative to 1986–2005 will *likely* be in the 5–95% range of 0.3–1.7 °C under RCP2.6 and 2.6–4.8 °C under RCP8.5 (Collins et al. 2013). Assessment of precipitation based on CMIP5 models indicates that it is virtually certain that global mean precipitation will increase by more than 0.05 mm day^{-1} and 0.15 mm day^{-1} by the end of the twenty-first century under the RCP2.6 and RCP8.5 scenarios, respectively (Collins et al. 2013). The median of the global mean sea-level rise for the period 2081–2100 is 0.47 m in RCP4.5 and 0.63 m in the

Table 1.1 Change in surface air temperature (TAS, °C) and precipitation (PR, mm day^{-1}) relative to 1850–1900 for the RCP4.5 and RCP8.5 scenarios from CMIP5 models for the Global and the Indian region during the historical (1951–2014), near future (2040–2069) and far future (2070–2099) periods

Variables	Estimates from CMIP5; (base period 1850–1900)									
	Global mean estimates					Indian region estimates				
	Historical	RCP4.5		RCP8.5		Historical	RCP4.5		RCP8.5	
	1951–2014	2040–2069	2070–2099	2040–2069	2070–2099	1951–2014	2040–2069	2070–2099	2040–2069	2070–2099
TAS (°C)	0.54 (0.28 to 0.68)	2.16 (1.43 to 2.75)	2.62 (1.80 to 3.16)	2.75 (1.94 to 3.48)	4.31 (3.08 to 5.25)	0.72 (0.47 to 1.28)	2.67 (1.72 to 3.70)	3.27 (2.25 to 4.27)	3.37 (2.32 to 4.68)	5.33 (3.70 to 6.11)
Precip. (mm day^{-1})	0.01 (−0.02 to 0.40)	0.09 (0.05 to 0.16)	0.13 (0.09 to 0.18)	0.12 (0.07 to 0.20)	0.20 (0.11 to 0.30)	−0.06 (−0.36 to 0.28)	0.10 (−0.32 to 0.33)	0.23 (−0.13 to 0.49)	0.22 (−0.09 to 0.43)	0.28 (−0.31 to 0.68)

Included are CMIP5 multi-model means and range (in parenthesis) among models. Models used for this analysis are listed in Table 1.3

RCP8.5 scenario (Church et al. 2013). The corresponding increase in the thermosteric sea level is 0.2 m for RCP4.5 and 0.27 m for the RCP8.5 scenario. A summary of projections for temperature and precipitation from CMIP5 models, averaged globally and over India, are provided in Table 1.1.

In the upcoming IPCC Sixth Assessment Report (IPCC AR6), projections are utilizing the a new range of scenarios known as Shared Socio-economic Pathways (SSPs) (O'Neill et al. 2017) introduced in the Coupled Model Intercomparison Project Phase 6 (CMIP6). The SSPs define five different ways in which the world might evolve in the absence of climate policy and how different levels of climate change mitigation could be achieved when the mitigation targets of RCPs are combined with the SSPs. These include SSP1: a world of sustainability-focused growth and equality; SSP2: a 'middle of the road' world where trends broadly follow their historical patterns; SSP3: a fragmented world of 'resurgent nationalism'; SSP4: a world of ever-increasing inequality and SSP5: a world of rapid and unconstrained growth in economic output and energy use.

Projections by the IITM-ESM of changes in key climate variables based on different SSP scenarios are discussed in detail in Sect. 1.2.3. A summary of these projections for temperature and precipitation, including those from other CMIP6 models, can be found in Table 1.2.

1.2.3 Contribution from IITM-ESM

In this section, we assess changes in the major global indicators of climate change using the IITM-ESM CMIP6 simulations. Change is assessed relative to the pre-industrial period (1850–1900). Simulations spanning over the historical period (1950–2014), near future (2040–2069)[3] and far future (2070–2099) are presented along with simulations from other available CMIP6 models. The projections are based on the Shared Socio-economic Pathways. Two of the priority SSPs, the SSP2-4.5 considered as middle-of-the-road and SSP5-8.5 as fossil-fuel-rich development, are presented in this section.

Changes in selected global climate indices including Global Mean Surface Temperature (GMST), global mean precipitation and Global Mean thermosteric Sea Level (GMSL) for the period 1900–2099 are shown in Fig. 1.3. These include historical simulations for the period 1900–2014 and projections from 2015 onwards. The historical simulations enable evaluation of model performance w.r.t observations and show whether the models are able to reproduce the observed aspects of climate change and variability. The time-series of GMST show an increase of more than 0.6 °C during 1951–2014 (with reference to a base period of 1900–1930). However, the increase is not steady; warming is more pronounced since 1970 and a slowing down of warming occurred during the recent period with a slow down in global warming (2000–2010). These fluctuations arise from natural variations within the climate system or internal climate variability, and are also driven by external forcings. The CMIP6 models simulate the observed warming trend but exhibit a wide range of warming levels especially in the far future (Fig. 1.3a). The higher warming level in the CMIP6 models may be associated with higher Equilibrium Climate Sensitivity (ECS). The GMST from the IITM-ESM for the historical period closely follows the observed warming (figure not shown), and the global mean temperature rise

---

[3]Throughout this report, climate change projections refer to changes in the 'near future' and 'far future' relative to a specified historical baseline period. 'Near future' refers to the climatological average over the period 2040–2069 and is used interchangeably with 'middle of the century'. Likewise 'far future' refers to the climatological average over the period 2070–2099 and is used interchangeably with 'end of the century'.

Table 1.2 Change in near-surface air temperature (TAS, °C) and precipitation (Precip, mm day^{-1}) relative to 1850–1900 for the SSP2-4.5 and SSP5-8.5 scenarios from IITM-ESM and CMIP6 models for the Global and the Indian region during the historical (1951–2014), near future (2040–2069) and far future (2070–2099) periods

Variables		Estimates from CMIP6 & IITM-ESM (base period 1850–1900)									
		Global mean estimates					Indian region estimates				
		Historical	SSP2-45		SSP5-85		Historical	SSP2-45		SSP5-85	
		1951 2014	2040– 2069	2070– 2099	2040– 2069	2070– 2099	1951– 2014	2040– 2069	2070– 2099	2040– 2069	2070– 2099
TAS (°C)	CMIP6	0.59 (0.32 to 1.07)	2.5 (1.60 to 3.28)	3.16 (1.96 to 4.05)	3.13 (1.87 to 4.11)	5.0 (2.93 to 6.55)	0.50 (0.29 to 0.87)	2.37 (1.67 to 3.16)	3.14 (2.11 to 4.48)	3.04 (1.92 to 4.53)	5.35 (3.26 to 7.30)
	IITM-ESM	0.60	1.91	2.22	2.28	3.36	0.54	1.67	2.11	2.08	3.26
Precip. (mm day^{-1})	CMIP6	0.01 (0.00 to 0.05)	0.11 (0.04 to 0.19)	0.15 (0.06 to 0.24)	0.13 (0.04 to 0.21)	0.21 (0.09 to 0.32)	0.01 (−0.30 to 0.31)	0.33 (−0.36 to 1.38)	0.45 (−0.60 to 1.60)	0.49 (−0.20 to 0.96)	0.84 (−0.25 to 1.89)
	IITM-E5M	0.02	0.09	0.11	0.09	0.15	−0.02	0.25	0.03	0.16	0.5

The estimates for CMIP6 constitute the multi-model mean and range (in parenthesis) across the considered models. The CMIP6 models used for this analysis are listed in Table 1.4

is within the range of warming shown by other CMIP6 models during the historical period (Fig. 1.3a).

A consistent feature across climate projections is a continuous rise in global mean temperature (Fig. 1.3a). The evolution of GMST for the future scenarios from the AR5 assessment concluded that GMST will continue to rise over the twenty-first century with the increase in GHGs. The GMST change at the end of the twenty-first century w.r.t. the pre-industrial period (1850–1900) from IITM-ESM under the highest emissions scenario (SSP5-8.5), is 3.4 °C and under the intermediate emissions scenarios (SSP2-4.5), is about 2.2 °C, respectively. The corresponding increase in

GMST in CMIP6 models show a range of about 2.9 °C–6.5 °C and 2 °C–4.1 °C under SSP5-8.5 and SSP2-4.5, respectively. The increase in the mean surface temperature over the Indian region projected by the IITM-ESM is 2.1 °C and 3.3 °C under the intermediate and high emissions scenarios, respectively, by the end of the twenty-first century.

Time-series of global mean precipitation from observations and CMIP6 simulations of IITM-ESM and other CMIP6 models are shown in Fig. 1.3b. Global mean precipitation in IITM-ESM and other CMIP6 models show an increase with an increase in temperature, with higher rate of increase in those models with higher levels of warming

Table 1.3 Climate models from the CMIP5 (Taylor et al. 2012) database used in this study. Historical, RCP4.5 and RCP8.5 forcing experiments have been used

Model ID	Institute, Country
ACCESS1.0	Commonwealth Scientific and Industrial Research Organization (CSIRO), Australia and Bureau of Meteorology (BOM), Australia
CMCC-CM	Euro-Mediterraneo sui Cambiamenti Climatici, Italy
CMCC-CMS	
CNRM-CM5	Centre National de Recherches Meteorologiques, Meteo-France, France
CSIRO-Mk3-6-0	Commonwealth Scientific and Industrial Research Organization (CSIRO), Australia
Inm-cm4	Institute for numerical mathematics, Russia
IPSL-CM5A-LR	Institute Pierre-Simon Laplace, France
IPSL-CM5A-MR	
MPI-ESM-LR	Max Planck Institute for Meteorology, Germany
MPI-ESM-MR	
Total no. models	10

Only the first available realization is used. (Expansions of acronyms are available online at http://www.ametsoc.org/PubsAcronymList). *All the models from CMIP5/CMIP6 used in the study are interpolated to the resolution of 1 × 1 degree

Table 1.4 Climate models from the CMIP6 (Eyring et al. 2016) database used in this study

Model ID	Institute, Country
BCC-CESM2-MR	Beijing Climate Center, China Meteorological Administration, China
CAMS-CSM1-0	Chinese Academy of Meteorological Sciences, China
CANESM5	Canadian Centre for Climate Modelling and Analysis, Canada
CESM2	National Science Foundation, Department of Energy, NCAR, USA
EC-Earth3	EC-Earth brings together 27 research institutes from 10 European countries, Europe
EC-Earth-Veg	
IPSL-CM6A-LR	Institute Pierre-Simon Laplace, France
MIROC6	Atmosphere and Ocean Research Institute (The University of Tokyo), National Institute for Environmental Studies and Japan Agency for Marine-Earth Science and Technology, Japan
MRI-ESM2-0	Meteorological Research Institute, Japan
IITM-ESM	Indian Institute of Tropical Meteorology, India
Total no. models	10

Other specifications are the same as in Table 1.3

(Fig. 1.3b). However, regional precipitation patterns can deviate significantly from the global mean, driven by different drivers of climate change like GHGs, aerosols, etc. Precipitation projections from IITM-ESM and CMIP6 models show a gradual increase in global precipitation over the twenty-first century (Fig. 1.3b). Global mean precipitation is projected to increase in IITM-ESM by more than 0.11 and 0.15 mm day^{-1} by the end of the twenty-first century (w.r.t 1850–1900) under the SSP2-4.5 and SSP5-8.5 scenarios. The precipitation is also projected to increase over the Indian region in IITM-ESM and CMIP6 models under both SSP5-8.5 and SS2-4.5 scenarios. The far future precipitation increase over the Indian region is 0.51 mm day^{-1} under SSP5-8.5 and 0.03 mm day^{-1} under SS2-4.5 as projected by IITM-ESM.

The AR5 assessment of changes in the ocean indicates that the global ocean will warm in all scenarios. Thermal expansion due to ocean warming and glacier melt have been the dominant contributors to the twentieth-century global mean sea-level rise (Church et al. 2013). The projected changes in global mean sea level due to thermal expansion (thermosteric sea level) are shown in Fig. 1.3c under SSP2-4.5 and SSP5-8.5 scenarios. The global mean thermosteric sea level (TSL) from CMIP6 models shows an increase, especially in the recent decades. The IITM-ESM shows an increase in global mean TSL which is similar to the TSL increase in other CMIP6 models. The global mean TSL rise during the far future (2070–2099) will be about 0.08 m–0.17 m under SSP2-4.5 and 0.11 m–0.23 m under SSP5-8.5, respectively. The TSL projections from

IITM-ESM and other CMIP6 models indicate that the rate of global mean sea-level rise for the twenty-first century will exceed the rate observed during the historical period for SSP2-4.5 and SSP5-8.5 scenarios.

We now assess changes in spatial patterns of surface temperature and precipitation for the historical period (1900–2014) and projections for far future (2070–2099) based on IITM-ESM historical simulation and SSP5-8.5 scenario. Spatial patterns of surface temperature change from IITM-ESM and NASA GISS surface temperature analysis (GISSTEMPv4) w.r.t pre-industrial period are shown in Fig. 1.4. Observations reveal increasing surface temperatures over most of the continental region with fastest-warming over the Arctic. Warming over land is higher as compared to oceans. The anthropogenic warming trend is reasonably well represented in IITM-ESM historical simulation. The model simulates the Arctic amplification and increasing temperature over west-central Asia while warming trends over Europe and east-central Asia are underestimated in the historical simulation. Spatial maps of annual mean surface temperature changes in the far future (2070–2099) in the SSP5-8.5 scenario show the largest warming over high latitudes, particularly over the Arctic. As the GMST continues to increase, it is *very likely* that by the end of the twenty-first century most of the global land and ocean areas will be warmer than during the historical period.

The spatial patterns of temperature change over the Indian region are well simulated in IITM-ESM historical simulation. Larger warming pattern is seen over the north and north-west India during the historical period

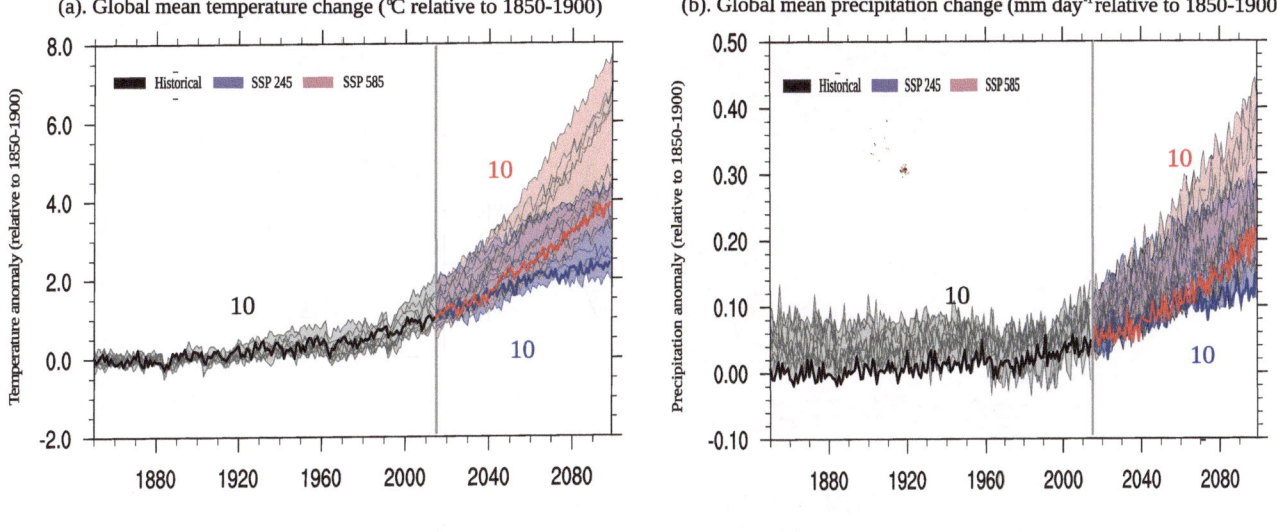

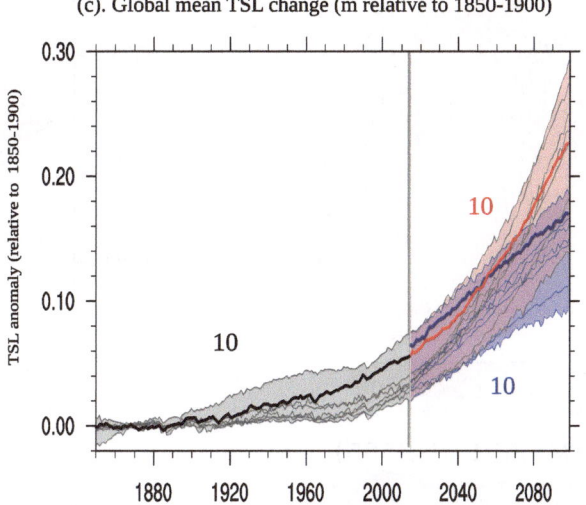

Fig. 1.3 Time series of CMIP6 simulations of the globally averaged annual mean **a** surface air temperature (°C), **b** precipitation (mm day^{-1}) and **c** thermosteric sea level (TSL; m). Plotted are the anomalies with respect to the pre-industrial baseline (1850–1900) for the historical period (black) and future projections following the SSP2-4.5 (blue) and SSP5-8.5 (red) scenarios. Numbers denote the number of models selected for each variable. Solid lines represent the IITM-ESM simulations and the shading represents the range across the models

(1951–2014), warming exceeding 0.5 °C over most part of the country. The warming pattern is projected to enhance in the far future, with warming level exceeding 3 °C as can be seen from the SSP5-8.5 scenario (Fig. 1.4).

Spatial patterns of change in observed and simulated summer monsoon (JJAS) precipitation for the period 1951–2014 are shown in Fig. 1.5. The precipitation pattern shows large spatial variability over the tropical regions. It is interesting to note that the IITM-ESM captures the observed decrease in summer monsoon precipitation over Central India during 1951–2014. A north–south pattern of monsoon rainfall decrease over Central India and an increase over the equatorial Indian Ocean is seen in the IITM-ESM

simulation. The observed weakening of the southwest summer monsoon post-1950s has been attributed to anthropogenic aerosol forcing, regional land use land cover changes as well as rapid warming of the equatorial Indian Ocean SST (Krishnan et al. 2016).

The global mean precipitation is projected to increase with an increase in global mean temperature. Large-scale increase in precipitation is seen over the tropical regions, especially over the monsoon domains and over the Indian landmass in the far future under SSP5-8.5 scenario (Fig. 1.5). Quantitiative estimates of global and regional changes in temperature and precipitation for the near and far future, based on the IITM-ESM and other CMIP6

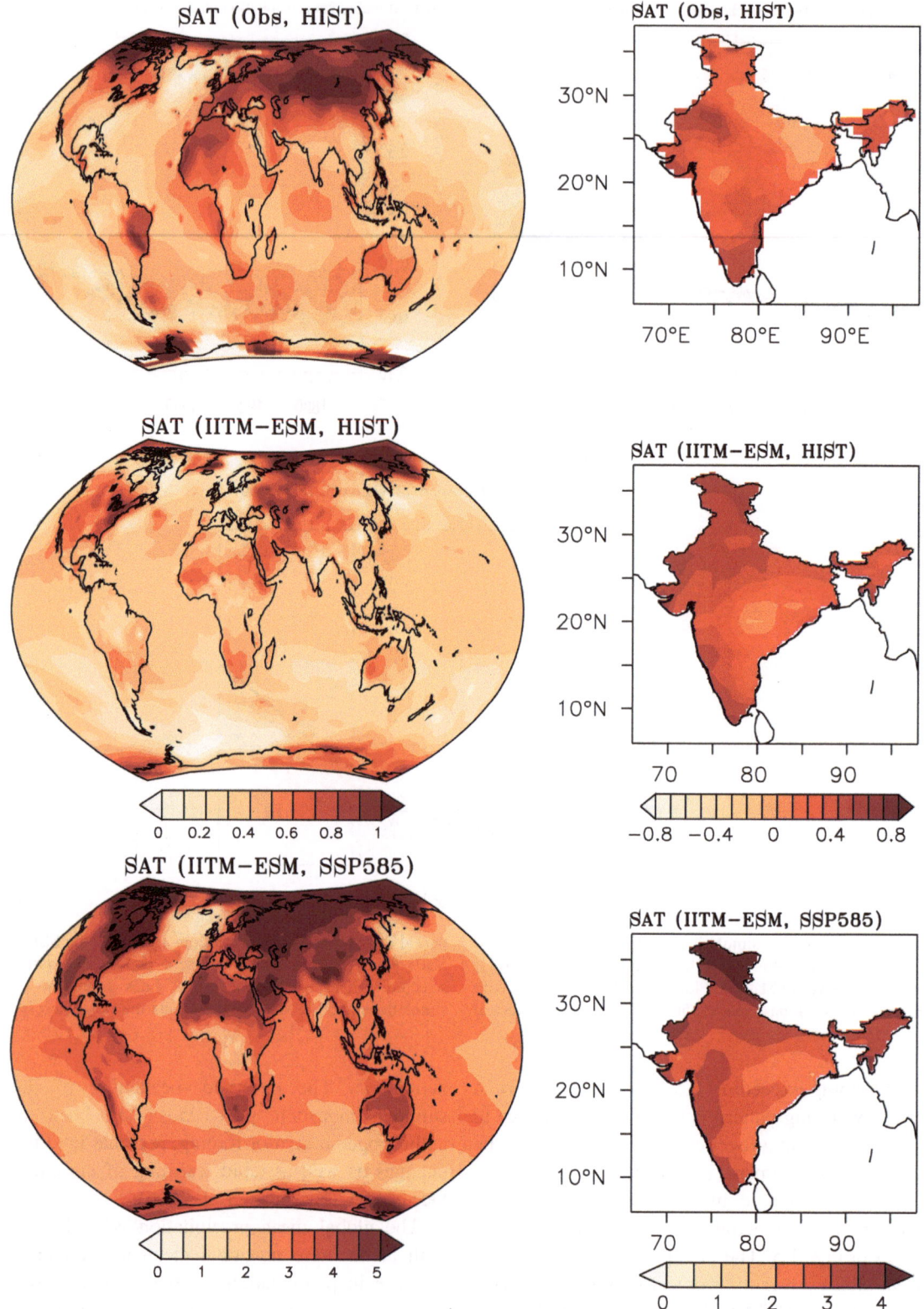

Fig. 1.4 Spatial patterns of change in surface air temperature (SAT; °C) over the globe in the left-hand column, and over India in the right-hand column. In the top row are plotted the observed changes over the globe based on the Climate Research Unit (CRU) dataset, and over India based on the India Meteorological Department (IMD) dataset. Plots in the middle row are from the IITM-ESM simulations for the historical period, and those in the last row are from the IITM-ESM projections following the SSP5-8.5 scenario. The historical changes in the global maps are shown for the period (1901–2014) and those over India are shown for the period (1951–2014). Changes in the historical period (first and middle rows) are calculated as linear trends and expressed in (°C per 114 years) for the global maps and (°C per 64 years) over India. Changes under the SSP5-8.5 scenario (last row) are plotted as the difference in mean temperature between the far future (2070–2099) and pre-industrial (1850–1900) periods

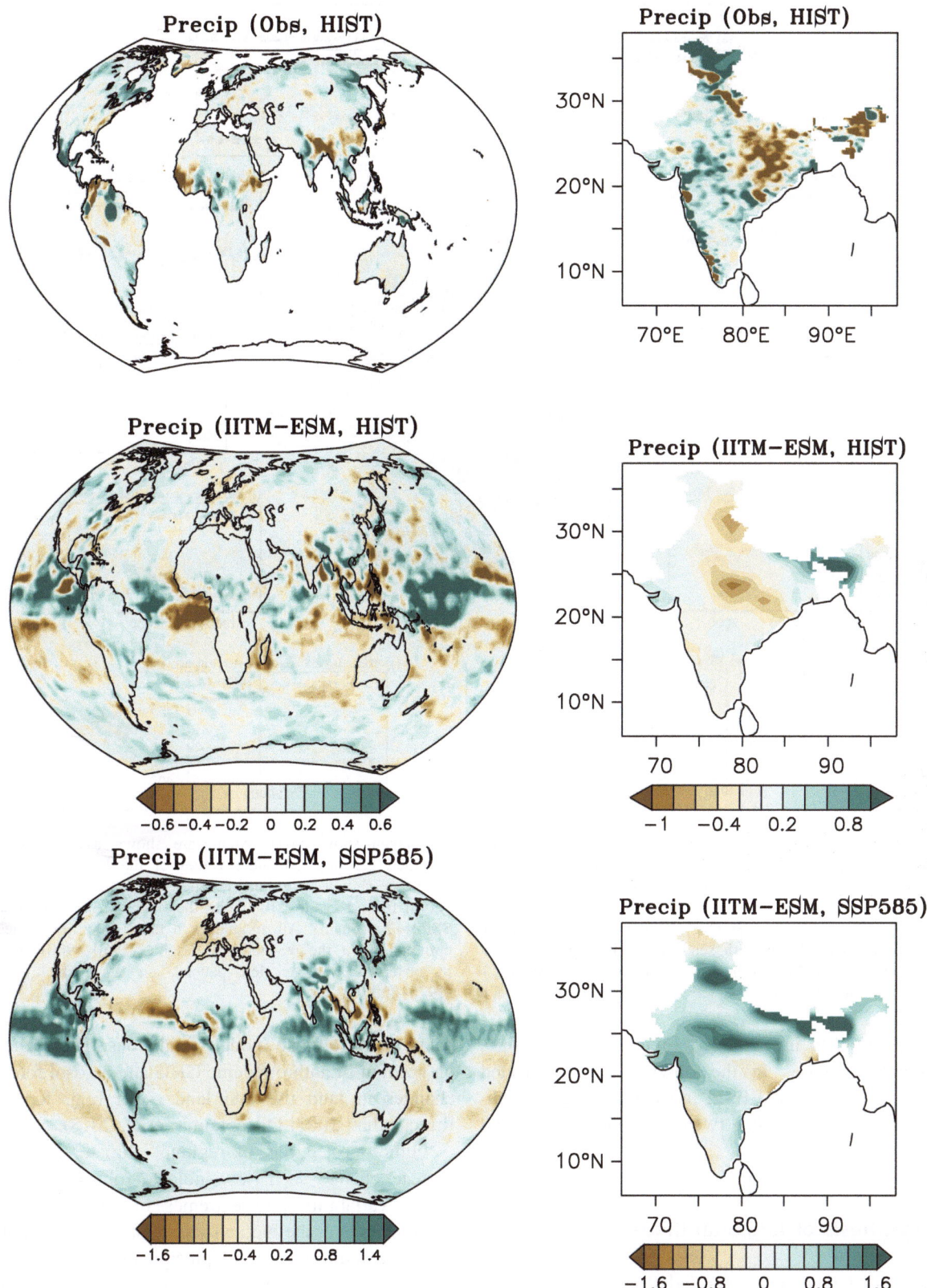

Fig. 1.5 Spatial patterns of change in the June–to-September seasonal precipitation (mm day^{-1}) over the globe in the left-hand column, and over India in the right-hand column. In the top row are plotted the observed changes for the period (1951–2014) relative to (1900–1930) over the globe based on the CRU dataset, and over India based on the IMD dataset. Plots in the middle row are from the IITM-ESM simulations for the historical period, and the plots in the last row are from the IITM-ESM projections following the SSP5-8.5 scenario. The IITM-ESM simulated changes in the historical period (first and middle rows) are plotted as difference for the period (1951–2014) relative to (1850–1900). Changes under the SSP5-8.5 scenario (last row) are plotted as difference between the far-future (2070–2099) relative to (1850–1900).

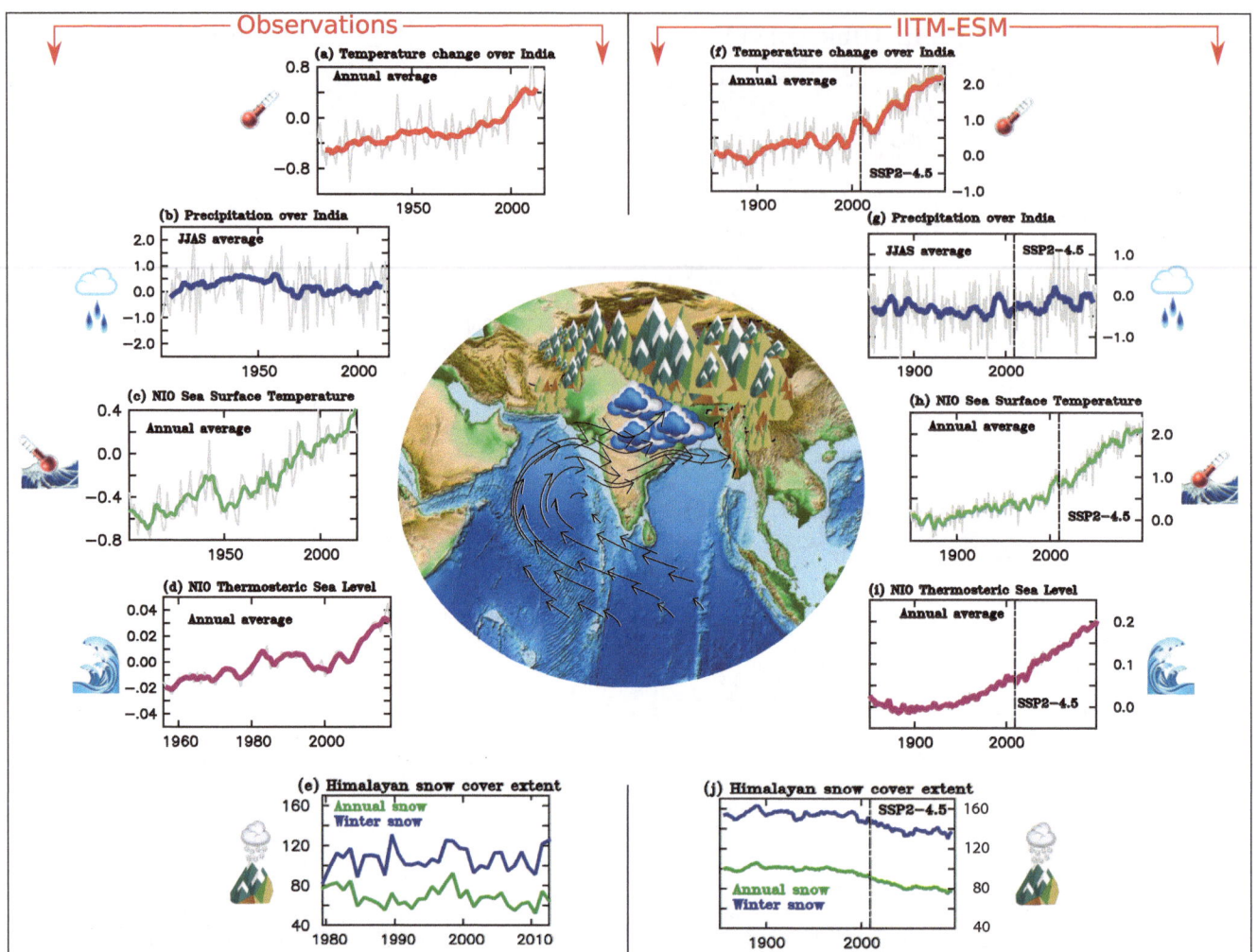

Fig. 1.6 Temporal variation of selected climate variables from observations (panels on the left) and IITM-ESM projections (panels on the right) : Near surface air temperature (°C) over India in (**a**) and (**f**); Summer monsoon precipitation (mm/day) over Indian land region in (**b**) and (**g**); Sea surface temperature of the north Indian Ocean (NIO; °C) in (**c**) and (**h**); Thermosteric sea level (m) in the north Indian Ocean in (**d**) and (**i**); and Himalayan snow cover extent (x 10000 km²) in (**e**) and (**j**). Note that the time-series of all variables, except Himalayan snow cover extent, are shown as anomalies w.r.t. a specified baseline. Absolute values of the Himalayan snow cover extent are shown in (**e**) and (**j**). The IITM-ESM projections cover the period from 1850 to 2100, with the future projections based on the SSP2-4.5 scenario. All anomalies in the IITM-ESM projections are computed w.r.t. the pre-industrial baseline (1850–1900) and anomalies in observed changes (left panels) are w.r.t. the recent period (1976–2005)

models, are tabulated in Table 1.2. Figure 1.6 summarizes the temporal variations in the observed and projected future changes by the IITM-ESM in some key regional climate variables during the twentieth and twenty-first centuries.

1.2.4 Synthesis of Regional Climate Change

The surface air temperature over India has risen by about 0.7 °C during 1901–2018. There are clear signatures of human-induced changes in climate over the Indian region in recent decades, as evidenced by observations, reanalysis datasets and climate model simulations. These changes are on account of anthropogenic GHG and aerosol forcings, and changes in land use and land cover (e.g. Krishnan and Ramanathan 2002; Dileepkumar et al. 2018).

Future projections of regional climate, under different climate change scenarios, indicate robust changes in the mean, variability and extremes of several key climatic parameters over the Indian subcontinent and adjoining areas, e.g. land temperature and precipitation, monsoons, Indian Ocean temperature and sea level, tropical cyclones, Himalayan cryosphere, etc.

A synthesis of the assessed past and projected changes in key climate variables pertinent to the Indian region is presented in Table 1.5 and also illustrated in Fig. 1.6 (IITM-ESM). The qualifiers to express certainty in the assessments are presented in Box 1.4.

Box 1.4: Qualifiers to Express Certainty in Assessment Findings

Two qualifiers have been used in this assessment report to express the degree of certainty in key findings.

- **Confidence** is a qualitative expert judgement by the authors based on an evaluation of available information in terms of 1. The amount, quality and consistency of evidence and 2. Agreement within the surveyed literature.

Confidence levels (based on IPCC):

- Very high: Robust evidence, high agreement.
- High: Robust evidence, medium agreement; Medium evidence, high agreement.
- Medium: Medium evidence, medium agreement.
- Low: Limited evidence, low agreement.

Broad criteria adopted in this Report to gauge 'evidence':

- >3 scientific papers: Robust evidence;
- 2–3 scientific papers: Medium evidence;
- 2 scientific papers: Limited evidence;
- 1 scientific paper: Insufficient evidence.

Broad criteria adopted in this Report to gauge 'agreement':

- >70% agreement in surveyed literature: High agreement;
- 50–70% agreement in surveyed literature: Medium agreement;
- 30–50% agreement in surveyed literature: Low agreement;
- <30% agreement in surveyed literature: No agreement.

Confidence cannot be interpreted statistically.

- **Likelihood** is a quantitative measure of uncertainty based on the probability of an outcome or result based on a statistical analysis of observational data or modelling outcomes or on the authors' expert judgement.

Likelihood levels (adopted from IPCC):

- Virtually certain: 99–100% probability;
- Very likely: 90–100%;
- Likely: 66–100%;
- About as likely as not: 33–66%;
- Unlikely: 0–33%;
- Very unlikely: 0–10%;
- Exceptionally unlikely: 0–1%.

See for more details: Mastrandrea, M. D., C. B. Field, T. F. Stocker, O. Edenhofer, K. L. Ebi, D. J. Frame, H. Held, E. Kriegler, K. J. Mach, P. R. Matschoss, G.-K. Plattner, G. W. Yohe and F. W. Zwiers 2010: Guidance Note for Lead Authors of the IPCC Fifth Assessment Report on Consistent Treatment of Uncertainties, Intergovernmental Panel on Climate Change (IPCC), Geneva, Switzerland, 4 pp.

1.3 Scope of the Report

This report provides a detailed overview and synthesis of the published scientific literature on climate change over India and adjoining regions. The objectives are to provide a state-of-the-art assessment of how and why India's climate is changing, changes that are projected for the future, uncertainties and knowledge gaps, and identification of areas that require greater research.

The scope of this report is the physical science basis of climate change with a focus on regional climate drivers specific to the Indian land area and the surrounding ocean. It is a region-focused analogue of the global scale assessment by Working Group I of the Intergovernmental Panel on Climate Change.

This report will be useful to advance public awareness of India's changing climate and to inform mitigation and adaptation decision making. While it is meant to be policy relevant, this report is not intended to be policy prescriptive.

This report is organized as follows: Chapters 2 and 3 focus on observed and projected trends in the mean and extremes of temperature and precipitation. Chapters 4 and 5 quantify the spatial and temporal variations in GHG fluxes and concentrations, and climate forcing due to anthropogenic aerosol emissions and trace gases over India. Chapter 6 discusses the observed and projected changes in

Table 1.5 Synthesis of the assessed past and projected changes in key climate variables pertinent to the Indian region

Warming over India and the Indian Ocean (Chaps. 2, 10)

The annual mean near-surface air temperature over India has warmed by around 0.7 °C during 1901–2018 (Srivastava et al. 2019), with the post-1950 trends attributable largely to anthropogenic activities (Dileepkumar et al. 2018) (*High confidence*). Atmospheric moisture content over the Indian region has also risen during this period (Krishnan et al. 2016; Mukhopadhyay et al. 2017; Mukherjee et al. 2018) (*High confidence*). The mean temperature rise over India by the end of the twenty-first century is projected to be in the range of 2.4–4.4 °C across greenhouse gas warming scenarios relative to the average temperature over 1976–2005.

The Indian Ocean has also experienced significant warming in recent decades in association with anthropogenic radiative forcing (Du and Xie 2008), as well as ocean–atmosphere coupled feedbacks arising from long-term changes in monsoonal wind patterns (Swapna et al. 2014) (*High confidence*). Sea surface temperature (SST) in the tropical Indian Ocean has risen by 1 °C on average over 1951–2015 and is projected to increase further during the twenty-first century.

Monsoon Precipitation (Chap. 3)

Warming due to increasing concentration of atmospheric GHGs and moisture content is generally expected to strengthen the Indian monsoon. Yet, the observational records show that there has been a declining trend in summer monsoon precipitation since 1950 (Kulkarni 2012), with particularly notable decreases in parts of the Indo-Gangetic plains and the Western Ghats (Krishnan et al. 2013; Roxy et al. 2015). Climate modelling studies suggest that the observed changes have resulted in response to the radiative effects of the northern hemispheric (NH) anthropogenic aerosols and regional LULC, which have more than offset the precipitation enhancing tendency of GHG warming in the past 6–7 decades (e.g. Bollasina et al. 2011; Krishnan et al. 2016; Sanap et al. 2015; Undorf et al. 2018) (*Medium confidence*).

In contrast, the frequency of localized heavy precipitation occurrences has risen significantly over Central India in the past 6–7 decades (Roxy et al. 2017; Mukherjee et al. 2018) (*High confidence*).

With anticipated reductions in NH aerosol emissions, future changes in the monsoon precipitation are expected to be prominently constrained by the effects of GHG warming. With the resultant increase in temperature and atmospheric moisture, climate models project a considerable rise in the mean, extremes and interannual variability of monsoon precipitation by the end of the century (Kitoh 2017).

Droughts and Floods (Chap. 6)

India has witnessed a higher frequency of droughts and expansion of drought-affected areas since 1950. While climate models project an enhancement of mean monsoon rainfall in the future, they concurrently project an increase in the occurrence, severity and area under drought. These changes are linked to increased variability of monsoon precipitation, and increase in water vapour demand in a warmer atmosphere that would tend to decrease soil moisture content (Menon et al. 2013; Scheff and Frierson 2014; Jayasankar et al. 2015; Sharmila et al. 2015; Krishnan et al. 2016; Preethi et al. 2019) (*High confidence*).

Flooding events over India have also increased since 1950, in part due to enhanced occurrence of localized, short-duration intense rainfall events and flooding occurrences due to intense rainfall are projected to increase in the future (Hirabayashi et al. 2013; Ali and Mishra 2018; Lutz et al. 2019) (*High confidence*). Higher rates of glacier and snowmelt in a warming world would enhance stream flow and compound flood risk over the Himalayan river basins. The Indus, Ganga and Brahmaputra basins are considered particularly at risk of enhanced flooding in the future in the absence of additional adaptation and risk mitigation measures (Lutz et al. 2014).

Sea-level rise in the North Indian Ocean (Chap. 9)

Sea-level rise is intimately related to thermal expansion due to rising ocean SST and heat content, and the melting of glaciers that add water to the world's oceans. Rates of sea-level variations differ from region to region.

The North Indian Ocean (NIO) rose at a rate of 3.3 mm year^{-1} during 1993–2017, similar to the global mean (Swapna et al. 2017). While thermal expansion (thermosteric) has dominated sea-level rise in the NIO) (*High confidence*), the major contribution to global mean sea-level rise is from glacier melt (IPCC AR5).

The thermosteric sea-level rise of the NIO during the recent 3–4 decades is closely linked to the weakening trend of summer monsoon winds and the associated slow down of heat transport out of the NIO (Swapna et al. 2017). Future changes in the strength of monsoon winds have implications on the NIO sea-level variations.

Tropical Cyclonic Storms (Chap. 8)

The intensity of tropical cyclones (TC) is closely linked to ocean SST and heat content, with regional differences in their relationships.

The frequency of very severe cyclonic storms (VSCS) over the NIO during the post-monsoon season has significantly increased in the past two decades, despite an overall reduction in the annual TC activity (*High confidence*). With continued global warming, the activity of VSCS over the NIO is projected to further increase during the twenty-first century.

Himalayan Cryosphere (Chap. 11)

The Hindukush Himalayas (HKH) underwent rapid warming at a rate of about 0.2°C per decade during the last 6–7 decades) (*High confidence*). Higher elevations of the Tibetan Plateau (> 4 km) experienced even stronger warming in a phenomenon alluded to as Elevation Dependent Warming (Liu et al. 2009; Krishnan et al. 2019b) (*High confidence*). With continued global warming, the temperature in the HKH is projected to rise substantially during the twenty-first century.

The HKH experienced a significant decline in snowfall (Ren et al. 2015; You et al. 2015) and glacial area (Kulkarni and Karyakarte 2014; Wester et al. 2019) in the last 4–5 decades (*Medium confidence*). With continuing warming, climate models project a continuing decline in snowfall over the HKH during the 21st century, but with wide inter-model spread. In contrast, parts of the Karakoram Himalayas have experienced increase in wintertime frozen precipitation in the recent decades, in association with enhanced amplitude variations of Western Disturbances (Kapnick et al. 2014; Kääb et al. 2015; Krishnan et al. 2019b).

the characteristics of floods and droughts in India associated with changes in precipitation. Chapter 7 discusses synoptic scale systems including trends in monsoon depressions and western disturbances. Chapters 8, 9, and 10 discuss the changes in the frequency and intensity of extreme storms, trends and projections of sea level rise and its variability along the Indian coast, and historical changes and projections of warming of the Indian Ocean and changes in its heat content. Chapter 11 discusses observed and projected changes in temperature, snowfall, and glaciers in the Hindu Kush Himalayas. Chapter 12 closes with a brief outline of the potential impacts of climate change pertinent to India and policies that may help advance adaption and mitigation efforts.

Comprehensive assessments of the societal and sectoral impacts of regional climate change will require targeted research and are beyond the scope of this report.

References

Allen, M.R., O.P. Dube, W. Solecki, F. Aragón-Durand, W. Cramer, S. Humphreys, M. Kainuma, J. Kala, N. Mahowald, Y. Mulugetta, R. Perez, M. Wairiu, and K. Zickfeld, 2018: Framing and Context. In: Global Warming of 1.5°C. An IPCC Special Report on the impacts of global warming of 1.5°C above pre-industrial levels and related global GHG emissions pathways, in the context of strengthening the global response to the threat of climate change, sustainable development, and efforts to eradicate poverty [Masson-Delmotte, V., P. Zhai, H.-O. Pörtner, D. Roberts, J. Skea, P.R. Shukla, A. Pirani, W. Moufouma-Okia, C. Péan, R. Pidcock, S. Connors, J.B. R. Matthews, Y. Chen, X. Zhou, M.I. Gomis, E. Lonnoy, T. Maycock, M. Tignor, and T. Waterfield (eds.)]

Ali H, Mishra V (2018) Increase in subdaily precipitation extremes in india under 1.5 and 2.0 °C warming worlds. Geophys Res Lett 45 (14):6972–6982

Bindoff NL, Stott PA, AchutaRao KM, Allen MR, Gillett N, Gutzler D, Hansingo K, Hegerl G, Hu Y, Jain S, Mokhov II, Overland J, Perlwitz J, Sebbari R, Zhang X (2013) Detection and attribution of climate change: from global to regional. In: Stocker TF, Qin D, Plattner G-K, Tignor M, Allen SK, Boschung J, Nauels A, Xia Y, Bex V, Midgley PM (eds) Climate change 2013: the physical science basis. Contribution of working group I to the fifth assessment report of the intergovernmental panel on climate change. Cambridge University Press, Cambridge

Bollasina MA, Ming Y, Ramaswamy V (2011) Anthropogenic aerosols and the weakening of the South Asian summer monsoon. Science 334:502–505

Boos WR, Kuang Z (2010) Dominant control of the South Asian monsoon by orographic insulation versus plateau heating. Nature 463:218–222. https://doi.org/10.1038/nature08707

Burkett VR, Suarez AG, Bindi M, Conde C, Mukerji R, Prather MJ, St. Clair AL, Yohe GW (2014) Point of departure. In: Field CB, Barros VR, Dokken DJ, Mach KJ, Mastrandrea MD, Bilir TE, Chatterjee M, Ebi KL, Estrada YO, Genova RC, Girma B, Kissel ES, Levy AN, MacCracken S, Mastrandrea PR, White LL (eds) Climate change 2014: impacts, adaptation, and vulnerability. Part A: global and sectoral aspects. Contribution of working group II to the fifth assessment report of the intergovernmental panel on

climate change. Cambridge University Press, Cambridge, pp 169–194

Calel R, Stainforth DA (2017) On the physics of three integrated assessment models. Bull Am Meteorol Soc 98:1199–1216. https://doi.org/10.1175/BAMS-D-16-0034.1

Choudhury AD, Krishnan R (2011) Dynamical response of the South Asian Monsoon trough to latent heating from stratiform and convective precipitation. J Atmos Sci 68(6):1347–1363. https://doi.org/10.1175/2011JAS3705.1

Choudhury AD, Krishnan R, Ramarao MVS, Vellore R, Singh M, Mapes B (2018) A phenomenological paradigm for midtropospheric cyclogenesis in the Indian summer monsoon. J Atmos Sci 75 (9):2931–2954. https://doi.org/10.1175/JAS-D-17-0356.1

Church JA, Clark PU, Cazenave A, Gregory JM, Jevrejeva S, Levermann A, Merrifield MA, Milne GA, Nerem RS, Nunn PD, Payne AJ, Pfeffer WT, Stammer D, Unnikrishnan AS (2013) Sea level change. In: Stocker TF, Qin D, Plattner G-K, Tignor M, Allen SK, Boschung J, Nauels A, Xia Y, Bex V, Midgley PM (eds) Climate change 2013: the physical science basis. Contribution of working group I to the fifth assessment report of the intergovernmental panel on climate change. Cambridge University Press, Cambridge

Collins M, Knutti R, Arblaster J, Dufresne J-L, Fichefet T, Friedlingstein P, Gao X, Gutowski WJ, Johns T, Krinner G, Shongwe M, Tebaldi C, Weaver AJ, Wehner M (2013) Long-term climate change: projections, commitments and irreversibility. In: Stocker TF, Qin D, Plattner G-K, Tignor M, Allen SK, Boschung J, Nauels A, Xia Y, Bex V, Midgley PM (eds) Climate change 2013: the physical science basis. Contribution of working group I to the fifth assessment report of the intergovernmental panel on climate change. Cambridge University Press, Cambridge

Cowtan K, Hausfather Z, Hawkins E, Jacobs P, Mann ME, Miller SK, Steinman BA, Stolpe MB, Way RG (2015) Robust comparison of climate models with observations using blended land air and ocean sea surface temperatures. Geophys Res Lett 42:6526–6534. https://doi.org/10.1002/2015GL064888

Dileepkumar R, AchutaRao K, Arulalan T (2018) Human influence on sub-regional surface air temperature change over India. Sci Rep 8 (1):8967. https://doi.org/10.1038/s41598-018-27185-8

Dimri AP, Niyogi D, Barros AP, Ridley J, Mohanty UC, Yasunari T, Sikka DR (2015) Western disturbances: a review. Rev Geophys. https://doi.org/10.1002/2014RG000460

Du Y, Xie S-P (2008) Role of atmospheric adjustments in the tropical Indian Ocean warming during the 20th century in climate models. Geophys Res Lett 35:L08712. https://doi.org/10.1029/2008GL033631

Eyring V, Bony S, Meehl G, Senior C, Stevens B, Stouffer R, Taylor K (2016) Overview of the coupled model intercomparison project phase 6 (CMIP6) experimental design and organization. Geosci Model Dev 9(5):1937–1958. https://doi.org/10.5194/gmd-9-1937-2016

Flato G, Marotzke J, Abiodun B, Braconnot P, Chou SC, Collins W, Cox P, Driouech F, Emori S, Eyring V, Forest C, Gleckler P, Guilyardi E, Jakob C, Kattsov V, Reason C, Rummukainen M (2013) Evaluation of climate models. In: Stocker TF, Qin D, Plattner G-K, Tignor M, Allen SK, Boschung J, Nauels A, Xia Y, Bex V, Midgley PM (eds) Climate change 2013: the physical science basis. Contribution of working group I to the fifth assessment report of the intergovernmental panel on climate change. Cambridge University Press, Cambridge

Gadgil S (2003) The Indian monsoon and its variability. Annu Rev Earth Planet Sci 31:429–467

Hartmann DL, Klein Tank AMG, Rusticucci M, Alexander LV, Brönnimann S, Charabi Y, Dentener FJ, Dlugokencky EJ, Easterling DR, Kaplan A, Soden BJ, Thorne PW, Wild M, Zhai PM (2013) Observations: atmosphere and surface. In: Stocker TF,

Qin D, Plattner G-K, Tignor M, Allen SK, Boschung J, Nauels A, Xia Y, Bex V, Midgley PM (eds) Climate change 2013: the physical science basis. Contribution of working group I to the fifth assessment report of the intergovernmental panel on climate change. Cambridge University Press, Cambridge

Haustein K, Allen MR, Forster PM, Otto FEL, Mitchell DM, Matthews HD, Frame DJ (2017) A real-time global warming index. Sci Rep 7:15417. https://doi.org/10.1038/s41598-017-14828-5

Hirabayashi Y, Mahendran R, Koirala S et al (2013) Global flood risk under climate change. Nat Clim Change 3:816–821. https://doi.org/10.1038/nclimate1911

Houze RA (1997) Stratiform precipitation in regions of convection: a meteorological paradox? Bull Am Meteor Soc 78(10):2179–2196. https://doi.org/10.1175/1520-0477(1997)078%3c2179:SPIROC%3e2.0.CO;2

Houze RA, Wilton DC, Smull BF (2007) Monsoon convection in the Himalayan region as seen by the TRMM precipitation radar. Quart J R Meteorol Soc. https://doi.org/10.1002/qj.106

Hunt KMR, Turner AG, Shaffrey LC (2018) The evolution, seasonality and impacts of western disturbances. R Meteorol Soc, Q.J. https://doi.org/10.1002/qj.3200

Jayasankar CB, Surendran S, Rajendran K (2015) Robust signals of future projections of Indian summer monsoon rainfall by IPCC AR5 climate models: role of seasonal cycle and interannual variability. Geophys Res Lett 42(9):3513–3520. https://doi.org/10.1002/2015GL063659

Kääb A, Treichler D, Nuth C, Berthier E (2015) Brief communication: contending estimates of 2003–2008 glacier mass balance over the Pamir–Karakoram–Himalaya. Cryosphere 9(2):557–564. https://doi.org/10.5194/tc-9-557-2015

Kapnick SB, Delworth TL, Ashfaq M, Malyshev S, Milly PCD (2014) Snowfall less sensitive to warming in Karakoram than in Himalayas due to a unique seasonal cycle. Nat Geosci 7:834–840

Kirtman B et al (2013) Near-term climate change: projections and predictability. In: Stocker TF, Qin D, Plattner G-K, Tignor M, Allen SK, Boschung J, Nauels A, Xia Y, Bex V, Midgley PM (eds) Climate change 2013: the physical science basis. Contribution of working group I to the fifth assessment report of the intergovernmental panel on climate change. Cambridge University Press, Cambridge, pp 953–1028

Kitoh A (2017) The Asian monsoon and its future change in climate models: a review. J Meteorol Soc Jpn. Ser. II 95(1):7–33. https://doi.org/10.2151/jmsj.2017-002

Konwar M, Das SK, Deshpande SM, Chakravarty K, Goswami BN (2014) Microphysics of clouds and rain over the Western Ghat. J Geophys Res Atmos 119(10):6140–6159. https://doi.org/10.1002/2014JD021606

Krishnamurti TN, Surgi N (1987) Observational aspects of summer monsoon. In: Chang C-P, Krishnamurti TN (eds) Chapter in book Monsoon meteorology. Oxford University Press, Oxford, pp 3–25

Krishnan R, Ramanathan V (2002) Evidence of surface cooling from absorbing aerosols. Geophys Res Lett 29:2002GL014687, 54 (1–4)

Krishnan R, Sabin TP, Ayantika DC, Kitoh A, Sugi M, Murakami H, Turner AG, Slingo JM, Rajendran K (2013) Will the South Asian monsoon overturning circulation stabilize any further? Clim Dyn 40:187–211

Krishnan R, Sabin TP, Vellore R, Mujumdar M, Sanjay J, Goswami BN, Hourdin F, Dufresne J-L, Terray P (2016) Deciphering the desiccation trend of the South Asian monsoon hydroclimate in a warming world. Clim Dyn 47:1007–1027. https://doi.org/10.1007/s00382-015-2886-5

Krishnan R, Sabin TP, Madhura RK, Vellore RK, Mujumdar M, Sanjay J, Nayak S, Rajeevan M (2019a) Non-monsoonal precipitation response over the western Himalayas to climate change. Clim Dyn 52:4091–4109

Krishnan R et al (2019b) Unravelling climate change in the Hindu Kush Himalaya: rapid warming in the mountains and increasing extremes. In: Wester P, Mishra A, Mukherji A, Shrestha A (eds) The Hindu Kush himalaya assessment. Springer, Cham

Kulkarni A (2012) Weakening of Indian summer monsoon rainfall in warming environment. Theoret Appl Climatol 109:447–459

Kulkarni AV, Karyakarte Y (2014) Observed changes in Himalayan glaciers. Curr Sci 106(2):237–244

Liu X, Cheng Z, Yan L et al (2009) Elevation dependency of recent and future minimum surface air temperature trends in the Tibetan Plateau and its surroundings Glob. Planet Change 68:164–174

Lutz AF, Immerzeel WW, Shrestha AB, Bierkens MFP (2014) Consistent increase in High Asia's runoff due to increasing glacier melt and precipitation. Nat Clim Chang 4:587–592. https://doi.org/10.1038/nclimate2237

Lutz AF, terMaat HW, Wijngaard RR et al (2019) South Asian river basins in a 1.5 °C warmer world. Reg Environ Chang 19:833–847. https://doi.org/10.1007/s10113-018-1433-4

Marcott SA, Shakun JD, Clark PU, Mix AC (2013) A reconstruction of regional and global temperature for the past 11,300 years. Science 339(6124):1198–1201. https://doi.org/10.1126/science.1228026

Meehl GA, Boer GJ, Covey C, Latif M, Stouffer RJ (2000) The coupled model intercomparison project (CMIP). Bull Amer Meteor Soc 81:313–318

Meehl GA, Hibbard KA (2007) A strategy for climate change stabilization experiments with AOGCMS and ESMS. WCRP Informal Rep. 3/2007, ICPO Publ. 112, IGBP Rep. 57, 35 pp

Menon A, Levermann A, Schewe J (2013) Enhanced future variability during India's rainy season. GRL 40:3242–3247

Mukherjee S, Aadhar S, Stone D, Mishra V (2018) Increase in extreme precipitation events under anthropogenic warming in India. Weather Clim Extremes 20:45–53

Mukhopadhyay P, Jaswal A, Deshpande M (2017) In: Rajeevan M, Nayak S (eds) Observed climate variability and change over the Indian region. Springer, Berlin, pp 129–144

Myhre G, Shindell D, Bréon F-M, Collins W, Fuglestvedt J, Huang J, Koch D, Lamarque J-F, Lee D, Mendoza B, Nakajima T, Robock A, Stephens G, Takemura T, Zhang H (2013) Anthropogenic and natural radiative forcing. In: Stocker TF, Qin D, Plattner G-K, Tignor M, Allen SK, Boschung J, Nauels A, Xia Y, Bex V, Midgley PM (eds) Climate change 2013: the physical science basis. Contribution of working group I to the fifth assessment report of the intergovernmental panel on climate change. Cambridge University Press, Cambridge

Nair VS, Babu SS, Manoj MR, Moorthy KK, Chin M (2016) Direct radiative effects of aerosols over South Asia from observations and modeling. Clim Dyn 49:1411–1428. https://doi.org/10.1007/s00382-016-3384-0

O'Neill BC et al (2017) The roads ahead: narratives for shared socioeconomic pathways describing world futures in the 21st century. Glob Environ Change 42:169–180. https://doi.org/10.1016/j.gloenvcha.2015.01.004

Otto FEL, Frame DJ, Otto A, Allen MR (2015) Embracing uncertainty in climate change policy. Nat Clim Change 5:1–5. https://doi.org/10.1038/nclimate2716

Pant GB, Parthasarathy B (1981) Some aspects of an association between the Southern Oscillation and Indian summer monsoon. Arch Meteor Geophys Biokl B29:245–252

Rajeevan M, Unnikrishnan CK, Bhate J, Niranjan Kumar K, Sreekala PP (2012) Northeast monsoon over India. Meteorol Appl 19:226–236

Rajendran K, Kitoh A (2008) Indian summer monsoon in future climate projection by a super high-resolution global model. Curr Sci 95 (11):1560–1569

Rajeevan M, Gadgil S, Bhate J (2010) Active and break spells of the Indian summer monsoon. J Earth Sys Sci 119:229–248

Rao YP (1976) Southwest Monsoons. Meteor Monogr 1. India Meteorological Department, pp 1–367

Ren YY, Parker D, Ren GY, Dunn R (2015) Tempo-spatial characteristics of sub-daily temperature trends in mainland China. Clim Dyn 46(9–10):2737–2748. https://doi.org/10.1007/s00382-015-2726-7

Richardson M, Cowtan K, Millar RJ (2018) Global temperature definition affects achievement of long-term climate goals. Environ Res Lett 13(5):054004. https://doi.org/10.1088/1748-9326/aab305

Romatschke U, Houze RA (2011) Characteristics of precipitating convective systems in the South Asian Monsoon. J Hydrometeorol 12(1):3–26. https://doi.org/10.1175/2010JHM1289.1

Roxy MK, Ghosh S, Pathak A, Athulya R, Mujumdar M, Raghu M, Pascal T, Rajeevan M (2017) A threefold rise in widespread extreme rain events over central India. Nat Commun 8:708

Roxy MK, Kapoor R, Terray P, Murtugudde R, Ashok K, Goswami BN (2015) Drying of Indian subcontinent by rapid Indian Ocean warming and a weakening land-sea thermal gradient. Nat Commun 6(7423)

Sabin TP, Krishnan R, Ghattas J, Denvil S, Dufresne J-L, Hourdin F, Pascal T (2013) High resolution simulation of the South Asian monsoon using a variable resolution global climate model. Clim Dyn 41(1):173–194. https://doi.org/10.1007/s00382-012-1658-8

Sanap SD, Pandithurai G, Manoj MG (2015) On the response of Indian summer monsoon to aerosol forcing in CMIP5 model simulations. Clim Dyn. https://doi.org/10.1007/s00382-015-2516-2

Scheff J, Frierson DMW (2014) Scaling potential evapotranspiration with greenhouse warming. J Clim 27:1539–1558. https://doi.org/10.1175/JCLI-D-13-00233.1

Schneider T, Bischoff T, Huag GH (2014) Migrations and dynamics of the intertropical convergence zone. Nature 45. DOI:10.1038/nature13636

Sharmila S, Joseph S, Sahai AK, Abhilash S, Chattopadhyay R (2015) Future projection of Indian summer monsoon variability under climate change scenario: an assessment from CMIP5 climate models. Global Planet Change 124:62–78. https://doi.org/10.1016/j.gloplacha.2014.11.004

Shige S, Nakano Y, Yamamoto MK (2017) Role of orography, diurnal cycle, and intraseasonal oscillation in summer monsoon rainfall over the Western Ghats and Myanmar Coast. J Clim 30(23):9365–9381. https://doi.org/10.1175/JCLI-D-16-0858.1

Sinha A, Kathayat G, Cheng H, Breitenbach SFM, Berkelhammer M, Mudelsee M, Biswas J, Edwards RL (2015) Trends and oscillations in the Indian summer monsoon rainfall over the last two millennia. Nat Commun. https://doi.org/10.1038/ncomms7309

Sperber KR, Annamalai H, Kang I-S, Kitoh A, Moise A, Turner A et al (2013) The Asian summer monsoon: an intercomparison of CMIP5 vs. CMIP3 simulations of the late 20th century. Clim Dyn 41:2711–2744. https://doi.org/10.1007/s00382-012-1607-6

Srivastava AK, Revadekar JV, Rajeevan M (2019) Regional climates: Asia: South Asia (in "State of the climate in 2018"). Bull Am Meteor Soc 100(9):S236–S240. https://doi.org/10.1175/2019BAMSStateoftheClimate.1

Stano G, Krishnamurti TN, Vijaya Kumar TSV, Chakraborty A (2002) Hydrometeor structure of a composite monsoon depression using the TRMM radar. Tellus A 54:370–381. https://doi.org/10.1034/j.1600-0870.2002.01330.x

Stocker TF, Qin D, Plattner G-K, Alexander LV, Allen SK, Bindoff NL, Bréon F-M, Church JA, Cubasch U, Emori S, Forster P, Friedlingstein P, Gillett N, Gregory JM, Hartmann DL, Jansen E, Kirtman B, Knutti R, Krishna Kumar K, Lemke P, Marotzke J, Masson-Delmotte V, Meehl GA, Mokhov II, Piao S, Ramaswamy V, Randall D, Rhein M, Rojas M, Sabine C, Shindell D, Talley LD, Vaughan DG, Xie S-P (2013) Technical summary. In: Stocker TF, Qin D, Plattner G-K, Tignor M, Allen SK, Boschung J, Nauels A, Xia Y, Bex V, Midgley PM (eds) Climate change 2013: the physical science basis. Contribution of working group I to the fifth assessment report of the intergovernmental panel on climate. Cambridge University Press, Cambridge

Swapna P, Jyoti J, Krishnan R, Sandeep N, Griffies SM (2017) Multidecadal weakening of Indian summer monsoon circulation induces an increasing Northern Indian Ocean Sea Level. Geophys Res Lett 44. DOI:10.1002/2017GL074706

Swapna P, Krishnan R, Sandeep N, Prajeesh AG, Ayantika DC, Manmeet S et al (2018) Long-term climate simulations using the IITM earth system Model (IITM-ESMv2) with focus on the South Asian Monsoon. J Adv Model Earth Syst 10. DOI:10.1029/2017MS001262

Swapna P, Krishnan R, Wallace JM (2014) Indian Ocean and monsoon coupled interactions in a warming environment. Clim Dyn 42(9–10):2439–2454. https://doi.org/10.1007/s00382-013-1787-8

Swapna P, Roxy MK, Aparna K, Kulkarni K, Prajeesh AG, Ashok K et al (2015) The IITM earth system model: transformation of a seasonal prediction model to a long-term climate model. Bull Am Meteorol Soc 96. DOI:10.1175/BAMS-D-13-00276.1

Taylor KE, Stouffer RJ, Meehl GA (2012) An overview of CMIP5 and the experiment design. Bull Am Meteor Soc 93(4):485–498. https://doi.org/10.1175/BAMS-D-11-00094.1

Turner AG, Annamalai H (2012) Climate change and the South Asian summer monsoon. Nat Clim Chang 2:587. Available at https://doi.org/10.1038/nclimate1495

Undorf S, Polson D, Bollasina MA, Ming Y, Schurer A, Hegerl GC (2018) Detectable impact of local and remote anthropogenic aerosols on the 20th century changes of West African and South Asian monsoon precipitation. J Geophy Res (Atmos) 123:4871–4889. https://doi.org/10.1029/2017JD027711

van Vuuren DP, Edmonds J, Kainuma M, Riahi K, Thomson A, Hibbard K et al (2011) The representative concentration pathways: an overview. Clim Change 109:5–31. https://doi.org/10.1007/s10584-011-0148-z

Vaughan DG, Comiso JC, Allison I, Carrasco J, Kaser G, Kwok R, Mote P, Murray T, Paul F, Ren J, Rignot E, Solomina O, Steffen K, Zhang T (2013) Observations: cryosphere. In: Stocker TF, Qin D, Plattner G-K, Tignor M, Allen SK, Boschung J, Nauels A, Xia Y, Bex V, Midgley PM (eds) Climate change 2013: the physical science basis. Contribution of working group I to the fifth assessment report of the intergovernmental panel on climate change

Xie S-P et al (2006) Role of narrow mountains in large-scale organization of Asian Monsoon Convection*. J Clim 19 (14):3420–3429. https://doi.org/10.1175/JCLI3777.1

You QL, Min J, Zhang W, Pepin N, Kang S (2015) Comparison of multiple datasets with gridded precipitation observations over the Tibetan Plateau. Clim Dyn 45:791–806

Wester P et al (eds) (2019) The Hindu Kush Himalaya Assessment. Springer International Publishing, Cham. https://doi.org/10.1007/978-3-319-92288-1

Coordinating Lead Authors

J. Sanjay, Indian Institute of Tropical Meteorology (IITM-MoES), Pune, India,
e-mail: sanjay@tropmet.res.in (corresponding author)
J. V. Revadekar, Indian Institute of Tropical Meteorology (IITM-MoES), Pune, India

Lead Authors

M. V. S. Ramarao, Indian Institute of Tropical Meteorology (IITM-MoES), Pune, India
H. Borgaonkar, Indian Institute of Tropical Meteorology (IITM-MoES), Pune, India
S. Sengupta, Indian Institute of Tropical Meteorology (IITM-MoES), Pune, India
D. R. Kothawale, Indian Institute of Tropical Meteorology (IITM-MoES), Pune, India
Jayashri Patel, Indian Institute of Tropical Meteorology (IITM-MoES), Pune, India
R. Mahesh, Indian Institute of Tropical Meteorology (IITM-MoES), Pune, India
S. Ingle, Indian Institute of Tropical Meteorology (IITM-MoES), Pune, India

Review Editors

K. AchutaRao, Centre for Atmospheric Sciences, Indian Institute of Technology Delhi, New Delhi, India
A. K. Srivastava, India Meteorological Department (IMD-MoES), Pune, India
J. V. Ratnam, Application Laboratory, Japan Agency for Marine Earth Science and Technology (JAMSTEC), Yokosuka, Japan

Corresponding Author

J. Sanjay, Indian Institute of Tropical Meteorology (IITM-MoES), Pune, India,
e-mail: sanjay@tropmet.res.in

The original version of this chapter was revised. On page 40, line 5 the word "business-as-usual" has been changed. The correction to this chapter is available at https://doi.org/10.1007/978-981-15-4327-2_13

© The Author(s) 2020, corrected publication 2021
R. Krishnan et al. (eds.), *Assessment of Climate Change over the Indian Region*,
https://doi.org/10.1007/978-981-15-4327-2_2

Key Messages

- Annual mean, maximum and minimum temperatures averaged over India during 1986–2015 show significant warming trend of 0.15 °C, 0.15 °C and 0.13 °C per decade, respectively (*high confidence*), which is consistent with dendroclimatic studies.
- Pre-monsoon temperatures displayed the highest warming trend followed by post-monsoon and monsoon seasons.
- The frequency of warm extremes over India has increased during 1951–2015, with accelerated warming trends during the recent 30 year period 1986–2015 (*high confidence*). Significant warming is observed for the warmest day, warmest night and coldest night since 1986.
- The CORDEX mean surface air temperature change over India for the mid-term (long-term) period 2040–2069 (2070–2099) relative to 1976–2005 is projected to be in the range of 1.39–2.70 °C (1.33–4.44 °C) across greenhouse gas warming scenarios. The ranges of these Indian mean temperature trends are broadly consistent with the CMIP5 based estimates.
- The frequency and intensity of warm days and warm nights are projected to increase over India in the next decades, while that of cold days and cold nights will decrease (*high confidence*). These changes will be more pronounced for cold nights and warm nights.
- The pre-monsoon season heatwave frequency, duration, intensity and areal coverage over India are projected to substantially increase during the twenty-first century (*high confidence*).

2.1 Introduction

Temperature is an essential climate quantity that directly affects human and natural systems. The global mean surface temperature is a key indicator of climate change because it increases quasi-linearly with cumulative greenhouse gas emissions as documented in multiple assessment reports of the Intergovernmental Panel on Climate Change (IPCC) including the most recent Fifth Assessment Report (AR5; IPCC 2013). This chapter assesses observed and projected changes in the mean and extreme temperature over India. The surface air temperature, typically measured at 2 m above the ground, varies from one region to another within India. This temperature also fluctuates naturally in interannual and decadal time scales in the background of human-induced changes in the climate. One of the important contributors of the observed changes in temperature not caused by human activities is the natural internal climate

variability, which refers to the chaotic short-term fluctuations around the mean climate over a region or at a location. This includes phenomena such as the variability in the El Nino–Southern Oscillation (ENSO). The presence of internal variability places fundamental limits on the accuracy with which future temperature can be projected. The internal variability becomes larger when averaging results over smaller areas, which may lead to larger uncertainty in projections at the Indian country scale, relative to that at a global scale (Collins et al. 2013).

The temperature projections are obtained by driving climate models with different future forcing scenarios. These projections include the response of the climate system to external forcing (e.g. changing greenhouse gas concentrations), internal variability and uncertainties associated with differences between models. The multi-model ensemble mean averages out the internal variability and model differences to a large extent and provides an estimate of the response of the climate system to forcing.

2.2 Observed Temperature Changes Over India

A significant observed warming trend in all India averaged annual mean surface air temperature for the long-term period 1901–2010 was assessed using the estimates derived from the India Meteorological Department (IMD) gridded monthly station data (Srivastava et al. 2017). This assessment also documented several past studies which reported the variability and trends in temperature over India. In this section, the observed temperature changes over India for the more recent three decades between 1986 and 2015 are assessed using the estimates derived from the IMD gridded daily station data.

2.2.1 Mean Temperature

The mean temperature over India has warmed from the mid-twentieth century (Fig. 2.1), with an increased rate of warming of 0.15 °C, 0.15 °C and 0.13 °C per decade for the annual mean, maximum and minimum temperatures, respectively, between 1986 and 2015 (Table 2.1).

The warming is not uniform across the seasons, with considerably more warming in the pre-monsoon season (March–May; MAM) than in other seasons. The rate of warming of 0.26 °C, 0.29 °C and 0.20 °C per decade for the pre-monsoon season mean, maximum and minimum temperatures, respectively, between 1986 and 2015 are relatively higher than that for the respective annual values (see

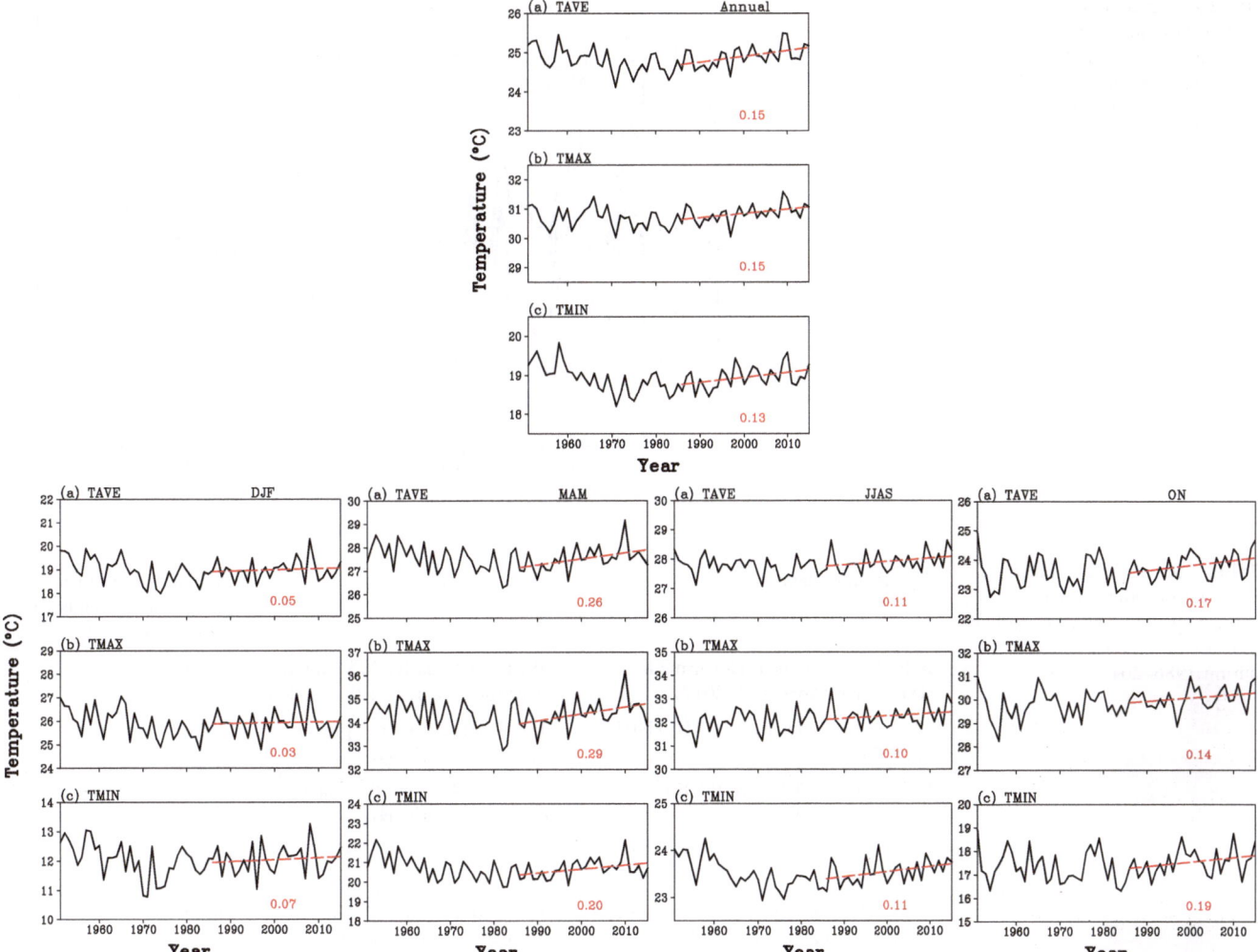

Fig. 2.1 Time-series of all India averaged (top panel) annual and (bottom panels) seasonal **a** mean, **b** maximum, and **c** minimum surface air temperatures between 1951 and 2015. Estimates are derived from the IMD daily gridded station data. Recent changes are computed based on linear trends (dashed red line) over the 30-year period 1986–2015. The rate of warming during this period in °C per decade is shown below the trend lines. The 90% confidence intervals for these trend estimates are assessed in Table 2.1

Table 2.1 Observed changes in India land mean annual and seasonal surface air temperature between 1986 and 2015

Season	Temperature trends 1986–2015 (°C per decade)		
	Mean	Maximum	Minimum
Annual	**0.15**[*] ± 0.09	**0.15**[*] ± 0.10	**0.13**[*] ± 0.10
Winter (Dec–Feb)	0.05 ± 0.16	0.03 ± 0.20	0.07 ± 0.18
Pre-monsoon (Mar–May)	**0.26**[*] ± 0.17	**0.29**[*] ± 0.20	**0.20**[*] ± 0.16
Monsoon (Jun–Sep)	0.11 ± 0.12	0.10 ± 0.17	**0.11**[*] ± 0.08
Post-monsoon (Oct–Nov)	0.17 ± 0.17	0.14 ± 0.22	0.19 ± 0.20

Estimates are derived from the IMD gridded station data. Changes are represented by linear trend[a] estimates (°C per decade) and 90% confidence intervals. Bold values with star sign (*) indicate that trend is significant (i.e. a trend of zero lies outside the 90% confidence interval)
[a]A linear trend model that allows for first-order autocorrelation in the residuals is adopted following IPCC AR5 (Hartmann et al. 2013)

Fig. 2.1 and Table 2.1). These changes in temperatures are statistically significant at the 90% confidence level. This behaviour of temperature changes is broadly consistent with the long-term variations and trends among the various global and regional gridded temperature estimates derived with different approaches (see Fig. 2.2 and Table 2.2).

Fig. 2.2 Indian annual average land surface air temperature anomalies relative to 1981–2010 climatology in the observed datasets (see details in Table 2.2)

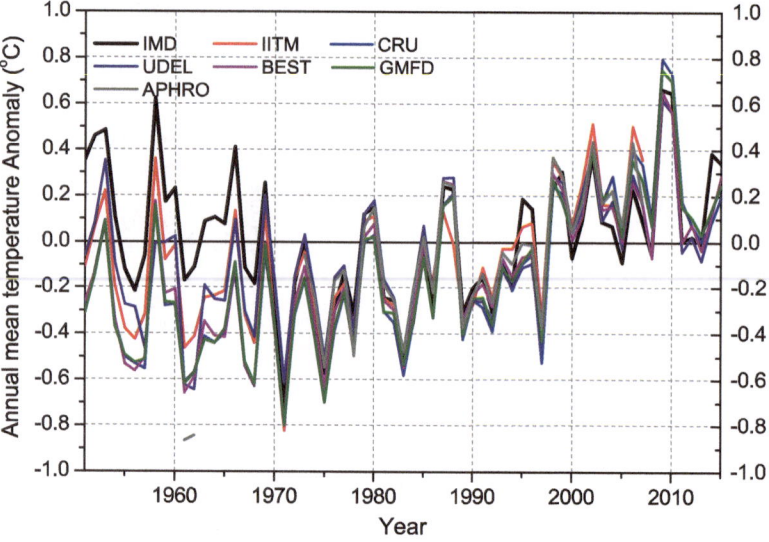

Table 2.2 Observed annual warming in India averaged surface air temperature in various global and regional datasets during 1986–2015

Temperature dataset	Data resolution	Indian annual mean temperature trends (°C per decade) 1986–2015
India Meteorological Department (IMD; Srivastava et al. 2017)	1951–2015; daily; 395 stations over India; 1.0° × 1.0° gridded	0.15
Climate Research Unit (CRU; Harris et al. 2014)	1901–2016; monthly; global; 0.5° × 0.5° gridded	0.20
University of Delaware (UDEL; Peterson et al. 1998)	1901–2014; monthly; global; 0.5° × 0.5° gridded	0.13
Berkeley Earth (BEST; Rhode et al. 2013)	1750–2017; monthly; global; 1.0° × 1.0° gridded	0.13
Global Meteorological Forcing Dataset (GMFD; Sheffield et al. 2006)	1948–2016; daily; global; blended reanalysis with observations; 0.25° × 0.25° gridded	0.19
		1986–2007
Indian Institute of Tropical Meteorology (IITM; Kothawale et al. 2010a, b)	1901–2007; monthly; 121 stations over India; 0.5° × 0.5° gridded	0.26
Asian Precipitation—Highly Resolved Observational Data Integration Towards Evaluation (APHRODITE; Yasutomi et al. 2011)	1961–2007; daily; over Asia; 0.25° × 0.25° gridded	0.21

These long-term changes in surface air temperature over India during the twentieth century also broadly agree with earlier assessments (e.g. Rupa Kumar et al. 1994; Sen Roy and Balling 2005; Kothawale and Rupa Kumar 2005; Srivastava et al. 2009; Kothawale et al. 2010a; Jain and Kumar 2012; Rai et al. 2012; Vinnarasi et al. 2017; Kothawale et al. 2016; Kulkarni et al. 2017; Srivastava et al. 2017, 2019). Indian annual mean land surface air temperatures have warmed by 0.6 °C century^{-1} between 1901 and 2018 (Srivastava et al.

2019). All India mean annual tropospheric temperature measured by radiosonde stations also showed an increasing trend from the surface to 500 hPa during the period 1971–2015 (Kothawale and Singh 2017). A similar warming trend is revealed by dendroclimatic studies over the eastern Himalaya including Sikkim and Bhutan in recent decades (Krusic et al. 2015; Yadava et al. 2015; Borgaonkar et al. 2018). Further details on the long-term changes of temperature over India based on paleoclimatic proxies are provided in Box 2.1.

Box 2.1 Trends Based on Tree-Ring Proxies

Palaeoclimatic records of temperature over monsoon Asia are limited and mainly based on the tree-ring proxies from the Himalayan region. Tree-ring based reconstructions of summer climate (temperature and rainfall) of Indian Himalaya, Nepal, Tibet, Karakoram region of Himalaya did not show significant increasing or decreasing trend during the past three to four centuries (Esper et al. 2002; Hughes 2001; Borgaonkar et al. 1994, 1996; Pant et al. 1998; Yadav et al. 1999; Cook et al. 2003; Thapa et al. 2015; Wu and Shao 1995). These reconstructions also indicated that the Little Ice Age (LIA) phenomenon was not prominent over this part of the Himalaya. However, few warm and cold epochs were observed over the region. A millennium-long mean summer temperature reconstruction from the monsoon-shadow zone in the western Himalaya (Yadav et al. 2011) indicated warming (Eleventh–fifteenth century) and cooling episodes (Fiftieth–nineteenth century) followed by a warming trend in the twentieth century. Higher growth in recent few decades detected in the high altitude tree-ring chronology has been noticed coinciding with the warming trend and rapid retreat of the Himalayan glaciers (Borgaonkar et al. 2009, 2011).

Dendroclimatic studies over the eastern Himalaya including Sikkim and Bhutan have indicated a warming trend in recent decades (Yadava et al. 2015; Borgaonkar et al. 2018; Krusic et al 2015). The reconstructed mean late-summer (July–August–September) temperature showed warming since the 1930s, with 1996–2005 being the warmest in context of the past ∼150 years (Yadava et al. 2015). Figure 2.3 shows the reconstructed late-summer temperature of Sikkim with a slight cooling trend since 1705 C.E. and noticeable increasing trend from 1850 C.E.

On a longer time scale, the first evidence of cooling during the Younger Dryas was provided by mineral magnetic susceptibility data and elemental concentrations that reveal a high around 13 ± 2 to 11 ± 1 ka (Juyal et al. 2009). The biochemical data of the Mansar Lake sediments, Lesser Himalaya indicated a hot and wet climate regime during the early Holocene and a dry and cold one during the late Holocene period (Das et al. 2010).

The observed warming is also unevenly distributed across India (Fig. 2.4). The largest increase in the annual mean temperature of more than 0.2 °C per decade are observed in some areas of north India between 1986 and 2015. The warming is much weaker in the southern peninsula, with mean temperature increase in some parts of the west coast lesser than 0.1 °C per decade. The winter warming is limited to peninsular India. The pre-monsoon season shows predominant warming of more than 0.5 °C per decade over north India. The summer monsoon season warming is confined to the eastern parts of the Indo-Gangetic plains and adjoining central India. The post-monsoon season warming pattern is similar to the pre-monsoon season, but with smaller magnitude and more uniformly distributed across the country than for other seasons. These estimates of warming across India based on simple linear trends are found to be, in general, similar to the earlier assessments of the temperature trends derived using non-stationary approach (Vinnarasi et al. 2017).

The all India averaged annual mean temperature increases due to greenhouse gas forcing outweighs the observed decrease in solar radiation (solar dimming; Padma Kumari et al. 2007). The radiative forcing is more effective in altering the strength of hydrological cycle than thermal forcing due to changes in the greenhouse gases (Padma Kumari and Goswami 2010; Soni et al. 2012; Padma Kumari et al. 2013).

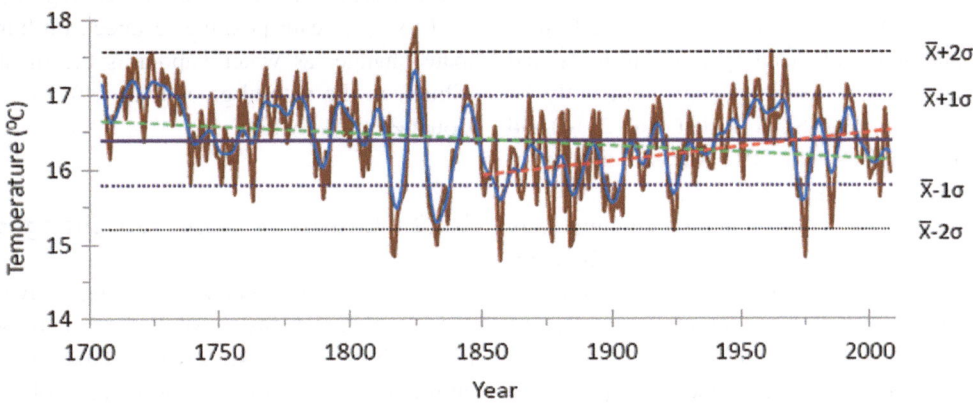

Fig. 2.3 Reconstructed late-summer (July–September) temperature of Sikkim from 1705–2008 C.E. (Brown line). The blue line indicates low-frequency variations at the decadal scale. Green- and red-dotted lines indicate a trend for full reconstructed period and for the period 1850–2008 C.E., respectively (Borgaonkar et al. 2018)

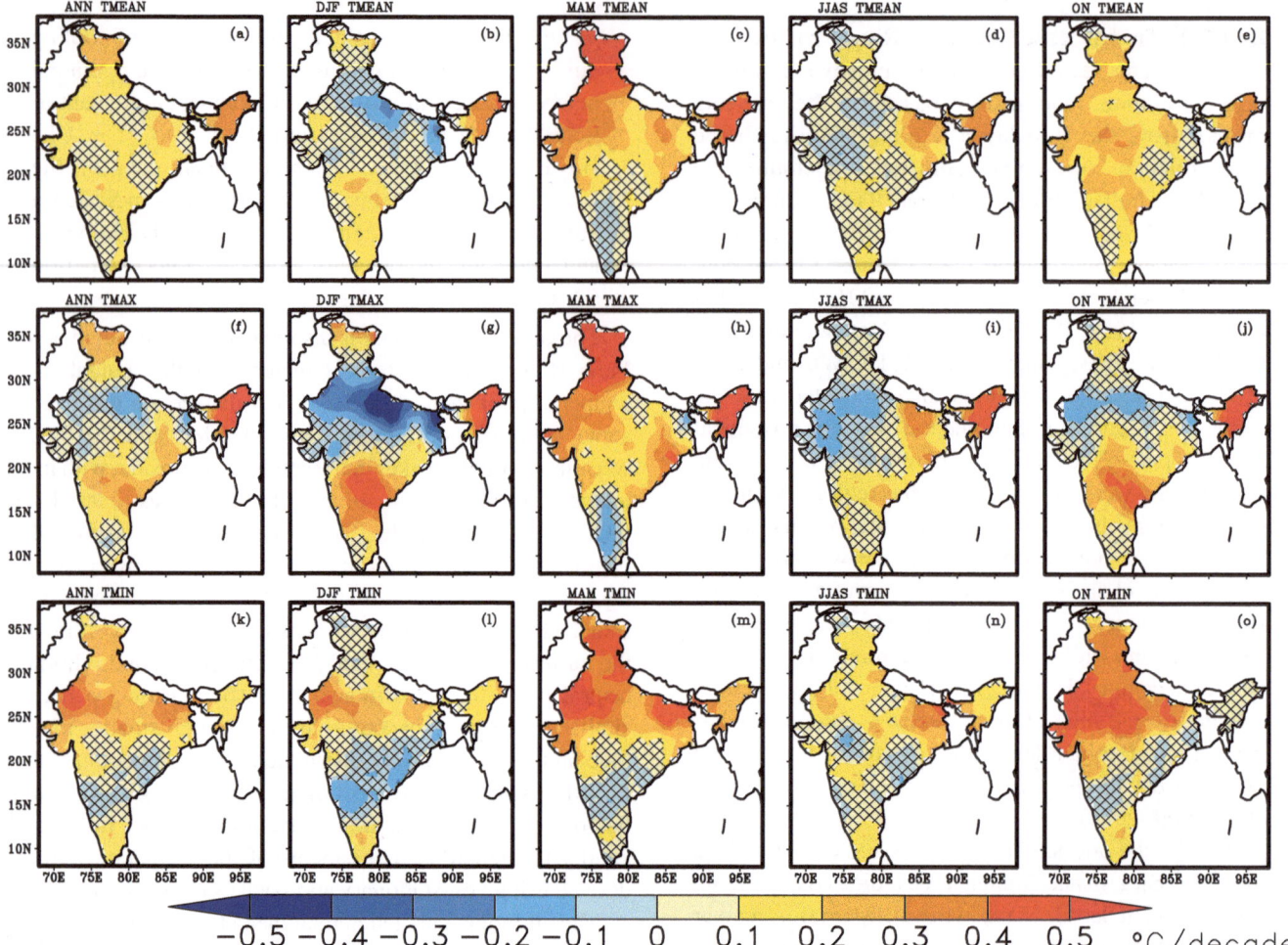

Fig. 2.4 Spatial distribution of observed annual and seasonal trends (°C per decade) for (top panel) mean, (middle panel) maximum and (bottom panel) minimum temperatures in (left to right panels) annual (ANN), winter (December–February, DJF), pre-monsoon (March–May, MAM), monsoon (June–September, JJAS) and post-monsoon (October–November, ON) seasons during the period 1986–2015. The grid boxes are hatched where the trends are insignificant (i.e. a trend of zero lies inside the 95% confidence interval)

The amount of water vapour in the atmosphere is expected to increase at a rate governed by the Clausius–Clapeyron ($\sim 7\%\ °C^{-1}$) under conditions of warming (Willett et al. 2007; Boucher et al. 2013). The all India averaged annual and seasonal mean specific humidity and relative humidity are found to be increasing significantly for both gridded observations (HadISDH, Willet et al. 2014) and reanalysis (ERA-Interim, Dee et al. 2011) datasets during the period 1979–2015 (Fig. 2.5 and Table 2.3).

The estimated magnitude of the annual specific humidity trend is similar for both datasets and is found to be comparable with the earlier assessment using IMD station data (Mukhopadhyay et al. 2017). The significant increasing trend in specific humidity assessed during the pre-monsoon season is consistent with the largest surface warming trend found for this season (see Table 2.1). Past studies had also reported a rise in the moisture content of the atmosphere

associated with warming over the Indian region (Krishnan et al. 2016; Mukhopadhyay et al. 2017). This increased water vapour under conditions of regional warming may lead to significant positive feedback on human-induced climate change, as water vapour is the most important contributor to the natural greenhouse effect (Willett et al. 2007; Boucher et al. 2013).

2.2.2 Causes of Observed Changes

The surface air temperature changes over India between 1956 and 2005 are attributed to anthropogenic forcing mostly by greenhouse gases and partially offset by other anthropogenic forcings including aerosols and land use land cover change (Dileepkumar et al. 2018). The observed changes in maximum temperature during the post-monsoon

Fig. 2.5 Time-series of all India averaged annual (left panels) specific humidity and (right panels) relative humidity from (top panels) HadISDH dataset and (bottom panels) ERA-Interim reanalysis for the period 1979–2015. The dashed blue lines indicate the linear trend for the period 1979–2015. The rate of change during this period is shown below the trend line. The 90% confidence intervals for these trend estimates are assessed in Table 2.3

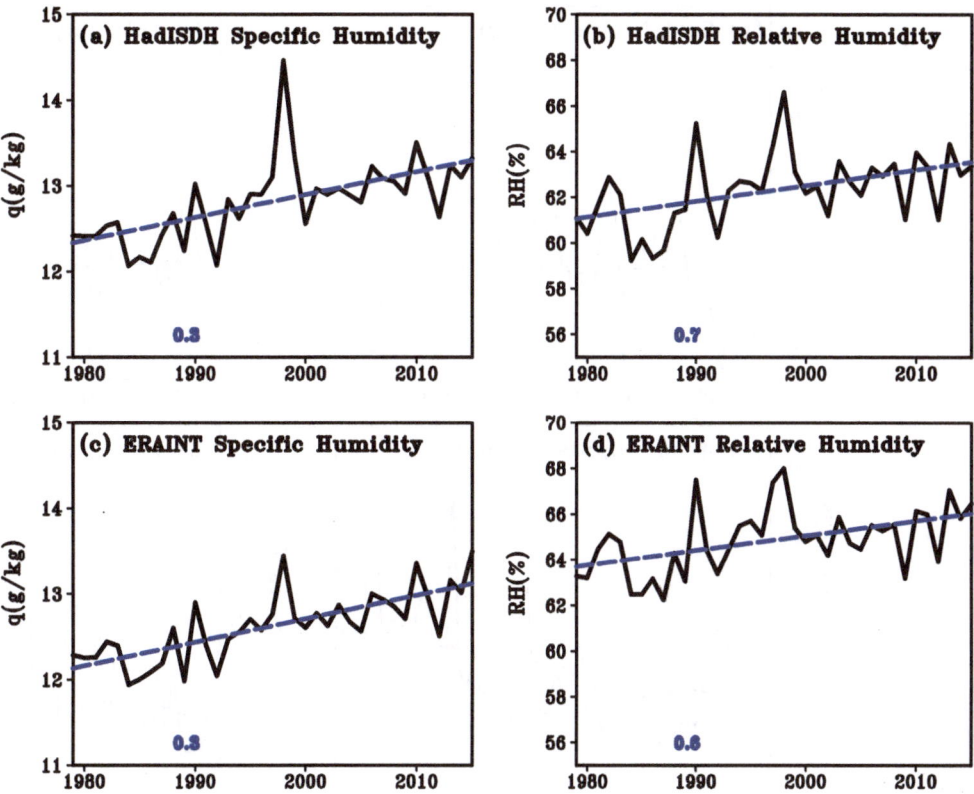

Table 2.3 Observed changes in India land mean annual and seasonal surface specific humidity and relative humidity between 1979 and 2015

Season	Specific humidity trends 1979–2015 (g kg^{-1} per decade)		Relative humidity trends 1979–2015 (% per decade)	
	HadISDH	ERA-interim	HadISDH	ERA-interim
Annual	**0.27**[*]± 0.13	**0.27**[*]± 0.07	**0.69**[*]± 0.48	**0.65**[*]± 0.40
Winter (Dec–Feb)	**0.20**[*]± 0.10	**0.20**[*]± 0.10	**0.86**[*]± 0.54	**0.75**[*]± 0.60
Pre-monsoon (Mar–May)	**0.37**[*]± 0.11	**0.36**[*]± 0.11	**0.79**[*]± 0.58	**0.85**[*]± 0.64
Monsoon (Jun–Sep)	**0.22**[*]± 0.11	**0.25**[*]± 0.09	0.47 ± 0.60	0.45 ± 0.51
Post-monsoon (Oct–Nov)	**0.33**[*]± 0.25	**0.37**[*]± 0.16	1.00 ± 1.08	**0.86**[*]± 0.79

Estimates are derived from the HadISDH dataset and ERA-Interim reanalysis. Trends and significance have been calculated as in Table 2.1. Bold values with star sign (*) indicate that trend is significant (i.e. a trend of zero lies outside the 90% confidence interval)

and minimum temperature during the pre-monsoon and monsoon seasons in South India during 1950–2005 are assessed to be detectably different from natural internal climate variability, and these temperature changes are attributed with confidence to climate change induced by anthropogenic effects (Sonali et al. 2018). These assessments are based on detection and attribution studies using observational datasets and the multiple atmosphere–ocean coupled general circulation model (AOGCM) outputs of historical simulation experiments conducted in the fifth phase of the Coupled Model Intercomparison Project

(CMIP5; Taylor et al. 2012), from which human and natural causes of climate change could be identified and quantified.

2.2.3 Temperature Extremes

The all India averaged annual frequency of warm days and nights have increased, and cold days and nights have decreased since 1951 (Fig. 2.6 and Table 2.4). These extreme temperature indices are defined from daily temperatures as days when daily maximum (daytime) and mini-

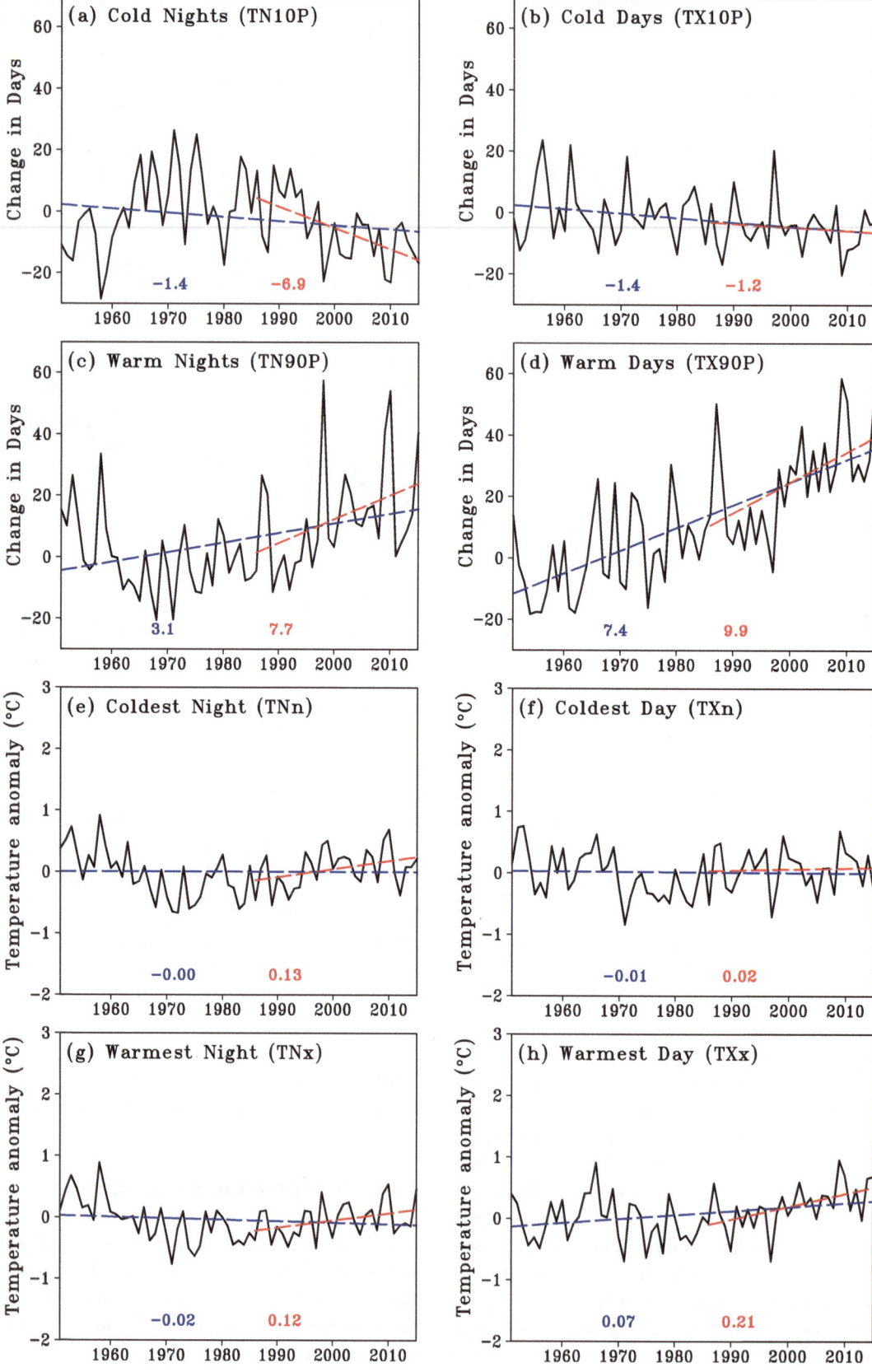

Fig. 2.6 Time-series of all India averaged annual frequency of **a** cold nights (TN10p), **b** cold days (TX10p), **c** warm nights (TN90p), and **d** warm days (TX90p); and annual intensity of **e** coldest night (TNn), **f** coldest day (TXn), **g** warmest night (TNx) and **h** warmest day (TXx) estimated from the IMD daily gridded maximum and minimum surface air temperature datasets for the period from 1951 to 2015. The dashed blue and red lines indicate the linear trends for the periods 1951–2015 and 1986–2015, respectively. The rates of changes during these two periods are shown below the trend lines. The 90% confidence intervals for these trend estimates are assessed in Tables 2.4 and 2.5

Table 2.4 Observed changes in India land mean annual and seasonal frequency indices of daily extreme temperatures for the periods 1951–2015 and 1986–2015

Season	Linear trends 1951–2015 (days per decade)				Linear trends 1986–2015 (days per decade)			
	Cold nights (TN10p)	Cold days (TX10p)	Warm nights (TN90p)	Warm days (TX90p)	Cold nights (TN10p)	Cold days (TX10p)	Warm nights (TN90p)	Warm days (TX90p)
Annual	−1.4 ± 2.3	**−1.4***± 1.1	**3.1***± 2.2	**7.4***± 1.7	**−6.9***± 3.8	−1.2 ± 2.8	**7.7***± 6.4	**9.9***± 6.4
Winter (Dec–Feb)	−0.2 ± 0.5	0.2 ± 0.4	0.6 ± 0.6	**2.2***± 0.7	−0.5 ± 1.2	0.5 ± 1.4	1.3 ± 1.4	**3.3***± 2.0
Pre-monsoon (Mar–May)	−0.2 ± 0.6	−0.4 ± 0.5	0.2 ± 0.7	**1.5***± 0.6	**−2.6***± 1.2	−1.0 ± 1.4	1.4 ± 2.2	**2.5***± 2.4
Monsoon (Jun–Sep)	−0.3 ± 0.9	**−0.6***± 0.4	**1.7***± 0.9	**2.4***± 0.8	**−3.0***± 1.5	−0.6 ± 1.1	**3.1***± 3.0	1.9 ± 3.3
Post-monsoon (Oct–Nov)	**−0.6***± 0.4	**−0.6***± 0.5	**0.7***± 0.6	**1.3***± 0.6	−0.8 ± 0.8	−0.1 ± 1.1	1.9 ± 1.9	2.2 ± 2.4

Estimates are derived from the IMD gridded station data. Trends and significance have been calculated as in Table 2.1. Bold values with star sign (*) indicate that trend is significant (i.e. a trend of zero lies outside the 90% confidence interval)

mum (nighttime) temperatures are above the 90th (warm) or below the 10th (cold) percentile (see more details in Box 2.2).

Box 2.2 Frequency and Intensity Indices of Daily Temperature Extremes

The all India averaged annual frequency and intensity indices of daily extreme temperatures observed over India during the periods 1951–2015 and 1986–2015 are estimated using the IMD daily maximum and minimum surface air temperature gridded datasets. The percentile indices representing the occurrence of cold nights (TN10p), cold days (TX10p), warm nights (TN90p) and warm days (TX90p) are defined following Zhang et al. (2011), and describe the threshold exceedance rate of days where daily minimum or daily maximum temperature is below the 10th or above the 90th percentile, respectively. The thresholds are based on the annual cycle of the percentiles calculated for a 5-day sliding window centered on each calendar day in the base period 1951–1980. The absolute indices representing the intensity of coldest night (TNn), coldest day (TXn), warmest night (TNx) and warmest day (TXx) are also defined following Zhang et al. (2011), and are presented as annual anomalies from the 1951–1980 climatological mean.

The significant annual increase of warm days (about 7.4 days per decade) is found to be higher than that of warm nights (about 3.1 days per decade) during the period 1951–2015. The magnitude of the significant annual decrease of cold days (about −1.4 days per decade) is weaker than that of the annual increase in the frequency of warm extremes for this long-term period. The seasonal frequency of warm days

and warm nights also increased significantly since 1951, except for warm nights in the pre-monsoon season (Table 2.4). The increase in frequency of warm days during monsoon (about 2.4 days per decade) and winter (about 2.2 days per decade) seasons contribute largely to the highest annual increase in the number of warm days over India since 1951. The recent 30-year period (1986–2015) shows a significant acceleration in the all India averaged annual increase of warm days (about 9.9 days per decade) and warm nights (about 7.7 days per decade) and decrease of cold nights (about −6.9 days per decade). In recent decades, the significant increase of winter season warm days (about 3.3 days per decade) largely contributes to the accelerated annual increase in number of warm days over India since 1986. The significant decreases in the frequency of cold nights during monsoon (about −3.0 days per decade) and pre-monsoon (about −2.6 days per decade) seasons contribute largely to the accelerated annual decrease in number of cold nights over India in this recent period (Table 2.4). The significant increase of pre-monsoon season warm days (about 2.5 days per decade) assessed for the recent period 1986–2015 is consistent with the earlier assessments of a gradual increasing trend in all India averaged warm day frequency during 1970–2005 period (Kothawale et al. 2010b; Revadekar et al. 2012). However, the decreasing trend in cold day frequency reported in these past studies is not significant between 1986 and 2015 (see Table 2.4).

The all India averaged annual intensity of warmest day shows a significant increasing trend (about 0.07 °C per decade) between 1951 and 2015 (Fig. 2.6 and Table 2.5). The recent 30-year period (1986–2015) shows significant acceleration for the annual increase in the intensity of warmest day (about 0.21 °C per decade) and warmest night (about 0.12 °C per decade), and decrease in the intensity of

Table 2.5 Observed changes in India land mean annual and seasonal intensity indices of daily extreme temperatures for the periods 1951–2015 and 1986–2015

Season	Linear trends 1951–2015 (°C per decade)				Linear trends 1986–2015 (°C per decade)			
	Coldest night (TNn)	Coldest day (TXn)	Warmest night (TNx)	Warmest day (TXx)	Coldest night (TNn)	Coldest day (TXn)	Warmest night (TNx)	Warmest day (TXx)
Annual	0.00 ± 0.07	-0.01 ± 0.06	-0.02 ± 0.05	$\mathbf{0.07^*} \pm 0.05$	$\mathbf{0.13^*} \pm 0.12$	0.02 ± 0.13	$\mathbf{0.12^*} \pm 0.10$	$\mathbf{0.21^*} \pm 0.11$
Winter (Dec–Feb)	-0.01 ± 0.08	$\mathbf{-0.09^*} \pm 0.08$	-0.01 ± 0.08	0.02 ± 0.09	-0.08 ± 0.19	-0.10 ± 0.28	$\mathbf{0.17^*} \pm 0.14$	$\mathbf{0.26^*} \pm 0.18$
Pre-monsoon (Mar–May)	-0.02 ± 0.09	-0.02 ± 0.10	$\mathbf{-0.09^*} \pm 0.07$	0.05 ± 0.07	$\mathbf{0.28^*} \pm 0.20$	-0.03 ± 0.32	0.10 ± 0.16	$\mathbf{0.29^*} \pm 0.18$
Monsoon (Jun–Sep)	-0.01 ± 0.05	0.04 ± 0.05	-0.02 ± 0.04	$\mathbf{0.09^*} \pm 0.06$	$\mathbf{0.15^*} \pm 0.09$	0.10 ± 0.15	0.05 ± 0.09	0.12 ± 0.20
Post-monsoon (Oct–Nov)	0.05 ± 0.09	0.04 ± 0.09	0.03 ± 0.08	0.10 ± 0.10	0.19 ± 0.19	0.01 ± 0.25	0.20 ± 0.22	0.17 ± 0.26

Estimates are derived from the IMD gridded station data. Trends and significance have been calculated as in Table 2.1. Bold values with star sign (*) indicate that trend is significant (i.e. a trend of zero lies outside the 90% confidence interval)

coldest night (about 0.13 °C per decade). The significant increase in the intensity of warmest day during pre-monsoon (about 0.29 °C per decade) and winter (about 0.26 °C per decade) seasons contribute largely to the accelerated annual increase in the intensity of the warmest day over India in the recent period (Table 2.5). The annual increase in the intensity of the warmest night is dominated by the significant increases in the winter season (about 0.17 °C per decade). The significant decrease in the intensity of coldest night during the pre-monsoon (about 0.28 °C per decade) and monsoon (about 0.15 °C per decade) seasons contribute to the accelerated annual decrease in the intensity of the coldest night over India during the recent period 1986–2015.

Significant increasing (decreasing) trends in heatwaves (cold waves) are observed during the hot (cold) weather season over most parts of India (Rohini et al. 2016, 2019; Ratnam et al. 2016; Pai et al. 2017). These periods containing consecutive extremely hot days (cold nights) are defined when departure in daily maximum (minimum) temperature exceeds (are below) the objectively defined threshold value (Pai et al. 2017). The observed frequency, total duration and maximum duration of heat waves during the hot summer months (April–June) are increasing over central and north-western parts of India (Rohini et al. 2016). The increase in the number of intensive heat waves between March and June in India over a recent-past decade was attributed to the presence of an upper-level cyclonic anomaly over the west of North Africa and a cooling anomaly in the Pacific (Ratnam et al. 2016). A significant decadal variation was observed in the frequency, spatial coverage and area of the maximum frequency of heat (cold) wave events over India (Pai et al. 2017). The variability of heat waves over India was found to be influenced by both the tropical Indian Ocean and central Pacific sea surface temperature anomalies. A noticeable increase (decrease) in the frequency of

heatwave days was observed during the El Nino (La Nina) events. It is also assessed that the spatial extent affected by concurrent meteorological droughts and heatwaves is increasing across India during the period 1981–2010 relative to the base period 1951–1980 (Sharma and Mujumdar 2017).

2.3 Projected Temperature Changes Over India

The projected future changes in temperature over India are assessed using the recently available high-resolution regional climate information from CORDEX South Asia and NEX-GDDP datasets generated by downscaling the CMIP5 AOGCM global-scale climate change projections using dynamical (i.e. regional climate modelling) and statistical (i.e. empirical) methods, respectively, (see more details in Box 2.3). The downscaled future projections in temperature are assessed over the Indian land area, by masking out the oceans and territories outside the geographical borders of India, and are reported for two 30-year future periods: 2040–2069 and 2070–2099 relative to the reference baseline period: 1976–2005, representing the mid-term and long-term changes in future climate over India.

Box 2.3 Downscaled High-Resolution CORDEX South Asia and NEX-GDDP Climate Change Projections

The coupled Atmospheric-Ocean General Circulation Models (AOGCMs) are the primary tools used to assess the nature and extent of the anthropogenic changes that are leading to global climate change since 1950s (Bindoff et al. 2013). The AOGCMs numerically represent the global climate system, and simulate historical and future climate projections. The most recent Fifth Assessment Report of IPCC (AR5; IPCC 2013) was based on multiple AOGCM outputs that participated in the fifth phase of the Coupled Model Intercomparison Project (CMIP5; Taylor et al. 2012) of the World Climate Research Program (WCRP). The CMIP5 AOGCMs projected distinct increases in temperature over South Asia during the twenty-first century, especially during the winter season (Christensen et al. 2013). These AOGCMs with coarse horizontal resolution (~ 100 km) were assessed to have good skill in simulating the regional synoptic-scale circulation pattern and smoothly varying climate variables like temperature. However, the temperature biases were assessed to be larger in few specific regions, particularly at high elevations over the Himalayas (Flato et al. 2013). Also, the assessment of a wide range of regional climate processes and features that are important for capturing the complexity of the Indian summer monsoon rainfall indicated that the performance of the individual AOGCMs varied in the CMIP5 historical experiments depending on which aspect of a model simulation was evaluated (Singh et al. 2017).

The recent developments to generate high-resolution regional-scale climate information by downscaling the CMIP5 AOGCM based global-scale climate change projections using statistical (i.e. empirical) and dynamical (i.e. regional climate modelling) methods are used to assess the future changes in temperature over India in Sect. 2.3.

The statistical downscaling approach derives empirical relationships linking large-scale atmospheric variables (predictors) and local/regional climate variables (predictands). These relationships are then applied to equivalent predictors from AOGCMs. The NASA Earth Exchange (NEX) Global Daily Downscaled Projections (GDDP) dataset uses the Bias-Correction Spatial Disaggregation (BCSD) method (Thrasher et al. 2012) to correct the systematic bias of the CMIP5 AOGCM daily maximum and minimum temperature historical data through comparisons performed against the Global Meteorological Forcing Dataset (GMFD; Sheffield et al. 2006), and spatially interpolates the adjusted AOGCM data to the finer resolution grid of the 0.25° GMFD data. The BCSD approach used in generating this downscaled dataset inherently assumes that the relative spatial patterns in temperature observed from 1950 through 2005 will remain constant for future climate change under the RCP4.5 and RCP8.5 emission scenarios. The limitation of the NEX-GDDP dataset is that other than the higher spatial resolution and bias-correction this dataset does not add information beyond what is contained in the original CMIP5 scenarios, and preserves the frequency of periods of anomalously high and low temperature (i.e. extreme events) within each individual CMIP5 scenario.

The dynamical downscaling derives regional climate information using physical–dynamical relationships by embedding a high-resolution regional climate model (RCM) within a coarse-resolution AOGCM. The WCRP regional activity Coordinated Regional climate Downscaling Experiment (CORDEX; http://www.cordex.org/) has generated an ensemble of regional climate change projections for South Asia with a high spatial resolution (50 km) by dynamically downscaling several CMIP5 AOGCM outputs using multiple RCMs. Section 2.3 assess the future changes in the annual mean, maximum and minimum surface air temperature over India using the CORDEX South Asia dynamically downscaled historical simulations and future projections of climate change till the end of the twenty-first century available from the CORDEX data archives on the Earth System Grid Federation (ESGF). This multi-RCM ensemble consists of six simulations with IITM-RegCM4 RCM and ten simulations with SMHI-RCA4 RCM, respectively, under the future RCP4.5 and RCP8.5 emission scenarios, and five simulations with SMHI-RCA4 RCM under the future RCP2.6 emission scenario (see more details in Table 2.6). These dynamically downscaled CMIP5 future temperature projections for India are also compared in Sect. 2.3 with the NEX-GDDP statistically downscaled daily maximum and minimum temperature projections under the RCP4.5 and RCP8.5 emission scenarios available from the NEX-GDDP data archives for the 10 CMIP5 host models that were used to provide lateral and ocean surface boundary conditions for the CORDEX South Asia RCMs (see Table 2.6).

The performance of the CORDEX South Asia multi-RCM historical temperature simulations have been evaluated in several studies (e.g. Mishra 2015; Sanjay et al. 2017a, b; Nengker et al. 2018; Hasson et al. 2018). These dynamically downscaled RCM simulations showed added value relative to their driving CMIP5 AOGCMs in simulating the climatological seasonal and annual spatial patterns of surface air temperature over the South Asia land region, and the climatological amplitude and phase of the annual cycle of monthly mean temperature over central India (Sanjay et al. 2017a). The spatial pattern of temperature climatology over the Himalayas for the present climate was simulated exceptionally well even though these RCMs showed a significant cold bias (Nengker et al. 2018). These RCMs showed larger uncertainty of 1–3.6 °C for simulated temperature in the CORDEX South Asia historical experiments than that of the observations in the Himalayan water towers (e.g. Indus, Ganges and Brahmaputra river basins; Mishra 2015). This evaluation also showed that the RCMs exhibited large cold bias (6–8 °C) and were not able to reproduce the observed warming in the Himalayan water towers. The downscaled seasonal mean temperature in this multi-RCM ensemble was found to have relatively larger cold bias than their driving CMIP5 AOGCMs over the hilly sub-regions within the Hindu Kush Himalayan region (Sanjay et al. 2017b). Also, these downscaled CORDEX South Asia RCMs and their driving CMIP5 AOGCM experiments consistently showed substantial cold (6–10 °C) biases for the observed climatology of temperature over the Himalayan watersheds of Indus basin (Jhelum, Kabul and upper Indus basin; Hasson et al. 2018).

2.3.1 Mean Temperature

The CORDEX South Asia multi-RCM ensemble mean projected long-term (2070–2099) annual warming exceeds 4 °C over most parts of India except the southern peninsula, relative to the reference period 1976–2005 under the high (RCP8.5) emission scenario, with relatively higher change exceeding 5 °C projected in the semi-arid north-west and north India (Fig. 2.7).

The geographical patterns of long-term change remain below 2 °C relative to the reference period under the low (RCP2.6) emission scenario over most parts of India. The projections of mid-term (2040–2069) change in these CORDEX South Asia multi-RCM ensemble mean indicate modest sensitivity to alternate RCP scenarios over the Indian land area. The projected mid-term annual warming exceed 1 °C over most parts of the country, with higher warming exceeding 2 °C projected in the north-west and north India under the medium (RCP4.5) emission scenario (Fig. 2.7). The summer monsoon temperature projections by a small subset of the CORDEX South Asia RCMs had indicated mean warming of more than 1.5 °C over the central and northern parts of India for the period 2031–2060 under this medium emission scenario (Sanjay et al. 2017a). The CORDEX South Asia multi-RCMs had provided

relatively better confidence than their driving CMIP5 AOGCMs in projecting the magnitude of seasonal warming for the hilly sub-region within the Karakoram and north-western Himalaya, with a higher projected change of 5.4 °C during winter than of 4.9 °C during summer monsoon season by the end of the twenty-first century under the high (RCP8.5) emissions scenario (Sanjay et al. 2017b). The CORDEX South Asia RCMs and their ensemble had projected statistically significant strong rate of warming (0.03–0.09 °C per year) across all seasons and RCPs over the Indian Himalayan region (Dimri et al. 2018a). The seasonal response to warming with respect to elevation was found to be substantial with December–January season followed by October–November showing the highest rate of warming at higher elevation sites such as the western Himalayas and northern part of central Himalayas.

The earlier assessment of temperature projections for India using the CMIP5 multi-model and multi-scenario ensemble had also suggested that generally in future the northern part of the country will experience higher warming compared to the southern peninsula (Chaturvedi et al. 2012). It was assessed that the areas in the Himalayas and Kashmir will be particularly subject to large warming to the tune of 8 °C in RCP8.5 by 2099 relative to the pre-industrial baseline (the 1880s). It was concluded that this assessment of a broad range of temperature projections for India, ranging from 1 to 8 °C during the period 1880–2099 under different RCP scenarios indicated that these regional climate change projections were associated with a range of limitations and uncertainties—driven mainly by the climate model and future scenario uncertainties (Chaturvedi et al. 2012). Also, an earlier study using multiple CMIP5 model outputs together with a single model ensemble assessed that for temperature in most regions within India the component of uncertainty due to model spread tends to be larger than that arising due to natural internal variability, and tends to grow with time (Singh and AchutaRao 2018).

A consistent and robust feature across the downscaled CORDEX South Asia RCMs is a continuation of warming over India in the twenty-first century for all the RCP scenarios (Fig. 2.8). The CORDEX South Asia historical RCM simulations capture the observed interannual variations and the warming trend reasonably well. The all India averaged annual surface air temperature increases are similar for all the RCP scenarios during the first decade after 2005. The warming rate depends more on the specified greenhouse gas concentration pathway at longer time scales, particularly after about 2050. The multi-RCM ensemble mean under RCP2.6 scenario stays around 1.5 °C above 1976–2005 levels throughout the twenty-first century, clearly demonstrating the potential of mitigation policies. The ensemble

Table 2.6 List of the 16 CORDEX South Asia RCM simulations driven with 10 CMIP5 AOGCMs

CORDEX South Asia RCM	RCM description	Contributing CORDEX modelling center	Driving CMIP5 AOGCM (see details at https://verc.enes.org/data/enes-model-data/cmip5/resolution)	Contributing CMIP5 modelling center
IITM-RegCM4 (6 members)	The Abdus Salam International Centre for Theoretical Physics (ICTP) Regional Climatic Model Version 4 (RegCM4; Giorgi et al. 2012)	Centre for Climate Change Research (CCCR), Indian Institute of Tropical Meteorology (IITM), India	CCCma-CanESM2	Canadian Centre for Climate Modelling and Analysis (CCCma), Canada
			NOAA-GFDL-GFDL-ESM2M	National Oceanic and Atmospheric Administration (NOAA), Geophysical Fluid Dynamics Laboratory (GFDL), USA
			CNRM-CM5	Centre National de RecherchesMe´te´orologiques (CNRM), France
			MPI-ESM-MR	Max Planck Institute for Meteorology (MPI-M), Germany
			IPSL-CM5A-LR	Institut Pierre-Simon Laplace (IPSL), France
			CSIRO-Mk3.6	Commonwealth Scientific and Industrial Research Organization (CSIRO), Australia
SMHI-RCA4 (10 members)	Rossby Centre Regional Atmospheric Model Version 4 (RCA4; Samuelsson et al. 2011)	Rossby Centre, Swedish Meteorological and Hydrological Institute (SMHI), Sweden	ICHEC-EC-EARTH	Irish Centre for High-End Computing (ICHEC), European Consortium (EC)
			MIROC-MIROC5	Model for Interdisciplinary Research On Climate (MIROC), Japan Agency for Marine-Earth Sci. & Tech., Japan
			NCC-NorESM1	Norwegian Climate Centre (NCC), Norway
			MOHC-HadGEM2-ES	Met Office Hadley Centre for Climate Change (MOHC), United Kingdom
			CCCma-CanESM2	CCCma, Canada
			NOAA-GFDL-GFDL-ESM2M	NOAA, GFDL, USA
			CNRM-CM5	CNRM, France
			MPI-ESM-LR	MPI-M, Germany
			IPSL-CM5A-MR	IPSL, France
			CSIRO-Mk3.6	CSIRO, Australia

mean annual India warming exceeds 2 °C within the twenty-first century under RCP4.5, and the warming exceeds 4 °C by the end of the twenty-first century under RCP8.5 scenario. The spread in the minimum to maximum range in the projected warming among the CORDEX South Asia RCMs for each RCP scenario (shown as shading in Fig. 2.8) provides a simple, but crude, measure of uncertainty.

A reliable quantitative estimate of all India averaged annual surface air temperature changes and the associated uncertainty range of future temperature projections for India are obtained by incorporating individual model performance and model convergence criteria within a reliability ensemble averaging (REA) methodology (see more details in Box 2.4).

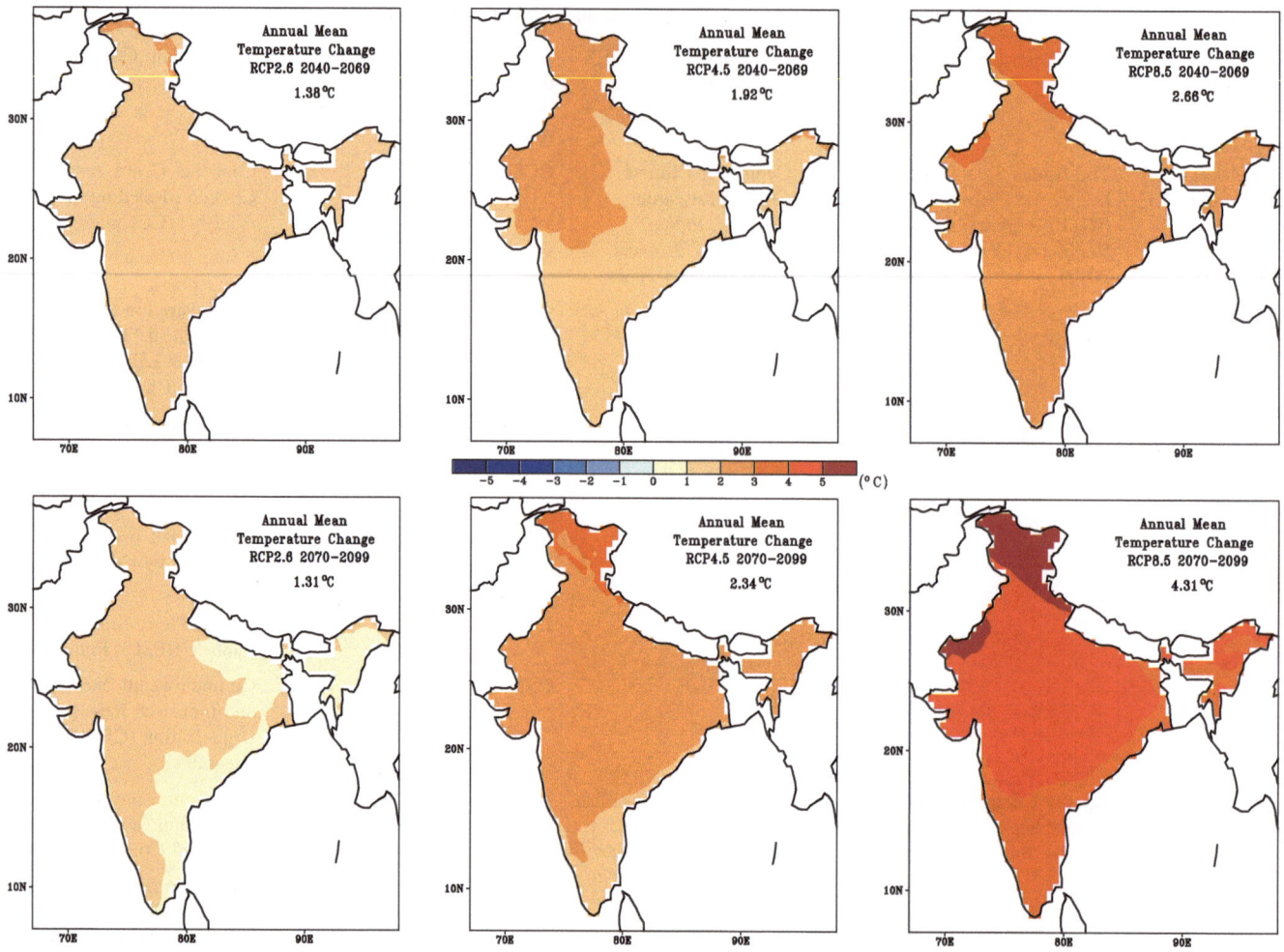

Fig. 2.7 CORDEX South Asia multi-RCM ensemble mean projections of annual average surface air temperature changes (in °C) over India for the mid-term (2040–2069) and long-term (2070–2099) climate relative to 1976–2005 under RCP2.6, RCP4.5 and RCP8.5 emission scenarios. The estimates of all India averaged ensemble mean projected changes are shown in each panel

---

Box 2.4 Climate Model Projections and Weighting

The climate change projections for the twenty-first century at the regional or subcontinental spatial scales are based on transient simulations with coupled atmosphere–ocean general circulation models (AOGCMs) including relevant anthropogenic forcings, for example, due to greenhouse gases (GHG) and atmospheric aerosols. These projections have been characterized by a low level of confidence and a high level of uncertainty deriving from different sources: estimates of future anthropogenic forcings, the response of a climate model to a given forcing, the natural variability of the climate system. One of the major sources of uncertainty in future temperature projections is that the different AOGCMs respond differently to the same forcing resulting in differences in the projected changes. These differences are due to

the differences in representing the real climate system through a set of mathematically approximated physical, chemical, and biological processes. A recent study assessed that the component of uncertainty due to CMIP5 model spread tends to be larger than that arising due to natural internal variability for temperature in most regions within India, and the model spread tends to grow with time (Singh and AchutaRao 2018). Therefore, a comprehensive assessment of regional change projection needs to be based on the collective information from the ensemble of AOGCM simulations.

A quantitative method called "reliability ensemble averaging" (REA) was introduced by Giorgi and Mearns (2002) for calculating the average, uncertainty range and collective reliability of regional climate change projections from ensembles of different

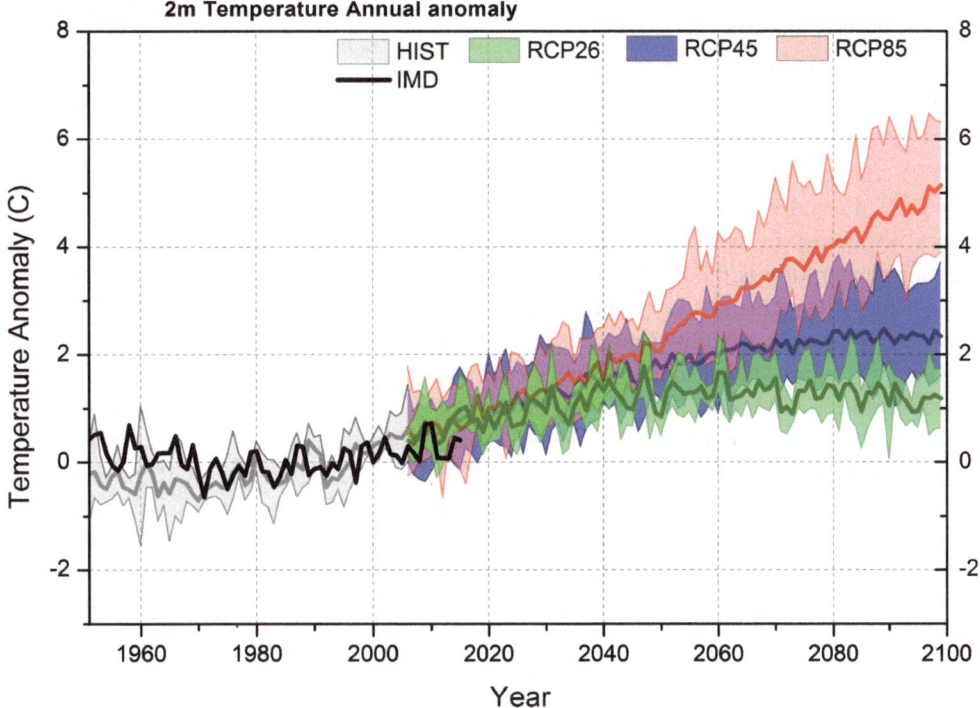

Fig. 2.8 Time series of Indian annual mean surface air temperature (°C) anomalies (relative to 1976–2005) from CORDEX South Asia concentration-driven experiments. The multi-RCM ensemble mean (solid lines) and the minimum to maximum range of the individual RCMs (shading) based on the historical simulations during 1951–2005 (grey), and the downscaled future projections during 2006–2099 are shown for RCP2.6 (green), RCP4.5 (blue) and RCP8.5 (red) scenarios. The black line shows the observed anomalies during 1951–2015 based on IMD gridded station data

AOGCM simulations. This method takes into account two reliability criteria: the performance of the model in reproducing present-day climate and the convergence of the simulated changes across models. The REA average was estimated as the weighted average of the AOGCM ensemble members. The uncertainty range was measured as the root-mean-square difference around the REA average. In the REA method, a model projection is reliable when both its present-day bias and distance from the ensemble average are within the natural variability. When compared to simpler approaches, the REA method estimated a reduction of the uncertainty range in the simulated seasonal temperature and precipitation changes over land regions of subcontinental scale (Giorgi and Mearns 2002). The REA method reduces the uncertainty range by minimizing the contribution of simulations that either performed poorly in the representation of present-day climate over a region or provided outlier simulations with respect to the other models in the ensemble, thus extracting only the most reliable information from each model (Giorgi and Mearns 2002). The use of REA methodology reduced the overall CMIP5 model

uncertainty range compared to simpler ensemble average approach for the future projections of surface air temperature and precipitation during the Indian summer monsoon season (Sengupta and Rajeevan 2013). The estimated REA average projected mean monsoon warming was also characterized by consistently high-reliability index in comparison with participating individual CMIP5 AOGCMs.

The results of applying this REA methodology to the dynamically downscaled CORDEX South Asia multi-RCMs and the statistically downscaled NEX-GDDP all India averaged annual surface air temperature changes under the three different RCP scenarios are assessed in Sect. 2.3.1. A measure of natural variability is estimated following Sengupta and Rajeevan (2013) as the difference between the maximum and minimum values of the 30 years moving average of the time series of all India averaged surface air temperature data available from the Indian Institute of Tropical Meteorology (IITM, Pune, http://www.tropmet.res.in) for the period 1901–2005, after linearly detrending the data (to remove century-scale trends). The natural variability in the observed all India

averaged annual mean, maximum and minimum surface air temperatures are estimated to be 0.347 °C,
0.513 °C and 0.213 °C respectively.

The projected REA estimates of warming for the
mid-term (2040–2069) ranges between 1.39 °C under
RCP2.6 scenario to 2.70 °C under RCP8.5 scenario (see
Table 2.7). The natural variability in the observed all India
annual mean surface air temperature is 0.347 °C (see more
details in Box 2.4) while the REA based temperature
increases are well above this natural variability estimate. The
uncertainty range defined by the root-mean square difference
varies in the mid-term between 0.17 and 0.37 °C for the
three RCP scenarios, with the RCP4.5 mid-term warming of
1.92 °C indicating the maximum uncertainty of 15.6% (see
Table 2.7). The proper weighting of individual CORDEX
South Asia RCMs based on their present-day performance
by the REA method has resulted in the REA warming under
RCP2.6 scenario to be 1.33 °C above 1976–2005 levels till
the end of the twenty-first century, which is slightly higher
than that defined by multi-RCM ensemble mean shown in

Fig. 2.8. However, this estimate of annual warming well
below 2 °C by the end of the twenty-first century for RCP2.6
is found to be associated with the highest uncertainty of 18%
among all the RCP scenarios. The REA estimates of
long-term (2070-2099) warming are 2.44 ± 0.41 °C and
4.44 ± 0.45 °C under RCP4.5 and RCP8.5 scenarios,
respectively (see Table 2.7). The assessment of 4.44 °C
warming by the end of the twenty-first century under
RCP8.5 scenario is highly reliable as it is associated with the
lowest uncertainty (of 10.1%) among the three RCP scenarios. These CORDEX South Asia multi-RCM
ensemble-based temperature projections for India are generally in line with earlier estimates. The CMIP5 ensemble
based on 18 models had projected warming of 1.5 °C, 2.8 °
C and 4.3 °C under the RCP2.6, RCP4.5 and RCP8.5 scenarios, respectively, for the 30 year average of 2071–2100
relative to the 1961–1990 baseline (Chaturvedi et al. 2012).

The forced signal of warming occurs not only in the
annual mean of daily mean surface air temperature but also
in the annual means of daily maximum and daily minimum
temperatures (see Tables 2.8 and 2.9). The CORDEX South

Table 2.7 CORDEX South Asia multi-RCM ensemble mean (CDX-ENS) and reliability ensemble average (CDX-REA) estimates of projected changes in annual mean surface air temperature over India relative to 1976–2005, and the associated uncertainty range

Emission scenario	Model ensemble (members)	Annual mean temperature (°C)	
		2040–2069	2070–2099
RCP2.6	CDX-ENS(5)	1.38 ± 0.17 (12.3%)	1.31 ± 0.24 (18.3%)
	CDX-REA(5)	1.39 ± 0.18 (12.9%)	1.33 ± 0.24 (18.0%)
RCP4.5	CDX-ENS(16)	1.92 ± 0.30 (15.6%)	2.34 ± 0.44 (18.8%)
	CDX-REA(16)	2.03 ± 0.28 (13.8%)	2.44 ± 0.41 (16.8%)
RCP8.5	CDX-ENS(15)	2.66 ± 0.37 (13.9%)	4.31 ± 0.56 (13.0%)
	CDX-REA(15)	2.70 ± 0.31 (11.5%)	4.44 ± 0.45 (10.1%)

The values in parenthesis of columns 3 and 4 show the uncertainty range (in %) measured as the root-mean-square difference around the respective ensemble mean

Table 2.8 CORDEX South Asia multi-RCM ensemble mean (CDX-ENS) and reliability ensemble average (CDX-REA), and NEX-GDDP reliability ensemble average (NEX-REA) estimates of projected changes in annual mean of daily maximum surface air temperature over India relative to 1976–2005, and the associated uncertainty range

Emission scenario	Model ensemble (members)	Annual maximum temperature (°C)	
		2040–2069	2070–2099
RCP2.6	CDX-ENS(5)	1.29 ± 0.14 (10.9%)	1.23 ± 0.22 (17.9%)
	CDX-REA(5)	1.29 ± 0.14 (10.9%)	1.25 ± 0.23 (18.4%)
RCP4.5	CDX-ENS(16)	1.79 ± 0.29 (16.2%)	2.14 ± 0.39 (18.2%)
	CDX-REA(16)	1.88 ± 0.26 (13.8%)	2.33 ± 0.37 (15.9%)
	NEX-REA(10)	1.91 ± 0.28 (14.7%)	2.35 ± 0.42 (17.9%)
RCP8.5	CDX-ENS(15)	2.44 ± 0.34 (13.9%)	3.93 ± 0.53 (13.5%)
	CDX-REA(15)	2.59 ± 0.36 (13.9%)	4.10 ± 0.45 (11.0%)
	NEX-REA(10)	2.51 ± 0.46 (18.3%)	4.38 ± 0.65 (14.8%)

The values in parenthesis of columns 3 and 4 show the uncertainty range (in %) measured as the root-mean-square difference around the respective ensemble mean

Table 2.9 CORDEX South Asia multi-RCM ensemble mean (CDX-ENS) and reliability ensemble average (CDX-REA), and NEX-GDDP reliability ensemble average (NEX-REA) estimates of projected changes in annual mean of daily minimum surface air temperature over India relative to 1976–2005, and the associated uncertainty range

Emission scenario	Model ensemble (members)	Annual minimum temperature (°C)	
		2040–2069	2070–2099
RCP2.6	CDX-ENS(5)	1.49 ± 0.28 (18.8%)	1.42 ± 0.31 (21.8%)
	CDX-REA(5)	1.45 ± 0.24 (16.6%)	1.33 ± 0.27 (20.3%)
RCP4.5	CDX-ENS(16)	2.09 ± 0.38 (18.2%)	2.58 ± 0.54 (20.9%)
	CDX-REA(16)	2.24 ± 0.29 (12.9%)	2.66 ± 0.38 (14.3%)
	NEX-REA(10)	2.10 ± 0.30 (14.3%)	2.38 ± 0.41 (17.2%)
RCP8.5	CDX-ENS(15)	2.92 ± 0.45 (15.4%)	4.77 ± 0.70 (14.7%)
	CDX-REA(15)	2.90 ± 0.25 (8.6%)	4.71 ± 0.35 (7.4%)
	NEX-REA(10)	2.79 ± 0.40 (14.3%)	4.87 ± 0.55 (11.3%)

The values in parenthesis of columns 3 and 4 show the uncertainty range (in %) measured as the root-mean-square difference around the respective ensemble mean

Asia REA estimate of warming for the two 30 year future periods are lower (higher) for the annual means of daily maximum (minimum) temperature than the respective warming assessed for the annual mean of daily mean temperature under all three RCP scenarios. The REA changes for annual minimum temperature of 4.71 ± 0.35 °C (see Table 2.9) is more pronounced than that of 4.10 ± 0.45 °C and 4.44 ± 0.45 °C increases estimated for all India annual maximum (see Table 2.8) and mean (see Table 2.7) temperatures respectively by the end of the twenty-first century under the high (RCP8.5) emission scenario. The assessment of 4.71 °C warming for annual mean of daily minimum surface air temperature by the end of the twenty-first century under RCP8.5 scenario is highly reliable as it is associated with the lowest uncertainty (of 7.4%) among not only the three RCP scenarios for this variable but also for the annual mean and maximum statistic shown in Tables 2.7 and 2.8. This finding illustrates that the dynamically downscaled CORDEX South Asia RCMs based regional climate projections do certainly bring more confidence to future temperature projections for India than the regional climate change information provided by the statistically downscaled NEX-GDDP dataset.

These CORDEX South Asia multi-RCMs had also projected statistically significant higher warming rate (0.23–0.52 °C/decade) for both minimum and maximum air temperatures over the Indian Himalayan region under RCP4.5 and RCP8.5 scenarios (Dimri et al. 2018b).

2.3.2 Temperature Extremes

The CORDEX South Asia multi-RCMs project that all India averaged annual frequency of warm nights and warm days will increase from about 10% in the reference base period (1976–2005) to 80% and 65%, respectively, by the end of

the twenty-first century under the high (RCP8.5) emission scenario (Fig. 2.9). The future changes in the percentile indices based on minimum temperature (warm nights and cold nights) are more pronounced than those based on maximum temperature (warm days and cold days). The downscaled future temperature projections under the high (RCP8.5) emission scenario also indicate that by the end of the twenty-first century there will be virtually no cold nights and cold days over India as defined for the reference base period (1976–2005). The spread among the RCMs (shading in Fig. 2.9) generally becomes smaller as the projection approaches the zero exceedance rates as more models simulate fewer cold nights and cold days. The largest decreases in cold nights and largest increases in warm nights projected over India are typical for tropical regions that are characterized by small day-to-day temperature variability so that changes in mean temperature are associated with comparatively larger changes in exceedance rates below the 10th and above the 90th percentiles.

The CORDEX South Asia multi-RCMs project robust increase (decrease) in the all India averaged annual intensity of warm (cold) temperature extremes by the end of twenty-first century, with the magnitude of the changes increasing with increased anthropogenic forcing (Fig. 2.9). The coldest night of the year warms (about 5.5 °C) more than the warmest day (about 4.7 °C) over India by the end of the twenty-first century relative to the reference base period (1976–2005) under the high (RCP8.5) emission scenario. This tendency is consistent with the assessment that the increases in the frequency of warm nights are greater than increases in the frequency of warm days.

The CORDEX South Asia multi-RCM ensemble simulate about one heatwave event with an average total duration of about 5 days per summer season (April to June) over India during the historical period 1976–2005 (Fig. 2.10). These heatwave characteristics are identified based on the 90th

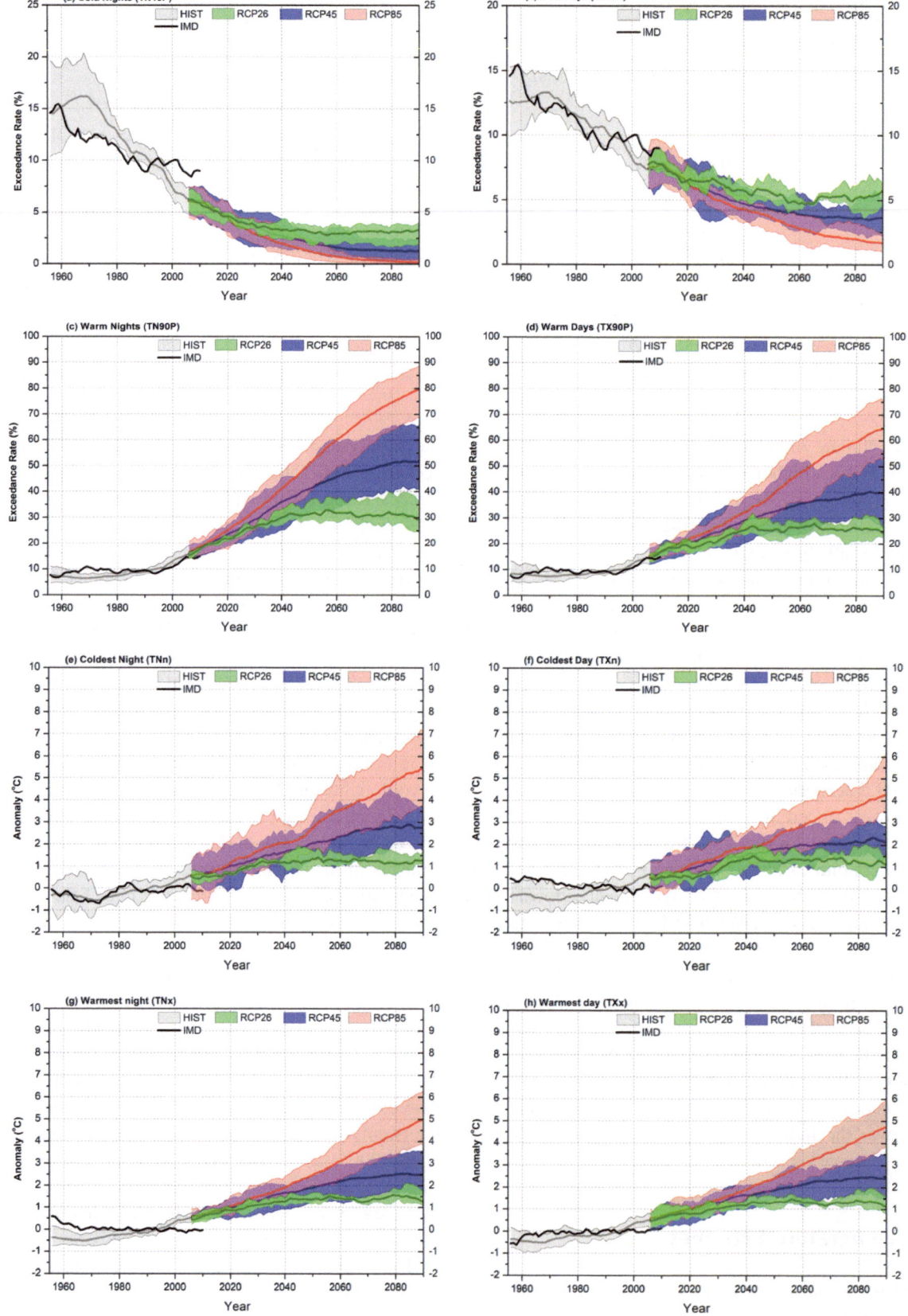

◄ **Fig. 2.9** India averages of temperature indices over land as simulated by the CORDEX South Asia multi-RCM ensemble (see more details in Table 2.6) for the RCP2.6 (green), RCP4.5 (blue), and RCP8.5 (red) displayed for the annual percentile frequency indices **a** cold nights (TN10p), **b** cold days (TX10p), **c** warm nights (TN90p), and **d** warm days (TX90p); and for the annual absolute intensity indices **e** coldest night (TNn), **f** coldest day (TXn), **g** warmest night (TNx) and **h** warmest day (TXx). Changes for the percentile frequency indices are displayed as absolute exceedance rates (in %). By construction the exceedance rate averages to about 10% over the base period 1976–2005. Changes for the absolute intensity indices are displayed as annual anomalies relative to the base period 1976–2005. Solid lines show the ensemble mean and the shading indicates the range among the individual RCMs. The black line shows the observed indices based on IMD gridded station data. Time series are smoothed with an 11-year running mean filter

Fig. 2.10 Time series of all India averaged CORDEX South Asia multi-RCM projections of the summer (April–June) heatwave **a** frequency (HWF; events per season) and **b** total average duration (HWD; days per season) for the CORDEX South Asia RCM ensemble mean (solid lines) and the minimum to maximum range of the individual RCMs (shading) based on the historical simulations during 1951–2005 (grey), and based on the future projections during 2006–2099 under RCP4.5 scenario (blue) and RCP8.5 scenario (red)

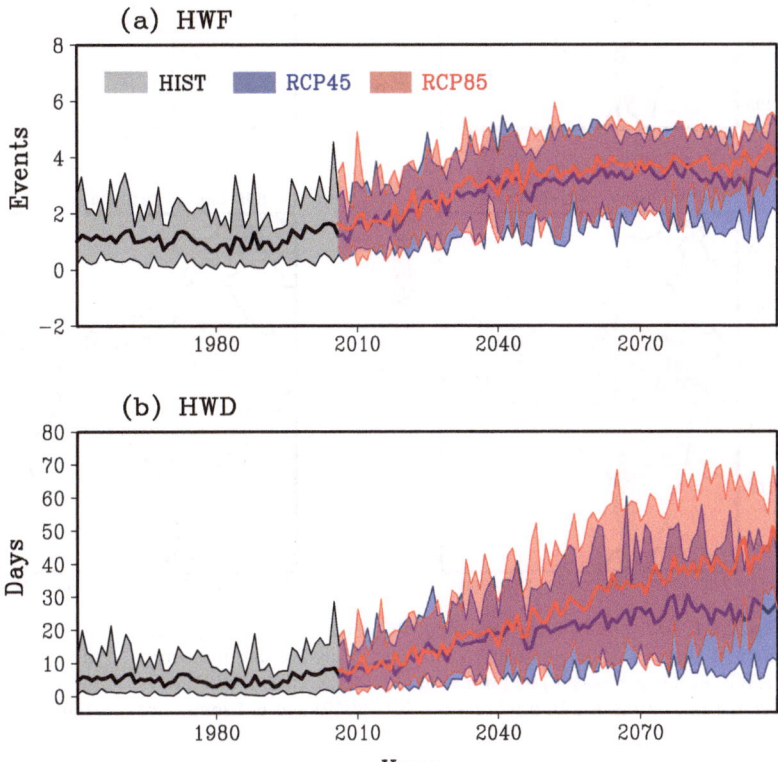

percentile threshold of daily maximum air temperature from each CORDEX model outputs after applying a quantile mapping bias-correction following Lafon et al. (2013), which was based on an empirical distribution correction method (Wood et al. 2004). These findings compare well with the heatwave climatology estimated using the CMIP5 multi-model ensemble and IMD observations over north-west India for the present period (Rohini et al. 2019).

The CORDEX multi-RCM ensemble mean project that all India averaged frequency of summer heatwaves will increase to about 2.5 events per season by the mid-twenty-first century (2040–2069), with a further slight rise to about 3.0 events by the end-twenty-first century (2070–2099) under the medium (RCP4.5) emission scenario (Fig. 2.10). The average total duration of summer heatwaves is projected to increase to about 15 and 18 days per season during the mid- and end-twenty-first century respectively under this future scenario. The projected increase in these all India averaged summer heatwave characteristics are similar to the assessment based on CMIP5 multi-model ensemble over north-west India in the period 2020–2064 under the RCP4.5 scenario (Rohini et al. 2019). The CORDEX ensemble mean projects that India averaged frequencies of summer heatwaves will increase to about 3.0 and 3.5 events per season during the mid- and end-twenty-first century, respectively under the high (RCP8.5) emission scenario (Fig. 2.10). The future rise in heatwave frequencies under this high emission scenario is marginally higher than the increase in the number of summer events under the RCP4.5 scenario for the corresponding periods. The average total duration of summer heatwaves in India under the RCP8.5 scenario is assessed to be substantially higher than that under RCP4.5 scenario, with about 25 and 35 heatwave days per season during the mid- and end-twenty-first century respectively (Fig. 2.10). The projected CORDEX ensemble mean change in the frequency of heatwaves for the mid- and end-twenty-first century under RCP8.5 scenario relative to the historical reference period (1976–2005) are higher over the north-west region (more than 3 days per summer season) compared to

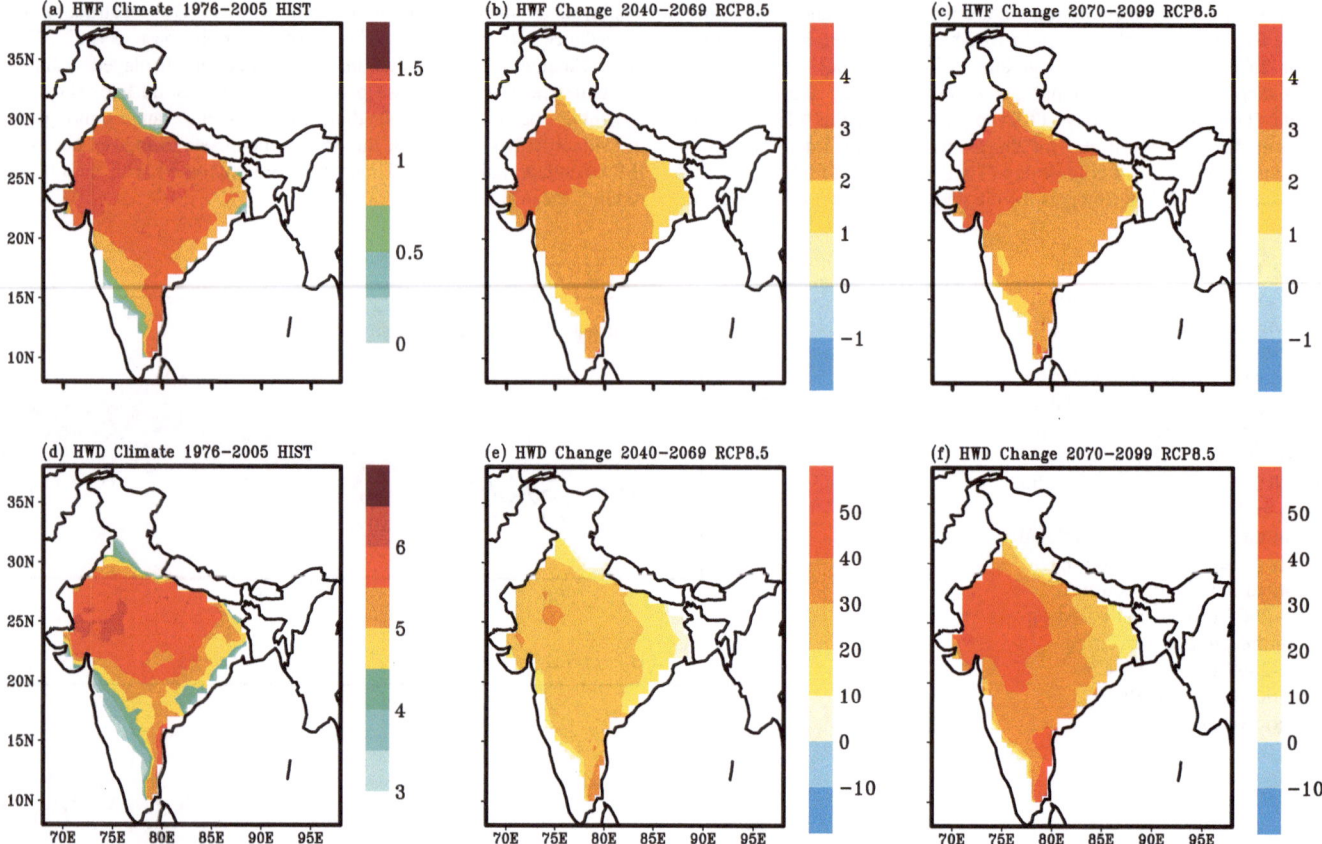

Fig. 2.11 CORDEX South Asia multi-RCM ensemble mean projections of changes in the summer (April–June) heatwave **a–c** frequency (HWF; events per season) and **d–f** total average duration (HWD; days per season) during the **b, e** mid-term (2040–2069) and **c, f** long-term (2070–2099) climate under RCP8.5 scenario relative to the **a, d** climate for a historical reference period 1976–2005. The projected ensemble mean changes based on 12 CORDEX South Asia RCM are statistically significant at *p*-value <0.05

other parts of India (Fig. 2.11). The CORDEX ensemble members consistently project a substantial increase in the average total duration of heat waves of more than 40 days per summer season relative to the historical reference period over north-west India by end of the twenty-first century under this high emission scenario.

These findings are consistent with results based on the CMIP5 multi-model ensemble assessments that the heatwave frequency, duration, intensity and areal coverage over India will substantially increase during the twenty-first century (Murari et al. 2015; Im et al. 2017; Mishra et al. 2017; Russo et al. 2017; Dosio et al. 2018; Rohini et al. 2019). The CMIP5 ensemble had projected that the future warm season (March to June) severe heat waves in India (based on the IMD criteria) will be more intense, with longer durations and will likely occur at a higher frequency and earlier in the year (Murari et al. 2015). The southern India, currently not influenced by heatwaves, is expected to be severely affected by the end of the twenty-first century. Rohini et al. (2016) reported that in near future more frequent and long lasting heatwave events are projected to

affect the Indian sub continent in response to the warming of the tropical Indian Ocean (Roxy et al. 2014, 2015) and the increasing frequency of extreme El Nino events (Cai et al. 2015). A CMIP5 multi-model ensemble-based assessment projected 40 times rise in the frequency of severe heat waves over India than in the present climate by the end of the twenty-first century under the RCP8.5 scenario with a heatwave index that captures both duration and magnitude estimated using CMIP5 model projections of all India averaged daily maximum air temperature without bias-correction (Mishra et al. 2017). This study also assessed using a global earth system model ensemble that the projected frequency of severe heatwaves by the end of the twenty-first century will be about 2.5 times lesser (than for RCP8.5 scenario) if the global mean temperature is limited to the low-warming scenario of 2.0 °C above pre-industrial conditions. However an assessment of the projected population exposure in India due to heat stress (Im et al. 2017) differed considerably from the findings of Mishra et al. (2017) based on a heatwave index without considering the effect of humidity in heatwave estimations.

The projected future heatwaves based on the extremes of wet-bulb temperature, which includes the effect of humidity, are found to be concentrated around densely populated agricultural regions of the Ganges and Indus river basins (Im et al. 2017). The human populations in these regions highly vulnerable to heat stress are assessed to experience maximum daily wet-bulb temperatures exceeding 31 °C by 2100 under RCP4.5 scenario, considered dangerous levels for most humans (Im et al. 2017). These extreme wet-bulb temperatures are likely to approach and, in few locations, exceed 35 °C by the 2070s under RCP8.5 scenario, considered as an upper limit on human survivability (Coffel et al. 2018). The annual probability of occurrence of heat waves with magnitude greater than the one in Russia in 2010 (the most severe of the present era) at 1.5 and 2 °C global warming above pre-industrial levels is assessed using CMIP5 ensemble to be different than zero in few parts of India when measured by means of an heatwave index, which takes into account both temperature and relative humidity (Russo et al. 2017). The yearly probability of occurrence of a heatwave at 4° global warming with magnitude greater than this most severe event in the present period will be greater than 10% in many parts of India, with the southern peninsula expected to experience such type of humid heatwaves with an annual probability greater than 50%, corresponding to an average return period of two years (Russo et al. 2017).

2.4 Knowledge Gaps

To improve the assessment of India's observed and projected warming and its impacts, the following gaps would need to be addressed:

- The uneven spatial distribution of temperature observation sites over India may lead to errors in the assessment of present-day temperature changes, particularly over the northern parts of the country with a very sparse network.
- Confidence in the assessed long-term temperature trends may be constrained by the data inhomogeneity due to changing observation site locations.
- There has been an increase in costly extreme temperature events (e.g. heat waves) across India. Hence urgent research studies are needed on event attribution that evaluates how the probability or intensity of a heatwave event, or more generally, a class of extreme temperature events, has changed as a result of increases in atmospheric greenhouse gases from human activity.
- The CMIP5 multi-model ensemble members sampling structural uncertainty and internal variability cannot be treated as purely independent because some climate models have been developed by sharing model components leading to shared biases. This implies a reduction in

the effective number of independent CMIP5 models. The CORDEX South Asia ensemble consists of two RCMs driven by a subset of CMIP5 AOGCMs, implying a very little effective number of independent members in this multi-RCM ensemble.

- The contribution of natural internally generated variability to the total uncertainty in the sub-regional/local temperature projections need to be quantitatively assessed using an ensemble of high-resolution future climate projections for India. The existing ensemble of dynamically downscaled temperature projections from CORDEX South Asia multi-RCMs does not sample initial conditions, which are needed to quantify the contribution of internal variability to the total uncertainty at smaller spatial scales.
- More research is needed to understand whether the increased water vapour under conditions of regional warming is leading to significant positive feedback on human-induced climate change, as water vapour is the most important contributor to the natural greenhouse effect.
- Assessment of joint projections of multiple variables over India are needed to understand the key processes relevant to future projected significant increases in temperature variability and extremes, for example, projected changes by combining mean temperature and precipitation; linking soil moisture, precipitation and temperature mean and variability; combining temperature, humidity, etc.

2.5 Summary

In summary, the annual mean, maximum and minimum temperatures averaged over India as a whole show significant warming trend of 0.15 °C, 0.15 °C and 0.13 °C per decade respectively since 1986 (*high confidence*). The maximum warming trend is seen during the pre-monsoon season for the recent 30-year period 1986–2015. It is very likely that all India averaged annual and seasonal near-surface air specific humidity have increased since the 1980s. The significant increasing trend in specific humidity assessed during the pre-monsoon season is consistent with the largest surface warming trend found for this season.

The observed surface air temperature changes over India are attributed to anthropogenic forcing (*medium confidence*). The maximum temperature during the post-monsoon and minimum temperature during the pre-monsoon and monsoon seasons are attributed with confidence to climate change induced by anthropogenic effects (*medium confidence*).

The all India averaged frequency of warm extremes has increased since 1951 with accelerated warming trends during the recent 30 year period 1986–2015. The annual increase in the number of warm days is higher than warm nights, with a relatively milder decrease in number of cold days since 1951. Increase in frequency of warm days during monsoon and winter seasons largely contributed to the observed highest annual increase in number of warm days over India from the mid-twentieth century. The accelerated annual increase in number of warm days and warm nights, and decrease in number of cold nights are observed during 1986–2015. Increase in frequency of warm days during winter season resulted in the highest annual increase in number of warm days over India during this recent 30 year period. Significant annual warming is observed for the warmest day, warmest night and coldest night since 1986. The warming during pre-monsoon and winter seasons contributed to the strongest warming of the warmest day over India after about 1986.

The all India mean surface air temperature change for the mid-term period 2040–2069 relative to 1976–2005 is projected to be in the range of 1.39–2.70 °C, and is larger than the natural internal variability. This assessment is based on a reliability ensemble average (REA) estimate incorporating each RCM performance and convergence and is associated with less than 13% uncertainty range. All India mean surface air temperature is projected to increase in the far future (2070–2099) by 1.33 ± 0.24 °C under RCP2.6, 2.44 ± 0.41 °C under RCP4.5 and 4.44 ± 0.45 °C under RCP8.5 scenario, respectively. These changes are relative to the period 1976–2005. The semi-arid north-west and north India will likely warm more rapidly than the all India mean. The REA changes for all India annual minimum temperature of 4.71 ± 0.35 °C is more pronounced than that of 4.10 ± 0.45 °C and 4.44 ± 0.45 °C increases estimated for the respective annual maximum and mean temperatures, respectively by the end of the twenty-first century under RCP8.5 scenario. The frequency and intensity of warm days and warm nights are projected to increase over India in the next decades, while that of cold days and cold nights will decrease (*high confidence*). The changes will be more pronounced for cold nights and warm nights. The pre-monsoon season heatwave frequency, duration, intensity and areal coverage over India are projected to increase substantially during the twenty-first century (*high confidence*).

References

Bindoff NL, Stott PA, AchutaRao KM, Allen MR, Gillett N, Gutzler D, Hansingo K, Hegerl G, Hu Y, Jain S, Mokhov II, Overland J, Perlwitz J, Sebbari R, Zhang X (2013) Detection and attribution of climate change: from global to regional. In: Stocker TF, Qin D, Plattner DG, Tignor M, Allen SK, Boschung J, Nauels A, Xia Y, Bex V, Midgley PM (eds) Climate change 2013: the physical science basis. Contribution of working group I to the fifth assessment report of the intergovernmental panel on climate change. Cambridge University Press, Cambridge, United Kingdom and New York, NY, USA. p 867–952. https://doi.org/10.1017/CBO9781107415324.022

Borgaonkar HP, Pant GB, Rupa Kumar K (1994) Dendroclimatic reconstruction of summer precipitation at Srinagar, Kashmir, India since the late 18th century. Holocene 4(3):299–306

Borgaonkar HP, Pant GB, Rupa Kumar K (1996) Ring-width variations in Cedrusdeodara and its climatic response over the western Himalaya. Int J Climatol 16(12):1409–1422

Borgaonkar HP, Somaru Ram, Sikder AB (2009) Assessment of tree-ring analysis of high elevation *Cedrusdeodara* D. Don from Western Himalaya (India) in relation to climate and glacier fluctuations. Dendrochronologia 27(1):59–69

Borgaonkar HP, Sikder AB, Ram Somaru (2011) High altitude forest sensitivity to the recent warming: a tree-ring analysis of conifers from Western Himalaya India. Quat Int 236(2011):158–166

Borgaonkar HP, Gandhi N, Ram Somaru, Krishnan R (2018) Tree-ring reconstruction of late summer temperatures in northern Sikkim (eastern Himalayas). Palaeogeogr Palaeoclimatol Palaeoecol 504(1):125–135

Boucher O, Randall D, Artaxo P, Bretherton C, Feingold G, Forster P, Kerminen VM, Kondo Y, Liao H, Lohmann U, Rasch P, Satheesh SK, Sherwood S, Stevens B, Zhang XY (2013) Clouds and aerosols. In: Stocker TF, Qin D, Plattner GK, Tignor M, Allen SK, Boschung J, Nauels A, Xia Y, Bex V, Midgley PM (eds) Climate change 2013: the physical science basis. Contribution of working group I to the fifth assessment report of the intergovernmental panel on climate change. Cambridge University Press, Cambridge, United Kingdom and New York, NY, USA. p 571–658. https://doi.org/10.1017/CBO9781107415324.016

Cai W, Santoso A, Wang G (2015) ENSO and greenhouse warming. Nat Clim Chang 5:849–859. https://doi.org/10.1038/nclimate2743

Chaturvedi RK, Joshi J, Jayaraman M, Bala G, Ravindranath NH (2012) Multimodel climate change projections for India under representative concentration pathways. Curr Sci 103(7):791–802

Christensen JH, Krishna Kumar K, Aldrian E, An SI, Cavalcanti IFA, de Castro M, Dong W, Goswami P, Hall A, Kanyanga JK, Kitoh A, Kossin J, Lau NC, Renwick J, Stephenson DB, Xie XP, Zhou T (2013) Climate phenomena and their relevance for future regional climate change. In: Stocker TF, Qin D, Plattner GK, Tignor M, Allen SK, Boschung J, Nauels A, Xia Y, Bex V, Midgley PM (eds) Climate change 2013: the physical science basis. Contribution of working group I to the fifth assessment report of the intergovernmental panel on climate change. Cambridge University Press, Cambridge, United Kingdom and New York, NY, USA. p 1217–1308. https://doi.org/10.1017/CBO9781107415324.028

Coffel ED, Horton RM, de Sherbinin A (2018) Temperature and humidity based projections of a rapid rise in global heat stress exposure during the 21st century. Environ Res Lett 13. https://doi.org/10.1088/1748-9326/aaa00e

Collins M, Knutti R, Arblaster J, Dufresne JL, Fichefet T, Friedlingstein P, Gao X, Gutowski WJ, Johns T, Krinner G, Shongwe M, Tebaldi C, Weaver AJ, Wehner M (2013) Long-term climate change: projections, commitments and irreversibility. In: Stocker TF, Qin D, Plattner GK, Tignor M, Allen SK, Boschung J, Nauels A, Xia Y, Bex V, Midgley PM (eds) Climate change 2013: the physical science basis. Contribution of working group I to the fifth assessment report of the intergovernmental panel on climate change. Cambridge University Press, Cambridge, United Kingdom and New York, NY, USA. p 1029–1136. https://doi.org/10.1017/CBO9781107415324.024

Cook ER, Krusic PJ, Jones PD (2003) Dendroclimatic signals in long tree-ring chronologies from the Himalayas of Nepal. Int J Climatol 23:707–732

Das BK, Gaye B, Malik MA (2010) Biogeochemistry and paleoclimate variability during the Holocene: a record from Mansar Lake lesser Himalaya. Environ Earth Sci 61(3):565–574

Dee DP et al (2011) The ERA-interim reanalysis: configuration and performance of the data assimilation system. Q J R Meteorol Soc 137(656):553–597

Dileepkumar R, AchutaRao K, Arulalan T (2018) Human influence on sub-regional surface air temperature change over India. Sci Rep 8:8967. https://doi.org/10.1038/s41598-018-27185-8

Dimri AP, Kumar D, Choudhary A, Maharana P (2018a) Future changes over the Himalayas: mean temperature. Glob Planet Chang 162:235–251

Dimri AP, Kumar D, Choudhary A, Maharana P (2018b) Future changes over the Himalayas: maximum and minimum temperature. Glob Planet Chang 162:212–234

Dosio A, Mentaschi L, Fischer EM, Wyser K (2018) Extreme heat waves under 1.5 and 2 °C global warming. Environ Res Lett 13. https://doi.org/10.1088/1748-9326/aab827

Esper J, Schweingruber FH, Winiger M (2002) 1300 years of climatic history for western central Asia inferred from tree-rings. Holocene 12(3):267–277

Flato G, Marotzke J, Abiodun B, Braconnot P, Chou SC, Collins W, Cox P, Driouech F, Emori S, Eyring V, Forest C, Gleckler P, Guilyardi E, Jakob C, Kattsov V, Reason C, Rummukainen M (2013) Evaluation of climate models. In: Stocker TF, Qin D, Plattner GK, Tignor M, Allen SK, Boschung J, Nauels A, Xia Y, Bex V, Midgley PM (eds) Climate change 2013: the physical science basis. Contribution of working group I to the fifth assessment report of the intergovernmental panel on climate change. Cambridge University Press, Cambridge, United Kingdom and New York, NY, USA. p 741–866. https://doi.org/10.1017/CBO9781107415324.020

Giorgi F, Coppola E, Solmon F et al (2012) RegCM4: model description and preliminary tests over multiple CORDEX domains. Climate Res 52:7–29

Giorgi F, Mearns LO (2002) Calculation of average, uncertainty range, and reliability of regional climate changes from AOGCM simulations via the 'reliability ensemble averaging' (REA) method. J Clim 15:1141–1158

Harris I, Jones PD, Osborn TJ, Lister DH (2014) Updated high-resolution grids of monthly climatic observations—the CRU TS3.10 Dataset. Int J Climatol 34(3):623–642

Hartmann DL, Klein Tank AMG, Rusticucci M, Alexander LV, Brönnimann S, Charabi Y, Dentener FJ, Dlugokencky EJ, Easterling DR, Kaplan A, Soden BJ, Thorne PW, Wild M, Zhai PM (2013) Observations: atmosphere and surface. In: Stocker TF, Qin D, Plattner GK, Tignor M, Allen SK, Boschung J, Nauels A, Xia Y, Bex V, Midgley PM (eds) Climate change 2013: the physical science basis. Contribution of working group I to the fifth assessment report of the intergovernmental panel on climate change. Cambridge University Press, Cambridge, United Kingdom and New York, NY, USA. p 159–254. https://doi.org/10.1017/CBO9781107415324.008

Hasson S, Böhner J, Chishtie F (2018) Low fidelity of CORDEX and their driving experiments indicates future climatic uncertainty over Himalayan watersheds of Indus basin. Clim Dyn 52(1):777–798

Hughes MK (2001) An improved reconstruction of summer temperature at Srinagar, Kashmir since 1660 AD based on tree-ring width and maximum latewood density of Abiespindrow (Royle) Spach. Palaeobotanist 50:13–19

Im E-S, Pal JS, Eltahir EAB (2017) Deadly heat waves projected in the densely populated agricultural regions of South Asia. Sci Adv 3 (8):1–7

IPCC (2013) Climate change 2013: the physical science basis. In: Stocker TF, Qin D, Plattner G-K, Tignor M, Allen SK, Boschung J, Nauels A, Xia Y, Bex V, Midgley PM (eds) Contribution of working group I to the fifth assessment report of the intergovernmental panel on climate change. Cambridge University Press, Cambridge, United Kingdom and New York, NY. p 1535. https://doi.org/10.1017/CBO9781107415324

Jain SK, Kumar V (2012) Trend analysis of rainfall and temperature data for India. Curr Sci 102(1):37–49

Juyal N, Pant RK, Basavaiah N, Bhushan R, Jain M, Saini NK, Yadava MG, Singhvi AK (2009) Reconstruction of Last Glacial to early Holocene monsoon variability from relict lake sediments of the Higher Central Himalaya, Uttrakhand, India. J Asian Earth Sci 34(3):437–449

Kothawale DR, Munot AA, Krishna Kumar K (2010a) Surface air temperature variability over India during 1901–2007, and its association with ENSO. Climate Res 42(2):89–104

Kothawale DR, Revadekar JV, Rupa Kumar K (2010b) Recent trends in pre-monsoon daily temperature extremes over India. J Earth Syst Sci 119:51–65

Kothawale DR, Deshpande NR, Kolli RK (2016) Long term temperature trends at major, medium, small cities and hill stations in India during the period 1901–2013. Am J Clim Chang 5 (3):383–398

Kothawale DR, Singh HN (2017) Recent trends in tropospheric temperature over India during the period 1971–2015. Earth Space Sci 4(5):240–246

Kothawale DR, Rupa Kumar K (2005) On the recent changes in surface temperature trends over India, Geophysical. Res Lett 32. http://doi.org/10.1029/2005GL023528

Krishnan R, Sabin TP, Vellore R, Mujumdar M, Sanjay J, Goswami BN, Hourdin F, Dufresne JL, Terray P (2016) Deciphering the desiccation trend of the South Asian monsoon hydroclimate in a warming world. Clim Dyn 47(3):1007–1027

Krusic PJ, Cook ER, Dukpa D, Putnam AE, Rupper S, Schaefer J (2015) Six hundred thirty-eight years of summer temperature variability over the Bhutanese Himalaya. Geophys Res Lett 42. http://dx.doi.org/10.1002/2015GL063566

Kulkarni A, Deshpande N, Kothawale DR, Sabade SS, Ramarao MVS, Sabin TP, Patwardhan S, Mujumdar M, Krishnan R (2017) Observed climate variability and change over India. In: Krishnan R, Sanjay J (eds) Climate change over India—an interim report. Published by Centre for Climate Change Research, IITM, p 38. http://cccr.tropmet.res.in/home/reports.jsp

Lafon T, Dadson S, Buysa G, Prudhomme C (2013) Bias correction of daily precipitation simulated by a regional climate model: a comparison of methods. Int J Climatol 33:1367–1381

Mishra V (2015) Climatic uncertainty in Himalayan water towers. J Geophys Res Atmos 120(7):2689–2705

Mishra V, Mukherjee S, Kumar R, Stone DA (2017) Heat wave exposure in India in current, 1.5 and 2.0 °C worlds. Environ Res Lett 12. https://doi.org/10.1088/1748-9326/aa9388

Mukhopadhyay P, Jaswal AK, Deshpande M (2017) Variability and trends of atmospheric moisture over the Indian region. In: Rajeevan MN, Nayak S (eds) Observed climate variability and change over the Indian region, pp 129–144. https://doi.org/10.1007/978-981-10-2531-0_8

Murari KK, Ghosh S, Patwardhan A, Daly E, Salvi K (2015) Intensification of future severe heat waves in India and their effect on heat stress and mortality. Reg Environ Change 15(4):569–579

Nengker T, Choudhary A, Dimri AP (2018) Assessment of the performance of CORDEX-SA experiments in simulating seasonal mean temperature over the Himalayan region for the present climate: part I. Clim Dyn 50(7):2411–2441

Padma Kumari B, Goswami BN (2010) Seminal role of clouds on solar dimming over the Indian monsoon region. Geophys Res Lett 37: L06703. https://doi.org/10.1029/2009GL042133

Padma Kumari B, Londhe AL, Daniel S, Jadhav DB (2007) Observational evidence of solar dimming: offsetting surface warming over India. Geophys Res Lett 34:L21810. https://doi.org/10.1029/2007GL031133

Padma Kumari B, Jaswal AK, Goswami BN (2013) Decrease in evaporation over the Indian monsoon region: implication on regional hydrological cycle. Clim Change 121:787–799. https://doi.org/10.1007/s10584-013-0957-3

Pant GB, Borgaonkar HP, Rupa Kumar K (1998) Climatic signals from tree-rings: a dendro climatic investigation of Himalayan spruce (piceasmithiana). Himalayan Geol 19(2):65–73

Pai DS, Srivastava AK, Smitha Anil Nair (2017) Heat and cold waves over India. In: Rajeevan MN, Nayak S (eds) Observed climate variability and change over the Indian region. Springer Geology, pp 51–71. https://doi.org/10.1007/978-981-10-2531-0_4

Peterson TC, Vose RS, Schmoyer R, Razuvaev V (1998) Global Historical Climatology Network (GHCN) quality control of monthly temperature data. Int J Climatol 18:1169–1179

Rai A, Joshi MK, Pandey AC (2012) Variations in diurnal temperature range over India: under global warming scenario. J Geophys Res 117, D02114. http://doi.org/10.1029/2011JD016697

Ratnam JV, Behera SK, Ratna SB, Rajeevan M, Yamagata T (2016) Anatomy of Indian heat waves. Sci Rep 6, 24395. https://doi.org/10.1038/srep24395

Revadekar J, Kothawale D, Patwardhan S, Pant G, Kumar K (2012) About the observed and future changes in temperature extremes over India. Nat Hazards 60(3):1133–1155

Rohde R, Muller R, Jacobsen R, Perlmutter S, Rosenfeld A et al (2013) Berkeley earth temperature averaging process. Geoinfor Geostat Overv 1:2. https://doi.org/10.4172/2327-4581.1000103

Rohini P, Rajeevan M, Srivastava AK (2016) On the variability and increasing trends of heat waves over India. Sci Rep 6, 26153. https://doi.org/10.1038/srep26153

Rohini P, Rajeevan M, Mukhopadhay P (2019) Future projections of heat waves over India from CMIP5 models. Clim Dyn 53:975–988. https://doi.org/10.1007/s00382-019-04700-9

Roxy MK, Ritika K, Terray P, Masson S (2014) The curious case of Indian ocean warming. J Clim 27(22):8501–8509

Roxy MK, et al (2015) Drying of Indian subcontinent by rapid Indian ocean warming and a weakening land-sea thermal gradient. Nature Comm 6, 7423. http://doi.org/10.1038/ncomms8423

Rupa Kumar K, Krishna Kumar K, Pant GB (1994) Diurnal asymmetry of surface temperature trends over India. Geophys Res Lett 21:677–680

Russo S, Sillmann J, Ster A (2017) Humid heat waves at different warming levels. Sci Rep 7, 7477. http://doi.org/10.1038/s41598-017-07536-7

Samuelsson P, Jones CG, Willen U et al (2011) The rossby centre regional climate model RCA3: model description and performance. Tellus 63(1):4–23

Sanjay J, Ramarao MVS, Mujumdar M, Krishnan R (2017a) Regional climate change scenarios. In: Rajeevan MN, Nayak S (eds) Observed climate variability and change over the Indian region. Springer Geology, pp 285–304. https://doi.org/10.1007/978-981-10-2531-0_16

Sanjay J, Krishnan R, Shrestha AB, Rajbhandari R, Ren G-Y (2017b) Downscaled climate change projections for the Hindu Kush Himalayan region using CORDEX South Asia regional climate models. Adv Clim Chang Res 8(3):185–198. https://doi.org/10.1016/j.accre.2017.08.003

Sen Roy S, Balling RC (2005) Analysis of trends in maximum and minimum temperature, diurnal temperature range, and cloud cover over India. Geophys Res Lett 32:L12702. http://doi.org/10.1029/2004GL022201

Sengupta A, Rajeevan M (2013) Uncertainty quantification and reliability analysis of CMIP5 projections for the Indian summer monsoon. Curr Sci 105:1692–1703

Sheffield J, Goteti G, Wood EF (2006) Development of a 50-year high-resolution global dataset of meteorological forcings for land surface modeling. J Clim 19(13):3088–3111

Sharma S, Mujumdar P (2017) Increasing frequency and spatial extent of concurrent meteorological droughts and heatwaves in India. Sci Rep 7, 15582. https://doi.org/10.1038/s41598-017-15896-3

Singh S, Ghosh S, Sahana AS, VittalH Karmakar S (2017) Do dynamic regional models add value to the global model projections of Indian monsoon? Clim Dyn 48(3):1375–1397

Singh R, AchutRao K (2018) Quantifying uncertainty in twenty-first century climate change over India. Clim Dyn 52(7):3905–3928

Sonali P, Nanjundiah RS, Kumar DN (2018) Detection and attribution of climate change signals in South India maximum and minimum temperatures. Climate Res 76(2):145–160

Soni VK, Pandithurai G, Pai DS (2012) Evaluation of long-term changes of solar radiation in India. Int J Climatol 32(4):540–551

Srivastava AK, Rajeevan M, Kshirsagar SR (2009) Development of a high resolution daily gridded temperature data set (1969–2005) for the Indian region. Atmos Sci Lett 10(4):249–254

Srivastava AK, Kothawale DR, Rajeevan MN (2017) Variability and long-term changes in surface air temperatures over the Indian subcontinent. In: Rajeevan MN, Nayak S (eds) Observed climate variability and change over the Indian region. Springer Geology, pp 17–35. https://doi.org/10.1007/978-981-10-2531-0_2

Srivastava AK, Revadekar JV, Rajeevan M (2019) South Asia in "State of the climate in 2018". Bull Amer Meteor Soc 100(9):S236–S240. https://doi.org/10.1175/2019BAMSStateoftheClimate.1

Thapa UK, Shah SK, Gaire NP et al (2015) Climate dynamics spring temperatures in the far-western Nepal Himalaya since AD 1640 reconstructed from Piceasmithiana tree-ring widths 45:2069–2081

Taylor KE, Stouffer RJ, Meehl GA (2012) An overview of CMIP5 and the experiment design. Bull Am Meteorol Soc 93:485–498

Thrasher B, Maurer EP, McKellar C, Duffy PB (2012) Technical note: bias correcting climate model simulated daily temperature extremes with quantile mapping. Hydrol Earth Syst Sci 16(9):3309–3314

Vinnarasi R, Dhanya CT, Chakravorty A, AghaKouchak A (2017) Unravelling diurnal asymmetry of surface temperature in different climate zones. Sci Rep 7(1):1–8. https://doi.org/10.1038/s41598-017-07627-5

Willett KM, Gillett NP, Jones PD, Thorne PW (2007) Attribution of observed surface humidity changes to human influence. Nature 449:710–712. https://doi.org/10.1038/nature06207

Willett KM, Dunn RJH, Thorne PW, Bell S, de Podesta M, Parker DE, Jones PD, Williams CN Jr (2014) HadISDH land surface multi-variable humidity and temperature record for climate monitoring. Clim Past 10(6):1983–2006

Wood AW, Leung LR, Sridhar V, Lettenmaier DP (2004) Hydrologic implications of dynamical and statistical approaches to downscaling climate model outputs. Clim Chang 62:189–216

Wu X, Shao X (1995) Status and prospects of dendrochronological study in Tibetan Plateau. Dendrochronologia 13:89–98

Yadav RR, Park WK, Bhattacharya A (1999) Spring temperature variations in the western Himalayan region as reconstructed from tree-rings: AD 1390–1987. Holocene 9(1):85–90

Yadav RR, Braeuning A, Singh J (2011) Tree ring inferred summer temperature variations over the last millennium in western Himalaya, India. Clim Dyn 36(7):1545–1554.

Yadava AK, Yadav RR, Misra KG, Singh J, Singh D (2015) Tree ring evidence of late summer warming in Sikkim, northeast India. Quat Int 371:175–180

Yasutomi N, Hamada A, Yatagai A (2011) Development of a long-term daily gridded temperature dataset and its application to rain/snow discrimination of daily precipitation. Glob Environ Res 15(2): 165–172

Zhang X, Alexander L, Hegerl GC, Jones P, Tank AK, Peterson TC, Trewin B, Zwiers FW (2011) Indices for monitoring changes in extremes based on daily temperature and precipitation data. WIREs Clim Chang 2(6):851–870

Coordinating Lead Authors

Ashwini Kulkarni, Indian Institute of Tropical Meteorology (IITM-MoES), Pune, India,
e-mail: ashwini@tropmet.res.in (corresponding author)
T. P. Sabin, Indian Institute of Tropical Meteorology (IITM-MoES), Pune, India
Jasti S. Chowdary, Indian Institute of Tropical Meteorology (IITM-MoES), Pune, India

Lead Authors

K. Koteswara Rao, Indian Institute of Tropical Meteorology (IITM-MoES), Pune, India
P. Priya, Indian Institute of Tropical Meteorology (IITM-MoES), Pune, India
Naveen Gandhi, Indian Institute of Tropical Meteorology (IITM-MoES), Pune, India
Preethi Bhaskar, Indian Institute of Tropical Meteorology (IITM-MoES), Pune, India
Vinodh K. Buri, Department of Earth and Atmospheric Sciences, National Institute of Technology,
Rourkela, India
S. S. Sabade, Indian Institute of Tropical Meteorology (IITM-MoES), Pune, India

Review Editors

D. S. Pai, Climate Research and Services, India Meteorological Department (IMD-MoES), Pune, India
K. Ashok, University Center for Earth and Space Sciences, University of Hyderabad, Hyderabad, India
A. K. Mitra, Modeling and Data Assimilation, National Center for Medium Range Weather
Forecasting (NCMRWF-MoES), New Delhi, India
Dev Niyogi, Purdue University, West Lafayette, IN and University of Texas at Austin, Austin, TX, USA
M. Rajeevan, Ministry of Earth Sciences, Government of India, New Delhi, India

Corresponding Author

Ashwini Kulkarni, Indian Institute of Tropical Meteorology (IITM-MoES), Pune, India,
e-mail: ashwini@tropmet.res.in

© The Author(s) 2020
R. Krishnan et al. (eds.), *Assessment of Climate Change over the Indian Region*,
https://doi.org/10.1007/978-981-15-4327-2_3

Key Messages

- There has been a decreasing trend in the all-India annual, as well as summer monsoon mean rainfall during 1951–2015, notably so over areas in the Indo-Gangetic Plains and the Western Ghats. Increasing concentrations of anthropogenic aerosols over the northern hemisphere appear to have played a role in these changes (medium confidence).

- The frequency of localized heavy rain occurrences over India has increased during 1951–2015 (high confidence). Urbanization and other land use, as well as aerosols, likely contribute to these localized heavy rainfall occurrence (medium confidence).

- With continued global warming and expected reductions of aerosol concentrations in the future, climate models project an increase in the annual and summer monsoon mean rainfall, as well as frequency of heavy rain occurrences over most parts of India during the twenty-first century (medium confidence).

- The interannual variability of summer monsoon rainfall is projected to increase through the twenty-first century (high confidence).

3.1 Introduction

Precipitation is an important component of the global water cycle, and the impacts of anthropogenic climate change on precipitation have significant implications on agricultural activities (Porter et al. 2014). Over India, the seasonal monsoon rains during the June-September months, which contribute to more than 75% of the annual rainfall (Fig. 3.1a), are vital for the country's agriculture and economy (Parthasarathy et al. 1988; Gadgil 2007). Abrupt changes in the Indian monsoon precipitation on decadal and centennial time-scales are evident from high-resolution climate proxy records, extending back several thousands of years (Berkelhammer et al. 2012; Sanyal and Sinha 2010). This chapter provides an assessment of historical changes in precipitation, as well as projections of future changes from climate models.

Box 3.1   Processes in the Monsoon System
The annual cycle of monsoon precipitation can be described as a manifestation of the seasonal migration of the Inter-Tropical Convergence Zone (ITCZ) and the associated shift in rain band from southern to northern Indian Ocean (Fig. 3.1a). Summer monsoon rainfall across the Indian Subcontinent shows

substantial spatial variability with heaviest rainfall along the Western Ghats and the Himalayan foothills due to orographic features, and over central India due to low-level convergence (Fig. 3.1; upper panel).

Mean Onset and Withdrawal
The onset of summer monsoon over India is characterized by the dramatic rise in mean daily rainfall over Kerala (Ananthkrishnan and Soman 1988; Soman and Kumar 1993). The onset of the Indian summer monsoon (ISM) is a key indicator characterizing the abrupt transition from the dry season to the rainy season and subsequent seasonal march (Koteswaram 1958; Ananthakrishnan et al. 1968; Krishnamurti and Ramanathan 1982; Ananthakrishnan and Soman 1988; Wang et al. 2001, 2009; Pai and Rajeevan 2009). The mean onset date of summer monsoon rainfall over India has been around 1 June. By the first week of July, the southwest monsoon is established over the entire subcontinent. The southwest monsoon starts withdrawing from the extreme northwestern portion of India by the beginning of September.

Intra-Seasonal Oscillations (ISOs)
The strength of summer monsoon rainfall is modulated by the intra-seasonal variability which is characterized by the active/break spells of enhanced/decreased precipitation over India. The active/break spells over India play an essential role in modulating mean monsoon rainfall (e.g. Ramamurthy 1969; Sikka and Gadgil 1980; Rodwell 1997; Webster et al. 1998; Krishnan et al. 2000; Krishnamurthy and Shukla 2000, 2007, 2008; Annamalai and Slingo 2001; Goswami and Ajayamohan 2001; Lawrence and Webster 2001; De and Mukhopadhyay 2002; Goswami et al. 2003; Waliser et al. 2003; Kripalani et al. 2004; Wang et al. 2005; Mandke et al. 2007; Goswami 2005; Waliser 2006). Active and break episodes, characteristic of subseasonal variations of the ISM, are associated with enhanced (decreased) rainfall over central and western India and decreased (enhanced) rainfall over the southeastern peninsula and eastern India (Singh et al. 1992; Krishnamurthy and Shukla 2000; Krishnan et al. 2000; Goswami et al. 2003).

ISOs are influenced by various factors such as variations in the position and strength of the continental monsoon trough, quasi-periodic oscillations of the monsoon, as well as synoptic systems such as lows and depressions (Shukla 1987 Yasunari 1979; Sikka and Gadgil 1980). These quasi-periodic fluctuations play a major role in determining the amplitude of seasonal mean of individual summer seasons by modulating the strength and duration of active/break

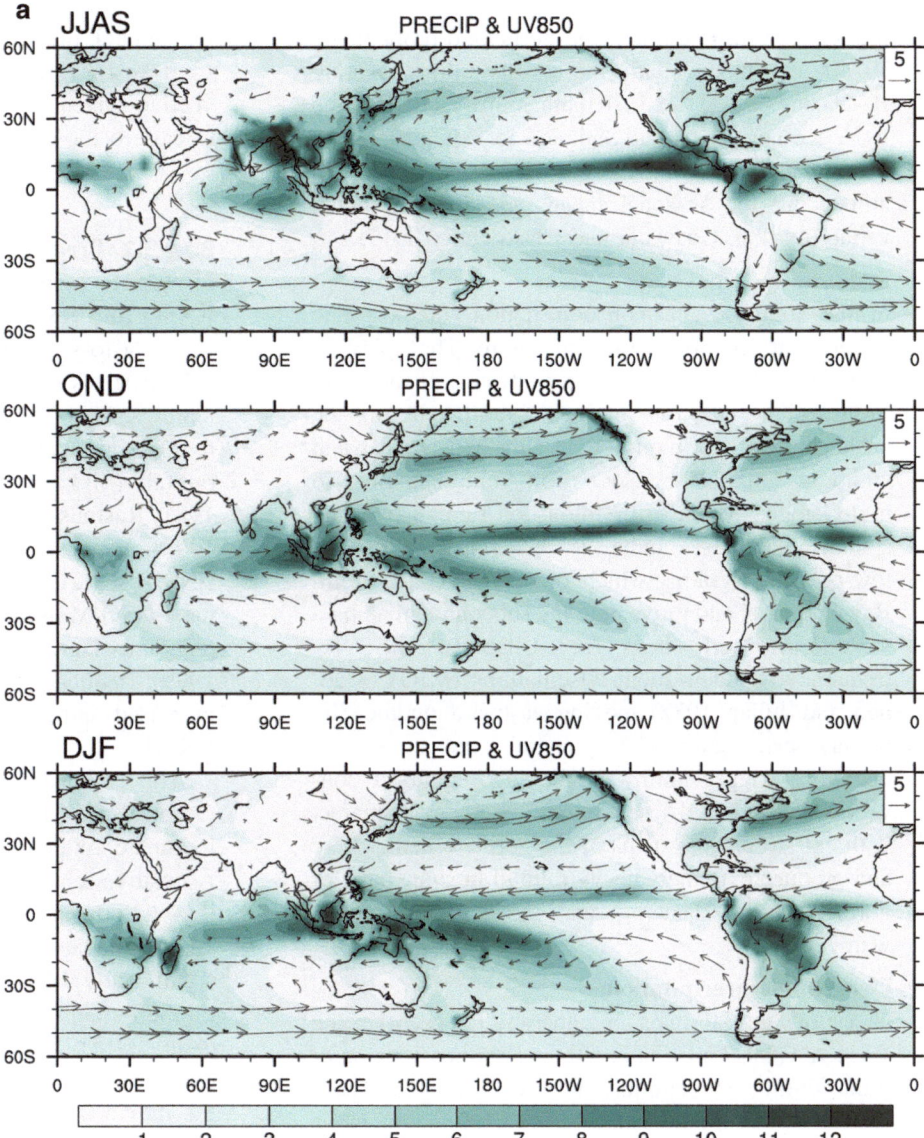

Fig. 3.1 **a** Precipitation (mm/day) and 850 hPa (m/s) wind climatology for southwest monsoon period (JJAS; upper panel), northeast monsoon period (OND; middle panel) and winter monsoon season (DJF; lower panel). GPCP rainfall and ERA-interim winds for the period of 1979–2017 are used. **b** Interannual variability of all-India summer monsoon rainfall (1901–2018). The excess (green), deficit (red) and normal (black) monsoons are also shown. *Data Source*: IMD 0.25° x 0.25° gridded daily rainfall data (Pai et al. 2014)

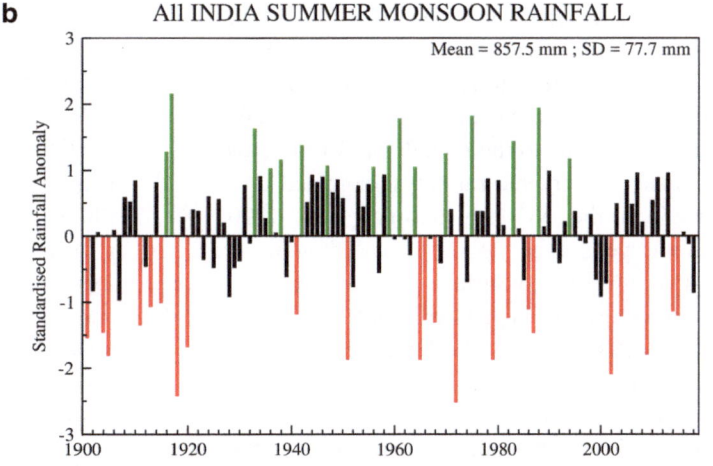

spells of the ISM through the northward-propagating 30–60-day mode and the westward-propagating 10–20-day mode (e.g. Krishnamurti and Bhalme 1976; Keshavamurty and Sankar Rao 1992). Monsoon intra-seasonal oscillation has a seminal role in influencing the seasonal mean, and its interannual variability (e.g. Goswami and Chakravorty 2017).

Synoptic systems

In the south Asian monsoon region, the synoptic systems are mainly of tropical nature. They play an important role for the onset and advance of the monsoon during June, distribution of rainfall during the peak phase of the monsoon in July and August and withdrawal of the monsoon from mid-September to mid-October. Much of the monsoon rainfall over the central plains of India is associated with the low-pressure systems which develop over the north Bay of Bengal and move onto the subcontinent along a west-northwesterly track (see Chap. 7). The low-frequency Madden–Julian Oscillations (MJO) (Madden and Julian 1972) moving eastward on the 30–40-day scale along near-equatorial belts, on several occasions trigger northward-moving organized convective episodes. These systems sometimes interact with extratropical systems of the northern hemisphere and produce extremely heavy rainfall in some parts of northern India (Pisharoty and Desai 1956; Ramaswamy 1962).

Orographic precipitation

Capacious rainfall rates are generally noticed over the Western Ghats (WG) and north and northeast region of India during the summer monsoon season. These regions have unique characteristics of mountainous terrain that acts as a barrier to southwesterly winds (e.g. Patwardhan and Asnani 2000a, b; Tawde and Singh 2015). The windward side of WG receives highest intense rainfall and leeward side of the WG is a strong rain shadow region. Rain shadow areas differ from one region to another along Karnataka, Maharashtra and Kerala due to the complexity of mountain terrains. Intense orographic rainfall is confined up to 800 m height in the WG (e.g. Rahman et al. 1990). High rainfall spells over the west coast of India are associated with warm Sea Surface Temperature (SST), low-level convergence, high CAPE and low convective inhibition (CIN) (e.g. Maheskumar et al. 2014).

The Himalaya mountain range acts as a barrier by blocking the warm moist monsoon air masses primarily on south-facing slopes and preventing their migration on the other side creating a prominent rain shadow contrast (see Chap. 11). Nearly 80% (20%) of the annual rainfall occur in the Himalayas due to southwest (winter) monsoon. The nature of the convective systems varies dramatically from the western to eastern foothills of Himalayas.

Box 3.2 Precipitation teleconnections with modes of climate variability
ENSO and IOD

The year-to-year variability of Indian monsoon rainfall (Fig. 3.1b) is governed by the slowly varying surface features. El Nino conditions in the Pacific play a major role in modulating the interannual variability of ISM rainfall (Sikka 1980, 1977; Pant and Parthasarathy 1981; Rasmusson and Carpenter 1983; Webster et al. 1998). Almost 50% of the droughts are associated with ENSO (see Chap. 6), however in the last few decades the ENSO-Monsoon relationship has been weakened (e.g. Kripalani and Kulkarni 1997; Krishna Kumar et al. 1999), frequency and intensity of droughts have been increased and some of them are not associated with ENSO.

The coupled mode in the Indian Ocean (Indian Ocean Dipole; IOD; Saji et al. 1999) is also known to modulate interannual variability of ISM rainfall. A positive relationship between IOD and ISM rainfall is well known (Ashok et al. 2004; Saji et al. 1999; Saji and Yamagata 2003). The positive (negative) IOD significantly dilutes the influence of El Nino (La Nina) on the Indian monsoon (Ashok et al. 2004; Chowdary et al. 2015). There are more frequent positive IOD events in recent decades due to the rapid warming of the Indian Ocean (e.g. Cai et al. 2018). In addition to IOD and ENSO, there is a strong link between ISM rainfall and the equatorial Indian Ocean oscillation (EQUINOO; Gadgil et al. 2004). In general, positive phase of the EQUINOO is favourable for a good monsoon. Association between EQUINOO and ENSO also determines the variations in ISM rainfall on the interannual time scale.

Eurasian snow cover

Eurasian snow cover also plays a major role in the year-to-year variability of ISM rainfall (Blanford 1884). Generally, positive Eurasian snow cover anomalies during winter and spring tend to be followed by an anomalous deficit rainfall over the Indian subcontinent in the subsequent summer monsoon season, while negative snow cover anomalies tend to be followed by abundant rainfall (Bhanu Kumar 1987; Bamzai and Kinter 1997). It has been observed that all non-ENSO related droughts over India have been associated with excessive snow depth over Eurasia (Kripalani and Kulkarni 1999).

Other than ENSO, the Atlantic, western North Pacific circulation changes (e.g. Chowdary et al. 2019; Srinivas et al. 2018) also play a role in monsoon interannual and decadal variability (Sankar et al. 2016; Yadav 2017). The variabilities in the rainfall, as well as the teleconnections of monsoon, could be natural, but there is an intriguing possibility of global warming to modulate these variations. It is suggested that the weakening linkage between ENSO and ISM, despite the increase in ENSO activity, could be due to global warming (Krishna Kumar et al. 1999). Also the anomalous warming over the Eurasian land mass and enhanced moisture conditions over the Indian region in a global warming scenario could have contributed to the weakening of the influence of warm ENSO events on ISM rainfall (e.g. Ashrit et al. 2001). Moreover, the warming of the Indian Ocean at a faster rate than the global oceans (Roxy et al. 2014) has implications on the variability of rainfall over India, by playing a major role in the declining trend of ISM rainfall (Preethi et al. 2017).

Decadal Variations

Variations in ISM rainfall are characterized by distinct epochs typically of about three decades, of above and below normal monsoon activity (e.g. Parthasarathy et al. 1991a, b; Kripalani and Kulkarni 1997). The observational, paleo-climatic and simulated datasets show increased (decreased) ISM rainfall during the positive (negative) phase of the Atlantic Multidecadal Oscillation (AMO) (e.g. Goswami et al. 2006; Joshi and Rai 2015; Krishnamurthy and Krishnamurthy 2015). The leading mode of SSTs in the North Pacific Ocean, Pacific Decadal Oscillation (PDO) with periodicities of 15–25 years and
50–70 years (e.g. Mantua and Hare 2002), could negatively impact ISM rainfall (e.g. Krishnan and sugi 2003; Krishnamurthy and Krishnamurthy 2013). The high correlation between the inter-decadal component of variability of ISM with that of Nino-3 SST highlights the importance of El Nino-Monsoon relationship (Parthasarathy et al. 1994; Kripalani et al. 1997; Kripalani and Kulkarni 1997; Mehta and Lau 1997; Krishnamurthy and Goswami 2000). This indicates that low-frequency modulation of summer monsoon could largely influence rainfall over the Indian subcontinent. Along with this, a strong multi-decadal variability with alternate wet (above normal) and dry (below normal) epochs of monsoon rainfall has been observed in the instrumental records extending back to 150 years (Kripalani and Kulkarni 2001; Joseph et al. 2016; Preethi et al. 2017).

Northeast and Winter Monsoon precipitation

While most parts of India receive the major share of the annual rainfall during southwest/summer monsoon season (from June to September), southeast peninsular India falls under the rain shadow region during this season. During the northeast monsoon season from October to December (OND), the zone of maximum rainfall migrates to southern India and the prevailing winds become northeasterly (Fig. 3.1a; middle panel) (e.g. Ramaswamy 1972; Dhar and Rakhecha 1983; Singh and Sontakke 1999; Balachandran et al. 2006; Rajeevan et al. 2012). The northeast monsoon rainfall shows strong interannual variability (28%), which is more than twice the variability of southwest monsoon rainfall (11%) (e.g. Nageswara Rao 1999; Sreekala et al. 2011). The normal date of the northeast monsoon onset is 20th October with a standard deviation of 7–8 days (Raj 1992).

The relationship between ENSO and northeast monsoon has been strengthened during 1979–2005 (Kumar et al. 2007) while it is weakened in the decade of 2001–2010 (Rajeevan et al. 2012). Local air-sea interaction within the Indian Ocean also modulates the northeast monsoon rainfall (Yadav 2013). IOD-related circulation is found to be an important local forcing mechanism for northeast monsoon (Kripalani and Kumar 2004).

During winter (December to February), cold air masses originating from the Siberian High move southward (Fig. 3.1; bottom panel), and lead to interaction between high northern latitudes and the tropics (Wang et al. 2003). Western disturbances from the Mediterranean to Central Asia transport moisture to the Indian winter monsoon contributing significantly to annual precipitation in the Himalaya region (Dimri 2013; Dimri et al. 2015).

3.2 Observed Changes in Mean Precipitation and Circulation

3.2.1 Precipitation Records in Paleo Time Scale—Inferences from Proxies

The paleoclimate proxy data of monsoonal record of the past 640,000 years suggest that the millennial-scale variability arose by the solar insolation changes which are caused by precession and obliquity (Cheng et al. 2016). In the last 11,000 years (The Holocene age), summer monsoon is declining with variability at the multi-decadal scale to centennial scales (Chao and Chen 2001). The decline of the summer monsoon is linked to, among other factors, the southward migration of the ITCZ as a result of the decrease in solar insolation (Fleitmann et al. 2007). Additionally, prolonged wet/drought periods of multi-decadal and century-scales have occurred during the last 4000 years (Sinha et al. 2011; Prasad et al. 2014) with notable century scale long declining trends 1550–2200 years BP (Roman Warm Period; RWP) and 100–550 years BP (Little Ice Age; LIA) and an increasing trend during 650–1050 year BP (Medieval warm period; MWP) in summer monsoon (Trends marked by arrows; Fig. 3.2). Abrupt changes in monsoon around 2800 years BP and 2350 years BP have been attributed to solar variability (Sinha et al. 2018). Tree ring-based studies from the Himalayan region reveal a declining trend in the summer monsoon over the last 200 years with the possible linkages to large-scale greenhouse warming, and anthropogenic aerosol emissions (Xu et al. 2013; Shi et al. 2017). Speleothem (cave deposits)-based 4000-year-long monsoon reconstructions from central, peninsular and northeast India

indicate that the monsoon has undergone multi-decadal changes of larger magnitude in the Holocene than in the last 200 years alone (Fig. 3.2). Therefore, as per the available paleo records, changes in the monsoon due to the regional forcing such as anthropogenic aerosol emissions are difficult to detect against the large multi-decadal natural variability.

3.2.2 Recent Changes in Precipitation

The mean summer monsoon rainfall (JJAS) over India from 1979 to 2005 from multiple observational datasets is shown in Fig. 3.3. It is noted that, in general, all datasets show a high rainfall zone over east-central India and low rainfall zones over northwest India, northern parts of Kashmir, and the rain shadow area of southeast India. The discrepancy in observations occurs mainly over northern parts of India, Himalayan region (e.g., Prakash et al. 2015) and northeast India (Bidyabati et al. 2017).

The annual rainfall averaged over Indian landmass does not show any trend over the period 1901–2015. However, in the recent period 1951–2015 as well as 1986–2015 the annual rainfall series shows decreasing trend (though not statistically significant, or evident in all the datasets). The summer monsoon rainfall series averaged over India landmass does not show any long-term trend on a century-scale where it has been found that the contribution from increasing heavy rain events has been offset by decreasing moderate rain events (Goswami et al. 2006). However, a downward trend of rainfall over the Indian subcontinent has been observed in the period 1951–2004 (Kulkarni 2012). The decreasing tendency of summer precipitation is found

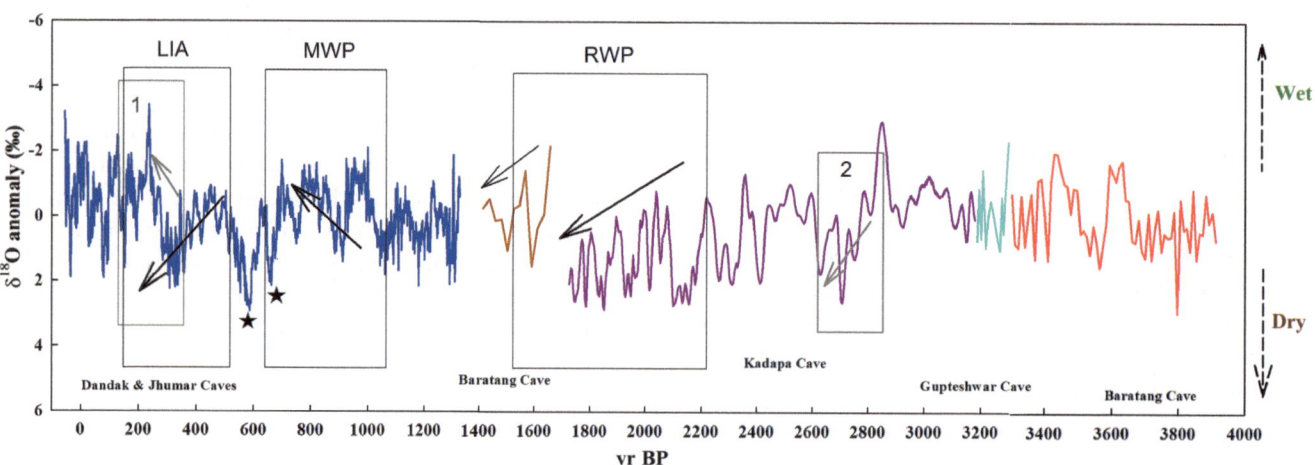

Fig. 3.2 Reconstruction of the ISM of the past 4000 years from a synthesis of records of δ¹⁸O of speleothems from Kadapa (Andhra Pradesh), Baratang (Andaman) and Gupteshwar (Orissa) (right y-axis) and Dandak-Jhumar (Chhattisgarh -Meghalaya) composite (left y-axis). Variations/trends in summer monsoon associated with major climatic events of the past such as LIA, MWP and RWP are highlighted. Abrupt changes in monsoonal strength (box 1 and 2), as well as major drought periods, are also shown (Sinha 2018; marked with stars). Some gaps in the time series are due to the poor resolution of available records for those particular periods, thus, unaccounted

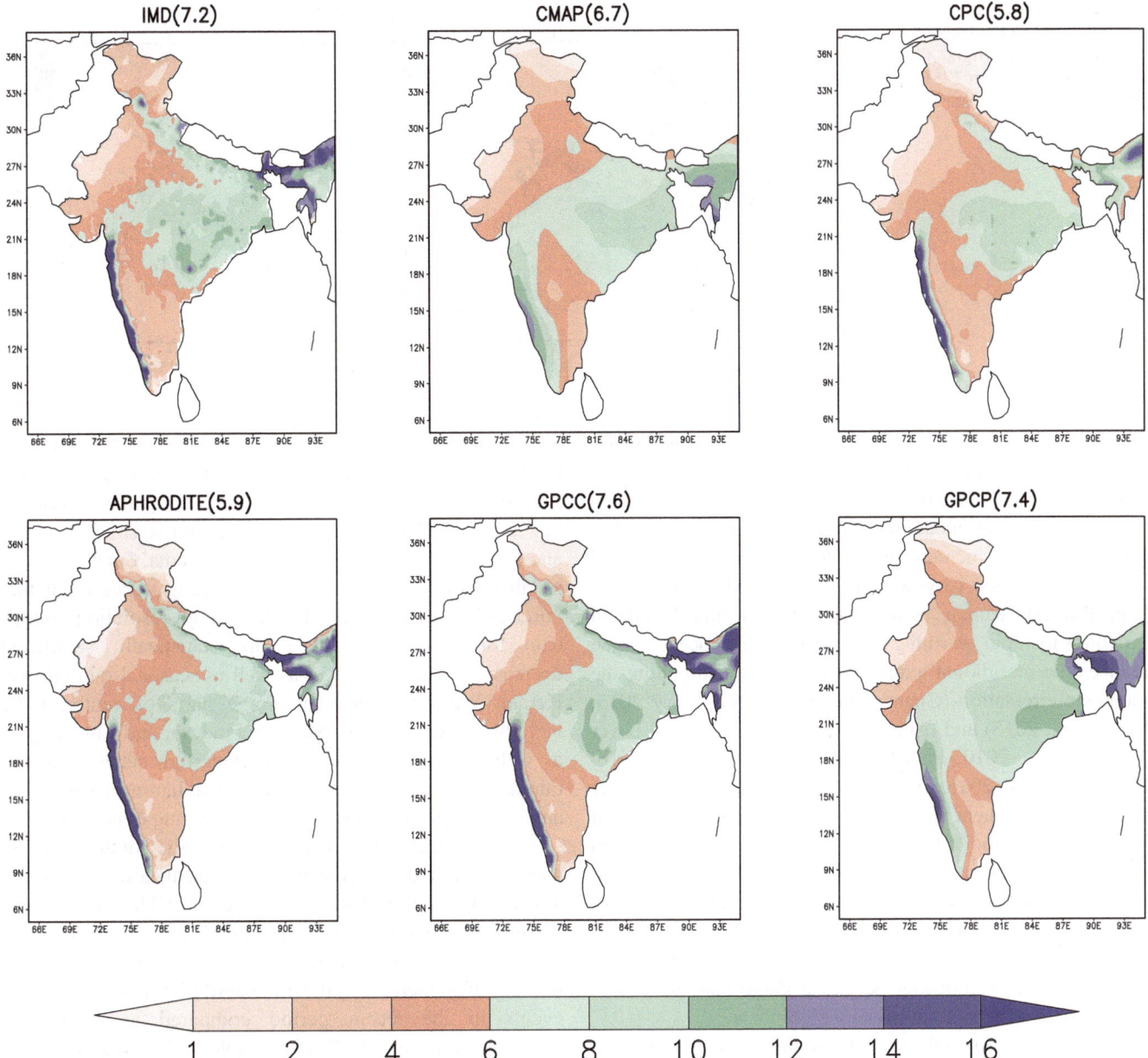

Fig. 3.3 Mean summer monsoon (June through September) rainfall over India from 1979 to 2005 in multiple observed datasets. The dataset and all-India mean rainfall (mm/day) are given in brackets

to have accelerated during 1971–2002 (Kothawale et al. 2008) and is evident in multiple datasets (Annamalai et al. 2013). This secular decline in mean rainfall is attributed to wakening monsoon Hadley circulation. A decreasing trend in rainfall may also be a result of multi-decadal epochal variability associated with an east-west shift in monsoon rainfall due to anomalous warming of the Indo-Pacific warm pool (Annamalai et al. 2013; Guhathakurta et al. 2015) or weakening of the land-ocean temperature gradient (Kulkarni 2012). The drying trend may be attributed to a number of factors such as: (a) increasing anthropogenic

aerosol concentration over northern hemisphere which may cool northern hemisphere and slowing of the tropical meridional overturning circulation (e.g. Ramanathan et al. 2005; Chung and Ramanathan 2006; Bollasina et al. 2011, 2014); (b) to increasing trend of the Pacific Decadal Oscillation (Salzmann and Cherian 2015); (c) to significant weakening of monsoon low-level southwesterly winds, the upper-tropospheric tropical easterlies from the out-flow aloft, the large-scale monsoon meridional overturning circulations (Rao et al. 2010; Joseph and Simon 2005; Sathiyamoorthy 2005; Fan et al. 2010; Krishnan et al.

Fig. 3.4 Linear trends (mm/day over 64 years) in the southwest (left) and northeast (right) monsoon rainfall from 1951 to 2015 based on IMD data

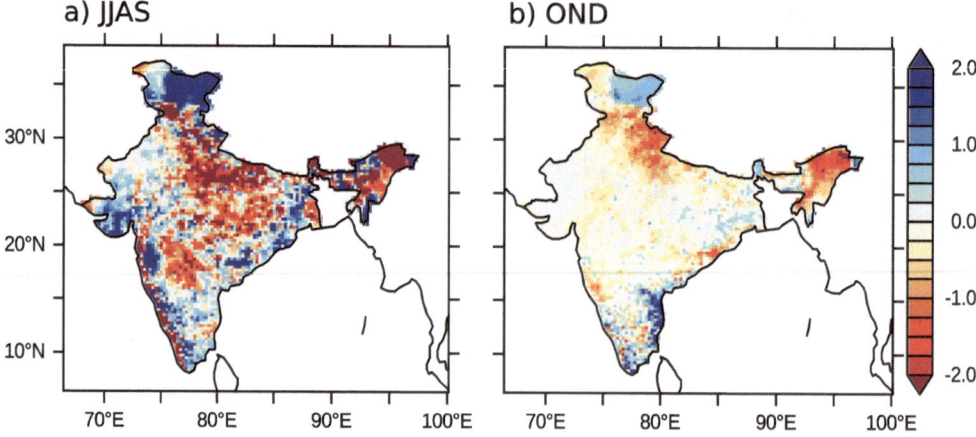

2013); (d) to strong Atlantic Multidecadal Oscillation (AMO) which weakens the meridional temperature gradient resulting in early withdrawal of monsoon over India and thus reducing the mean rainfall (Goswami et al. 2006); (e) to significant increase in the duration and frequency of 'monsoon- breaks' (dry spells) over India since the 1970s (e.g. Ramesh Kumar et al. 2009; Turner and Hannachi 2010);(f) to rapid warming of western Indian ocean which reduces the meridional temperature gradient dampening the monsoon circulation (Gnanaseelan et al. 2017; Roxy 2015; Roxy et al. 2015) and (g) to changes in land use/land cover (Niyogi et al. 2010; Pathak et al. 2014; Paul et al. 2016; Krishnan et al. 2016). However, enhanced warming over Indian subcontinent and comparatively slower rate of warming over India Ocean has favoured the land-ocean temperature gradient in the recent decade (2002–2014) and helped a possible short-term revival of the monsoon over India at the rate of 1.34 mm/day/decade (Jin and Wang 2017). Thus, over the recent three decades (1986–2015) all-India summer monsoon shows a decreasing tendency, but the decline is not statistically significant.

There is considerable spatial variability in precipitation changes. As compared to the period 1901–1975, rainfall has reduced by 1–5 mm/day during 1976–2015 over central parts of India (the core monsoon zone), Kerala and extreme northeastern parts and increased over the Jammu and Kashmir region as well as in parts of western India (Kulkarni et al. 2017). Regional anthropogenic forcings such as from aerosols and land-use change from urbanization and agricultural intensification could be dominant contributors to this recent spatial variability (Paul et al. 2018).

Trends in Indian rainfall records have been extensively studied, but the subject remains complicated by the high spatiotemporal variability of rainfall arising from complex atmospheric dynamics and, to some extent, differences that emerge from the methods used in creating the datasets. Monsoon rainfall has shown moderate increasing trends in 27 (out of 36) subdivisions across India (Guhathakurta and

Rajeevan 2008). The linear trend in annual as well as seasonal rainfall shows a statistically significant decreasing trend over Jharkhand, Chhattisgarh, and Kerala, and eight subdivisions, viz. Gangetic WB, West UP, Jammu and Kashmir, Konkan and Goa, Madhya Maharashtra, Rayalaseema, Coastal AP and North Interior Karnataka show increasing trends (Guhathakurta and Rajeevan 2008). Based on high-resolution gridded data for 1901–2015, there are statistically significant decreasing trends in annual as well as seasonal rainfall over Kerala, Western Ghats and some parts of central India including Uttar Pradesh, Madhya Pradesh, and Chhattisgarh as well as some parts of the northeastern states. Whereas rainfall over Gujarat, Konkan coast, Goa, Jammu and Kashmir and east coast shows a significant increasing trend (Fig. 3.4).

Climate change is not just affecting the southwest monsoon but is also driving changes in the northeastern monsoon. The variability of northeast monsoon rainfall has increased in the period 1959–2016. Seasonal rainfall has increased over Tamil Nadu, Rayalaseema, as well as south peninsular India because of an increase in the number of high-intensity rainfall events in the recent period compared to 1901–1958 (Nageswararao et al. 2019). Table 3.1 gives summary statistics for rainfall over India based on 1951-2015.

3.2.3 Understanding the Observed Changes in the Summer Monsoon Precipitation

3.2.3.1 Anthropogenic Causes of Observed Precipitation Changes

In general, we can summarize that the unprecedented increase in atmospheric greenhouse gases (GHGs) is responsible for the global rise in temperature, which as feedback to atmospheric dynamics and convection has also led to changes in rainfall characteristics globally (Alexander 2016). Some regional forcings, such as aerosols and land-cover changes, have additionally detectable and notable impact on the monsoon rainfall changes.

Anthropogenic aerosols modulate regional precipitation patterns, over monsoon regions (Bollasina et al. 2011; Krishnan et al. 2016; Undorf et al. 2018). Aerosols play an important role in the earth-atmosphere system through their interactions with solar radiation, clouds and the cryosphere aerosol solar absorption over the Indian monsoon region has a potential role in influencing the monsoon circulation and rainfall distribution (Chap. 5 *provides a summary*). Observed patterns of regional changes in precipitation are missing from the CMIP5 (Coupled Model Intercomparison Project 5) assessments–primarily due to the coarse resolution of models and also due to missing local features that can be important for such regional variability. The current generation of coupled models shows very substantial dry bias in simulating Indian monsoon precipitation over the core monsoon zones of central India, and the Western Ghats. Sabin et al. (2013) used a variable resolution global atmospheric model with telescopic zooming over south Asia (~35 km) and demonstrated that the high resolution provides particular value addition in simulating better monsoon rainfall over the Indian region. Using the same set of model analysis, recent changes in observed mean monsoon over India (1951–2005) have been attributed to a combined effect of anthropogenic aerosol, equatorial Indian Ocean warming and land-use/land-cover change (Krishnan et al. 2016, Fig. 3.5).

3.2.3.2 Changes in Circulation Features

The core of the Tropical Easterly Jet (TEJ) has been shrinking over the South Asian region (Pattanaik and Satyan 2000). The strength of TEJ has been found to have been decreasing before 2000 (Sathiyamoorthy 2005), but since 2000 has increased at the rate of 1 m/s per year (Roja Raman et al. 2009; Venkat Ratnam et al. 2013). The weakening of the TEJ may be attributed to increase in convection due to the excessive warming of Indian Ocean SST (Joseph and Sabin 2008), cooling of upper-tropospheric temperature over the Tibetan anticyclone region, and a significant warming over the equatorial Indian Ocean which might have resulted in decreasing trend of the upper-tropospheric meridional temperature gradient. These changes have caused a reduction in the strength of the easterly thermal wind at the core region of the TEJ, after the weakening of the TEJ. Further, the weakening of TEJ and associated decrease of easterly shear is attributed to the reduced north-south temperature gradient between the equator and $20°N$ for the longitude belt of $40°E–100°E$, that is, the air temperature on the equator side is increasing compared to the north. These variations are particularly high above 500 hPa (Rai and Dimri 2017).

In response to the global warming, the intensity of the summer monsoon overturning circulation (monsoon Hadley cell) and the associated southwesterly monsoon flow (LLJ) have significantly weakened from the 1950s (Joseph and Simon 2005; Krishnan et al. 2013). An ultra-high-resolution global general circulation model (about 20 km resolution) also shows a stabilization (weakening) of the summer monsoon Hadley-type circulation in response to global warming which has resulted in a weakened large-scale monsoon flow (Rajendran et al. 2012; Krishnan et al. 2013). The weakening of Asian monsoon circulation (Fig. 3.5) may be due to relatively smaller warming in Asia compared to the surrounding regions which make the landmass a 'heat sink' (Zuo et al. 2012). Indeed, the tropospheric temperature over Asia has lowered in recent decades. As a consequence, the meridional and zonal land-sea thermal contrasts are reduced, and the Asian summer monsoon becomes weaker.

3.2.3.3 Observed Changes in Active/Break Spells

The seasonal monsoon strength is mainly modulated by the intra-seasonal variability of the summer monsoon rainfall

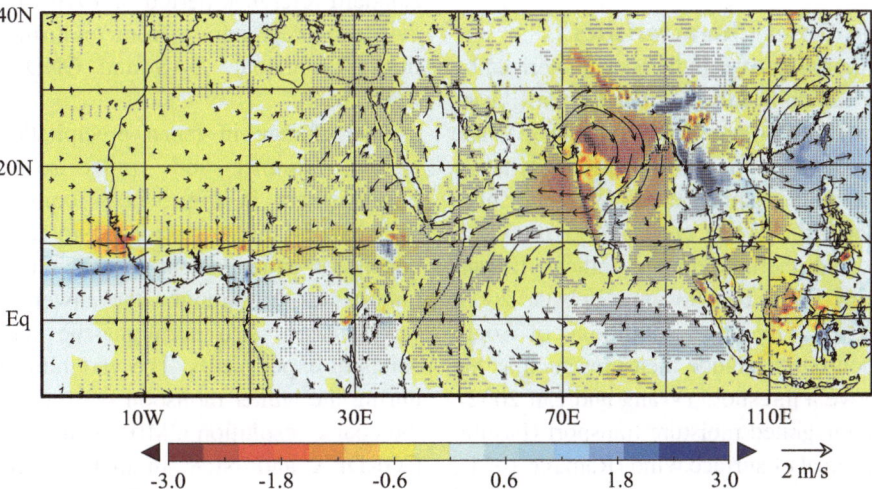

Fig. 3.5 Attribution of the decline in monsoonal rainfall: The difference in mean precipitation (JJAS; mm/day) and low-level circulation at 850 hPa (m/s) between the Historical and Historical natural simulations for the period (1951–2005) from a high-resolution simulation. Regions with a significance level above 95% level are shown with grey dots. Reprinted with permission from Krishnan et al. 2016

which is characterized by the active/break spells of enhanced and decreased precipitation over India (e.g. Ramamurthy 1969; Sikka and Gadgil 1980; Rodwell 1997; Webster et al. 1998; Krishnan et al. 2000; Krishnamurthy and Shukla 2000, 2007, 2008; Annamalai and Slingo 2001; Goswami and Ajayamohan 2001; Lawrence and Webster 2001; De and Mukhopadhyay 2002; Goswami et al. 2003; Waliser et al. 2003; Kripalani et al. 2004; Wang et al. 2005; Mandke et al. 2007; Goswami 2005 and Waliser 2006). The relative strength of the northward-propagating low-frequency (20–60 days) modes has a significant decreasing trend during 1951–2010, possibly due to the weakening of large-scale circulation in the region during the monsoon season. This reduction is compensated by a gain in synoptic-scale (3–9 days) variability. The decrease in low-frequency ISO variability is associated with a significant decreasing trend in the percentage of extreme events during the active phase of the monsoon. However, this decrease is likely balanced by significant increasing trends in the percentage of extreme events in the break and transition phases. These changes are accompanied by a weakening of low-frequency variability (Karmakar et al. 2015, 2017). Also, while there is no change in the distribution of the break events, the number of active spells shows an increase of about 12% in the period 1951–2010; the increase was mainly in the short duration (3–6 days) spells (Pai et al. 2016). A statistically significant increase in the frequency of dry spells (27% higher during 1981–2011 relative to 1951–1980) and intensity of wet spells and statistically significant decreases in the intensity of dry spells have been observed in recent six decades (Singh et al. 2014). The changes in frequency, intensity and speed of intra-seasonal oscillations have been attributed to Indian ocean warming (Sabeerali et al. 2015); developing and decaying phase of ENSO (Pillai and Chowdary 2016); increase in convective available potential energy, low-level moisture convergence and changes in large-scale circulation in upper atmosphere (Singh et al. 2014).

3.2.3.4 Changes in Onset Characteristics

The onset of summer monsoon over India is characterized by the dramatic rise in mean daily rainfall over Kerala (Ananthkrishnan and Soman 1988; Soman and Kumar 1993). The onset of the ISM has been defined with various dynamic (Koteswaram 1958; Ananthakrishnan et al. 1968; Krishnamurti and Ramanathan 1982; Wang et al. 2001; 2009; Pai and Rajeevan 2009) and thermodynamic indices (Ananthakrishnan and Soman 1988; Fasullo and Webster 2003; Janowiak and Xie 2003). Objective definitions of South Asian summer monsoon onset include measures such as the increase of rainfall above a threshold (Wang and Lin 2002), transition in vertically integrated moisture transport (Fasullo and Webster 2003), reversal of surface wind (Ramage 1971), and intensification of the lower level Somali jet (Taniguchi

and Koike 2006; Wang et al. 2009). As per these different definitions, the mean onset date of summer monsoon rainfall over India has been stable around 1 June. In recent decades, the monsoon onset over India is seen to be delayed to 5th June since 1976 (Sahana et al. 2015), which can be attributed to the net decrease in moisture supply from the Arabian Sea in the post-1976 period. The interannual variability of the onset date is associated with ENSO with early onsets preceded by La Nina, and late onsets preceded by El Nino (e.g. Noska and Mishra 2016).

3.3 Projected Changes in Precipitation Over India

Understanding the projected future changes in precipitation has a profound importance for policy. In this report, the assessment of rainfall changes over India is carried out based on the multiple ensemble member simulations from CMIP5, CORDEX-SA (COordinated Regional Downscaling EXperiment-South Asia) and NEX-GDDP (Nasa earth Exchange-Global Daily Downscaled Products) in which CMIP5 is the parent GCM, CORDEX is dynamically downscaled to 50 km × 50 km grid resolution, and NEX-GDDP is statistically downscaled to 25 km × 25 km grid resolution.

Historical and projected changes until the end of the twenty-first century based on various simulations (28 from CMIP5, 16 from CORDEX and 19 from NEX; see the list of models in Tables 3.2 and 3.3) are provided in this section. The future changes are mostly quantified as percentage changes in the near future (2040–2069) and far future (2070–2099) epochs. We provide our analysis for annual, summer (JJAS) and winter (OND) seasons in all cases. Mostly the analysis is restricted to the Indian landmass, by masking out the seas, and regions outside the geographical area of India. Projections are stated with respect to the standard reference period of 1976–2005.

Mean precipitation from multi-model ensemble simulations for annual, JJAS and OND seasons is shown in Fig. 3.6.

The change in mean precipitation over India for the annual, summer and winter seasons is presented as box-whiskers in Fig. 3.7. A comparison of the various sources of climate data used in this assessment shows a consistent enhancement in precipitation across the Indian landmass throughout the twenty-first century. The box-whiskers also highlight the spread among the three suites of experiments. The variability is comparatively high during the winter monsoon season (OND). Comparing with the coarse resolution CMIP5 simulation, the high-resolution CORDEX and NEX simulations show higher variability irrespective of seasons. This increased variability in the

Table 3.1 Statistics of rainfall over the Indian landmass

	Monsoon	Post-monsoon	Annual
Seasons	JJAS	OND	
Mean (mm)	858.0	119.1	1142.1
Std Dev	80.2	29.5	98.8
% Contribution to annual average	75.1	10.4	
Max RF (Year)	1011.7 (1988)	205.1 (1956)	1359.6 (1990)
Min RF (Year)	665.2 (1972)	63.5 (2011)	922.4 (1972)

Table 3.2 List of CORDEX South Asia simulations used

CORDEX -SA	RCM used	Contributing RCM modelling centre	Driving CMIP5 model	CMIP5 modelling centre
IITM-RegCM4	ICTP, regional climate model version 4 (Giorgie et al. 2012)	CCCR, IITM, India	CCCma-CanESM2	*Canadian Centre for Climate Modelling and Analysis (CCCma), Canada*
			NOAA-GFDL-ESM2M	National Oceanic and Atmospheric Administration-Geophysical Fluid Dynamics Laboratory, USA
			CNRM-CM5	Centre National de Recherches Météorologiques, France
			MPI-ESM-MR	Max Planck Institute for Meteorology, Germany
			IPSL-CM5A-LR	The Institute Pierre Simon Laplace, France
			CSIRO-Mk.6	Commonwealth Scientific and Industrial Research Organisation, Australia
SMHI-RCA4	Rossby Centre regional atmospheric model version 4 (Samuelsson et al. 2011)	Rossby Centre, Swedish Meteorological and Hydrological Institute, Sweden	ICHEC-EC-EARTH	Irish Centre for High End Computing, European Consortium
			MIROC-MIROC5	Model for Interdisciplinary Research On Climate, Japan Agency for Marine-Earth Science and Technology, Japan
			NCC NorESM1	Norwegian Climate Centre, Norway
			MOHC-HadGEMEM2-ES	Met Office Hadley Centre for Climate Change, UK
			CCCma-CanESM2	*Canadian Centre for Climate Modelling and Analysis (CCCma), Canada*
			NOAA-GFDL-ESM2m	National Oceanic and Atmospheric Administration-Geophysical Fluid Dynamics Laboratory, USA
			CNRM-CM5	Centre National de Recherches Météorologiques, France
			MPI-CM5A-MR	Max Planck Institute for Meteorology, Germany
			IPSL-CM5A-MR	The Institute Pierre Simon Laplace, France
			CSIRO-Mk.6	Commonwealth Scientific and Industrial Research Organisation, Australia

Table 3.3 List of CMIP5 and NEX Models used in this study. The availability of NEX-GDDP statistically downscaled product for the respective model is depicted with "†"

CMIP5	NEX	CMIP5 modelling centre
ACCESS1-0	†	Australian Community Climate and Earth System Simulator, Australia
BNU-ESM	†	Beijing Normal University Earth System Model, China
BCC-CSM1-1	†	Beijing Climate Centre Climate System Model, China
CCSM4 CESM1-BGC	† †	Community Climate System Model, NCAR, USA
CNRM-CM5	†	Meteo-France/Centre National de Recherches Meteorologiques, France
CanESM2	†	*Canadian Centre for Climate Modelling and Analysis (CCCma), Canada*
CMCC-CM		Centro Euro-Mediterraneo sui Cambiamenti Climatici, Italy
CSIRO-Mk3-6-0	†	Commonwealth Scientific and Industrial Research Organisation (CSIRO), Australia
GFDL-ESM2M GFDL-ESM2G GFDL-CM3	† † †	Geophysical Fluid Dynamics Laboratory, National Oceanic and Atmospheric Administration (NOAA), USA
GISS-E2H GISS-E2-R		National Aeronautics and Space Administration (NASA)/Goddard Institute for Space Studies (GISS), USA
HadCM3 HadGEM2-ES HadGEM2 CC		Met Office Hadley Centre, UK
INMCM4	†	Institute for Numerical Mathematics, Russia
IPSL-CM5A-MR IPSL-CM5A-LR	† †	Institute Pierre Simon Laplace, France
MIROC5 MIROC-ESM-CHEM MIROC-ESM	†	Centre for Climate System Research (University of Tokyo), National Institute for Environmental Studies and Frontier Research Center for Global Change (JAMSTEC), Japan
MPI-ESM-LR MPI-ESM-MR MRI-CGCM3	† † †	Max Planck Institute for Meteorology, Germany
MRI-CGCM3		Meteorological Research Institute, Japan
NorESM1-M	†	Norwegian Climate Centre, Norway

future climate is noted in all three experiments, especially for the RCP8.5 scenario as noted by Singh and Achutarao 2018. The quantitative estimate of future changes in annual mean precipitation from different Reliability Ensemble Average (REA) for projected change in precipitation (mm/day) along with uncertainty range is summarized in Table 3.4.

The percentage change in projected mean precipitation pattern for near and far future from RCP4.5 and RCP 8.5 is shown in Fig. 3.8. Multi-model mean change suggests wetter condition over India in near and far future on average. Slightly different scenario is projected in the CORDEX simulations over the northwest Indian region with a 10% drier condition than its present-day mean for near future in RCP4.5 simulations. During the winter months, northeast India is projected to witness a moderate deficit condition in the near future (both in CMIP5 and NEX); in addition to this, CORDEX models suggest a potential reduction over the Himachal and Jammu belt. From the extreme scenario (RCP8.5, both near and far future) and far future in RCP4.5, consistent pattern emerges among the three sources, irrespective of variations in their spatial resolution and methodologies followed. The changes in annual mean precipitation surpass 10% above baseline over the west coast and southern locale of the Indian landmass in the RCP4.5 scenario in the near future, and exceed 20% in far future (Figs. 3.8 and 3.9). Over the rest of India, the precipitation changes are not significant for the near future up to the mid-twenty-first century, yet in the long-term; increment surpasses 10% over northwest and the adjoining territory of the nation. The long-term projected annual precipitation increment surpasses 10% over most parts the Indian landmass.

3.3.1 Future Changes in the Summer Monsoon

ENSO and IOD typically exert an offsetting impact on Indian summer monsoon rainfall (ISMR), with an El Niño event tending to lower, whereas a positive IOD tending to increase ISM rainfall (Ashok and Saji 2007). In a recent study, Li et al. (2017) showed that CMIP5 models simulate an unrealistic present-day IOD-ISMR correlation due to an

Fig. 3.6 Mean precipitation (mm/day, 1976–2005) from multi-model ensemble simulations for annual, JJAS, and OND seasons from CMIP5, CORDEX-SA, and NEX-GDDP experiments

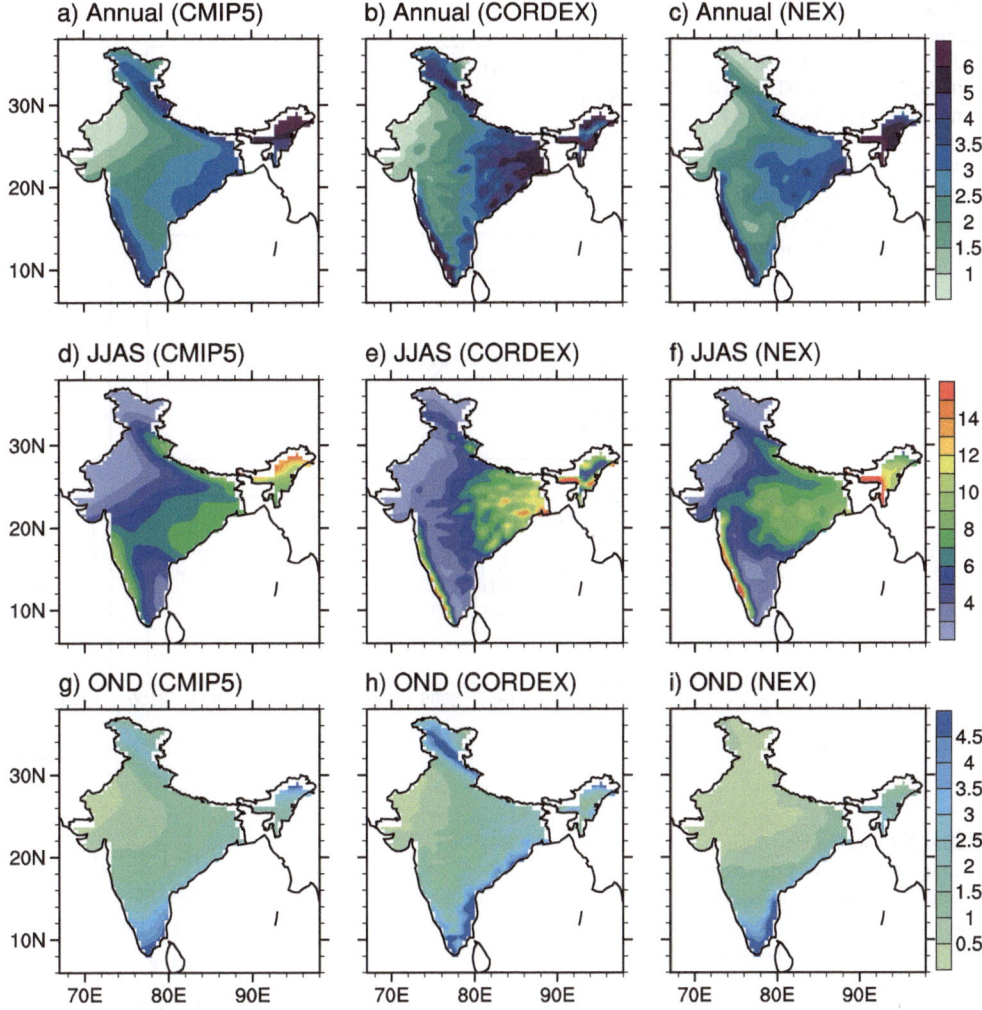

overly strong control by ENSO. As such, a positive IOD is associated with an ISM rainfall reduction in the simulated present-day climate. They further highlight that the uncertainties in ISM rainfall projection can be in part due to the present-day simulation of ENSO, the IOD, their relationship, and their rainfall correlations. Thus, the natural variability also plays a dominant role in the diverse ENSO-monsoon relationship during the twentieth century (Li and Ting 2015). From CMIP5 models, the analysis by Li et al. (2017) further showed that in future, the enhance SST warming could likely lead to a weak ENSO-monsoon relation as well.

Multi-model average changes considered in the different model sources, in general, suggest wetter future conditions. CMIP5 models project an increase of 6% (RCP4.5) and 8% (RCP8.5) in the near future over the central Indian region (core monsoon zone defined by Rajeevan et al. 2008). Projected changes in rainfall by the end of the twenty-first century are 10% (RCP4.5) and 14% (RCP8.5), respectively. The vast majority of the CMIP models shows enhanced monsoon precipitation due to global warming (e.g. Kitoh et al. 1997; Douville et al. 2000; Ueda et al. 2006; Cherchi

et al. 2011; Rajendran et al. 2012; Krishnan et al. 2013); however, they indicate a likely weakening of large-scale monsoonal circulation (Krishnan et al. 2016). Many studies noted that the poor skill in simulating monsoon amplifies the ambiguities in understanding the future changes in projected monsoon rainfall (e.g. Chaturvedi et al. 2012; Saha et al. 2014; Sharmila et al. 2015; Krishnan et al. 2016). Studies have highlighted the wide inter-model spread in the simulated precipitation changes over South Asia, which therefore makes the assessment of regional hydroclimatic response a bit ambiguous in reality (e.g. Kripalani et al. 2007; Annamalai et al. 2007; Turner and Slingo 2009; Sabade et al. 2011; Fan et al. 2010; Hasson et al. 2013; Saha et al. 2014).

Applying the Clausius–Clapeyron equation, the water vapour holding capacity of the atmosphere is expected to increase by about 7% per degree of warming. The enhanced availability of moisture can naturally lead to more precipitation over different parts of the globe (Trenberth 1998; Meehl et al. 2005). Indeed, there is a considerable multiscale feedback through large-scale circulation and various

Fig. 3.7 Percentage changes in mean precipitation over the Indian land across CMIP5, CORDEX-SA, and NEX-GDDP simulations from the **a)** RCP4.5 **b)** RCP8.5 scenario for annual, JJAS and OND seasons. Changes in 10-year means, with respect to the reference period (1976–2005), are shown as box-whiskers

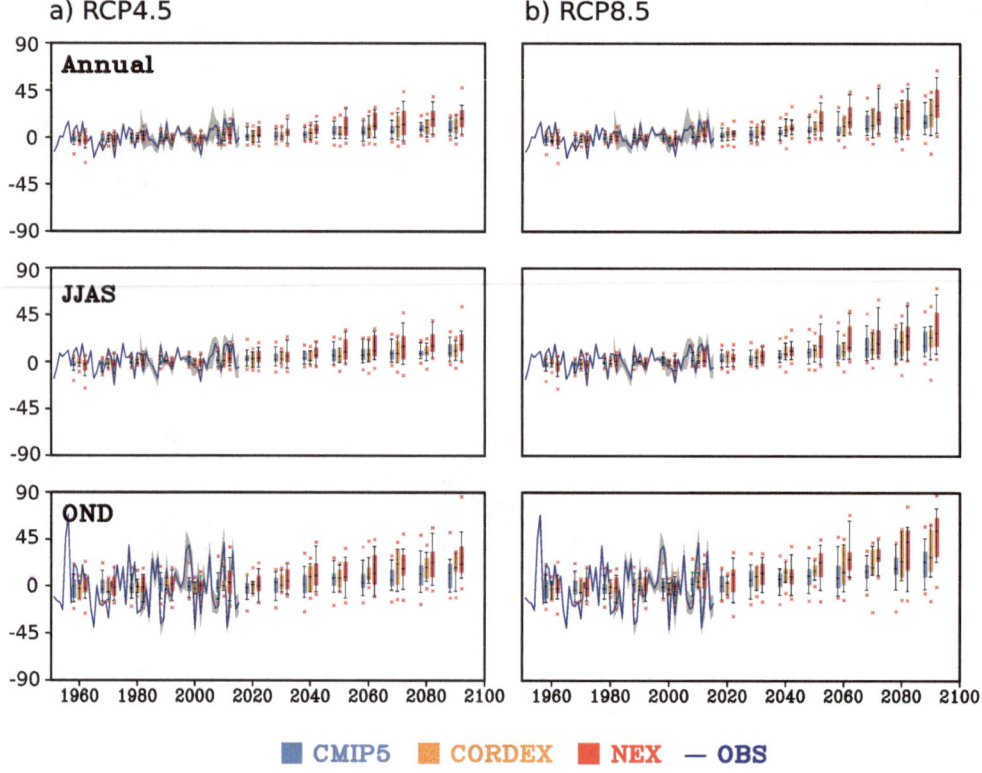

Table 3.4 Reliability ensemble average estimate of projected change in mean precipitation (mm/day) over Indian land associated with uncertainty range

	Scenario	Experiments	Annual
Near future	RCP 4.5	CMIP5	0.12 ± 0.11 *(91%)*
		CORDEX	0.16 ± 0.19 *(118%)*
		NEX	0.20 ± 0.21 *(105%)*
	RCP 8.5	CMIP5	0.20 ± 0.17 *(85%)*
		CORDEX	0.28 ± 0.18 *(64%)*
		NEX	0.38 ± 0.20 *(52%)*
Far future	RCP 4.5	CMIP5	0.25 ± 0.18 *(72%)*
		CORDEX	0.32 ± 0.22 *(68%)*
		NEX	0.40 ± 0.23 *(57%)*
	RCP 8.5	CMIP5	0.45 ± 0.21 *(46%)*
		CORDEX	0.58 ± 0.32 *(55%)*
		NEX	0.63 ± 0.31 *(49%)*

thermodynamic and dynamic conditions locally that also contribute to rainfall occurrence. In general, climate simulations hint that the global warming is expected to enhance the Indian summer monsoon precipitation by 5–10%, albeit some climate models suggest less. In the Indian monsoon region, the seasonal movement of the ITCZ brings rainfall over land, and the convection is strongly coupled with large-scale circulation (Goswami 2006; Joseph and Sabin 2008; Gadgil 2018). However, this convective-coupling is relatively weak in most of the CMIP5 models and mostly simulating an increase in precipitation irrespective of the anomalous

decrease in monsoon circulation (Krishnan et al. 2016). Sabeerali et al. (2015) postulated that the increase in future ISMR simulated in CMIP5 models is a result of unrealistic local convective precipitation enhancement that is not related to large-scale monsoon dynamics, which is mainly due to the unrealistic representation of stratiform and convective cloud ratio in the coupled model. Using CMIP5 outputs, Sabeerali and Ajaymohan (2018) showed that there is a possibility of a shorter rainy season (defined using tropospheric temperature gradient as outlined by Goswami and Xavier (2005)) by the end of the twenty-first century in the RCP8.5 scenario.

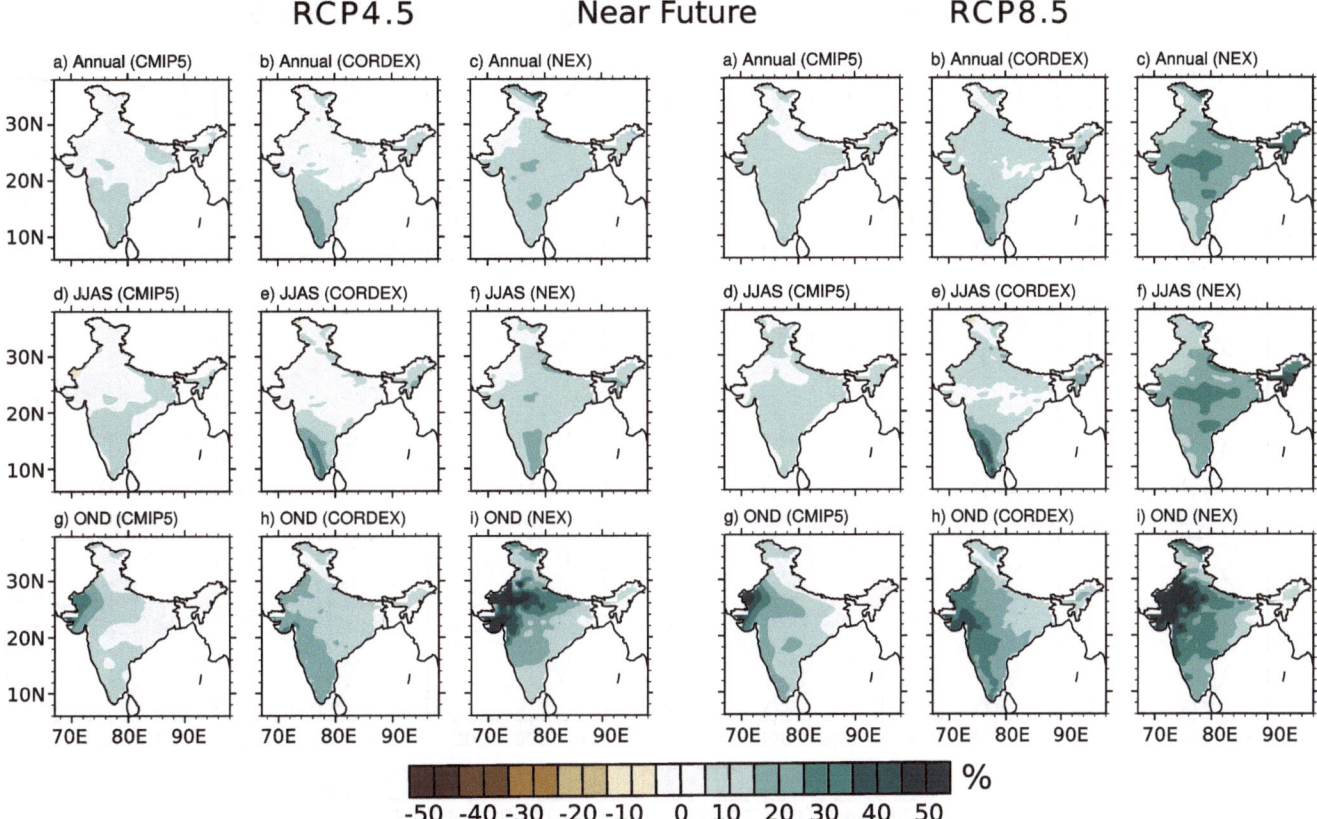

Fig. 3.8 Multi-Model Ensemble (MME) change (%) in annual, JJAS and OND rainfall as projected by CMIP5, CORDEX-SA and NEX-GDDP models for near future with respect to 1976–2005 from RCP4.5 and RCP8.5 scenario

Even though there is ambiguity in changes in projected seasonal mean precipitation, future projections from CMIP5 models show a significant increase in interannual variation during the summer monsoon season (Menon et al. 2013). Studies such as Sarita and Rajeevan (2016) have suggested that the periodicity of El Ninos is likely to shift to a shorter period (2.5–3 year) compared to present day (3–5 year) which will have serious implication on monsoon in interannual timescale. The increase in summer monsoon precipitation over South Asian region projected by CMIP5 models is mostly sustained by an increase in moisture supply due to enhanced warming (Mei et al. 2015). By analysing 14 CMIP models, they further showed that towards the end of the century the precipitable water over Indian landmass may increase by 8–16 mm/day, the evapotranspiration by around 0.6 mm/day while the change in moisture convergence around 2.4 mm/day or more under RCP8.5 scenario. Figure 3.10 shows the change in vertically integrated moisture convergence towards the end of the century under RCP4.5 and RCP8.5.

3.3.2 Future Changes in Northeast Monsoon

Interannual variations in the northeast monsoon are much larger compared to the southwest monsoons (Fig. 3.7). Projected changes for the near future and far future from CORDEX, NEX, and CMIP5 show a moderate increase in rainfall over the Indian landmass. Note that significant change is seen only over the region where northeast monsoon is not much significant (Figs. 3.8 and 3.9). Over major areas in Tamil Nadu and coastal Andhra, only modest changes are projected for both RCP 4.5 and RCP 8.5. Parvathi et al. (2017) reported that CMIP5 models project a robust reduction of the wind intensity during the northeast monsoon season especially over the Arabian Sea by the end of the twenty-first century (a reduction of 3.5% for RCP4.5 and 6.5% for RCP8.5, on an average). Despite a decrease in the winter monsoon winds, they noted an increased rainfall (10 ± 2%) in the winter monsoon rain zones in the equatorial Indian Ocean. Over Indian landmass, the precipitation enhancement is minimal in the near future in the RCP4.5

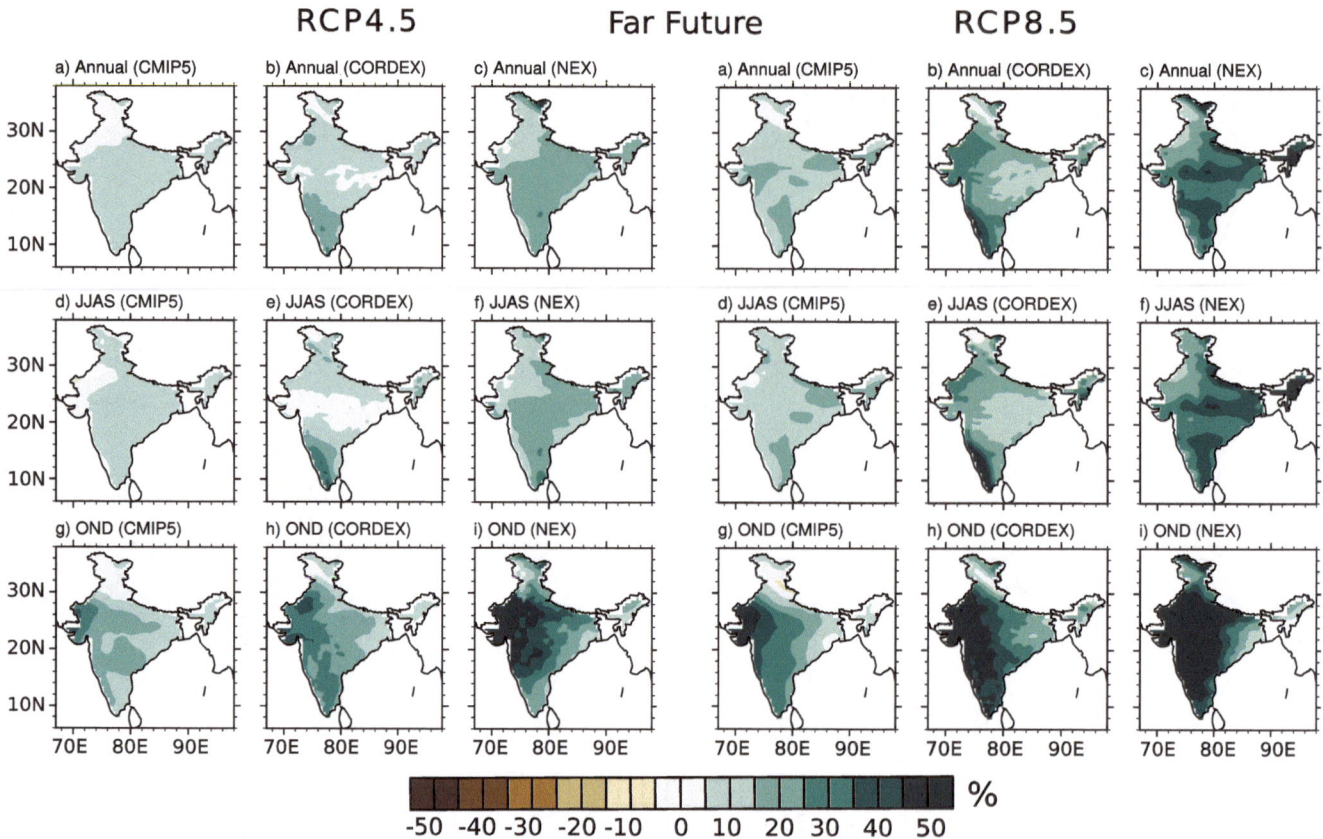

Fig. 3.9 Multi-Model Ensemble (MME) change (%) in annual, JJAS and OND rainfall as projected by CMIP5, CORDEX-SA and NEX-GDDP models in far future with respect to 1976–2005 from RCP4.5 and RCP8.5 scenario

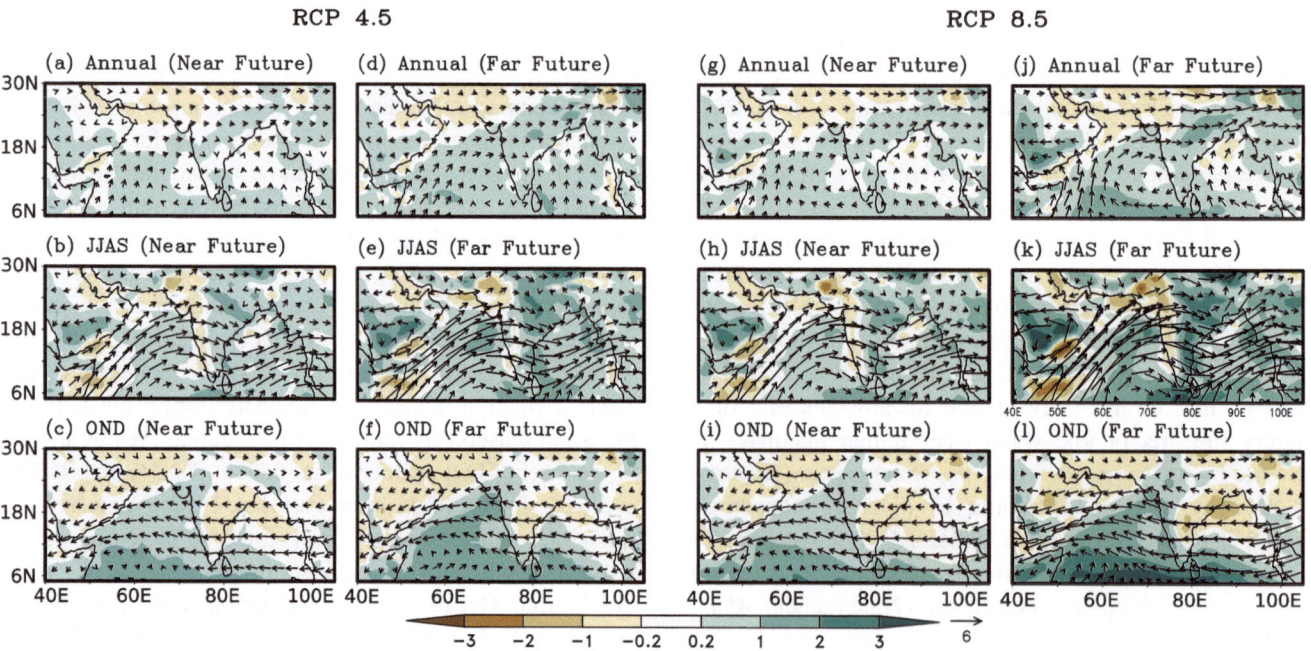

Fig. 3.10 Changes in vertically integrated moisture transport (vectors; Kg m^{-1} s^{-1}) and moisture convergence (shadings; 10^{-6} kg m^{-2} s^{-1}) from surface pressure level to 300 hPa for (**a, d**) Annual, (**b, e**) JJAS and (**c, f**) OND seasons, projected by multi-model ensemble of CMIP5 models for near future (2040–2069) and far future (2070–2099) for RCP4.5 and RCP8.5 scenarios, with respect to the reference period (1976–2005)

scenario; however, both projections (RCP4.5 and RCP8.5) shows an increase in precipitation [CMIP5 (10–20%), CORDEX (10–25%) and NEX (15–35%)] by the end of the twenty-first century.

3.4 Changes in Daily Precipitation Extremes

3.4.1 Observed Changes and Their Attribution

Detecting changes in the characteristics of extreme rain events is an important issue in view of their large impacts on human society (Ghosh et al. 2016). It is difficult to attribute a specific extreme event during the monsoon is owing to anthropogenic climate change—like the Uttarakhand surges of 2015, or recent flood in Kerala during 2018 monsoon— yet it is robustly anticipated that a warming atmosphere will result in more severe weather. It has been observed that from 1950 onwards there has been a significant rising trend in the frequency and intensity of extreme heavy rain events over central India, along with a decreasing trend in the moderate rain events (Goswami et al. 2006; Dash et al. 2009; Kulkarni et al. 2017; Krishnan et al. 2016; Roxy et al. 2017) (Fig. 3.11a). Consecutive dry days with minimum spell length of 5 days show significant increase of about 4 days in the period 1951–2015, while consecutive wet days show decrease of about 10 days in this period. Prolonged break spells appear to be more frequent in 1951–2015. Roxy et al. (2016) showed that the widespread changes in extreme rain events are mainly dominated by dynamic response of the atmosphere rather than thermodynamic factor alone. Krishnan et al. (2016) showed that, the enhancement of such deep localized convection, leading to heavy rainfall events, are more likely to happen in an atmosphere with weak vertical

shear (Romatschke and Houze 2011). Increased variability of low-level monsoon westerlies (Mishra et al. 2018; Roxy et al. 2017) and warming of north Arabian sea lead to increased moisture supply and thus enhance such events (Roxy et al. 2017). By examining the changes in the distribution of moderate and heavy monsoon precipitation in Historical and GHG, only simulations Krishnan et al. (2016) have shown that along with increase of atmospheric moisture the decrease of easterly vertical shear of the SAM circulation is also pivotal for favouring localized heavy rainfall over the Indian region (Fig. 3.11b). In a recent study, Singh et al. (2014) found statistically significant increase in the intensity and frequency of extreme wet and dry spells during the ISM during the 1951–2011 period.

3.4.2 Future Projections of Precipitation Extremes

The IPCC Special Report on Extremes (SREX; Intergovernmental Panel on Climate Change 2012) appraisal reported that extreme precipitation events globally are certain to rise in the future. From 1950 onwards, the number of extreme precipitation events over Indian landmass has also become more significant than it before (Sillmann et al. 2013; Goswami et al. 2006; Rao et al. 2014). A recent study by Mukherjee et al. (2017) showed that 1–5-day precipitation maxima at 5–500 year return period will increase (10–30%) with anthropogenic warming in RCP8.5 scenario. They further showed that the frequency of precipitation extremes is projected to rise more prominently in the RCP8.5 scenario over southern and central India by the middle and end of the twenty-first century. The analyses of select precipitation-based indices, defined by the Expert Team on Climate

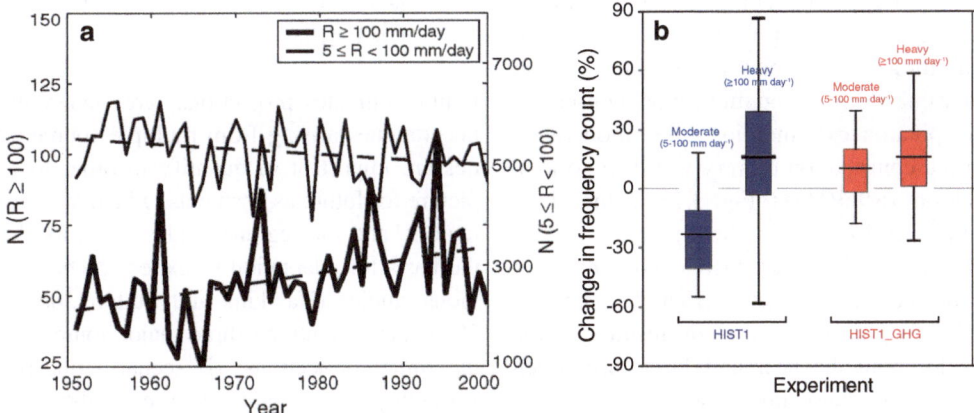

Fig. 3.11 Observed frequency of **a** heavy ($R \geq 100$ mm/day, bold line) and moderate ($5 \leq R < 100$ mm/day, thin line) daily rain events (Goswami et al. 2006). **b** Attribution of changes in moderate and heavy precipitation. *Box-whisker* plot of percentage distributions of yearly count of moderate (5–100 mm day^{-1}) and heavy (≥ 100 mm day^{-1})

events over Central India (74.5°–86.5° E, 16.5°–26.5° N) during the period (1951–2000) from Historical and GHG experiments with respect to the natural only simulation (Krishnan et al. 2016). Permission taken from American Association for the Advancement of Science for Fig. 3.11a.

Fig. 3.12 Precipitation indices averaged over Indian land area **a** contribution of very wet days to total wet day precipitation (R95PTOT), **b** simple daily intensity index (SDII) and **c** maximum 5-day precipitation (RX5day) based on CORDEX South Asia multi-model ensemble. Changes are displayed relative to the reference period 1976–2005 (in %). Solid lines show the ensemble mean and the shading indicates the range among the individual RCMs. Time series are smoothed with a 20-year running mean

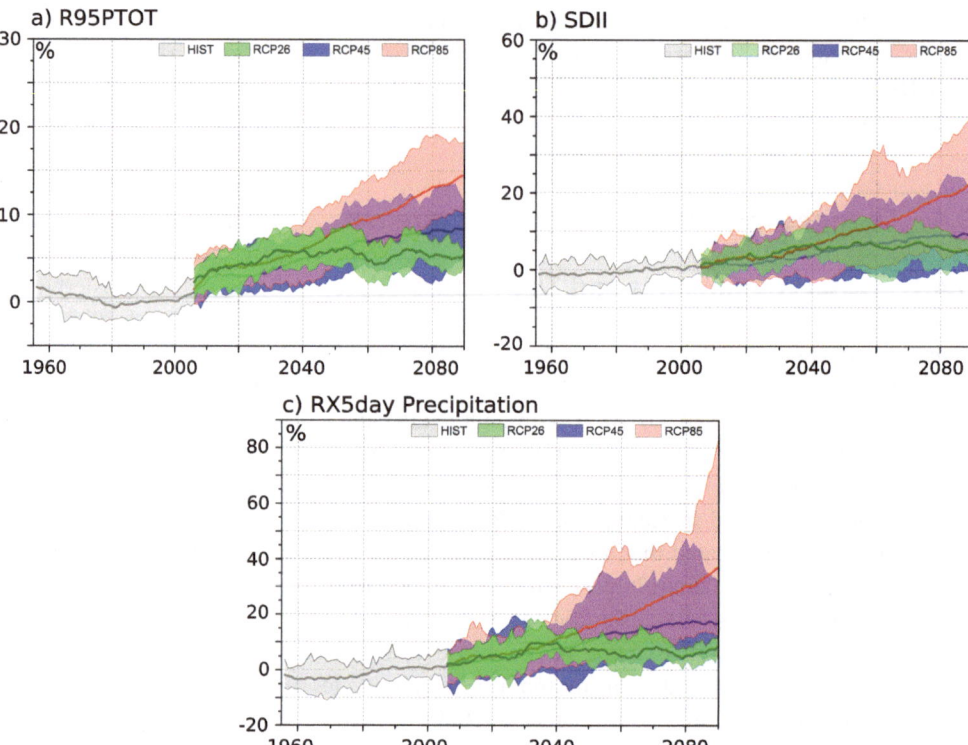

Change Detection and Indices (ETCCDI), computed with a consistent methodology for climate change simulations for different emission scenarios are discussed below.

Relative changes in the contribution of very wet days to total wet day precipitation (R95PTOT), the daily intensity index (SDII) and maximum 5-day precipitation (RX5day) with respect to 1976–2005 reference period are shown in Fig. 3.12. Similar to most of the global tropics, over Indian landmass as well, extreme precipitations are projected to increase throughout the twenty-first century. In RCP8.5, R95PTOT and SDII are projected to rise by 15 and 21%, by the end of the twenty-first century, whereas RX5day is projected to rise by 38%. The spread among ensemble members (shading in Fig. 3.12) is more in RCP8.5 comparing to RCP4.5 scenario all through the twenty-first century.

The spatial pattern of the projected multi-model ensemble means of the precipitation extremes identifies moderately higher increase in the contribution of very wet days to total wet day precipitation (R95PTOT; Fig. 3.13a), the daily intensity (SDII; Fig. 3.13b), and in the maximum 5-day precipitation (RX5day; Fig. 3.13c) are visible along the west coast, central and northern Indian states. Both the scenarios (RCP4.5 and RCP8.5) showed consistent results in the projected changes for both the near and far future. Even though the number of consecutive dry days (CDD; Fig. 3.13d) is increasing over various parts of India, the experiments provide a consensus only over the Indian peninsular region throughout the twenty-first century in the RCP8.5 scenario. The simultaneous increase in both CDD

and RX5day indicates an increase of both dry and wet epochs along the west coast and the peninsular region of India. This analysis is in agreement with the study by Mukherjee et al. (2017), who noted the frequency of extreme precipitation from CMIP5-GCM shows an increasing trend prominent over southern India (Fig. 3.14). This result enhances our confidence in assessing a likelihood of an increase in future precipitation extremes over the Indian peninsula throughout the twenty-first century, under the propensity of global warming signals.

3.5 Uncertainties in Projected Precipitation Changes

Future climate projections are inherently saddled with uncertainties arising from multiple sources. These uncertainties are important to quantify in order to convey a realistic picture for future assessments, which are particularly useful at regional and sub-regional scales where local actions may form the basis for adaptation to expected changes in climate. Previous studies (Hawkins and Sutton 2009; Terray and Boe 2013) have identified three major sources of uncertainties in the future projections: (i) scenario uncertainty (ii) internal variability from chaotic nature of the climate system, and (iii) model related, i.e. how different climate models respond to the same forcing. Kirtman et al. (2013) showed that the uncertainty in near-term projections is mostly dominated by internal variability and model spread. This provides some of

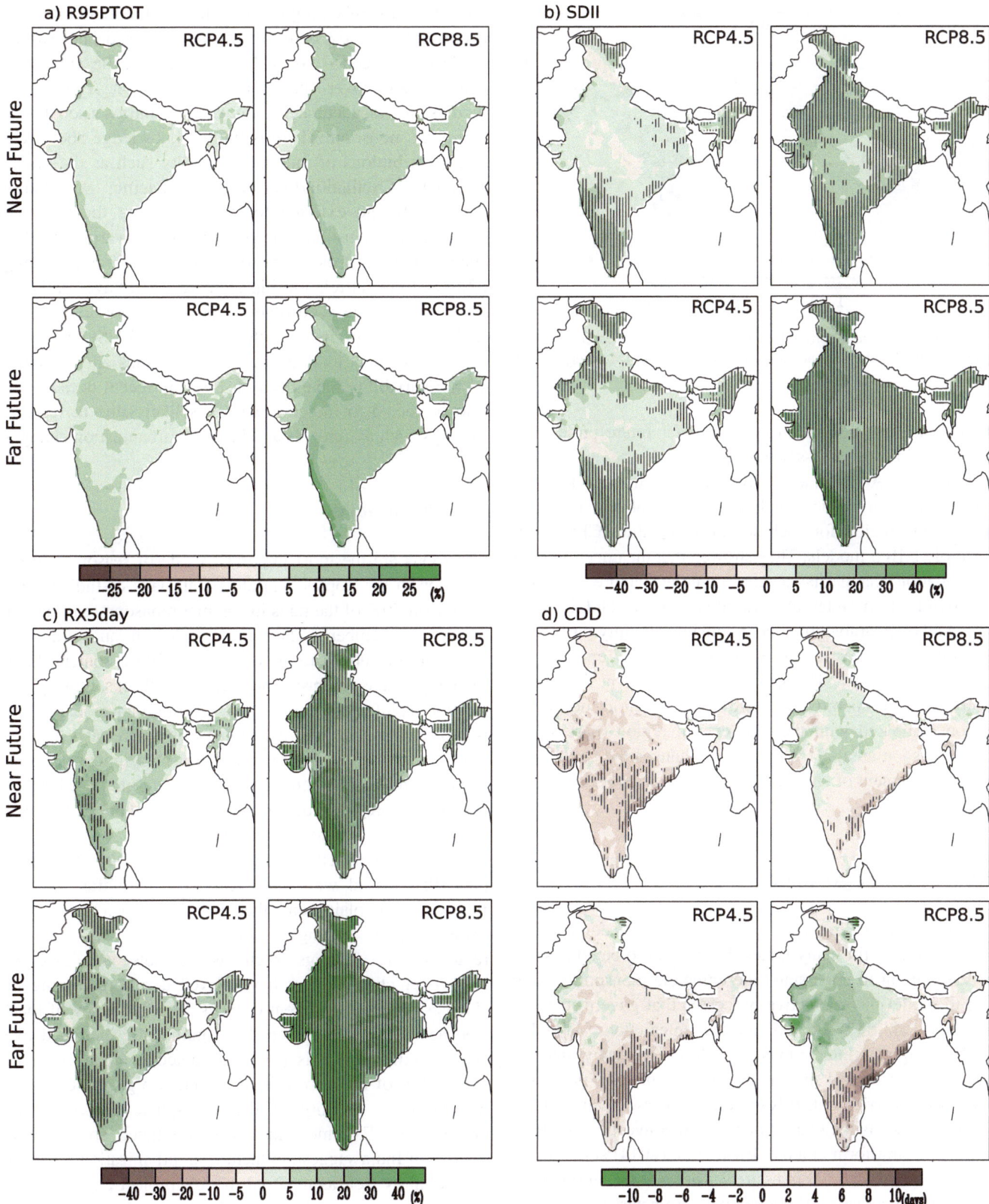

Fig. 3.13 Changes (%) in **a** contribution of very wet days to total wet day precipitation (R95PTOT) **b** simple daily intensity index (SDII), **c** maximum 5-day precipitation (RX5day), and **d** maximum number of consecutive dry days (CDD) for near future and far future from CORDEX South Asia multi-model ensemble mean with respect to the reference period 1976–2005. Striping indicates where at least 70% of the RCM realizations concur on the increase (vertical) or decrease (horizontal) in the future scenarios. The stapling is not plotted for R95PTOT as the RCM realizations have more than 70% consensus for increase

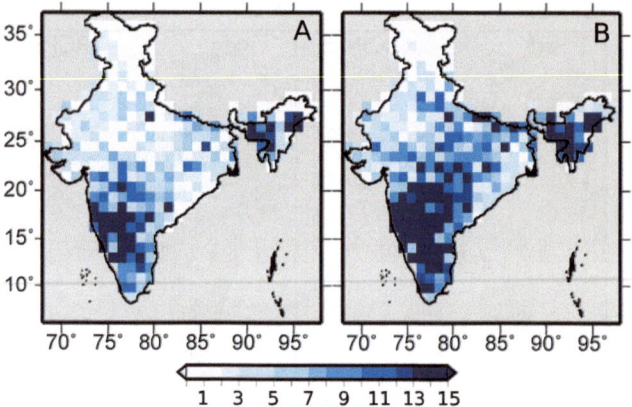

Fig. 3.14 Multi-model ensembles mean frequency of extreme precipitation events for the **a** Near Future and **b** Far future (Mukherjee et al. 2017)

the rationale for considering near and far future projections separately in most assessments.

Singh and AchutaRao (2018) quantified the uncertainties associated with models and internal variability over the Indian land region for each season using the RCP8.5 scenario of CMIP5 models. They showed that the uncertainty in precipitation change has a more complex picture such that uncertainty from internal variability persists and is quite large and comparable to model related uncertainty during the southwest and post-monsoon season by the end of the twenty-first century. The spatial heterogeneity in the uncertainty was comparatively more over the arid northwest region compared with the west-central region (part of the core monsoon area). This in a way enhances the confidence in the assessment of changes in precipitation over the central Indian region.

3.6 Knowledge Gaps

Though the modelling of the climate system has come a long way, there are still many issues that remain to be addressed. It is important to understand the complete monsoon system and to model the interactive processes that govern it. Identification of the coupled air-sea interactions, coupled land-atmosphere interactions, and flow-orography interactions that are critical in shaping the precipitation processes needs to be carried out. It is important to assess whether the model represents the phase transition in convection (shallow to deep to stratiform clouds). Bush et al. (2014) suggest that monsoon precipitation biases are sensitive to the entrainment and detrainment rates of convective parameterization. From observations and in models, it is important to identify the required thermodynamic conditions for convective phase transitions over the Asian monsoon region. There is lack of high-quality observations (atmosphere and ocean) over the

monsoon-influenced regions to constrain the model physics. More realistic biogeophysical processes-based land surface models are needed to realistically assess the impact of land-use/land-cover change on the monsoon. The present models fail to adequately simulate the intra-seasonal variability of monsoon. There is limited knowledge about relative contributions of internal variability such as the Pacific Decadal Oscillation/Inter-decadal Pacific Oscillation (PDO/IPO) and external forcing in driving the historical evolution of monsoons and other modes of variability. It is challenging to project the behaviour of climate forcing. ENSO-Monsoon relationship is observed to have weakened in recent decades; however, models do not capture the ENSO behaviour in the future and how ENSO–monsoon relationship may evolve. Increased extremes and increased spatial variability have been observed in recent decades and also projected to increase, and it is likely due to regional forcings such as aerosols and land-use/urbanization changes.

3.7 Summary

Monsoons are the most important mode of seasonal climate variation in the tropics, and almost all parts of India receive more than 70% of the rains in summer monsoon season (June through September). The intensity, length and timing of monsoon are related to atmospheric moisture content, land–sea temperature contrast, land surface feedbacks, atmospheric aerosol loading and other factors. Overall, monsoonal rainfall is projected to become more intense in future, and to affect larger areas mainly due to increase in atmospheric moisture content with temperature. The temperature gradient between land and sea, regional distribution of land and ocean as well as topography play major role in monsoon.

Summer monsoon rainfall has decreased over India in the post-1950 period with more reduction in rainfall over the Indo-Gangetic plains and the Western Ghats. Global-scale anthropogenic forcings such as GHGs as well as regional-scale forcings such as aerosols and land-use/land-cover changes may have played a role in driving the changes observed in recent decades. The frequency of localized heavy rain occurrences has significantly increased over central India, which is partly attributed to changes in the availability of moisture due to greenhouse gas-based warming, aerosols, stability of the atmosphere and increasing urbanization. Extreme rains are concentrated around urban regions of India suggesting there is an urbanization feedback.

Global as well as regional models project an increase in seasonal mean rainfall over India while also projecting a weakening monsoon circulation. However, this weakening of circulation is compensated by increased atmospheric moisture content leading to more precipitation. Frequency of extreme precipitation events may increase all over India, and more prominently so over the central and southern parts as a response to enhanced warming. Monsoon onset dates are likely to be early or not to change much, and the monsoon retreat dates are likely to be delayed, resulting in lengthening of the monsoon season.

References

Alexander LV (2016) Global observed long-term changes in temperature and precipitation extremes: a review of progress and limitations in IPCC assessments and beyond. Weather Clim Extrem 11:4–16. https://doi.org/10.1016/j.wace.2015.10.007

Ananthakrishnan R, Srinivasan V, Ramakrishnan AR, Jambunathan R (1968) Synoptic features associated with onset of southwest monsoon over Kerala, Forecasting Manual Report No. IV-18.2, India Meteorological Department, Pune, India

Ananthakrishnan R, Soman MK (1988) The onset of the southwest monsoon over Kerala: 1901–1980. J Clim 8(3):283–296

Annamalai H, Hafner J, Sooraj KP, Piilai P (2013) Global warming shifts the monsoon circulation, drying south Asia. J Clim 26 (9):2701–2718. https://doi.org/10.1175/JCLI-D-12-00208.1

Annamalai H, Slingo JM (2001) Active/break cycles: diagnosis of the intraseasonal variability of the Asian Summer Monsoon. Clim Dyn 18(1–2):85–102. https://doi.org/10.1007/s003820100161

Annamalai H, Hamilton K, Sperber KR (2007) The south Asian summer monsoon and its relationship with ENSO in the IPCC AR4 simulations. J Clim 20(6):1071–1092. https://doi.org/10.1175/JCLI4035.1

Ashok K, Guan Z, Saji NH, Yamagata T (2004) Individual and combined influences of ENSO and the Indian Ocean dipole on the Indian summer monsoon. J Clim 17:3141–3155. https://doi.org/10.1175/1520-0442(2004)017%3c3141:IACIOE%3e2.0.CO;2

Ashok K, Saji N (2007) On the impacts of ENSO and Indian Ocean dipole events on sub-regional Indian summer monsoon rainfall. Nat Hazards: J Int Soc Prev Mitig Nat Hazards. Springer; Int Soc Prev Mitig Nat Hazards 42(2):273–285. https://doi.org/10.1007/s11069-006-9091-0

Ashrit R, Kumar KR, Kumar KK (2001) ENSO-monsoon relationships in a greenhouse warming scenario. Geophys Res Lett 28:1727–1730

Balachandran S, Asokan A, Sridharan S (2006) Global surface temperature in relation to northeast monsoon rainfall over Tamil Nadu. J Earth Syst Sci 115(3):349–362. https://www.ias.ac.in/article/fulltext/jess/115/03/0349-0362

Bamzai AS, Kinter JL (1997) Climatology and interannual variability of Northern Hemisphere snow cover and depth based on satellite observations. Center for ocean land atmosphere studies report 52, COLA, Calverton, 48pp

Berkelhammer M, Sinha A, Stott L, Cheng H, Pausata FSR, Yoshimura K (2012) An abrupt shift in the Indian monsoon 4000 years ago. In: Giosan Let al (eds) Climates, landscapes, and civilizations, vol 198. American Geophysical Union Geophysical Monograph, pp 75–87. https://doi.org/10.1029/2012GM001207

Bhanu Kumar OSRU (1987) Seasonal variation of Eurasian snow cover and its impact on the Indian summer monsoon. In: Large scale effects of seasonal snow cover. Proceedings of Vancouver symposium, Publication 166. International Association of Hydrological Sciences, pp 51–60

Bidyabati S, Ashok K, Pai DS (2017) Uncertainties in observations and climate projections for the North East India. Glob Planet Chang 160:96–108. https://doi.org/10.1016/j.gloplacha.2017.11.010

Blanford HF (1884) On the connection of Himalayan snowfall and seasons of drought in India. Proc R Soc London 37(232–234):3–22. https://doi.org/10.1098/rspl.1884.0003

Bollasina M, Ming Y, Ramaswamy V (2011) Anthropogenic aerosols and the weakening of the South Asian summer monsoon. Science 334(6055):502–505

Bollasina M, Ming Y, Ramaswamy V, Schwarzkopf MD, Naik V (2014) Contribution of local and remote anthropogenic aerosols to the 20th century Weakening of the South Asian Monsoon. Geophys Res Lett 41(2):680–687. https://doi.org/10.1002/2013GL058183

Bush SJ, Turner AG, Woolnough SJ, Martin GM, Klingaman NP (2014) The effect of increased convective entrainment on Asian monsoon biases in the MetUM general circulation model. Q J R Meteorol Soc. https://doi.org/10.1002/qj.2371

Cai W, Wang G, Gan B, Wu L, Santoso A, Lin X, Chen Z, Jia F, Yamagata T (2018) Stabilised frequency of extreme positive Indian Ocean Dipole under 1.5 °C warming. Nat Commun 9(1):1419. https://doi.org/10.1038/s41467-018-03789-6

Chao WC, Chen B (2001) The origin of monsoons. J AtmosSci 58:3497–3507

Chaturvedi RK, Joshi J, Jayaraman M, Bala G, Ravindranath NH (2012) Multi-model climate change projections for India under representative concentration pathways. Curr Sci 103(7):791–802. http://dspace.library.iitb.ac.in/jspui/handle/100/15902

Cheng H, Edwards RL, Sinha A, Spötl C, Yi L, Chen S, Kelly M, Kathayat G, Wang X, Li X, Kong X, Wang Y, Ning Y, Zhang H (2016) The Asian monsoon over the past 640,000 years and ice age terminations. Nature 534(7609):640–646. https://doi.org/10.1038/nature18591

Cherchi A, Alessandri A, Masina S, Navarra A (2011) Effects of increased CO_2 levels on monsoons. Clim Dyn 37(1–2):83–101. https://doi.org/10.1007/s00382-010-0801-7

Chowdary JS, Bandgar AB, Gnanaseelan C, Luo JJ (2015) Role of tropical Indian Ocean air–sea interactions in modulating Indian summer monsoon in a coupled model. Atmos Sci Lett 16:170–176

Chowdary JS, Patekar D, Srinivas G, Gnanaseelan C, Parekh A (2019) Impact of the Indo-Western Pacific Ocean capacitor mode on South Asian summer monsoon rainfall. Clim Dyn 53:2327. https://doi.org/10.1007/s00382-019-04850-w

Chung CE, Ramanathan V (2006) Weakening of North Indian SST gradients and the monsoon rainfall in India and the Sahel. J Clim 19:2036–2045.https://doi.org/10.1175/JCLI3820.1

Dash SK, Kulkarni MA, Mohanty UC, Prasad K (2009) Changes in the characteristics of rain events

De US, Mukhopadhyay RK (2002) Breaks in monsoon and related precursors. Mausam 53:309–318

Dhar ON, Rakhecha PR (1983) Forecasting northeast monsoon rainfall over Tamil Nadu, India. Mon Weather Rev 111:109–112. https://doi.org/10.1175/1520-0493(1983)111%3c0109:FNMROT%3e2.0.CO;2

Dimri AP (2013) Intraseasonal oscillation associated with the Indian winter monsoon. J Geophys Res Atmos 118(3):1189–1198. https://doi.org/10.1002/jgrd.50144

Dimri AP, Niyogi D, Barros AP, Ridley J, Mohanty UC, Yasunari T, Sikka DR (2015) Western disturbances: a review. Rev Geophys 53 (2):225–246. https://doi.org/10.1002/2014RG000460

Douville H, Royer J-F, Polcher J, Cox P, Gedney N, Stephenson DB, Valdes PJ (2000) Impact of CO_2 doubling on the Asian summer monsoon: robust versus model-dependent responses. J Meteor Soc Japan 78(2000):421–439. https://doi.org/10.2151/jmsj1965.78.4_421

Fan F, Mann ME, Lee S, Evans JL (2010) Observed and modeled changes in the South Asian monsoon over the historical period. J Clim 23:5193–5205. https://doi.org/10.1175/2010JCLI3374.1

Fasullo J, Webster PJ (2003) A hydrological definition of Indian monsoon onset and withdrawal. J Clim 16:3200–3211. https://doi.org/10.1175/1520-0442(2003)016%3c3200a:AHDOIM%3e2.0.CO;2

Fleitmann D, Burns SJ, Mangini A (2007) Holocene ITCZ and Indian monsoon dynamics recorded in stalagmites from Oman and Yemen (Socotra). Quat Sci Rev 26:170–188. https://doi.org/10.1016/j.quascirev.2006.04.012

Gadgil S (2007) The Indian monsoon: 3. Physics of the monsoon. Resonance 12(5):4–20. https://doi.org/10.1007/s12045-007-0045-y

Gadgil S (2018) The monsoon system: land–sea breeze or the ITCZ?. J Earth Sys Sci 127(1)

Gadgil S, Vinayachandran PN, Francis PA (2004) Extremes of the Indian summer monsoon rainfall, ENSO and equatorial Indian ocean oscillation. Geophys Res Lett 31:L12213

Ghosh S, Vittal H, Sharma T, Karmakar S, Kasiviswanathan K, Dhanesh Y, Sudheer K, Gunthe S (2016) Indian summer monsoon rainfall: implications of contrasting trends in the spatial variability of means and extremes. PLoS ONE 11(7):e0158670. https://doi.org/10.1371/journal.pone.0158670

Giorgi F et al (2012) RegCM4: model description and preliminary tests over multiple CORDEX domains. Clim Res 52:7–29

Gnanaseelan C, Roxy MK, Deshpande A (2017) Variability and trends of sea surface temperature and circulation in the Indian Ocean. In: Nayak S, Rajeevan M (eds) Observed climate variability and change over the Indian Region. Springer, Singapore

Goswami BN (2005) Intraseasonal variability (ISV) of south Asian summer monsoon; In: Lau K, Waliser D (eds) Intraseasonal variability of the atmosphere-ocean climate system. Springer, Chichester

Goswami BN (2006) The annual cycle, intraseasonal oscillations and the roadblock to seasonal predictability of the Asian summer monsoon. J Clim 19:5078–5098. https://doi.org/10.1175/JCLI3901.1

Goswami BN, Ajayamohan RS (2001) Intraseasonal oscillations and interannual variability of the Indian summer monsoon. J Clim 14:1180–1198. https://doi.org/10.1175/1520-0442(2001)014%3c1180:IOAIVO%3e2.0.CO;2

Goswami BN, Ajayamohan RS, Xavier PK, Sengupta D (2003) Clustering of low pressure systems during the Indian summer monsoon by intra-seasonal oscillations. Geophys Res Lett 30(8). https://doi.org/10.1029/2002GL016734

Goswami BN, Venugopal V, Sengupta D, Mdhusoodanan MS, Xavier PK (2006) Increasing trend of extreme rain events over india in a warming environment. Science 314(5804):1442–1445. https://doi.org/10.1126/science.1132027

Goswami BN, Xavier PK (2005) Dynamics of "internal" interannual variability of the Indian summer monsoon in a GCM. J Geophys Res 110(D24). https://doi.org/10.1029/2005JD006042

Goswami BN, Chakravorty S (2017) Dynamics of the Indian summer monsoon climate. Oxford Research Encyclopedia. https://doi.org/10.1093/acrefore/9780190228620.013.613

Guhathakurta P, Rajeevan M (2008) Trends in rainfall patterns over India. Int J Climatol 28:1453–1469. https://doi.org/10.1002/joc.1640

Guhathakurta P, Rajeevan M, Sikka DR, Tyagi A (2015) Observed changes in southwest monsoon rainfall over India during 1901-2011. Int J Climatol 35(8):1881–1898. https://doi.org/10.1002/joc.4095

Hasson S, Lucarini V, Pascale S (2013) Hydrological cycle over South and Southeast Asian river basins as simulated by PCMDI/CMIP3 experiments. Earth Syst Dyn 4:199–217. https://doi.org/10.5194/esd-4-199-2013

Hawkins E, Sutton R (2009) The potential to narrow uncertainty in regional climate predictions. Bull Am Meteor Soc 90:1095–1107. https://doi.org/10.1175/2009BAMS2607.1

IPCC (2013) Climate Change (2013) The physical science basis. Contribution of working group I to the Fifth assessment report of the intergovernmental panel on climate change (Stocker TF, Qin D, Plattner G-K, Tignor M, Allen SK, Boschung J, Nauels A, Xia Y, Bex V, Midgley PM (eds)). Cambridge University Press, Cambridge, 1535 pp

Janowiak JE, Xie P (2003) The global scale examination of monsoon related precipitation. J Clim 16:4121–4133

Jin Q, Wang C (2017) A revival of Indian summer monsoon rainfall since 2002. Nat Clim Change 7:585–597. https://doi.org/10.1038/nclimate3348

Joseph PV, Sabin TP (2008) An Ocean-atmosphere interaction mechanism for the active break cycle of the Asian summer monsoon. Clim Dyn 30(6):553–566. https://doi.org/10.1007/s00382-007-0305-2

Joseph PV, Simon A (2005) Weakening trend of the southwest monsoon current through peninsular India from 1950 to the present. Curr Sci 89(4):687–694

Joseph PV, Bindu G, Preethi B (2016) Impact of the upper tropospheric cooling trend over Central Asia on the Indian summer monsoon rainfall and the Bay of Bengal cyclone tracks. Curr Sci 110:2105–2113

Joshi MK, Rai A (2015) Combined interplay of the Atlantic multidecadal oscillation and the Interdecadal Pacific Oscillation on rainfall and its extremes over Indian subcontinent. Clim Dyn 44 (11–12):3339–3359

Karmakar N, Chakraborty A, Nanjundiah R (2015) Decreasing intensity of monsoon low-frequency intraseasonal variability over India. Env Res Lett 10(5):054018. https://doi.org/10.1088/1748-9326/10/5/054018

Karmakar N, Chakraborty A, Nanjundiah R (2017) Increased sporadic extremes decrease the intraseasonal variability in the Indian summer monsoon rainfall. Sci Rep 7:7824. https://doi.org/10.1038/s41598-017-07529-6

Keshavamurty RN, Sankar Rao M (1992) The physics of monsoons. Allied Publishers Ltd, Bombay, p 199

Kirtman B et al (2013) Near-term climate change: projections and predictability. In: Stocker TF, Qin D, Plattner G-K, Tignor M, Allen SK, Boschung J, Nauels A, Xia Y, Bex V, Midgley PM (eds) Climate change 2013: the physical science basis. Contribution of working group I to the fifth assessment report of the intergovernmental panel on climate change. Cambridge University Press, Cambridge

Kitoh A, Yukimoto S, Noda A, Motoi T (1997) Simulated changes in the Asian summer monsoon at times of increased atmospheric CO_2. J Meteorol Soc Japan 75(6):1019–1031. https://doi.org/10.2151/jmsj1965.75.6_1019

Koteswaram P (1958) The easterly jet stream in the tropics. Tellus 10:43–57. https://doi.org/10.1111/j.2153-3490.1958.tb01984.xU

Kothawale DR, Munot AA, Borgaonkar HP (2008) Temperature variability over the Indian Ocean and its relationship with Indian summer monsoon rainfall. Theor Appl Climotol 92(1–2):31–45. https://doi.org/10.1007/s00704-006-0291-z

Kripalani RH, Kulkarni A (2001) Monsoon rainfall variations and teleconnections over South and East Asia. Int J Clim 21:603–616

Kripalani R, Oh JH, Kulkarni A, Sabade SS, Chaudhari HS (2007) South Asian summer monsoon precipitation variability: coupled climate model simulations and projections under IPCC AR4. Theor Appl Climatol 90(3–4):133–159. https://doi.org/10.1007/s00704-006-0282-0

Kripalani RH, Kulkarni A (1997) Climatic impact of El Niño/La Niña on the Indian monsoon: a new perspective. Weather 52:39–46. https://doi.org/10.1002/j.1477-8696.1997.tb06267.x

Kripalani RH, Kulkarni A (1999) Climatology and variability of historical Soviet snow depth data: some new perspectives in snow—Indian monsoon teleconnections. Clim Dyn 15:475–489. https://doi.org/10.1007/s003820050294

Kripalani RH, Kulkarni A, Sabade SS, Revadekar J, Patwardhan S, Kulkarni JR (2004) Intraseasonal oscillations during monsoon 2002 and 2003. Curr Sci 87:325–351

Kripalani RH, Kulkarni A, Singh SV (1997) Association of the Indian summer monsoon with the Northern Hemisphere mid-latitude circulation. Int J Climatol 17:1055–1067. https://doi.org/10.1002/(SICI)1097-0088(199708)17:10%3c1055:AID-JOC180%3e3.0.CO;2-3

Kripalani RH, Kumar P (2004) Northeast monsoon rainfall variability over south peninsular India vis-à-vis Indian Ocean dipole mode. Int J Climatol 24:1267–1282

Krishna Kumar K, Rajagopalan B, Cane MA (1999) On the weakening relationship between the Indian monsoon and ENSO. Science 284:2156–2159. https://doi.org/10.1126/science.284.5423.2156

Krishnamurthy L, Krishnamurthy V (2013) Influence of PDO on South Asian summer monsoon-ENSO relation. Clim Dynam 42(9–10):2397–2410. https://doi.org/10.1007/s00382-013-1856-z

Krishnamurthy L, Krishnamurthy V (2015) Teleconnection of Indian monsoon rainfall with AMO and Atlantic Tripole. Clim Dynam 46(7):2269–2285. https://doi.org/10.1007/s00382-015-2701-3

Krishnamurthy V, Goswami BN (2000) Indian monsoon-ENSO relationship on interdecadal time scales. J Clim 13:579–595. https://doi.org/10.1175/1520-0442(2000)013%3c0579:IMEROI%3e2.0.CO;2

Krishnamurthy V, Shukla J (2000) Intra-seasonal and inter-annual variability of rainfall over India. J Clim 13:4366–4377. https://doi.org/10.1175/1520-0442(2000)0132.0.CO;2

Krishnamurthy V, Shukla J (2007) Intraseasonal and seasonally persisting patterns of Indian monsoon rainfall. J Clim 20:3–20. https://doi.org/10.1175/JCLI3981.1

Krishnamurthy V, Shukla J (2008) Seasonal persistence and propagation of intraseasonal patterns over the Indian summer monsoon region. Clim Dyn 30:353–369. https://doi.org/10.1007/s00382-007-0300-7

Krishnamurti TN, Bhalme HN (1976) Oscillations of monsoon system. Part I. Observational aspects. J Atmos Sci 33(10):1937–1954. https://doi.org/10.1175/1520-0469(1976)033%3c1937:OOAMSP%3e2.0.CO;2

Krishnamurti TN, Ramanathan Y (1982) Sensitivity of the monsoon onset to differential heating. J Atmos Sci 39(6):1290–1306. https://doi.org/10.1175/1520-0469(1982)039%3c1290:SOTMOT%3e2.0.CO;2

Krishnan R, Sabin TP, Ayantika DC, Kitoh A, Sugi M, Murakami H, Turner AG, Slingo JM, Rajendran K (2013) Will the South Asian monsoon overturning circulation stabilize any further? Clim Dyn 40:187–211. https://doi.org/10.1007/s00382-012-1317-0

Krishnan R, Sugi M (2003) Pacific decadal oscillation and variability of the Indian summer monsoon rainfall. Clim Dyn 21(3):233–242. https://doi.org/10.1007/s00382-003-0330-8

Krishnan R, Zhang C, Sugi M (2000) Dynamics of breaks in the Indian summer monsoon. J Atmos Sci 57(9):1354–1372. https://doi.org/10.1175/1520-0469(2000)057%3c1354:DOBITI%3e2.0.CO;2

Krishnan R, Sabin TP, Vellore R, Mujumdar M, Sanjay J, Goswami BN, Hourdin F, Dufresne J-L, Terray P (2016) Deciphering the desiccation trend of the South Asian monsoon hydroclimate in a warming world. Clim Dyn 47(3–4):1007–1027. https://doi.org/10.1007/s00382-015-2886-5

Kulkarni A (2012) Weakening of Indian summer monsoon rainfall in warming environment. Theor Appl Climatol 109(3–4):447–459. https://doi.org/10.1007/s00704-012-0591-4

Kulkarni A, Deshpande N, Kothawale DR, Sabade SS, RamaRao MVS, Sabin TP, Patwardhan S, Mujumdar M, Krishnan R (2017) Observed climate variability and change over India. In: Krishnan R, Sanjay J (eds) Climate change over India—an interim report. Centre for Climate Change Research, IITM, Pune

Kumar P, Rupa Kumar K, Rajeevan M, Sahai AK (2007) On the recent strengthening of the relationship between ENSO and northeast monsoon rainfall over South Asia. Clim Dyn 8:649–660

Lawrence DM, Webster PJ (2001) Interannual variations of the intraseasonal oscillation in the south Asian summer monsoon region. J Clim 14:2910–2922. https://doi.org/10.1175/1520-0442(2001)014%3c2910:IVOTIO%3e2.0.CO;2

Li X, Ting M (2015) Recent and future changes in the Asian monsoon-ENSO relationship: natural or forced? Geophys Res Lett 42(9):3502–3512. https://doi.org/10.1002/2015GL063557

Li Z, Lin X, Cai W (2017) Realism of modelled Indian summer monsoon correlation with the tropical Indo-Pacific affects projected monsoon changes. Sci Rep 7(1). https://doi.org/10.1038/s41598-017-05225-z

Madden R, Julian P (1972) Description of global-scale circulation cells in the tropics with a 40–50 day period. J Atmos Sci 29(6):1109–1123

Maheskumar RS et al (2014) Mechanism of high rainfall over the Indian west coast region during the monsoon season. Clim Dyn 43:1513–1529. https://doi.org/10.1007/s00382-013-1972-9

Mandke S, Sahai AK, Shinde MA, Joseph S, Chattopadhyay R (2007) Simulated changes in active/break spells during the Indian summer monsoon due to enhanced CO_2 concentrations: assessment from selected coupled atmosphere–ocean global climate models. Int J Climatol 27(7):837–859. https://doi.org/10.1002/joc.1440

Mantua NJ, Hare SR (2002) The Pacific decadal oscillation. J Oceanogr 58:35–44

Meehl GA, Arblaster JM, Tebaldi C (2005) Understanding future patterns of increased precipitation intensity in climate model simulations. Geophys Res Lett 32:L18719. https://doi.org/10.1029/2005gl023680

Mehta VM, Lau KM (1997) Influence of solar irradiance on the Indian monsoon-ENSO relationship at decadal-multidecadal time scales. Geophys Res Lett 24(2):159–162. https://doi.org/10.1029/96GL03778

Mei R, Ashfaq M, Rastogi D, Leung L, Dominguez F (2015) Dominating controls for wetter south Asian Summer Monsoon in the twenty-first century. J Clim 28(8):3400–3419. https://doi.org/10.1175/JCLI-D-14-00355.1

Menon A, Levermann A, Schewe J (2013) Enhanced future variability during India's rainy season. Geophys Res Lett 40(12):3242–3247. https://doi.org/10.1002/grl.50583

Mishra AK, Nagaraju V, Rafiq M, Chandra S (2018) Evidence of links between regional climate change and precipitation extremes over India. Weather Roy Meteorol Soc 74(6):218–221. https://doi.org/10.1002/wea.3259

Mukherjee S, Saran A, Stone D, Mishra V (2017) Increase in extreme precipitation events under anthropogenic warming in India. Weather Clim Extrem 20:45–53. https://doi.org/10.1016/j.wace.2018.03.005

Nageswara Rao G (1999) Variations of the SO relationship with summer and winter monsoon rainfall over India. 1872–1993. J Clim 12:3486–3495

Nageswararao MN, Sannan MC, Mohanty UC (2019) Characteristics of various rainfall events over South Peninsular India during northeast monsoon using high-resolution gridded dataset (1901–2016). Theor Appl Climatol 137:2573–2973. https://doi.org/10.1007/s00704-018-02755-y

Niyogi D, Kishtawal C, Tripathi S, Govindaraju RS (2010) Observational evidence that agricultural intensification and land use change may be reducing the Indian summer monsoon rainfall. Water Res Res 46:W03533. https://doi.org/10.1029/2008WR007082

Noska R, Mishra V (2016) Characterizing the onset and demise of the Indian summer monsoon. Geophys Res Lett 43:4547–4554. https://doi.org/10.1002/2016GL068409

Pai DS, Rajeevan MN (2009) Summer monsoon onset over Kerala: new definition and prediction. J Earth Syst Sci 118(2):123–135. https://doi.org/10.1007/s12040-009-0020-y

Pai DS, Shridhar L, Rajeevan M, Sreejith OP, Satbhai NS, Mukhopadhyay B (2014) Development of a new high spatial resolution (0.25° × 0.25°) long period (1901–2010) daily gridded rainfall data set over India and its comparison with existing data sets over the region. Mausam 65:1–18

Pai DS, Shridhra L, Ramesh Kumar MR (2016) Active and break events of Indian summer monsoon during 1901–2014. Clim Dyn 46:3921–3939. https://doi.org/10.1007/s00382-015-2813-9

Pant GB, Parthasarathy B (1981) Some aspects of an association between the southern oscillation and Indian summer monsoon Archives for Meteorology. Arch Met Geoph Biocl B29:245–252. https://doi.org/10.1007/BF02263246

Parthasarathy B, Diaz HF, Eischeld JK (1988) Prediction of all-India summer monsoon rainfall with regional and large-scale parameters. J Geophys Res 93:5341–5350. https://doi.org/10.1029/JD093iD05p05341

Parthasarathy B, Rupa Kumar K, Deshpande VR (1991a) Indian summer monsoon rainfall and 200-mb meridional wind index: application for long-range prediction. Int J Climatol 11:165 ± 176

Parthasarathy B, Rupa Kumar K, Munot AA (1991b) Evidence of secular variations in Indian monsoon rainfall ± circulation relationships. J Climate 4(9):927 ± 938

Parthasarathy B, Munot AA, Kothawale DR (1994) All-India monthly and seasonal rainfall. Series 1871–1993. Theor Appl Climatol 49:217–224

Parvathi V, Suresh I, Lengaigne M, Izumo T, Vialard J (2017) Robust projected weakening of winter monsoon winds over the Arabian Sea under climate change. Geophys Res Lett 4:9833–9843. https://doi.org/10.1002/2017GL075098

Pathak A, Ghosh S, Kumar P (2014) Precipitation recycling in the Indian subcontinent during summer monsoon. J Hydrometeor 15:2050–2066. https://doi.org/10.1175/JHM-D-13-0172.1

Pattanaik D, Satyan V (2000) Fluctuations of Tropical Easterly Jet during contrasting monsoons over India: a GCM study. Meteorol Atmos Phys 75(1–2):51–60. https://doi.org/10.1007/s007030070015

Patwardhan SK, Asnani GC (2000a) Meso-scale distribution of summer monsoon rainfall near the Western Ghats (India). Int J Climatol 20 (5):575–581

Patwardhan SK, Asnani GC (2000b) Meso-scale distribution of summer monsoon rainfall near the Western Ghats (India). Int J Climatol J R Meteorol Soc 20(5):575–581

Paul S, Ghosh S, Oglesby R, Pathak A, Chadrasekharan A, Ramasankaran R (2016) Weakening of Indian summer monsoon rainfall due to changes in land use land cover. Sci Rep 6:32177. https://doi.org/10.1038/srep32177

Paul S, Ghosh S, Rajendran K, Murtugudde R (2018) Moisture supply from the Western Ghats Forests to water deficit East Coast of India. Geophys Res Lett 45(9):4337–4344. https://doi.org/10.1029/2018GL078198

Pillai P, Chowdary JS (2016) Indian summer monsoon intra-seasonal oscillation associated with the developing and decaying phase of El Niño. Int J Climatol 36:1846–1862. https://doi.org/10.1002/joc.4464

Pisharoty PR, Desai BN (1956) Western disturbances and Indian weather. Indian J Met Geophys 7:333–338

Porter JR, Xie L, Challinor AJ, Cochrane K, Howden SM, Iqbal MM, Lobell DB, Travasso MI (2014) Food security and food production systems. In: Field et al (eds) Climate change 2014: impacts, adaptation, and vulnerability. Part A: global and sectoral aspects. Contribution of working group II to the fifth assessment report of the intergovernmental panel on climate change. Cambridge University Press, Cambridge, pp 485–533

Prakash S, Mitra AK, Momin IM, Rajgopal EN, Basu S, Collins M, Turnar AG, Rao KA, Ashok K (2015) Seasonal intercomparison of observational rainfall datasets over India during the southwest monsoon season. Int J Climatol 35:2326–2338. https://doi.org/10.1002/joc.4129

Prasad S, Anoop A, Riedel N, Sarkar S, Menzel P, Basavaiah N, Krishnan R, Fuller D, Plessen B, Gaye B, Röhl U, Wilkes H, Sachse D, Sawant R, Wiesner MG, Stebich M (2014) Prolonged monsoon droughts and links to Indo-Pacific warm pool: a Holocene record from Lonar Lake, central India. Earth Planet Sci Lett 391:171–182. https://doi.org/10.1016/j.epsl.2014.01.043

Preethi B, Mujumdar M, Kripalani RH, Prabhu A, Krishnan R (2017) Recent trends and tele-connections among South and East Asian summer monsoons in a warming environment. Clim Dyn 48:2489–2505

Rahman SB, Templeman TS, Holt T, Murthy AB, Singh MP, Agarwal P, Nigam S, Prabhu A, Ameenullah S (1990) Structure of the Indian southwesterly pre-monsoon and monsoon boundary layers: observations and numerical simulation. Atmos Environ 24A (4):723–734

Rai P, Dimri A (2017) Effect of changing tropical easterly jet, low level jet and quasi-biennial oscillation phases on Indian summer monsoon. Atmos Sci Lett 18(2):52–59. https://doi.org/10.1002/asl.723

Raj YEA (1992) Objective determination of northeast monsoon onset dates over coastal Tamil Nadu for the period 1901-190. Mausam 43:273–282

Rajeevan M, Bhate J, Jaswal AK (2008) Analysis of variability and trends of extreme rainfall events over India using 104 years of gridded daily rainfall data. Geophy Res Lett 35:L18707. https://doi.org/10.1029/2008GL035143

Rajeevan M, Unnikrishnan CK, Bhate J, Niranjan Kumar K, Sreekala PP (2012) Northeast monsoon over India: variability and prediction. Meteorol Appl 19(2):226–236. https://doi.org/10.1002/met.1322

Rajendran K, Kitoh A, Srinivasan J, Mizuta R, Krishnan R (2012) Monsoon circulation interaction with Western Ghats orography under changing climate. Theor Appl Climatol 110(4):555–571. https://doi.org/10.1007/s00704-012-0690-2

Ramage CS (1971) Monsoon meteorology. In: International geophysics series, vol 15. Academic Press, San Diego California, p 296

Ramamurthy K (1969) Monsoon of India: some aspects of the 'break' in the Indian southwest monsoon during July and August. Forecast Man 1–57. (18.3 India Met. Dep Pune, India)

Ramanathan VC, Chung DK, Bettge T, Buja L, Kiehl JT, Washington WM, Fu Q, Sikka DR, Wild M (2005) Atmospheric brown clouds: impact on South Asian climate and hydrologic cycle. Proc Natl Acad Sci USA 102:5326–5333. https://doi.org/10.1073/pnas.0500656102

Ramaswamy C (1962) Breaks in the Indian summer monsoon as a phenomenon of interaction between the easterly and subtropical westerly jet streams. Tellus 14(3):337–349

Ramaswamy C (1972) The severe drought over Tamil Nadu during the retreating monsoon period of 1968 and its associations with anomalies in the upper level flow patterns over the northern hemisphere. Indian J Meteorol Geophys 23:303–316

Ramesh Kumar MR, Krishnan R, Syam S, Unnikrishnan AS, Pai DS (2009) Increasing trend of "break-monsoon" conditions over India-role of ocean-atmosphere processes in the Indian Ocean. IEEE Geosci Rem Sens Lett 6:332–336. https://doi.org/10.1109/LGRS.2009.2013366

Rao SA, Choudhari H, Pokhrel S, Goswami BN (2010) Unusual central Indian drought of summer monsoon 2008: role of southern tropical Indian Ocean warming. J Clim 23:5164–5173. https://doi.org/10.1175/2010JCLI3257.1

Rao KK, Patwardhan SK, Kulkarni A, Kamala K, Sabade SS, Krishna Kumar K (2014) Projected changes in mean and extreme precipitation indices over India using PRECIS. Global Planet Change 113:77–90

Rasmusson EM, Carpenter TH (1983) The relationship between eastern equatorial Pacific SSTs and rainfall over India and Sri Lanka. Mon Wea Rev 111:517–528. https://doi.org/10.1175/1520-0493(1983)111%3c0517:TRBEEP%3e2.0.CO;2

Rodwell MJ (1997) Breaks in the Asian monsoon: the influence of Southern Hemisphere weather systems. J Atmos Sci 54:2597–2611. https://doi.org/10.1175/15200469(1997)054%3c2597:BITAMT%3e2.0.CO;2

Roja Raman M, Jagannadha Rao V, Venkat Ratnam M, Rajeevan M, Rao S, Narayana Rao D, Prabhakara Rao N (2009) Characteristics of the Tropical Easterly Jet: long-term trends and their features during active and break monsoon phases. J Geophy Res 114:D19. https://doi.org/10.1029/2009JD012065

Romatschke U, Houze RA Jr (2011) Characteristics of precipitating convective systems in the South Asian monsoon. J Hydrometeorol 12:3–26. https://doi.org/10.1175/2010JHM1289.1

Roxy M (2015) Sensitivity of precipitation to sea surface temperature over the tropical summer monsoon region—and its quantification. Clim Dyn 43(5–6):1159–1169

Roxy MK, Ghosh S, Pathak A, Athulya R, Mujumdar M, Raghu M, Pascal T, Rajeevan M (2017) A threefold rise in widespread extreme rain events over central India. Nat Commun 8:708. https://doi.org/10.1038/s41467-017-00744-9

Roxy MK, Kapoor R, Terray P, Murtugudde R, Ashok K, Goswami BN (2015) Drying of Indian subcontinent by rapid Indian Ocean warming and a weakening land-sea thermal gradient. Nat Commun 6:7423. https://doi.org/10.1038/ncomms8423

Roxy MK, Modi A, Murtugudde R, Valsala V, Panickal S, Prasanna Kumar S, Ravichandran M (2016) A reduction in marine primary productivity driven by rapid warming over the tropical Indian Ocean. Geophys Res Lett 43:826–833. https://doi.org/10.1002/2015gl066979

Roxy MK, Ritika K, Terray P, Masson S (2014) The curious case of Indian Ocean warming. J Clim 27:8501–8509. https://doi.org/10.1175/JCLI-D-14-00471.1

Sabade S, Kulkarni A, Kripalani R (2011) Projected changes in South Asian summer monsoon by multi-model global warming experiments. Theor Appl Climatol 103(3–4):543–565. https://doi.org/10.1007/s00704-010-0296-5

Sabeerali CT, Ajayamohan RS (2018) On the shortening of Indian summer monsoon season in a warming scenario. Clim Dyn 50(5–6):1609–1624. https://doi.org/10.1007/s00382-017-3709-7

Sabeerali CT, Rao SA, George G, Rao DN, Mahapatra S, Kulkarni A, Murtugudde R (2015) Modulation of monsoon intraseasonal oscillations in the recent warming period. J Geophys Res Atmos 119:5185–5203. https://doi.org/10.1002/2013JD021261

Sabin TP, Ghattas J, Denvil S, Dufresne JL, Hourdin F, Pascal T (2013) High resolution simulation of the South Asian monsoon using a variable resolution global climate model. Clim Dyn 41(1):173–194. https://doi.org/10.1007/s00382-012-1658-8

Saha A, Ghosh S, Sahana AS, Rao EP (2014) Failure of CMIP5 climate models in simulating post-1950 decreasing trend of Indian monsoon. Geophys Res Lett 41:7323–7330. https://doi.org/10.1002/2014GL061573

Sahana A, Ghosh S, Ganguly A, Murtugudde R (2015) Shift in Indian summer monsoon onset during 1976/1977. Environ Res Lett 10 (5):054006. https://doi.org/10.1088/1748-9326/10/5/054006

Saji NH, Goswami BN, Vinayachandran PN, Yamagata T (1999) A dipole mode in the tropical Indian Ocean. Nature 401:360–363. https://doi.org/10.1038/43854

Saji NH, Yamagata T (2003) Possible impacts of Indian Ocean dipole mode events on global climate. Clim Res 25(2):151–169. https://doi.org/10.3354/cr025151

Salzmann M, Cherian R (2015) On the enhancement of the Indian summer monsoon drying by Pacific multidecadal variability during the latter half of the twentieth century. J Geophys Res 120 (18):9103–9118. https://doi.org/10.1002/2015JD023313

Samuelsson P, Jones CG, Willén U et al (2011) The Rossby centre regional climate model RCA3: model description and performance. Tellus A Dyn Meteorol Oceanogr 63:4–23. https://doi.org/10.1111/j.1600-0870.2010.00478.x

Sankar S, Svendsen L, Gokulapalan B, Joseph PV, Johannessen OM (2016) The relationship between Indian summer monsoon rainfall and Atlantic multidecadal variability over the last 500 years. Tellus A 68:1–14. https://doi.org/10.3402/tellusa.v68.31717

Sanyal P, Sinha R (2010) Evolution of the Indian summer monsoon: synthesis of continental records. Geol Soc Lond Spec Publ 342:153–183. https://doi.org/10.1144/SP342.11

Sarita A, Rajeevan M (2016) Possible shift in the ENSO-Indian monsoon rainfall relationship under future global warming. Sci Rep 6(1):20145. https://doi.org/10.1038/srep20145

Sathiyamoorthy V (2005) Large scale reduction in the size of the Tropical Easterly Jet. Geophys Res Lett 32:L14802. https://doi.org/10.1029/2005GL022956

Sharmila S, Joseph S, Sahai AK, Abhilash S, Chattopadhyay R (2015) Future projection of Indian summer monsoon variability under climate change scenario: an assessment from CMIP5 climate models. Glob Planet Chang 124:62–78. http://dx.doi.org/10.1016/j.gloplacha.2014.11.004

Shi F, Fang KY, Xu C, Guo ZT, Borgaonkar HP (2017) Interannual to centennial variability of the South Asian summer monsoon over the past millennium. Clim Dyn 49:2803–2814

Shukla J (1987) Interannual variability of monsoons. In: Lighthill J, Pearce RP (eds) Monsoon dynamic. Cambridge University Press, Cambridge, pp 399–463

Sikka DR (1977) Some aspects of the life history, structure and movement of monsoon depressions. Pure appl Geophys 115:1501–1529. https://doi.org/10.1007/BF00874421

Sikka DR (1980) Some aspects of the large scale fluctuations of summer monsoon rainfall over India in relation to fluctuations in the planetary and regional scale circulation parameters. Earth Planet Sci 89:179–195. https://doi.org/10.1007/BF02913749

Sikka DR, Gadgil S (1980) On the maximum cloud zone and the ITCZ over India longitude during the southwest monsoon. Mon Wea Rev 108:1840–1853. https://doi.org/10.1175/1520-0493(1980)108%3c1840:OTMCZA%3e2.0.CO;2

Sillmann J, Kharin VV, Zwiers FW, Zhang X, Bronaugh D (2013) Climate extremes indices in the CMIP5 multimodel ensemble: part 2. Future climate projections. J Geophys Res Atmos 118:2473–2493. https://doi.org/10.1002/jgrd.50188

Singh D, Tsiang M, Rajaratnam B, Diffenbaugh NS (2014) Observed changes in extreme wet and dry spells during the South Asian summer monsoon season. Nat Clim Change 4:456–461. https://doi.org/10.1038/nclimate2208

Singh N, Sontakke NA (1999) On the variability and prediction of post-monsoon rainfall over India. Int J Climatol 19:309–339. https://doi.org/10.1002/(SICI)1097-0088(19990315)19:3%3c309:AID-JOC361%3e3.0.CO;2

Singh SV, Kripalani RH, Sikka DR (1992) Interannual variability of the Madden–Julian oscillations in the Indian summer monsoon rainfall. J. Climate 5:973–978

Singh R, AchutaRao K (2018) Quantifying uncertainty in twenty-first century climate change over India. Clim Dyn 52(7–8):3905–3928. https://doi.org/10.1007/s00382-018-4361-6

Sinha A, Stott LD, Berkelhammer M, Cheng H, Edwards RL, Aldenderfer M, Mudelsee M, Buckley B (2011) A global context for monsoon mega droughts during the past millennium. Quat Sci Rev 30:47–62. https://doi.org/10.1016/j.quascirev.2010.10.005

Sinha, N (2018) Isotopic studies of rainfall and its reconstruction using speleothems from the Indian subcontinent. Ph.D. Thesis (unpublished) Savitribai Phule Pune University, Pune, pp 146. https://figshare.com/s/7a4296d67493f07c32d5

Sinha, N, Gandhi N, Chakraborty S, Krishnan R, Yadava MG, Ramesh R (2018) Abrupt climate change at ~2800 year BP evidenced by a stalagmite record from the peninsular India. Holocene 28(11):1720–1730. https://doi.org/10.1177/0959683618788647

Soman MK, Kumar K (1993) Space-time evolution of meteorological features associated with the onset of Indian summer monsoon. Mon Wea Rev 121(4):1177–1194. https://doi.org/10.1175/1520-0493(1993)121%3c1177:STEOMF%3e2.0.CO;2

Sreekala P, Rao S, Rajeevan M (2011) Northeast monsoon rainfall variability over south peninsular India and its teleconnections. Theoret Appl Climatol 108(1–2):73–83

Srinivas G, Chowdary JS, Kosaka Y, Gnanaseelan C, Parekh A, Prasad K (2018) Influence of the Pacific-Japan pattern on Indian summer monsoon rainfall. J Clim 31(10):3943–3958. https://doi.org/10.1175/JCLI-D-17-0408.1

Taniguchi K, Koike T (2006) Comparison of definitions of Indian summer monsoon onset: better representation of rapid transitions of atmospheric conditions. Geophys Res Lett 33:L02709. https://doi.org/10.1029/2005GL024526

Tawde SA, Singh C (2015) Investigation of orographic features influencing spatial distribution of rainfall over the Western Ghats of India using satellite data. Int J Climatol 35(9):2280–2293. https://doi.org/10.1002/joc.4146

Terray L, Boe J (2013) Quantifying 21st-century France climate change and related uncertainties. CR Geosci 345:136–149. https://doi.org/10.1016/j.crte.2013.02.003

Trenberth K (1998) Atmospheric moisture residence times and cycling: implications for rainfall rates and climate change. Clim Change 39:667–694. https://doi.org/10.1023/A:1005319109110

Turner AG, Hannachi A (2010) Is there regime behavior in monsoon convection in the late 20th century? Geophys Res Lett 37:L16706. https://doi.org/10.1029/2010GL044159

Turner AG, Slingo JM (2009) Subseasonal extremes of precipitation and active-break cycles of the Indian summer monsoon in a climate change scenario. Q J R Meteorol Soc 135:549–567. https://doi.org/10.1002/qj.401

Ueda H, Iwai A, Kuwako K, Hori ME (2006) Impact of anthropogenic forcing on the Asian summer monsoon as simulated by eight GCMs. Geophys Res Lett 33:L06703. https://doi.org/10.1029/2005GL025336

Undorf S, Polson D, Bollasina MA, Ming Y, Schurer A, Hegerl GC (2018) Detectable impact of local and remote anthropogenic aerosols on the 20th century changes of West African and South Asian monsoon precipitation. J Geophys Res 123(10):4871–4889. https://doi.org/10.1029/2017JD027711

Venkat Ratnam M, Krishna Murthy B, Jayaraman A (2013) Is the trend in TEJ reversing over the Indian subcontinent? Geophys Res Lett 40(13):3446–3449. https://doi.org/10.1002/grl.50519

Waliser D (2006) Intraseasonal variability. In: Wang B (ed) Asian monsoons. Springer, Chichester, pp 203–257

Waliser DE, Lau KM, Stern W (2003) Potential predictability of the Madden-Julian Oscillation. Bull Am Meteor Soc 84:33–50. https://doi.org/10.1175/BAMS-84-1-33

Wang B, Clemens SC, Liu P (2003) Contrasting the Indian and East Asian monsoons: implications on geologic timescales. Mar Geol 201:5–21. https://doi.org/10.1016/S0025-3227(03)00196-8

Wang B, Ding Q, Joseph PV (2009) Objective definition of the Indian summer monsoon onset. J Clim 22(12):3303–3316. https://doi.org/10.1175/2008JCLI2675.1

Wang B, Lin H (2002) Rainy season of the Asian-Pacific summer monsoon. J Climate 15:386–398. https://doi.org/10.1175/1520-0442(2002)015%3c0386:RSOTAP%3e2.0.CO;2

Wang B, Webster PJ, Teng HH (2005) Antecedents and self-induction of active–break south Asian monsoon unraveled by satellites. Geophys Res Lett 32:GL020996. https://doi.org/10.1029/2004

Wang B, Wu R, Lau KM (2001) Interannual variability of Asian summer monsoon: contrasts between the Indian and western North Pacific-East Asian monsoons. J Clim 14:4073–4090. https://doi.org/10.1175/1520-0442(2001)014%3c4073:IVOTAS%3e2.0.CO;2

Webster PJ, Magana VO, Palmer TN, Shukla J, Tomas RA, Yanai M, Yasunari T (1998) Monsoons: processes, predictability, and the prospects for prediction. J Geophy Res 103(14):451–510. https://doi.org/10.1029/97JC02719

Xu C, Zheng H, Nakatsuka T, Sano M (2013) Oxygen isotope signatures preserved in tree-ring cellulose as a proxy for April–September precipitation in Fujian, the subtropical region of southeast China. J Geophys Res 118:12805–12815

Yadav RK (2013) Emerging role of Indian ocean on Indian northeast monsoon. Clim Dyn 41:105–116. https://doi.org/10.1007/s00382-012-1637-0

Yadav RK (2017) On the relationship between east equatorial Atlantic SST and ISM through Eurasian wave. Clim Dyn 48:281–295. https://doi.org/10.1007/s00382-016-3074-y

Yasunari T (1979) Cloudiness fluctuations associated with the Northern Hemisphere summer monsoon. J Meteorol Soc Jpn 57:227–242

Zuo Z, Yang S, Kumar A, Zhang R, Xue Y, Jha B (2012) Role of thermal condition over Asia in the weakening Asian summer monsoon under global warming background. J Clim 25(9):3431–3436

Coordinating Lead Authors

Supriyo Chakraborty, Indian Institute of Tropical Meteorology (IITM-MoES), Pune, India,
e-mail: supriyo@tropmet.res.in (corresponding author)
Yogesh K. Tiwari, Indian Institute of Tropical Meteorology (IITM-MoES), Pune, India

Lead Authors

Pramit Kumar Deb Burman, Indian Institute of Tropical Meteorology (IITM-MoES), Pune, India
Somnath Baidya Roy, Centre for Atmospheric Sciences, Indian Institute of Technology Delhi, India
Vinu Valsala, Indian Institute of Tropical Meteorology (IITM-MoES), Pune, India

Contributing Authors

Smrati Gupta, Indian Institute of Tropical Meteorology (IITM-MoES), Pune, India
Abirlal Metya, Indian Institute of Tropical Meteorology (IITM-MoES), Pune, India
Shilpa Gahlot, Centre for Atmospheric Sciences, Indian Institute of Technology Delhi, India

Review Editors

V. K. Dadhwal, Indian Institute of Space Science and Technology, Thiruvananthapuram, India
Govindsamy Bala, Centre for Atmospheric and Oceanic Sciences, Indian Institute of Science, Bengaluru, India
Shamil Maksyutov, National Institute for Environmental Studies, Tsukuba, Japan
Prabir Patra, Japan Agency for Marine-Earth Science and Technology, Yokohama, Japan
P. Mahesh, National Remote Sensing Center, Indian Space Research Organization, Hyderabad, India

Corresponding Author

Supriyo Chakraborty, Indian Institute of Tropical Meteorology (IITM-MoES), Pune, India,
e-mail: supriyo@tropmet.res.in

The original version of this chapter was revised. On page 88, the last sentence has been changed. The correction to this chapter is available at
https://doi.org/10.1007/978-981-15-4327-2_13

Key Messages

- The surface CO_2 concentration observed at Sinhagad site, located in the western part of India, shows higher seasonal cycle amplitude as compared to the observations at Mauna Loa in the Pacific region (Fig. 4.1). The higher amplitude is caused by strong local–regional biospheric activity.

- The surface CO_2 concentration amplitude in the western Indian region is increasing with time, likely driven by the enhanced biospheric activities as well as the changes in nearby oceanic fluxes. To ascertain the driving mechanism behind this as well as long-term variability at country scale, a strategically designed network of long-term surface CO_2 concentration and associated flux observations is essential over India.

- Recent studies using flux tower measurements show that the carbon fluxes in Indian forests vary widely across the ecosystems. The Kaziranga forest in Northeast India sequesters maximum carbon during pre-monsoon season, whereas the forests in Haldwani and Barkot in northern India sequester maximum carbon during the summer monsoon season. The forests in Betul in central India, mangroves in Sundarbans in east India and forests in Kosi-Katarmal in north India sequester maximum carbon during post-monsoon, whereas mangroves in Pichavaram in the east coast of south India sequester maximum carbon during the winter season (Fig. 4.5).

- Satellite-derived vegetation indices indicate increasing vegetation productivity over India during recent decades.

- Modeling studies show that even though the Indian terrestrial ecosystem has not historically been a strong source or sink of carbon, it is behaving as a carbon sink since the 1980s. The terrestrial carbon sink is maintained primarily by the carbon fertilization effect aided to some extent by forest conservation, management, and reforestation policies in recent decades (Sect. 4.4.1).

- Surface GHGs measurement sites in India are sparse in nature. In the absence of long-term observational records and a comprehensive modeling of biogeochemical processes, there is a limited understanding of the dynamics of GHGs variability in India. Expansion of observational network as well as development of process-based biogeochemical and coupled climate–carbon models may fill this knowledge gap. Such expanded capabilities would help improve the assessments of the mitigation potential of Indian ecosystem in the future (Sect. 4.5).

Box 4.1 Preamble

In order to study the effect of primary drivers of climate change, observational data of at least a few decades ("climate timescale") are required. Such knowledge is also useful for a meaningful interpretation of a reliable estimate of future projection of the drivers. The primary drivers of climate change in the industrial era are the greenhouse gases (GHGs) which absorb the terrestrial radiation strongly. It is now well established that increase in the GHGs concentrations during the past century has been due to the anthropogenic activities, such as the fossil fuel consumption and land use land cover changes. The largest instrumental data of CO_2 concentration, reported as dry-air mole fraction, are available since the international geophysical year (1957–1958) from South Pole and Mauna Loa, Hawaii. The latter is considered as a reference to global-mean CO_2 concentration. In India however, the longest CO_2 record is available only for a couple of decades. Thus, most studies are limited to

Fig. 4.1 Seasonal variation of CO_2 mixing ratio at two tropical sites. The amplitude of CO_2 mixing ratio variability at Sinhagad, India (blue line), is much higher than that observed at a comparable latitudinal range (18.3–19.4° N), Mauna Loa in the Pacific region (red line). The data were obtained from Tiwari et al. (2014) re-plotted and extended till 2015

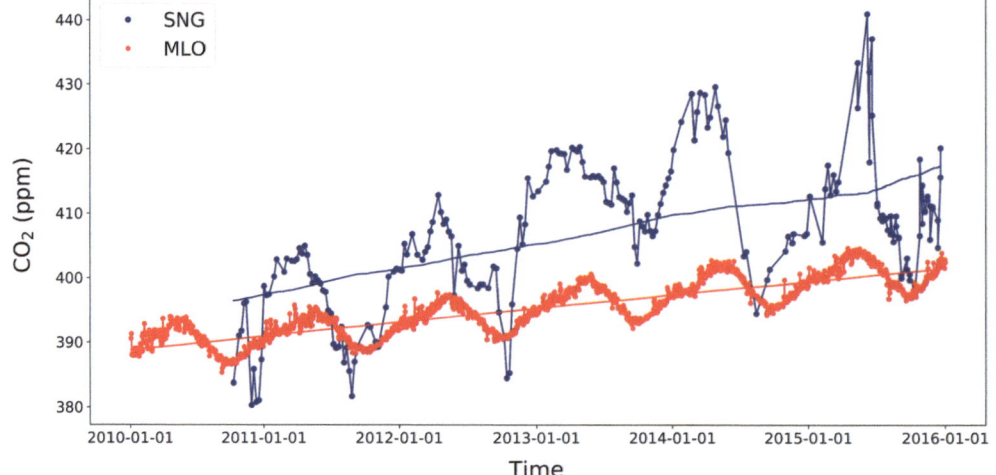

short-term changes that span over diurnal to inter-annual timescale. The lack of robust observational records is a severe constraint to make a reliable assessment of the trends in GHGs fluxes. In this chapter, we provide an overview of the measurement of a few GHGs concentration, how carbon is sequestered in natural ecosystems as determined by the eddy covariance technique and some model-based results for India. The chapter provides a summary assessment of GHGs (mainly CO_2, CH_4 and N_2O)-related research in India. Other trace gases, such as ozone, SO_2 and CFCs, have been discussed in Chap. 5. A brief discussion on the upper ocean carbon cycle in the Indian Ocean has been presented in Chap. 10.

4.1 Introduction

There is a general scientific consensus that radiative processes associated with increasing concentrations of greenhouse gases (GHGs) and related feedback processes are responsible for the global warming and climate change (IPCC 2013). The important GHGs in the earth's environment are carbon dioxide (CO_2), methane (CH_4), nitrous oxide (N_2O), halocarbons and ozone in the lower atmosphere. The concentrations of CO_2, CH_4 and N_2O have significantly increased since the beginning of the industrial era (Ciais et al. 2013). Among the GHGs of anthropogenic origin, the increase of atmospheric carbon is of primary concern because CO_2 has a long lifetime in the atmosphere (~ 100 years). The concentration of atmospheric CO_2 was ca. 280 ppm during the pre-industrial period. This has exceeded 400 ppm in recent time (https://www.esrl.noaa.gov/gmd/ccgg/trends). The average growth rate has been estimated at 2.11 ppm per year during the last decade, and a value of 410 ppm was observed in 2018 (https://www.esrl.noaa.gov/gmd). However, the CO_2 mixing ratio shows considerable variability on local to regional scale. For example, the amplitude of CO_2 variations observed in the west peninsular India (~ 25 ppm) was much higher than that observed in the Mauna Loa observatory in the Pacific region (~ 6 ppm) (see Fig. 4.1), though both the sites are situated within a narrow latitude band (Sinhagad: $18.35°$ N; Mauna Loa: $19.4°$ N). The measurement in India is likely affected by regional terrestrial biospheric processes and marine sources (Tiwari et al. 2014).

The concentration of CH_4 has increased from 700 to 1857 ppb and that of N_2O from 270 to 321 ppb since pre-industrial era (IPCC 2007). Although the concentrations of CH_4 and N_2O are small compared to that of CO_2, their global warming potential (GWP) in terms of radiative forcing is several times (28 and 265 for CH_4 and N_2O, respectively, to 100-year time horizon) higher than that of an equivalent amount of CO_2. Residence time of CO_2 in the atmosphere is ~ 100 years, whereas that of CH_4 and N_2O is about 10 years and 120 years, respectively. The increase in CH_4 concentration is mainly dominated by agricultural and animal husbandry operations, and that of N_2O is due to agricultural practices (Liu et al. 2019).

4.2 Observational Aspects of GHGs Research in India

In India, there are very few research organizations involved in the observational aspects of GHGs research with a long-term perspective. One of the earliest attempts was made by a group at the National Institute of Oceanography (NIO) in Goa and Physical Research Laboratory (PRL) in Ahmedabad, which took a major initiative in measuring GHGs at a coastal site in western India, Cabo de Rama, also known as Cape Rama (acronym CRI: $15.08°$ N, $73.83°$ E, 50 m ASL) in association with the Commonwealth Scientific and Industrial Research Organisation (CSIRO), Australia (Bhattacharya et al. 2009). The CRI site is located closer to the shoreline, free from any major vegetation and away from habitation (Tiwari et al. 2011). Routine measurement of the concentrations of CO_2, CH_4, CO, N_2O and H_2 was carried out at the bimonthly time intervals. Additionally, carbon ($\delta^{13}C$) and oxygen ($\delta^{18}O$) isotopic ratios of CO_2 were also measured. The observational record is available for about a decade (1992–2002) and also for a period of 2009–2013.

In the next phase of the carbon cycle study in India, the Indian Space Research Organisation (ISRO) started the National Carbon Project (NCP) under the auspices of the ISRO Geosphere-Biosphere Program (IGBP) in the early phase of the 2010s. The NCP endeavors to understand the GHGs dynamics and estimate their budget by means of a robust observational network across the country (Chanda et al. 2013, 2014; Sharma et al. 2013, 2014; Patel et al. 2011; Mahesh et al. 2014, 2016, 2019; Sreenivas et al. 2016, 2019; Jha et al. 2013, 2014; Rodda et al. 2016). At about the same time, another GHGs observational program (CO_2 and CH_4) at a semi-urban hilly site near Pune, called Sinhagad (SNG), was initiated by the Indian Institute of Tropical Meteorology, Pune, under the patronage of the Ministry of Earth Sciences (MoES), Government of India. The SNG site is 200 km east of the Arabian Sea ($73.75°$ E, $18.35°$ N, 1600 m ASL) situated over the Western Ghats mountainous terrain of southwestern peninsular region of the Indian subcontinent. The location is relatively free from major vegetation and local habitation disturbances for GHGs measurement on a long-term perspective. Routine air sampling at SNG, from a 10 m meteorological tower at weekly intervals, has been operational since November 2009 (Tiwari et al. 2014). The main objectives were (a) to investigate the GHGs transport mechanisms pertaining to the Indian subcontinent, (b) to understand the causes behind the seasonality of atmospheric CO_2 concentrations and (c) to develop a modeling framework (forward and inverse) for estimations of carbon sources and sinks over the Indian domain.

Among other studies, significant effort was made by the CSIR Fourth Paradigm Institute, Bengaluru (formerly CSIR Centre for Mathematical Modeling and Computer Simulation), in collaboration with a few other national and international research organizations, such as Laboratoire des sciences du climat et de l'environnement (LSCE) in Paris, France; National Institute of Ocean Technology in Chennai and Port Blair; and Indian Institute of Astrophysics, Bengaluru. Scientists from these institutes reported GHGs monitoring results at three sites Hanle (32.78° N, 78.96° E, 4500 m ASL), Pondicherry (11.91° N, 79.81° E, 6 m ASL) and Port Blair (11.62° N, 92.72° E, 17 m ASL) during the period 2007–2011 (Lin et al. 2015). Hanle is a high-altitude site and is considered as northern hemispheric midlatitude representative location. Pondicherry and Port Blair are coastal and oceanic sites, respectively, located at similar latitudes. Measurements were based on weekly flask sampling, and the flask samples were analyzed at the LSCE, France, under the Indo-French collaborative project. The observed concentration of CH_4 at Hanle peaked during the summer monsoon months, in contrast to the seasonal cycle at

other sites including Pondicherry and Port Blair. High CH_4 concentration during monsoon could be due to enhanced biogenic emissions from wetlands and rice fields.

Sreenivas et al. (2016) presented CO_2 and CH_4 measurements over Shadnagar (17.03° N, 78°18 E) at Telangana India. But unlike the Hanle observations, these authors found minimum CH_4 during Indian summer monsoon and maximum during post-monsoon, indicating spatial heterogeneity in methane emissions and hence concentrations across the country. Maximum CO_2 occurred during pre-monsoon in conformity with the other observations, such as SNG (Tiwari et al. 2014; Metya et al. 2020). The Physical Research Laboratory in Ahmedabad has also made significant contributions in this regard, especially in analyzing the GHGs dynamics in an urban environment of India. For example, Chandra et al. (2019) observed low concentration of CO_2 at Ahmedabad during the monsoon season in conformity with the result obtained by Tiwari et al. (2014) for the Sinhagad station.

4.2.1 Carbon Dioxide (CO_2)

Tiwari et al. (2014) presented a detailed account of the SNG record. According to these authors, the regional source contributions at SNG and also at CRI arise from the horizontal flow within the planetary boundary layer. Greater

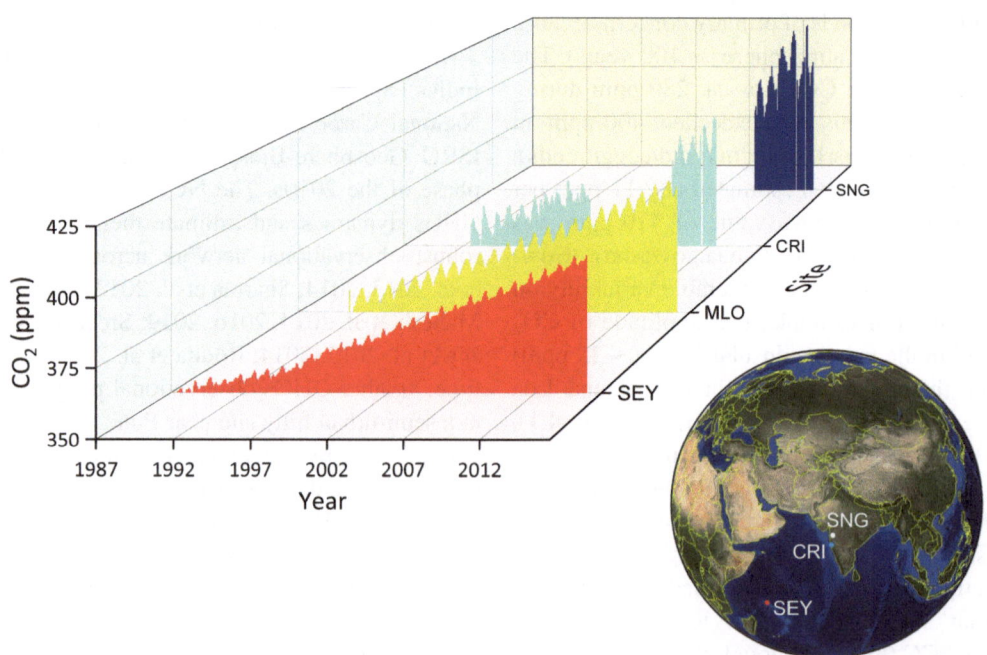

Fig. 4.2 Long-term records of CO_2 (ppm) concentration variability observed in the western peninsular region of India and selected global background locations. The Indian observing sites Sinhagad (SNG) and Cape Rama (CRI) CO_2 records are shown in blue and cyan bars, respectively. The CO_2 record of an Indian Ocean island site Seychelles (SEY) is shown in red, and the Mauna Loa (MLO) CO_2 record is shown in yellow. These are to compare the Indian CO_2 record with that of the global "background" record. The three sites (SNG, CRI, SEY) are also shown separately on the globe (globe credit: Google Earth)

CO_2 variability, >5 ppm, is observed during winter, while it is reduced nearly by half during the summer (Tiwari et al. 2014). Unlike the long-term Mauna Loa observational records, the Indian GHGs record, as mentioned earlier, is short. Bose et al. (2014) developed a fractionation model and used it to reconstruct the CO_2 variability using carbon isotopic analysis of tree rings from a western Himalayan region. These authors demonstrated that the tree ring-derived CO_2 record for the last one hundred year had a good match with the ice core records of CO_2 variability. Considering these observational records, we attempt to make a composite diagram of the CRI and the SNG CO_2 records to examine the regional features of the CO_2 variability in the Indian context. Figure 4.2 shows the CO_2 concentration variability approximately for the timescale of 1992–2013 with an intermittent gap from October 2002 to July 2009.

The CRI record is shown in cyan, while the blue bars represent the variability for the SNG site. The CRI and the SNG data show close resemblance, though a small difference is apparent. The CO_2 amplitude in the case of the SNG site is slightly higher than the CRI site. In case of CRI, the sampling was done once in two months, whereas in case of SNG, the sampling was done on weekly time interval. Since samples collected at lower temporal resolution impart a smoothing effect, the variability is expected to be subdued in the case of the CRI site compared to the SNG site which had a sampling frequency of about eight times higher than the CRI site. Nevertheless, we may assume the combined dataset as a representative of the CO_2 for this region and since there is no other known similar observational dataset available from the Indian region, we may call it as the "Indian CO_2 record."

When this record is compared with the Mauna Loa CO_2 (MLO) variability (see Fig. 4.2), one distinct feature is apparent. The Indian record shows higher amplitude, which has also been discussed earlier and illustrated in Fig. 4.1. In the earliest record, during the 1990s and 2000s, the CRI CO_2 data also show slightly larger amplitude than the MLO record. But, in the second phase of measurement during the early 2010s, this amplitude in the case of the Indian record shows an increasing trend. The amplitudes are larger than that observed during the 1990s and 2000s, and much larger than the Mauna Loa record. As mentioned earlier, Bhattacharya et al. (2009) also measured $\delta^{13}C$ of CO_2 during 1993–2002 and observed a strong inverse correlation between CO_2 and $\delta^{13}C$. This anti-correlation suggests a land biospheric control rather than the oceanic modulation on the seasonal behavior of CO_2 in this region. But according to a recent study, high-frequency (1 Hz) measurements of CO_2 and CH_4 concentrations using a laser-based GHG analyzer at SNG reveal that the oceanic emission of CO_2 and/or background CO_2 transported from the distant marine environment may also be playing a moderate role in determining the

seasonal pattern of CO_2 in this region (Metya et al. 2020). Though the marine influence needs to be further investigated to ascertain its role, the increasing amplitude may be attributed to enhanced land biospheric activities. It is important to mention that Barlow et al. (2015) observed similar behavior at a high latitude region, namely Barrow (71.3° N, 156.6° W) in Alaska. They argue that the observed change in amplitude is partially due to an increase in "peak respiration" (contributes to increase the maxima in CO_2 value) and a larger increase in "peak uptake" (contributes to extend the minima in CO_2 value). Since respiration and photosynthetic uptake are by and large driven by vegetation, an increased vegetation cover seems to be the most likely reason for increased CO_2 amplitude. One of the reasons could be that the vegetation and/or the forest cover are probably increasing, at least in the western part of India. In the last twenty years, the surface characteristics may have changed significantly, and interestingly, the forest cover has increased in India (MoEFCC 2015, 2018). On the contrary, some studies show that the Indian forest cover has been decreasing, at least for the last couple of decades (Meiyappan et al. 2017). Needless to say that an accurate estimate of Indian forest cover and its characteristic behavior for different climatic zones/geographical locations in terms of its carbon sequestration potential are essential in order to better understand the carbon dynamics. For example, the Northeast Indian region is characterized by high forest cover (ca. 64%; Jain et al. 2013) but its carbon dynamical characteristics are quite different from the rest of the country. Firstly, it shows a very different characteristic in terms of the timing of the maximum CO_2 uptake on a seasonal scale compared to the rest of India (discussed later), and secondly it shows a large CO_2 variability. Deb Burman et al. (2017) reported large amplitude of the CO_2 concentration in a deciduous forest at Kaziranga in Assam (380–460 ppm) during the year of 2016. Though one year of data may not be generalized as representative values, it may be indicative of distinct spatial characteristics of CO_2 seasonal pattern of different ecosystems in India. Analysis of such records provides useful information on the amplitude and phase of the time series by means of spectral decomposition technique (Barlow et al. 2015). Additionally, tree-ring based measurement of radiocarbon activity of the atmospheric air on sub-seasonal timescale would provide information about the fossil fuel component (Chakraborty et al. 2008) of the CO_2 emission and may help to partition the CO_2 emitted by the agricultural practices (Berhanu et al. 2017).

4.2.2 Methane (CH_4)

Methane is the second most important anthropogenic GHG after atmospheric CO_2 (on the scale of radiative forcing, which is about +0.48 Wm^{-2}; Stocker et al. 2013). The pre-industrial CH_4 concentration has been estimated as 700 ppb, but increased anthropogenic activities, such as the agricultural practices and industrial activities, have resulted in a steady increase of atmospheric CH_4, up to 1803 ppb in 2011 (Dlugokencky et al. 2011; Etheridge et al. 1998). It also plays an active role in tropospheric chemistry. CH_4 is mainly lost by reaction with the OH radical, and hence it directly contributes to stratospheric water vapor increase (Keppler et al. 2006). CH_4 emissions from the anthropogenic sources in India have grown from 18.85 to 20.56 Tg yr^{-1} during 1985–2008 (Garg et al. 2011). Patra et al. (2013) made a comprehensive analysis to estimate CH_4 emission from the entire South Asian region which turned out to be 37 ± 3.7 TgC-CH_4 yr^{-1} during the 2000s. Unlike CO_2, CH_4 has a relatively short lifetime of 9–10 years. Shorter lifetime, compared to CO_2, N_2O or CFCs, is potentially useful for its sources and sinks to achieve a steady state or even a decline, thus reducing its impact on global climate change. A quasi-steady state was observed in the late 1990s and the first half of the 2000s (Dlugokencky et al. 2011). But increased emission of CH_4, driven by the anthropogenic activities, later perturbed the equilibrium state (Rigby et al. 2008; Patra et al. 2016). The source and sink mechanism of CH_4 is complex, and several components of its budget remain poorly constrained. Surface observation of CH_4 in India is limited to a very few places, and those too provide only sporadic data; long time measurement is mostly absent. The CRI data provide one of the first long-term measurements of CH_4 in India. Figure 4.3 shows the time series of

the methane concentration at CRI (1993–2002; in cyan) in comparison with the MLO (in yellow) and the Seychelles site (SEY, an island in the equatorial Indian Ocean; in red). The data were obtained from the published results of Bhattacharya et al. (2009). It is noted that the seasonality in CH_4 is much stronger at CRI than at MLO and SEY as shown in the inset. Another feature is that the CH_4 mixing ratios at MLO were higher compared to the CRI values during the SW monsoon season (July–August; Fig. 4.3-inset). This provides a strong evidence of seasonality in CH_4 fluxes in the Indian Ocean sector than in the Pacific (Bhattacharya et al. 2009; Patra et al. 2009). Analysis of the high-frequency data at the Sinhagad site also revealed a strong correlation between the CO_2 and CH_4 mixing ratios during the monsoon season but weak correlation in other seasons (Metya et al. 2020), indicating that the concentrations of both these gases during the monsoon season are also controlled by the monsoon circulation that originates in the Indian Ocean.

Methane shows distinct diurnal- and seasonal-scale variations (Sreenivas et al. 2016), which, unlike CO_2, is characterized by a large variation on spatial domain. The seasonality of CH_4 is found to be varying differently over different parts of India due to the complex interaction between the surface emissions and monsoonal transport patterns (Patra et al. 2009). For example, the eastern Himalayan station Darjeeling (27.03° N, 88.26° E, 2000 m ASL) captures episodes of increased CH_4 concentrations throughout the year (Ganesan et al. 2013), but a northwestern Himalayan station Hanle experiences high values during summer monsoon season (Lin et al. 2015).

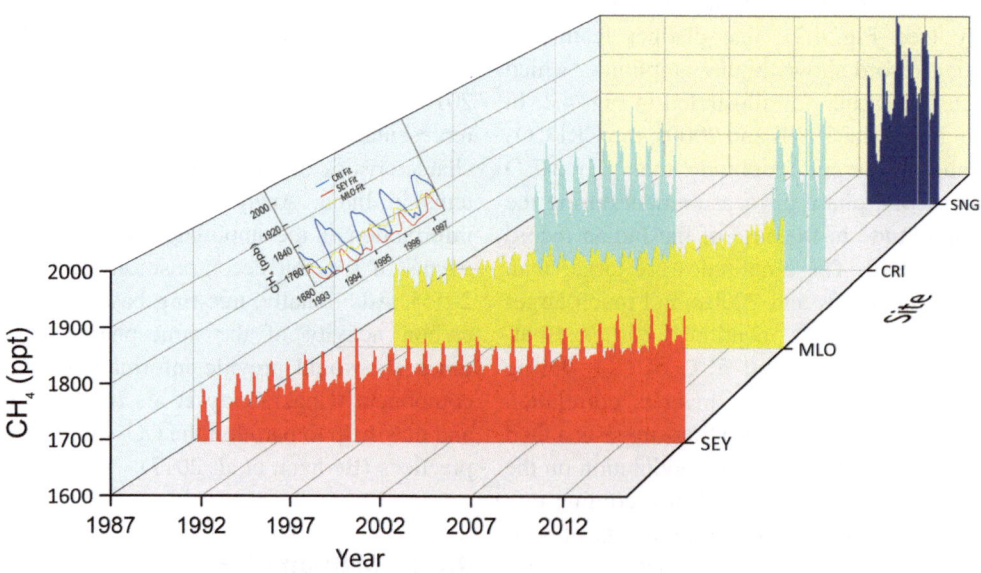

Fig. 4.3 CH_4 mixing ratio in air samples observed at SNG (deep blue) and CRI (since 1993, cyan). The Mauna Loa (yellow) and the SEY data (red) have also been shown for comparison. The inset shows a zoomed version of the CRI data emphasizing the seasonal cycles. The data are available in Bhattacharya et al. (2009) and redrawn with modification

Darjeeling experiences highest CH$_4$ values during October–November, while Pondicherry at the southeastern coastal site and Port Blair in Andaman Islands register comparatively lower values (Lin et al. 2015). A dense observational network is required for understanding the spatial and temporal variations of CH$_4$ over India, in particular the central Indian region which is characterized by sparse and intermittent observational network. Chandra et al. (2017) studied the variability of column dry-air mole fractions of methane (XCH$_4$) over India using GOSAT satellite retrievals. The satellite observation of the XCH$_4$ is an integrated measure of CH$_4$ densities at all altitudes from the surface to the top of the atmosphere. The Indian region was divided into eight subregions, and relationship between XCH$_4$ variability from surface to the upper troposphere with the surface emissions was discussed. More often, the XCH$_4$ variabilities are strongly linked with the transport of air mass from outside the domain of interests and monsoon anticyclone in the middle–upper troposphere. Variations of XCH$_4$ are controlled by both surface emissions and atmospheric transport largely driven by the monsoonal dynamics. Various observational techniques have been used to make estimates of top-down CH$_4$ emissions in India (Schuck et al. 2010; Ganesan et al. 2013; Parker et al. 2015). It was shown that there is a little growth in the CH$_4$ emissions in India during the period 2010–2015. The reported emissions in Ganesan et al. (2017) are 30% lower than the bottom-up global inventory Emission Database for Global Atmospheric

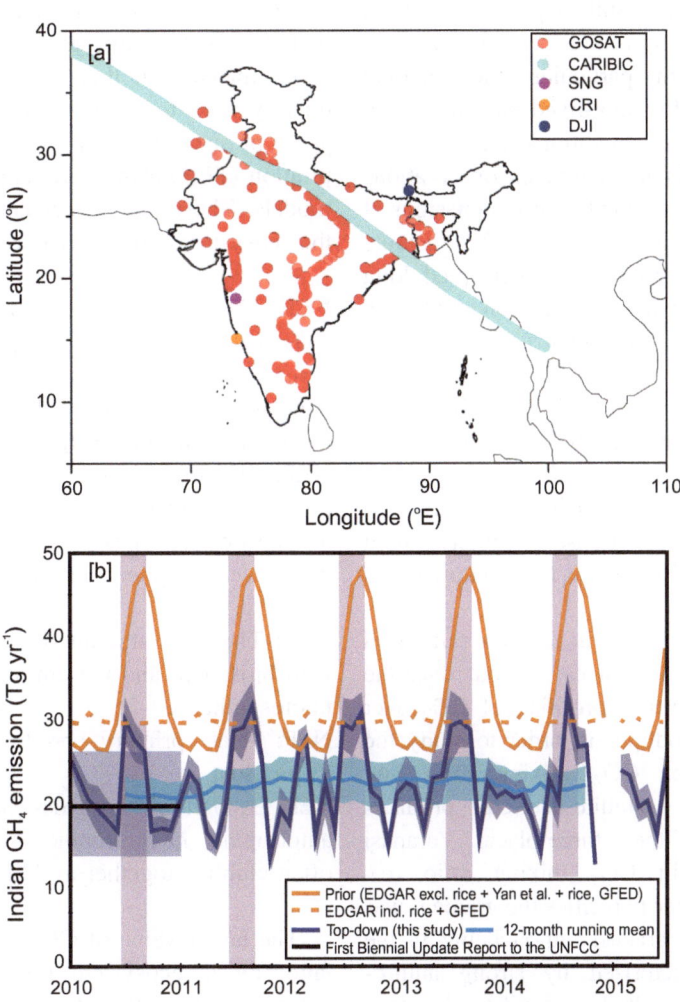

Fig. 4.4 **a** Map of observations used in top-down CH$_4$ emission estimations. Typical monthly coverage from GOSAT satellite retrievals (red), CARIBIC aircraft's flight path (light blue), surface sites Darjeeling, India (dark blue), Cape Rama (CRI) Goa, India (orange), Sinhagad (SNG), Pune, India (purple). **b** Monthly Indian emissions estimated by EDGAR (orange line), EDGAR2010 including rice, GFED and natural emissions (dashed orange line), top-down CH$_4$ estimations presented in Ganesan et al. (2017) (dark blue line) and 5th–95th percentile range (dark blue line shading), 12-month running mean along with the 5th–95th percentile range of the running mean from this study (light blue line and shading, respectively), 2010 emissions submitted to the UNFCCC by Government of India (solid black line and uncertainties as shaded line). Ref: Ganesan et al. (2017); the figure is reproduced through an open license

Research (EDGAR) during the same period (Fig. 4.4) and consistent with the emissions reported by India to the UNFCCC (NATCOM BUR-1 and BUR-2). The data used in Ganesan et al. (2013) are XCH_4 based on the CO_2 proxy method from the GOSAT satellite; flask-based measurements of dry-air CH_4 mole fraction from Sinhagad (73.75° E, 18.35° N, 1600 m ASL), Cape Rama, India (73.83° E, 15.08° N, 60 m ASL), in situ measurements from Darjeeling, India (88.25° E, 27.03° N, 2200 m ASL); and upper-atmospheric in situ measurements from the CARIBIC aircraft (Fig. 4.4a). Further, the Lagrangian particle dispersion model, Numerical Atmospheric dispersion Modeling Environment (NAME), was used to provide a quantitative relationship between atmospheric mole fractions and emissions. However, the observation of Ganesan et al. (2017) is subjected to further verification for certain reasons. It is well known that the inverse modeling results depend strongly on the selection of chemistry transport model and treatment of atmospheric measurements. In particular, the regional transport and inverse models for long-lived gas simulation suffer from the use of boundary conditions; e.g., a recent study shows the emission estimates based on observations over Siberia are greatly affected by the trends in emissions of CH_4 over Europe and Asia (Sasakawa et al. 2017). Accurate estimation of emission trends requires uninterrupted long term atmospheric observations from the region. Although Ganesan et al. (2017) used multiple streams of in situ data over India, none of the measurement locations continuously measured CH_4 during the period of their analysis (2010–2015). The remote-sensing measurements from GOSAT do not retrieve XCH_4 for the regions covered by clouds at any thickness, leading to a seasonal bias in inverse model calculation. To overcome these limitations, a well-structured GHGs long-term observational network is to be set up across India (Nalini et al. 2019).

Apart from the above activities, a few aircraft-based GHGs measurements have also been carried out over the Indian subcontinent for the vertical profiling of CO_2 over Bhubaneswar, Varanasi and Jodhpur in order to study the spatiotemporal distribution of CO_2 mixing ratio and inter-comparison with the satellite-derived products (Sreenivas et al. 2019). Out of these three places, Varanasi showed a strong gradient in CO_2 mixing ratio (ca. 3.12 ppm/km), while in the other two cities the gradient was much smaller (<2 ppm/km). Varanasi region being in the Indo-Gangetic Plain is characterized by strong anthropogenic loading, which has resulted in a relatively steep vertical gradient. The Cloud Aerosol Interaction and Precipitation Enhancement Experiment (CAIPEEX) Project of the Indian Institute of Tropical Meteorology also carried out similar observations during the Indian summer monsoon months of 2014, 2015 and 2018, 2019 over India and adjacent oceanic regions. The vertical profiling of methane

mixing ratio revealed a strong peak at about 4.5 km. It is believed that this kind of mid-tropospheric peak occurs due to strong convective activities (Chandra et al. 2017). Among other work, uptake of winter time carbon by the agricultural practices around Delhi has been discussed in Umezawa et al. (2016).

4.3 Greenhouse Gas Flux Measurements in Natural Ecosystems

GHGs fluxes are monitored over India primarily by two major measurement networks that corroborate efforts from multiple research institutes, universities and government organizations. Two important GHG components, viz., CO_2 and CH_4, are being monitored by the chamber-based or eddy covariance (EC) systems along with the other scalar fluxes including water vapor and energy.

Estimated CO_2 flux from these measurements is subsequently used to calculate net ecosystem exchange (NEE), gross primary productivity (GPP) and total ecosystem respiration (TER) using various process-based biogeoscientific models. These parameters are the different components of the ecosystem carbon cycle. Some definitions are as follows:

NEE: The net amount of carbon exchanged (in the form of CO_2) between the land biosphere and the atmosphere at a given location over a particular period of time; negative and positive values of NEE denote uptake and release of carbon by the land biosphere, respectively.
NEP: Net Ecosystem Productivity; opposite of NEE i.e. positive and negative values of NEP denote carbon uptake and loss by the biosphere, respectively.
NBP: Net biospheric productivity. NEP–disturbance fluxes (fire, etc.).
GPP: The total amount of carbon exchanged between the land biosphere and the atmosphere, through the process of photosynthesis.
NPP: Net primary productivity; GPP - autotrophic respiration.
TER: A part of the gross carbon uptake is lost through autotrophic, heterotrophic, microbial and soil respiration, often clubbed together as TER.

The larger value of GPP signifies greater carbon assimilation by the ecosystem; however, the net carbon uptake is defined by the NEE. In general, the carbon sequestration potential of an ecosystem depends on climatological conditions, soil moisture, texture and nutrient content, and vegetation type. Hence, for a regional or country-scale estimation of carbon budget, GHGs fluxes must be measured over different ecosystems scattered across the length and breadth of the region or the country.

4.3.1 Carbon Dioxide Fluxes and Net Ecosystem Exchange

Several micrometeorological flux towers were erected for the long-term monitoring of GHGs fluxes over different ecosystems in India by the Indian Space Research Organisation (ISRO) as part of the Geosphere-Biosphere Program. These sites include a mixed deciduous teak forest in Betul over central India (Jha et al. 2013); a mangrove forest in Sundarbans on the east coast of India, flanked by the Bay of Bengal (Jha et al. 2014; Rodda et al. 2016); a mixed deciduous forest in Haldwani over north India (Watham et al. 2014; Ahongshangbam et al. 2016); and a tropical moist deciduous sal forest in Barkot over north India (Watham et al. 2017; Pillai et al. 2019).

Being deciduous in nature, the carbon exchange at the Betul forest is observed to be strongly influenced by the availability of the leaves. It is seen to act as a sink in winter with a maximum value of NEE being $-300 \ \mu gC \ m^{-2} \ s^{-1}$ in the daytime. However, the sink strength is drastically reduced to $-24 \ \mu gC \ m^{-2} \ s^{-1}$ in the dry months of summer due to leaf dehiscence. Annual NEE, GPP and TER of the Betul forest were not available at the time of writing this report.

The Sundarbans mangrove acted as a net sink of CO_2 with annual NEE of $-249 \pm 20 \ gC \ m^{-2} \ y^{-1}$. Soil–$CO_2$ emission from this ecosystem was found to vary between 18 and 28 $\mu gC \ m^{-2} \ s^{-1}$. It remains positive throughout the year with a close dependency on the soil temperature (Chanda et al. 2014). Annual GPP and TER of this ecosystem were 1271 $gC \ m^{-2} \ y^{-1}$ and 1022 $gC \ m^{-2} \ y^{-1}$, respectively. Sink strength varied widely between $-72 \ \mu gC \ m^{-2} \ s^{-1}$ in summer and $-120 \ \mu gC \ m^{-2} \ s^{-1}$ in winter.

The 9-year-old man-made mixed forest at Haldwani acted as a net CO_2 source during the leafless months of winter, quite contradictory to Betul and Sundarbans (Watham et al. 2014). Subsequently, it transformed into a sink during the growing months from April to September. The monthly mean daily total NEE of this ecosystem was 0.35 $gC \ m^{-2} \ d^{-1}$, in January. It became slightly negative at $-0.38 \ gC \ m^{-2} \ d^{-1}$ in February and grew farther negative continuously to $-5.74 \ gC \ m^{-2} \ d^{-1}$ in September. The annual NEE of this ecosystem was unavailable at the moment to quantify the yearlong total carbon exchange. The GPP of Barkot forest was observed to vary from 5.38 $gC \ m^{-2} \ d^{-1}$ in December to 12.42 $gC \ m^{-2} \ d^{-1}$ in September. This shows the gross carbon exchange to be stronger in monsoon compared to winter. The annual mean daily GPP of this ecosystem was 7.98 $gC \ m^{-2} \ d^{-1}$. However, annual NEE is required for estimating the actual carbon sequestration by this ecosystem. Based on a three-year-long EC measurement during 2014–2016, the mixed forest at Kosi-Katarmal, Almora (20.05° N,

79.05° E, 1217 m amsl), acted as a net sink of atmospheric CO_2 with the average daily NEE being $-3.21 \ gC \ m^{-2} \ d^{-1}$ (Mukherjee et al. 2018). In another study, CO_2 exchange of a wheat field was measured by an EC system at Modipuram, north India, by the ISRO (Patel et al. 2011). Maximum daytime uptake and nighttime release of CO_2 by this ecosystem were observed to vary markedly during different stages of the crop growth. During the anthesis stage of the crop, maximum midday uptake and nighttime release were $-26.78 \ gC \ m^{-2} \ d^{-1}$ and $3.45 \ gC \ m^{-2} \ d^{-1}$, respectively. Due to reduced leaf area index (LAI), midday uptake drastically reduced to $-19.00 \ gC \ m^{-2} \ d^{-1}$ in the senescence stage. However, the nighttime release remained unaltered. Subsequently, in the mature stage daytime uptake further plummeted below $-11.22 \ gC \ m^{-2} \ d^{-1}$. The annual estimates of NEE, GPP and TER will be required to draw conclusion on the actual carbon sequestration of this wheat field.

The Indian Institute of Tropical Meteorology (IITM), Pune, also established a few flux towers over different ecosystems under the aegis of the MetFlux India project, initiated and funded by the Ministry of Earth Sciences (MoES), Government of India. These sites include a semievergreen moist deciduous forest in Kaziranga National Park over Northeast India (Deb Burman et al. 2017, 2019; Sarma et al. 2018), a mangrove forest in Pichavaram (Gnanamoorthy et al. 2019) on the southeast coast of India by the Bay of Bengal and an evergreen coniferous forest over the eastern Himalayan range in Darjeeling (Chatterjee et al. 2018) in Northeast India. In addition to the GHG fluxes, CO_2 concentration in seawater and atmosphere is also measured in two islands, in Agatti in the Lakshadweep Islands over the Arabian Sea (Kumaresan et al. 2018) and Port Blair in the Andaman and Nicobar Islands over the Bay of Bengal as part of the MetFlux India network. The ecosystem at Kaziranga has an annual GPP of 2110 $gC \ m^{-2} \ y^{-1}$ with a prominent seasonal variation primarily governed by the plant leaf phenology (Deb Burman et al. 2017). It is also reported that cloudiness during the Indian summer monsoon months drastically reduces the availability of the photosynthetically active radiation at this ecosystem which in turn affects its carbon sequestration activity negatively (Deb Burman et al. 2020b). The mangrove forest of Pichavaram in Tamil Nadu registered a NEE maximum (ca. $-11.40 \ gC \ m^{-2} \ d^{-1}$) during the month of January. Annual GPP and ecosystem respiration (TER) were estimated to be 1466 $gC \ m^{-2} \ y^{-1}$ and 1283 $gC \ m^{-2} \ y^{-1}$, respectively (Gnanamoorthy et al. 2020). Based on a two-month observation at a high-altitude Himalayan evergreen forest near Darjeeling, Chatterjee et al. (2018) estimated springtime CO_2 flux for the year 2015. These authors observed a maximum NEE reaching up to $-10.37 \ gC \ m^{-2} \ d^{-1}$.

The analysis of the available annual NEE data from across the country reveals that the maximum carbon uptake (maximum negative NEE value on a diurnal scale) takes place typically in monsoon to early winter in most of the places, except in the Kaziranga forest in Northeast India. This site is characterized by maximum uptake (large negative NEE values) during the pre-monsoon time, while the minimum values typically occur in the winter. Kosi-Katarmal in north India sequesters maximum carbon during the late monsoon, though significant amount of carbon is also sequestered during the pre-monsoon time. The reason for this unusual behavior is thought to be arising partly due to the leaf phenology and partly driven by the regional climate variability. For example, the northeast region of India receives a significant amount of rainfall (driven by nor'westers) during the early summer and high sunshine (compared to the cloudy days during the monsoon time) days resulting in the increased uptake of carbon by the vegetation. The kharif (rainfed) sesame crop cultivated at Barkachha, Uttar Pradesh, in the Indo-Gangetic Plain over north India sequesters maximum CO_2 during monsoon, whereas the sequestration activity is seen to be severely affected by the drought (Deb Burman et al. 2020a).

4.3.2 Methane Fluxes

The Sundarbans mangrove forest acted as a net source of atmospheric CH_4 with average daily flux being 150.2 ± 248.9 mg m^{-2} d^{-1} (Jha et al. 2014). Methane source was enhanced at Sundarbans during summer months due to the elevated temperature and moisture contents. Another study by Mukhophadhya et al. (2001) estimates the CH_4 source from Sundarbans mangroves to lie within 4.5–8.9 μg m^{-2} s^{-1}. The methane flux has also been measured in the Pichavaram mangrove forest. Purvaja and Ramesh (2001) used static chamber and reported methane emission in the range of 47.2–324.5 mg m^{-2} d^{-1}.

Methane flux from Indian rice paddy fields is reported to vary from 2.4 to 660.0 mg m^{-2} d^{-1} (Parashar et al. 1991; Lal et al. 1993; Adhya et al. 1994). Total CH_4 emission from an irrigated rice paddy field in New Delhi was reported to be 0.275–0.372 g m^{-2}. Fertilizers such as urea, ammonium sulfate and potassium nitrate increased the CH_4 emission, while dicyandiamide helped reduce it (Ghosh et al. 2003). Intermittently irrigated rice paddy fields were seen to emit less CH_4 compared to the continuously flooded fields (Jain et al. 2000). A dry land rice cultivation in Varanasi, north India, was reported to be a sink of CH_4 with an average growing season uptake of 8.4 mg m^{-2} d^{-1} (Singh et al. 1997). According to these authors in dry tropical ecosystems, N availability is remarkably low, and this may be the reason for high CH_4 uptake rates in these ecosystems.

Furthermore, these dry soils are well drained and permeable, and it has been clearly demonstrated that CH_4 consumption is diffusion-limited (Dörr et al. 1993). A recent study showed potential for CH_4 emission by changing rice cultivation practice from conventional transplanting (CT) to system of rice intensification (SRI) in India (Oo et al. 2018). Inland water bodies such as lakes, ponds, open wells, rivers, springs and canals are known to emit CO_2 and CH_4 to the atmosphere. Total CH_4 flux from multiple such ecosystems in India measured using flux chambers ranged from 0.16 to 834 mg m^{-2} d^{-1} (Selvam et al. 2014). Additionally, the CO_2 flux from these systems was measured to lie between 0.34 and 3.15 gC m^{-2} d^{-1} (Selvam et al. 2014).

4.3.3 Carbon Inventory of the Indian Forests

The total carbon stored in Indian forests, including forest soil, is estimated to be in the range of 8.58–9.6 PgC (Ravindranath et al. 2008). According to Chhabra and Dadhwal (2004), 3.8–4.3 PgC is stored as Indian forest phytomass, approximately 10% of the global forest phytomass carbon pool. Several studies assessed the influence of land use change on forest carbon (Ravindranath et al. 1997; Kaul et al. 2009). Kaul et al. (2009) estimated that net carbon flux attributable from land use change decreased from a source level of 5.65 Tg C yr^{-1} during 1982–1992 to a sink level of 1.09 Tg C yr^{-1} during 1992–2002. It indicates that Indian forests became a sink of atmospheric carbon due to the regeneration and afforestation efforts.

4.3.4 Agricultural Ecosystem

Apart from the natural vegetation, agricultural ecosystems also contribute significantly to the GHGs fluxes. Specifically, the rice paddy fields are known to emit significant amounts of CH_4 owing to the anaerobic conditions prevalent during the times when the fields are inundated with potentially warm water (Adhya et al. 2000). Moreover, the net emission from the agricultural ecosystems depends strongly on the agricultural practices such as applications of fertilizers, manure, pesticides, crop residue, straw and flooding of the agricultural field (Debnath et al. 1996; Singh et al. 1996; Bhatia et al. 2005). Indian Agricultural Research Institute (IARI) has several flux measurement systems across India for measuring the emissions from different agricultural fields.

4.3.5 Other Observations

Several satellites provide derived values of GPP, NEE and TER from space observations. Such products have also been compared with the ground-based estimates (Dadhwal 2012). Using SPOT VEGETATION 10-day NPP composites, Chhabra and Dadhwal (2004) estimated the total net carbon uptake by the Indian landmass to be 2.18 PgC y^{-1}. Bala et al. (2013) studied the trends and variability of the terrestrial net primary productivity over India for the period of 1982–2006 using the Advanced Very-High-Resolution Radiometer (AVHRR)-derived data. These authors found an increasing trend of 3.9% per decade over India, indicating an increased rate of carbon fixation by the terrestrial ecosystems during the past two decades. NPP estimates based on Dynamic Land Ecosystem Model for the 1901–2000 period also show an increasing trend, from 1.2 to 1.7 PgC yr^{-1} (see Sect. 4.4.1). These observations also support our hypothesis of increasing CO_2 amplitude due to increasing biospheric activity (discussed in Sect. 4.3.1). According to Watham et al. (2017), the MOD17 product by MODIS is a severe underestimation of the actual GPP

calculated from both surface measurements and remote-sensing-driven models. Deb Burman et al. (2017) showed that MODIS LAI fails to capture the seasonality in the leaf phenology which can lead to significant error in the GPP estimate if used as input in process-based models. Ground observations over different ecosystems and satellite products are combined together for a countrywide estimate of carbon sequestration by Nayak et al. (2013). According to their study, broadleaf evergreen forests are the largest contributor to the annual carbon storage followed by broadleaf deciduous forests. Annual NEE of these ecosystems is -1057 gC m^{-2} y^{-1} and -658 gC m^{-2} y^{-1}, respectively. Additionally, this upscaling study also reports an estimated growth rate of 0.005 PgC y^{-1} in the annual carbon sequestration over the Indian region.

Apart from the surface flux measurements, satellite observations are also used to estimate the GHGs fluxes. Valsala et al. (2013) studied the intra-seasonal oscillations in

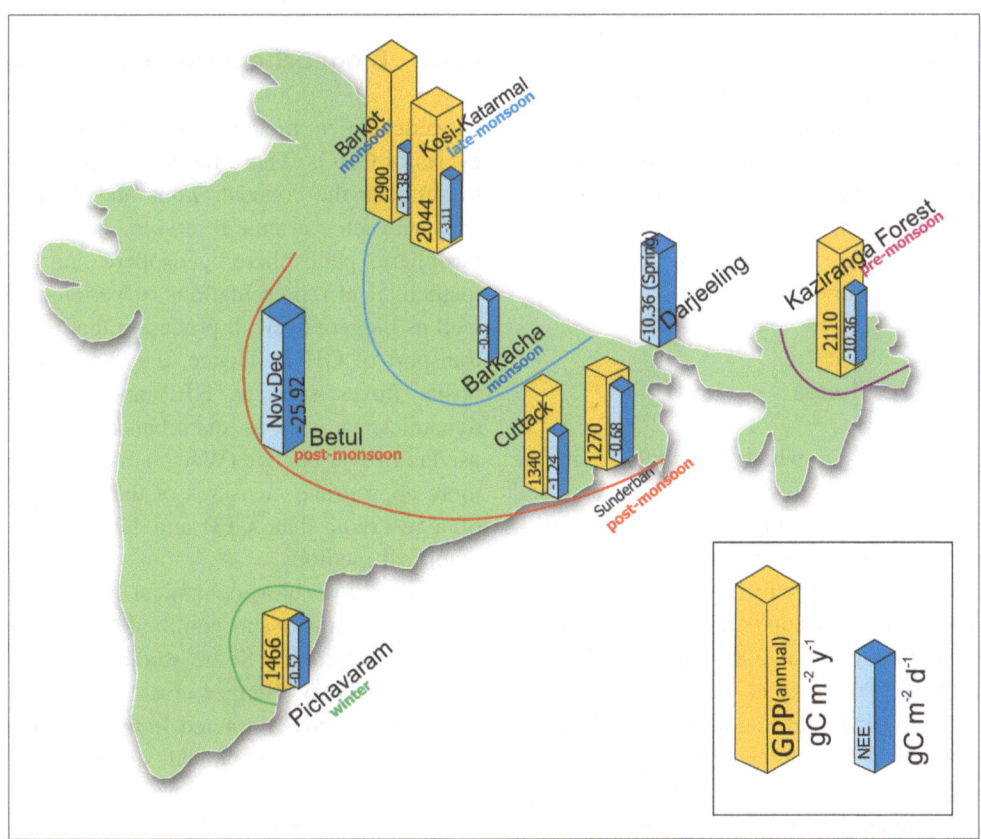

Fig. 4.5 A schematic representation of CO_2 fluxes measured by the eddy covariance technique across the different ecosystems over India. The blue bars depict the maximum CO_2 uptake by the vegetation on the diurnal timescale averaged for a particular season as mentioned in the figure (unit: gC m^{-2} d^{-1}). The yellow bar represents the GPP value for a particular ecosystem on an annual timescale (unit: gC m^{-2} y^{-1}). Maximum carbon sequestration varies greatly with season and geography: the northeast during the pre-monsoon (represented by magenta), the Gangetic plains in monsoon (shown in blue), the north, central and east India during post-monsoon (shown in orange) and the east coast of south India during the winter (shown in green). The curved lines very approximately show the geographical variation of carbon uptake for the specific season. No data are available for the whole of the western region. The total carbon uptake by the Indian landmass is estimated to be 2.18 PgC yr^{-1} (Chhabra and Dadhwal 2004)

terrestrial NEE during the Indian summer monsoon. These authors showed that while the terrestrial ecosystems act as a net source of CO_2 during June and July, they transform into the net CO_2 sink during August and September. However, due to spatial variability in GHGs distribution and dynamics, this characteristic feature may differ in specific regions. For example, the eddy covariance-based results show that the deciduous forest (Kaziranga in Assam) in the Northeast India acted as a strong sink of carbon during the pre-monsoon period (May–June; Sarma et al. 2018). On the other hand, as mentioned earlier, most of the forests in mainland India sequester significant carbon in the monsoon but maximum carbon during the post-monsoon to early winter. Hence, the study by Valsala et al. (2013) points out the limitations of satellite measurements (GHG column concentrations) in illustrating the GHGs dynamics in the Indian landmass. However during monsoon season and due to cloud cover, GHG absorption bands are obscured; hence, ground-based direct measurements of GHG are also necessary to complement the satellite measurements.

Figure 4.5 schematically shows the carbon sequestration by some Indian forests measured by means of the eddy covariance technique. The blue bar represents the maximum value of the net ecosystem exchange on the diurnal timescale for a particular month as indicated. The height is proportional to the amount of carbon uptake by the vegetation having a unit of $gC\ m^{-2}\ d^{-1}$. The gross primary productivity (expressed in $gC\ m^2\ yr^{-1}$), available only for a few ecosystems, is shown as yellow bar.

4.3.6 Nitrous Oxide Fluxes

Another potent GHG is N_2O, which is produced in the natural biological processes occurring in soil, ocean and inland water by the microbes (Thomson et al. 2012; Davidson and Kanter 2014). Various anthropogenic activities such as agriculture, energy production, heavy industries and waste management also contribute to the rising N_2O flux into the atmosphere (Wassman et al. 2004; Malla et al. 2005; Datta and Adhya 2014). Prasad et al. (2003) studied N_2O emissions from India's agricultural sector between 1961 and 2000 and suggested that the total N_2O emission had increased approximately 6 times over 40 years. Presently, agricultural activities alone account for more than 90% of the total anthropogenic N_2O emissions in India, with 65% of the total N_2O emission being attributable to the chemical fertilizers. In a study done by Ghosh et al. (2003) over an upland rice ecosystem grown during the summer monsoon months in New Delhi, N_2O fluxes were found to vary within 4.32–$2400\ \mu g\ m^{-2}\ d^{-1}$, whereas seasonal N_2O loss by the ecosystem varied between 0.037 and $0.186\ kg\ ha^{-1}$. An agricultural denitrification and decomposition model was calibrated for predicting the crop yield and GHG emissions and validated for Indian conditions by Pathak et al. (2005). According to their study, continuous flooding of rice fields results in annual net emissions as follows: 21.16–60.96 TgC in the form of CO_2, 1.07–1.10 TgC in the form of CH_4 and 0.04–0.05 TgN in the form of N_2O, whereas intermittent flooding changes the emissions to 16.66–48.80 TgC in the form of CO_2, 0.12–0.13 TgC in the form of CH_4 and 0.05–0.06 TgN in the form of N_2O. Noticeably, the agricultural practice of intermittent flooding of the rice paddy field had opposite effects on carbon and nitrogen emissions. An analysis using atmospheric observations and models suggests acceleration in N_2O emission per unit of nitrogen fertilizer use, thus a global emission factor of $2.3 \pm 0.6\%$, which is significantly larger than the IPCC default for combined direct and indirect emissions of 1.375% (Thompson et al. 2019).

4.4 Model Simulation of GHGs

4.4.1 Biogeochemical Model Study

Biogeochemical models are widely used to simulate the life cycles of GHGs. These models adopt a bottom-up approach to simulate the concentrations and fluxes of GHGs within and between various reservoirs such as the atmosphere, biosphere, pedosphere, geosphere and hydrosphere using mathematical representations of various biogeochemical, as well as biogeophysical, processes that affect the storage and transport of GHGs. Hence, they are able to simulate composite fluxes such as NEP and NBP that include soil dynamics as well as disturbances such as fire and land use/land cover change (Watson et al. 2000). Hence, these are more appropriate measures of the carbon source/sink status of a reservoir than GPP and NPP obtained using typical empirical methods.

Studies with biogeochemical models are very limited in the Indian context and confined to the carbon cycle in terrestrial ecosystems. Some studies have conducted simulations over India (Banger et al. 2015; Gahlot et al. 2017), while others have extracted India-specific information from global-scale biogeochemical simulations (Cervarich et al. 2016; Gahlot et al. 2017; Rao et al. 2019). Banger et al. (2015) used the Dynamic Land Ecosystem Model (DLEM) to estimate NPP patterns for the 1901–2000 period. They found that it has increased from 1.2 to 1.7 $PgC\ yr^{-1}$ during this period. Using an ensemble average of nine different dynamical vegetation models, Cervarich et al. (2016) found NEP and NBP values for India to be in the $200.6 \pm 137.7\ TgC\ yr^{-1}$ and $185.9 \pm 145.6\ TgC\ yr^{-1}$ range, respectively, for the 2000–2013 period. These values are much larger than other estimates. Gahlot et al. (2017)

Table 4.1 Comparison of net biome productivity (TgC yr^{-1}) for the terrestrial ecosystems of India estimated using different approaches based on Gahlot et al. (2017)

Decade	1980s	1990s	2000s
Process-based models	1.04 ± 46.75	34.65 ± 54.16	18.50 ± 48.40
Inverse models	–	45.22 ± 69.53	42.12 ± 67.40
ISAM	27.17	34.39	23.70

conducted a comprehensive study using the Integrated Science Assessment Model (ISAM). They found that the NBP in India changed from 27.17 TgC yr^{-1} in the 1980s to 34.39 TgC yr^{-1} in the 1990s to 23.70 TgC yr^{-1} in the 2000s indicating that the terrestrial ecosystems of India are a net carbon sink but the magnitude of the sink may be decreasing in recent years. Their estimates are comparable with results from the models involved in the TRENDY (Trends in net land carbon exchange) project (Table 4.1). Very importantly, their results show that there is a large uncertainty between different estimates of the terrestrial carbon sink.

Banger et al. (2015) and Gahlot et al. (2017) also conducted numerical experiments to quantify the impacts of various natural and anthropogenic forcings on the dynamics of the carbon cycle. They found that the net positive carbon sink is maintained mostly by the carbon fertilization effect, aided to some extent by forest conservation, management and reforestation policies in the past decade. In an idealized modeling study, Bala et al. (2011) showed that CO_2-fertilization has the potential to alter the sign of terrestrial carbon uptake over India. They found that modeled carbon stocks in potential vegetation increased by 17 GtC with unlimited fertilization for CO_2 levels and climate change corresponding to the end of the 21st century. However, the carbon stock declined by 5.5 GtC when fertilization is limited at 1975 levels of CO_2 concentration. Thus, the benefits from CO_2 fertilization could be partially offset by land use/land cover change and climate change (Bala et al. 2011). Further, the model simulations of Bala et al. (2011) also implied that the maximum potential terrestrial sequestration over India, under equilibrium conditions and best-case scenario of unlimited CO_2 fertilization, is only 18% of the twenty-first-century SRES A2 scenario emissions from India. The limited uptake potential suggests that reduction of CO_2 emissions and afforestation programs should be top priorities for India.

The broader trends, variability and drivers of the carbon cycle dynamics over India were comprehensively addressed by Rao et al. (2019) using the multi-model dataset TRENDY for the period 1900–2010. Their analysis showed that the TRENDY multi-model mean NPP shows a positive trend of 2.03% per decade over India during this period which is consistent with a global greening in the last two decades (Chen et al. 2019) and other studies such as Bala et al. (2013) which showed an NPP increase of about 4% per decade during the satellite era. Rao et al. (2019) also

analyzed the trends in water-use efficiency (WUE) of ecosystem in India and found that WUE has increased by 25% during the period 1900–2010. Further, it was found that the inter-annual variation in NPP and NEP over India is strongly driven by precipitation, but remote drivers such as El Nino–Southern Oscillation (ENSO) and Indian Ocean Dipole (IOD) may not have a strong influence. The multi-model-based estimate of the cumulative NEE is only 0.613 ± 0.1 PgC during 1901–2010, indicating that the Indian terrestrial ecosystem was neither a strong source nor a significant sink during this period. Among other studies, Chakraborty et al. (1994) used proxy-based atmospheric and surface ocean radiocarbon records to estimate the CO_2 exchange rate in the coastal region of Gujarat. Using a box model approach, these authors have estimated CO_2 exchange rate in the tune of 12 mol m^{-2} yr^{-1}.

4.4.2 Greenhouse gases Emission and Projections

In this chapter, we present projections of future changes based on the SSPs (Shared Socioeconomic Pathways)—a suite of future forcing scenarios being used for the latest generation of climate model (CMIP6) experiments. The SSPs describe five possible future emissions trajectories based on different narratives of socio-economic developments in the future. It may be noted that the future forcing scenarios used by the previous generation of climate models (CMIP5) were based on the Representative Concentration Pathways (RCPs; see Chap. 1).

The future projection of GHGs emissions may be understood in light of the shared socioeconomic pathways (SSPs) by climate change researcher community, which takes an account of qualitative and quantitative trends in population growth, economic development, urbanization and education leading to future development of nations. The narratives of these are discussed in detail in O'Neill et al. (2017). These SSPs are developed using integrated assessment models. These are considered to be new scenarios designed under 5 pathways, namely SSP1, SSP2, SSP3, SSP4 and SSP5 out of which SSP1, SSP3, SSP4 and SSP5 are based on different levels of challenges to climate adaptation and mitigation while SSP2 is a median pathway considering a moderate challenge to both (O'Neill et al. 2014, 2017). Briefly, these SSPs are named and understood as:

SSP1: Sustainability, in which the world gradually shifts toward a more sustainable path where the development happens considering the environmental boundaries.

SSP2: Middle of the road pathway, where social, economic and technological developments follow a historical trend, leading to uneven progress among the countries and slowly attaining sustainable goals.

SSP3: Regional rivalry, where there is increased competitiveness among the countries to attain development and policies are mostly oriented toward empowering national securities and local issues.

SSP4: Inequalities, where uneven investments and policies enhance the inequalities among and within the countries' increase. Energy demand increases so as the consumption of carbon as well as low-carbon energy sources.

SSP5: Fossil-fueled development; investment in technological progress and human capital is considered a way to achieve sustainable development. To cater the increasing energy demand for achieving economic development, this world still primarily relies on the fossil fuel resources.

Further details about these SSPs can be found in Calvin et al. (2016), Fricko et al. (2016), Fujimori et al. (2016), Hasegawa et al. (2018), Kriegler et al. (2016), van Vuuren et al. (2016). Emission projections are made using these SSPs in MAGICC6 (Model for the Assessment of Greenhouse Gas-Induced Climate Change) climate model and are available as 9 CMIP6 emission scenarios, 4 of which are in line with the radiative forcings of 2.6, 4.5, 6.0 and 8.5 W m^{-2} of RCPs of CMIP5 projections as SSP1-2.6, SSP2-4.5, SSP4-6.0 and SSP5-8.5. The 5 additional intermediate scenarios cover vivid possibilities of forcing targets during emergence of these scenarios in future as SSP1-1.9, SSP3-7.0, SSP3-low near-term climate forcing (LowNTCF), SSP4-3.4 and SSP5-3.4 Overshoot (OS). Out of these, SSP3-7.0 and SSP5-8.5 are baseline scenarios, considering no additional policies are brought into the action to mitigate GHGs emission other than the existing policies. Considering the base year as 2015, all these scenarios were then harmonized for smooth transition between the historical data and these projections and downscaled at country level (Gidden et al. 2018).

Future projections for India's GHG emissions, primarily CO_2 and CH_4, under different SSPs up to the year 2100 are shown in Fig. 4.6 where CO_2 emissions span over a large range of 2–16 Gt-CO_2 yr^{-1} under different scenarios and CH_4 emissions span over 10–90 Mt yr^{-1}. The data were sourced from Riahi et al. (2017). In India, emission projections under four SSPs and a total of 8 scenarios are available, SSP1-1.9, SSP1-2.6, SSP3-7.0, SSP3-LowNTCF, SSP4-3.4, SSP4-6.0, SSP5-3.4 OS and SSP5-8.5. SSP1-1.9 is

considered a greener world where energy consumption is based on more green and renewable energy sources, restricting the radiative forcing to 1.9 W m^{-2} which is in accordance with the recent Paris Agreement. Trajectories represent the immediately decreasing CO_2 emissions. In SSP1-2.6 which also implies a greener world, but the radiative forcing reaches a level of 2.6 W m^{-2}, CO_2 emission tends to decrease, but slightly less than that of SSP1-1.9 in the future due to investment in green energy. India's CO_2 emissions are projected to peak by the year 2030 and gradually decrease thereafter. CH_4 emissions are projected to decrease gradually by the year 2100 to its lowest value hovering around one-third of present emission. India's (intended) nationally determined contribution under the Paris Agreement commits to reduce the emission intensity of its GDP by 33–35% by 2030 from 2005 level and to achieve about 40% cumulative electric power installed capacity from non-fossil fuel-based energy sources by 2030. These SSP projections are in line with the green and sustainable world with least GHGs emissions from India as well in such scenario.

In SSP3-70 and SSP3-LowNTCF, there is a consistent increase in CO_2 emissions till the end of the century at almost the same rate. For CH_4, there is an increase in SSP3-70 scenario, but in the SSP3-LowNTCF scenario, where the policies are to reduce the forcing due to short-lived species, a sharp decrease after 2020 and then a slight increase are projected till the end of the century. CO_2 increase is attributed to the increased power consumption to meet the growing energy demand mostly by relying on fossil fuels and carbon-based energy sources along with the change in land use, whereas CH_4 emissions are dominated by agriculture (cultivation and livestock) as of now, also contributed by energy sector and waste management. Under the SSP3-70 scenario while CO_2 emission tends to increase about four times the present emission, increase in CH_4 emission is projected to rapidly increase after 2020 and shall be tripled by the end of the century; in fact, CH_4 emissions are seen to reach higher values in this scenario among all the cases.

SSP4 projection is in the world of unequal development. In SSP4-6.0, where radiative forcing is restricted to 6 W m^{-2}, India's emission of both CO_2 and CH_4 is projected to peak around the middle of the century, after which it stabilizes and shows a decreasing tendency around the year 2060 probably due to the inclusion of investments in better technologies and sustainable energy sources in the future. While CO_2 emissions tend to stabilize toward the end of this century around today's emission rate, CH_4 emissions are on the higher side, even after an anticipated decline in emission in the latter half of the century. In SSP4-3.4, CO_2

Fig. 4.6 Emission trajectories of (**a**) CO₂, and (**b**) CH₄ for India under different CMIP6 emission scenarios

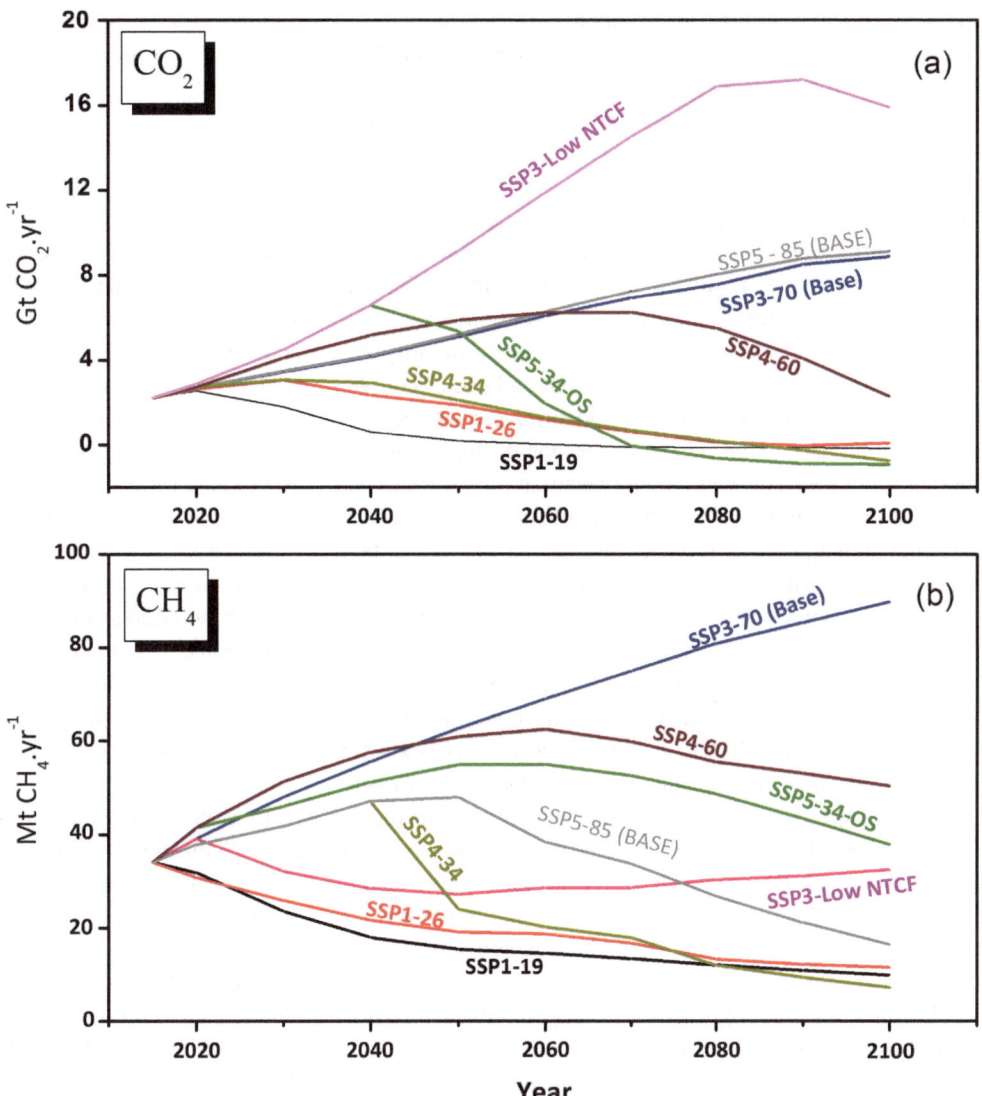

emission trajectories are toward decreasing trend in the near future and CH_4 emission trajectories following slightly lesser than the SSP4-6.0 trend.

In SSP5, CO_2 emissions are projected to follow an increasing trend for the next few decades and stabilize itself around the year 2080. It then slightly decreases by 2100, but yields the highest emission of CO_2 as a result of socioeconomic development powered by fossil fuel. Projected emissions hover at the highest rate of 16,000 Mt CO_2 yr^{-1}. CH_4 emission is projected to increase till 2040 and stabilize itself in a decade, and by 2050 it starts decreasing and ends up to the values near the emission values as projected in SSP1. This decrease in CH_4 emission may be attributed to better agriculture practices, livestock management and land usage as a result of investment in technologies.

4.5 Knowledge Gaps

The seasonal-scale atmospheric CO_2 concentrations typically show high amplitude in the high latitude region and low in the tropical areas. Despite being a tropical country, the Indian CO_2 variability shows unusually large seasonal cycle and an increasing trend in the CO_2 values. So, whether the natural and/or the anthropogenic emission of carbon is controlling the CO_2 amplitude needs to be properly understood.

A lack of long-term record and sparse observational data limits our ability to comprehensively assess the biogeochemical cycle of carbon over India. The eddy covariance-based data of NEE are available only for a few forested and agricultural environments. It is also believed that trees emit significant amount of methane, especially in forested environment; no study to ascertain and quantify the plant-derived methane emission exists in India. This poses a serious constraint in making an accurate estimation of the net ecosystem productivity and in turn the carbon sequestration potential of the Indian forests. A large amount of vapor is generated through the process of transpiration, but their role in modulating the monsoon processes remains largely unknown. Apart from insufficient observational data, there are no significant studies for Indian region to quantify this effect by climate modeling. Development and use of coupled climate–carbon models could help identify the sources and sinks of both carbon dioxide and water vapor fluxes. Inverse models have been shown to reliably estimate the sources and sinks of the GHGs, but insufficient observed data pose a serious challenge to achieve the task in the Indian context. Likewise, the role of the oceanic GHG emission on the terrestrial carbon cycle also remains poorly constrained.

One of the important parameters is the carbon isotopic values of certain GHGs, such as CO_2 and CH_4, which are known to provide valuable information about the sources, but their use in the Indian context is almost nonexistent. With the advent of new technology (i.e., laser-based cavity ring-down spectroscopy), real-time in situ monitoring of GHGs concentrations and their isotopes are possible (Mahesh et al. 2015; Chakraborty et al. 2020). Use of outputs from such instruments would greatly enhance our capability to characterize the GHGs source and sink patterns on a higher temporal and regional scale.

With increasing population and rapid industrialization, energy demand for the country is increasing at a rapid pace resulting in more GHGs emissions. But there is no network of observations available to monitor these emissions and to have a better understanding of sources and sink pattern over the country. In this context, there is an urgent need to develop a countrywide surface GHGs concentration observational network and their fluxes at all the major ecosystems and urban hotspots.

Other issues, such as the sensitivity of the photosynthetic sink and the respiration-driven sources of carbon to increased warming, relation between the CO_2 and other green house gas fluxes with the intra-seasonal variation of rainfall across the ecosystems, need to be investigated.

The scope of research needs to be expanded using process-based modeling of carbon cycle to quantify the role of different natural and anthropogenic factors. These models do not rely on the remote-sensing data and hence can be used to study the pre-satellite era as well as develop future scenarios. Development of a coupled climate and GHG cycle model constrained and validated by an extensive observational network would strengthen the effort in unraveling the regional sources and sinks of the GHGs and develop realistic projections of the future.

4.6 Summary

The observational records of CO_2 and CH_4 concentration are available from a West Indian location (Sinhagad) for the last several years. This is one of the very few flask-based measurements that is currently underway in India on a long-term perspective.

The observation shows that the amplitude of CO_2 mixing ratio has been increasing progressively for the past several years. The mechanism responsible for producing such variability is not fully understood, but is likely to be linked to the changes in vegetation and forest cover.

Reports available on Indian forest cover expansion (contraction) are not coherent. Systematic efforts are required to address these issues. The measurement of surface CO_2 concentration over a wide geographical area must be carried out on a long-term basis. Analysis of such kind of records has been proven to be an effective means to assess the biospheric activity.

Use of satellite-derived vegetation indices (proxy of terrestrial biosphere) indicates an increasing trend (ca. 4% per decade) during the last two decades in India. Also, the NPP estimates based on Dynamic Land Ecosystem Model for the 1901–2000 period show an increasing trend, from 1.2 to 1.7 $PgC.yr^{-1}$.

The available surface GHGs concentration observations, albeit from a limited area, indicate the role of marine processes, especially during the summer monsoon season, in determining the seasonal pattern of CO_2 and CH_4 fluxes. However, quantification of this process needs to be done for a better understanding of the GHGs dynamics.

The atmosphere–biosphere exchange of CO_2 tends to be active during summer monsoon and maximum during post-monsoon seasons of India. However, the Kaziranga forest in Northeast India appears to sequester maximum carbon during the pre-monsoon season.

The net ecosystem exchange (NEE) of carbon derived from satellite retrievals gives only a gross estimate over a wide region. Considerable differences exist with the eddy covariance-based in situ observations at several places.

A robust network of eddy covariance-based observations consisting of a large number of micrometeorological tower setup over the diverse ecosystems across the country can lead to a better understanding of the biogeochemical cycles.

CO_2 and CH_4 projections for the Indian environment have been made under different CMIP6 emission scenarios. Without rapid mitigation policies, atmospheric CO_2 and CH_4 loading will continue to increase for the next several decades.

However, with more sustainable approaches using green technologies these rising trends can be controlled.

Carbon cycle study in India using the biogeochemical models is still in its infancy. Process-based modeling of GHGs cycle coupled with general circulation models should be developed to estimate the sources and sinks, and identify the transport of the GHGs contributed by different natural and anthropogenic factors. This could be carried out in association with the radiocarbon analysis of atmospheric CO_2 in order to partition the emission due to fossil fuel burning and agricultural practices and/or natural processes.

References

Adhya T, Rath A, Gupta P et al (1994) Methane emission from flooded rice fields under irrigated conditions. Biol Fertil Soils 18:245–248

Adhya TK, Mishra SR, Rath AK et al (2000) Methane efflux from rice-based cropping systems under humid tropical conditions of eastern India. Agr Ecosyst Environ 79:85–90

Ahongshangbam J, Patel NR, Kushwaha SPS, Watham T, Dadhwal VK (2016) Estimating gross primary production of a forest plantation area using eddy covariance data and satellite imagery. J Indian Soc Remote Sens 44:895–904

Bala G, Gopalakrishnan R, Jayaraman M, Nemani R, Ravindranath NH (2011) CO_2-fertilization and potential future terrestrial carbon uptake in India. Mitig Adapt Strat Gl 16:143–160

Bala G, Joshi J, Chaturvedi RK, Gangamani HV, Hashimoto H, Nemani R (2013) Trends and variability of AVHRR-derived NPP in India. Remote Sens 5:810–829. https://doi.org/10.3390/rs5020810

Banger K, Tian H, Tao B, Ren W, Pan S, Dangal S, Yang J (2015) Terrestrial net primary productivity in India during 1901–2010: contributions from multiple environmental changes. Clim Change 132(4):575–588. https://doi.org/10.1007/s10584-015-1448-5

Barlow JM, Palmer PI, Bruhwiler LM, Tans P (2015) Analysis of CO_2 mole fraction data: first evidence of large-scale changes in CO_2 uptake at high northern latitudes. Atmos Chem Phys 15:13739–13758. https://doi.org/10.5194/acp-15-13739-2015

Berhanu TA, Szidat S, Brunner D, Satar E, Schanda R, Nyfeler P, Battaglia M, Steinbacher M, Hammer S, Leuenberger M (2017) Estimation of the fossil fuel component in atmospheric CO_2 based on radiocarbon measurements at the Beromünster tall tower, Switzerland. Atmos Chem Phys 17:10753–10766

Bhatia A, Pathak H, Jain N, Singh PK, Singh AK (2005) Global warming potential of manure amended soils under rice-wheat system in the Indo-Gangetic plains. Atmos Environ 39:6976–6984

Bhattacharya SK, Borole DV, Francey RJ, Allison CE, Steele LP, Krummel P, Langenfelds R, Masarie KA, Tiwari YK, Patra PK (2009) Trace gases and CO_2 isotope records from Cabo de Rama, India. Curr Sci 97:1336–1344

Bose T, Chakraborty S, Borgaonkar HP, Sengupta S, Ramesh R (2014) Estimation of past atmospheric carbon dioxide levels using tree-ring cellulose d13C. Curr Sci 107(6):971–982

Calvin K, Bond-Lambert B, Clarke L, Edmonds J, Eom J, Hartin C, Kim S, Kyle P, Link R, Moss R, McJeon H, Patel P, Smith S, Waldhoff S, Wise M (2016) SSP4: a world of inequality. Glob Environ Change. https://doi.org/10.1016/j.gloenvcha.2016.06.010

Cervarich M et al (2016) The terrestrial carbon budget of South and Southeast Asia. Environ Res Lett 11(10). https://doi.org/10.1088/1748-9326/11/10/105006

Chakraborty S, Ramesh R, Krishnaswami S (1994) Air-sea exchange of CO_2 in the Gulf of Kutch, northern Arabian Sea based on bomb-carbon in corals and tree rings. Proc Indian Acad Sci (Earth Planet Sci) 103:329–340

Chakraborty S, Dutta K, Bhattacharyya A, Nigam M, Schuur EAG, Shah S (2008) Atmospheric ^{14}C variability recorded in tree rings from Peninsular India: implications for fossil fuel CO_2 emission and atmospheric transport. Radiocarbon 50(3):321–330

Chakraborty S, Metya A, Datye A, Deb Burman PK, Sarma D, Gogoi N, Bora A (2020) Eddy covariance and CRDS based techniques of GHGs measurements provide additional constraint in partitioning the net ecosystem exchange. EGU General Assembly 2020 Abstract volume; https://doi.org/10.5194/egusphere-egu2020-901

Chanda A, Akhand A, Manna S, Dutta S, Hazra S et al (2013) Characterizing spatial and seasonal variability of carbon dioxide and water vapour fluxes above a tropical mixed mangrove forest canopy, India. J Earth Sys Sci 122:503–513

Chanda A, Akhand A, Manna S et al (2014) Measuring daytime CO_2 fluxes from the inter-tidal mangrove soils of Indian Sundarbans. Environ Earth Sci 72:417–427

Chandra N, Hayashida S, Saeki T, Patra PK (2017) What controls the seasonal cycle of columnar methane observed by GOSAT over different regions in India? Atmos Chem Phys 17:12633–12643. https://doi.org/10.5194/acp-17-12633-2017

Chandra N, Venkataramani S, Lal S, Patra PK, Ramonete M, Lin X, Sharma SK (2019) Observational evidence of high methane emissions over a city in western India. Atmos Environ 202:41–52

Chatterjee A, Roy A, Chakraborty S, Sarkar C, Singh S, Karipot AK, Ghosh SK, Mitra A, Raha S (2018) Biosphere atmosphere exchange of CO_2, H_2O vapour and energy during spring over a high altitude Himalayan forest in eastern India. Aerosol Air Qual Res. https://doi.org/10.4209/aaqr.2017.12.0605

Chhabra A, Dadhwal VK (2004) Estimating terrestrial net primary productivity over India using satellite data. Curr Sci 86:269–271

Chen C, Park T, Wang X, Piao S, Xu B, Chaturvedi RK, Fuchs R, Brovkin V, Ciais P, Fensholt R, Tømmervik H, Bala G, Zhu Z, Nemani RR, Myneni RB (2019) China and India lead in greening of the world through land-use management. Nat Sustain 2:122–129

Ciais P, et al (2014) Carbon and other biogeochemical cycles. Climate change 2013: the physical science basis. Contribution of Working Group I to the Fifth Assessment Report of the Intergovernmental Panel on Climate Change. Cambridge University Press, pp 465–570

Dadhwal VK (2012) Assessment of Indian carbon cycle components using earth observation systems and ground inventory. ISPRS Int Arch Photogram Remote Sens Spat Inf Sci XXXIX-B8:249–254

Datta A, Adhya TK (2014) Effects of organic nitrification inhibitors on methane and nitrous oxide emission from tropical rice paddy. Atmos Environ 92:533–545

Davidson EA, Kanter D (2014) Inventories and scenarios of nitrous oxide emissions. Environ Res Lett 9:105012. https://doi.org/10.1088/1748-9326/9/10/105012

Deb Burman PK, Sarma D, Williams M, Karipot A, Chakraborty S (2017) Estimating gross primary productivity of a tropical forest ecosystem over north-east India using LAI and meteorological variables. J Earth Syst Sci 126:99. https://doi.org/10.1007/s12040-017-0874-3

Deb Burman PK, Sarma D, Morrison R, Karipot A, Chakraborty S (2019) Seasonal variation of evapotranspiration and its effect on the surface energy budget closure at a tropical forest over north-east India. J Earth Syst Sci 128:127. https://doi.org/10.1007/s12040-019-1158-x

Deb Burman PK, Shurpali NJ, Chowdhuri S, Karipot A, Chakraborty S, Lind SE, Martikainen PJ, Arola A, Tiwari YK, Murugavel P,

Gurnule D, Todekar K, Prabha TV (2020a) Eddy covariance measurements of CO_2 exchange from agro-ecosystems located in subtropical (India) and boreal (Finland) climatic conditions. J Earth Syst Sci 129:43. https://doi.org/10.1007/s12040-019-1305-4

Deb Burman PK, Sarma D, Chakraborty S, Karipot A, Jain AK (2020b) The effect of Indian summer monsoon on the seasonal variation of carbon sequestration by a forest ecosystem over north-east India. SN Appl Sci 2:154. https://doi.org/10.1007/s42452-019-1934-x

Debnath G, Jain MC, Kumar S, Sarkar K, Sinha SK (1996) Methane emissions from rice fields amended with biogas slurry and farm yard manure. Clim Change 33:97–109

Dlugokencky EJ, Nisbet EG, Fisher R, Lowry D (2011) Global atmospheric methane: budget, changes and dangers. Philos Trans R Soc A Math Phys Eng Sci. https://doi.org/10.1098/rsta.2010.0341

Dörr H, Katruff L, Levin I (1993) Soil texture parameterization of the methane uptake in aerated soils. Chemosphere 26:698–713. https://doi.org/10.1016/0045-6535(93)90454-D

Etheridge DM, et al (1998) Atmospheric methane between 1000 AD and present: evidence of anthropogenic emissions and climatic variability. J Geophys Res Atmos 103.D13(1998):15979–15993. https://doi.org/10.1029/98JD00923

Fricko O, Havlik P, Rogelj J, Riahi K, Klimont Z, Gusti M, Johnson N, Kolp P, Strubegger M, Valin H, Amann M, Ermolieva T, Forsell N, Herrero M, Heyes C, Kindermann G, Krey V, McCollum DL, Obersteiner M, Pachauri S, Rao S, Schmid E, Schoepp W (2016) SSP2: a middle-of-the-road scenario for the 21st century. Glob Environ Change. https://doi.org/10.1016/j.gloenvcha.2016.06.004

Fujimori S, Hasegawa T, Masui T, Takahashi K, Herran DS, Dai H, Hijioka Y, Kainuma M (2016) AIM implementation of shared socioeconomic pathways. Glob Environ Change. https://doi.org/10.1016/j.gloenvcha.2016.06.009

Gahlot S, Shu S, Jain AK, Baidya Roy S (2017) Estimating trends and variation of net biome productivity in India for 1980–2012 using a land surface model. Geophys Res Lett 44. https://doi.org/10.1002/2017gl075777

Ganesan AL, Chatterjee A, Prinn RG, Harth CM, Salameh PK, Manning AJ, Hall BD, Mühle J, Meredith LK, Weiss RF, O'Doherty S, Young D (2013) The variability of methane, nitrous oxide and sulfur hexafluoride in Northeast India. Atmos Chem Phys 13:10633–10644. https://doi.org/10.5194/acp-13-10633-2013

Ganesan AL, Rigby M, Lunt MF, Parker RJ, Boesch H, Goulding N, Umezawa T, Zahn A, Chatterjee A, Prinn RG, Tiwari YK, van der Schoot M, Krummel PB (2017) Atmospheric observations show accurate reporting and little growth in India's methane emissions. Nat Commun 8. Article 836. https://doi.org/10.1038/s41467-017-00994-7

Garg A, Kankal B, Shukla PR (2011) Methane emissions in India: sub-regional and sectoral trends. Atmos Environ 45(28):4922–4929. https://doi.org/10.1016/j.atmosenv.2011.06.004

Ghosh S, Majumdar D, Jain MC (2003) Methane and nitrous oxide emissions from an irrigated rice of north India. Chemosphere 51:181–195

Gidden MJ, Riahi K, Smith SJ, Fujimori S, Luderer G, Kriegler E, van Vuuren DP et al (2018) Global emissions pathways under different socioeconomic scenarios for use in CMIP6: a dataset of harmonized emissions trajectories through the end of the century. Geosci Model Dev Discuss 2018:1–42. https://doi.org/10.5194/gmd2018-266

Gnanamoorthy P, Selvam V, Ramasubramanian R, Chakraborty S, Deb Burman PK, Karipot A (2019) Diurnal and seasonal patterns of soil CO_2 efflux from south east coastal Indian mangrove. Environ Monit Assess 191:258. https://doi.org/10.1007/s10661-019-7407-2

Gnanamoorthy P, Selvam V, Deb Burman PK, Chakraborty S, Nagarajan R, Karipot A, Ramasubramanian R, Grace J (2020) Seasonal variations of carbon dioxide, water vapor and energy fluxes in Indian tropical mangrove forest of Pichavaram using eddy covariance technique. Estuar Coast Shelf Sci (in press)

Hasegawa T, Havlik P, Hilaire J, Hoesly R, Horing J, Popp A, Stehfest E, Takahashi K (2018) Global emissions pathways under different socioeconomic scenarios for use in CMIP6: a dataset of harmonized emissions trajectories through the end of the century. Geosci Model Dev Discuss (in review). https://doi.org/10.5194/gmd-2018-266

IPCC (2007) Forster P, Ramaswamy V, Artaxo P, Berntsen T, Betts R, Fahey DW, Haywood J, Lean J, Lowe DC, Myhre G, Nganga J, Prinn R, Raga G, Schulz M, Van DR. Changes in atmospheric constituents and in radiative forcing. In: Solomon S, Qin D, Manning M, Chen Z, Marquis M, Averyt KB, Tignor M, Miller HL (eds) Climate change 2007: the physical science basis. Contribution of working group I to the fourth assessment report of the Intergovernmental Panel on Climate Change. Cambridge University Press, Cambridge, New York

IPCC, Climate Change (2013) The physical science basis. In: Stocker TF, Qin D, Plattner G-K, Tignor M, Allen SK, Boschung J, Nauels A, Xia Y, Bex V, Midgley PM (eds) Contribution of working group I to the fifth assessment report of the Intergovernmental Panel on Climate Change. Cambridge University Press, Cambridge, New York, 1535 pp

Jain MC, Kumar S, Wassmann R et al (2000) Methane emissions from irrigated rice fields in northern India (New Delhi). Nutr Cycl Agroecosyst 58:75–83

Jain SK, Kumar V, Saharia M (2013) Analysis of rainfall and temperature trends in northeast India. Int J Climatol 33:968–978

Jha CS, Thumaty KC, Rodda SR, Sonakia A, Dadhwal VK (2013) Analysis of carbon dioxide, water vapour and energy fluxes over an Indian teak mixed deciduous forest for winter and summer months using eddy covariance technique. J Earth Syst Sci 122:1259–1268

Jha CS, Rodda SR, Thumaty KC, Raha AK, Dadhwal VK (2014) Eddy covariance based methane flux in Sundarbans mangroves, India. J Earth Syst Sci 123:1089–1096

Kaul M, Dadhwal VK, Mohren GMJ (2009) Land use change and net C flux in Indian forests. For Ecol Manage 258:100–108

Keppler F, Hamilton JTG, Brass M, Röckmann T (2006) Methane emissions from terrestrial plants under aerobic conditions. Nature 439:187–191. https://doi.org/10.1038/nature04420

Kriegler E, Bauer N, Popp A, Humpenöder F, Leimbach M, Strefler J, Baumstark L, Bodirsky B, Hilaire J, Klein D, Mouratiadou I, Weindl I, Bertram C, Dietrich J-P, Luderer G, Pehl M, Pietzcker R, Piontek F, Lotze-Campen H, Biewald A, Bonsch M, Giannousakis A, Kreidenweis U, Müller C, Rolinski S, Schwanitz J, Stefanovic M (2016) Fossil-fueled development (SSP5): an energy and resource intensive scenario for the 21st century. Glob Environ Change. https://doi.org/10.1016/j.gloenvcha.2016.05.015

Kumaresan S, Shekhar S, Chakraborty S, Sundaramanickama A, Kuly N (2018) Environmental variables and nutrients in selected islands of Lakshadweep Sea; addressing coral bleaching. Reg Stud Mar Sci 22:38–48

Lal S, Venkataramani S, Subbaraya B (1993) Methane flux measurements from paddy fields in the tropical Indian region. Atmos Environ 27:1–3

Lin X, Indira NK, Ramonet M, Delmotte M, Ciais P, Bhatt BC, Reddy MV, Angchuk D, Balakrishnan S, Jorphail S, Dorjai T, Mahey TT, Patnaik S, Begum M, Brenninkmeijer C, Durairaj S, Kirubagaran R, Schmidt M, Swathi PS, Vinithkumar NV, Yver Kwok C, Gaur VK (2015) Long-lived atmospheric trace gases measurements in flask samples from three stations in India. Atmos Chem Phys 15:9819–9849. https://doi.org/10.5194/acp-15-9819-2015

Liu DN, Guo XD, Xiao BW (2019) What causes growth of global greenhouse gas emissions? Evidence from 40 countries. Sci Total Environ 661:750–766. https://doi.org/10.1016/j.scitotenv.2019.01.197

Mahesh P, Sharma N, Dadhwal VK, Rao PVN, Apparao BV, Ghosh AK, Mallikarjun K, Ali MM (2014) Impact of land-sea breeze and rainfall on CO_2 variations at a coastal station. J Earth Sci Clim Change 5:6

Mahesh P, Sreenivas G, Rao PVN, Dadhwal VK, Sai Krishna SVS, Mallikarjun K (2015) High-precision surface-level CO_2 and CH_4 using off-axis integrated cavity output spectroscopy (OA-ICOS) over Shadnagar, India. Int J Remote Sens 36(22):5754–5765. https://doi.org/10.1080/01431161.2015.1104744

Mahesh P, Sreenivas G, Rao PVN, Dadhwal VK (2016) Atmospheric CO_2 retrieval from ground based FTIR spectrometer over Shadnagar, India. Atmos Meas Tech Discuss. https://doi.org/10.5194/amt-2016-177

Mahesh P, Gaddamidi S, Gharai B, Mullapudi Venkata Rama SS, Kumar Sundaran R, Wang W (2019) Retrieval of CO_2, CH_4, CO and N_2O using ground-based FTIR data and validation against satellite observations over the Shadnagar, India. Atmos Meas Tech Discuss. https://doi.org/10.5194/amt-2019-7

Malla G, Bhatia A, Pathak H, Prasad S, Jain N, Singh J (2005) Mitigating nitrous oxide and methane emissions from soil in rice-wheat system of the Indo-Gangetic plain with nitrification and urease inhibitors. Chemosphere 58:141–147

Meiyappan P et al (2017) Dynamics and determinants of land change in India: integrating satellite data with village socioeconomics. Reg Environ Change 17:753–766. https://doi.org/10.1007/s10113-016-1068-2

Metya A, Datye A, Chakraborty S, Tiwari Y (2020) Diurnal and seasonal variability of CO_2 and CH_4 concentrations in a semi-urban environment of western India. Theor Appl Climatol (under review)

MoEFCC (2015) India: first biennial update report to the United Nations framework convention on climate change. Ministry of Environment, Forest and Climate Change, Government of India

MoEFCC (2018) India: second biennial update report to the United Nations framework convention on climate change. Ministry of Environment, Forest and Climate Change, Government of India

Mukherjee S, Sekar KC, Lohani P, Kumar K, Patra P, Ishijima K (2018) Investigation of scale interaction between rainfall and ecosystem carbon exchange of Western Himalayan Pine dominated vegetation. Biogeosci Discuss. https://doi.org/10.5194/bg-2018-299

Mukhophadhya SK, Biswas H, Das KL, Jana TK (2001) Diurnal variation of carbon dioxide and methane exchange above Sundarbans mangrove forest, in NW coast of India. Indian J Mar Sci 30:70–74

Nalini K, Sijikumar S, Valsala V, Tiwari YK, Ramachandran R (2019) Designing surface CO_2 monitoring network to constrain the Indian land fluxes. Atmos Environ 218:117003

Nayak RK, Patel NR, Dadhwal VK (2013) Inter-annual variability and climate control of terrestrial net primary productivity over India. Int J Climatol 33:132–142

O'Neill BC, Kriegler E, Riahi K, Ebi KL, Hallegatte S, Carter TR, Mathur R, Van Vuuren DP (2014) A new scenario framework for climate change research: the concept of shared socioeconomic pathways. Clim Change 122:387–400

O'Neill BC, Kriegler E, Ebi KL, Kemp-Benedict E, Riahi K, Rothman DS, Van Ruijven BJ, Van Vuuren DP, Birkmann J, Kok K, Levy M, Solecki W (2017) The roads ahead: narratives for shared socioeconomic pathways describing world futures in the 21st century. Glob Environ Change 42:169–180

Oo AZ, Sudo S, Inubushi K, Mano M, Yamamoto A, Ono K, Osawa T, Hayashida S, Patra PK, Terao Y, Elayakumar P, Vanitha K, Umamageswari C, Jothimani P, Ravi V (2018) Methane and nitrous oxide emissions from conventional and modified rice cultivation systems in South India. Agric Ecosyst Environ 252:148–158

Parashar D, Rai J, Gupta P, Singh N (1991) Parameters affecting methane emission from paddy fields. Indian J Radio Space Phys 20:12–17

Parker RJ et al (2015) Assessing 5 years of GOSAT Proxy XCH_4 data and associated uncertainties. Atmos Meas Tech 8:4785–4801

Patel NR, Dadhwal VK, Saha SK (2011) Measurement and scaling of carbon dioxide (CO_2) exchanges in wheat using flux-tower and remote sensing. J Ind Soc Remote Sens 39:383–391

Pathak H, Li C, Wassmann R (2005) Greenhouse gas emissions from Indian rice fields: calibration and upscaling using the DNDC model. Biogeosciences 2:113–123

Patra PK et al (2009) Growth rate, seasonal, synoptic, diurnal variations and budget of methane in the lower atmosphere. J Meteorol Soc Jpn 87:635–663. https://doi.org/10.2151/jmsj.87.635

Patra PK et al (2013) The carbon budget of South Asia. Biogeosciences 10:513–527. https://doi.org/10.5194/bg-10-513-2013

Patra PK, Saeki T, Dlugokencky EJ, Ishijima K, Umezawa T, Ito A, Aoki S, Morimoto S, Kort EA, Crotwell A, Ravikumar K, Nakazawa T (2016) Regional methane emission estimation based on observed atmospheric concentrations (2002–2012). J Meteorol Soc Jpn 94:91–113

Pillai ND, Nandy S, Patel NR, Srinet R, Watham T (2019) Integration of eddy covariance and process-based model for the intra-annual variability of carbon fluxes in an Indian tropical forest. Biodivers Conserv. https://doi.org/10.1007/s10531-019-01770-3

Prasad VK, Lata M, Badarinath KVS (2003) Trace gas emissions from biomass burning from northeast region in India—estimates from satellite remote sensing data and GIS. Environmentalist 23:229–236

Purvaja R, Ramesh R (2001) Natural and anthropogenic methane emission from coastal wetlands of south India. Environ Manage 27:547–557

Rao AS, Bala G, Ravindranath NH, Nemani R (2019) Multi-model assessment of trends, variability and drivers of terrestrial carbon uptake in India. J Earth Syst Sci 128:1–19. https://doi.org/10.1007/s12040-019-1120-y

Ravindranath NH, Somashekhar BS, Gadgil M (1997) Carbon flow in Indian forests. Clim Change 35:297–320

Ravindranath NH, Chaturvedi RK, Murthy IK (2008) Forest conservation, afforestation and reforestation in India: implications for forest carbon stocks. Curr Sci 95:216–222

Riahi K, Van Vuuren DP, Kriegler E, Edmonds J, O'Neill B, Fujimori SNB, Calvin K, Dellink R, Fricko O, Lutz W, Popp A, Crespo Cuaresma J, Leimbach M, Kram T, Rao S, Emmerling J, Hasegawa T, Havlik P, Humpenöder F, Aleluia Da Silva L, Smith SJ, Stehfest E, Bosetti V, Eom J, Gernaat D, Masui T, Rogelj J, Strefler J, Drouet L, Krey V, Luderer G, Harmsen M, Takahashi K, Wise M, Baumstark L, Doelman J, Kainuma M, Klimont Z, Marangoni G, Moss R, Lotze-Campen H, Obersteiner M, Tabeau A, Tavoni M (2017) The shared socio-economic pathways and their energy, land use and greenhouse gas emissions implications: an overview. Glob Environ Change 42:148–152

Rigby M, Prinn RG, Fraser PJ, Simmonds PG, Langenfelds RL, Huang J, Cunnold DM, Steele LP, Krummel PB, Weiss RF, O'Doherty S, Salameh PK, Wang HJ, Harth CM, Muhle J, Porter LW (2008) Renewed growth of atmospheric methane. Geophys Res Lett 35:L22805. https://doi.org/10.1029/2008GL036037

Rodda S, Thumaty K, Jha C, Dadhwal V (2016) Seasonal variations of carbon dioxide, water vapor and energy fluxes in tropical Indian mangroves. Forests 7:35

Sarma D, Baruah KK, Baruah R, Gogoi N, Chakraborty S, Karipot AK (2018) Carbon dioxide, water vapour and energy fluxes over a semi evergreen forest in Assam, Northeast India using eddy covariance technique. J Earth Syst Sci. https://doi.org/10.1007/s12040-018-0993-5

Sasakawa M, Machida T, Ishijima K, Arshinov M, Patra PK, Ito A, Aoki S, Petrov V (2017) Temporal characteristics of CH_4 vertical profiles observed in the West Siberian Lowland over Surgut from 1993 to 2015 and Novosibirsk from 1997 to 2015. J Geophys Res (Atmos). https://doi.org/10.1002/2017JD026836

Schuck TJ et al (2010) Greenhouse gas relationships in the Indian summer monsoon plume measured by the CARIBIC passenger aircraft. Atmos Chem Phys 10:3965–3984

Selvam B, Natchimuthu S, Arunachalam L et al (2014) Methane and carbon dioxide emissions from inland waters in India—implications for large scale greenhouse gas balances. Glob Change Biol 20:3397–3407

Sharma N, Nayak RK, Dadhwal VK, Kant Y, Ali MM (2013) Temporal variations of atmospheric CO_2 in Dehradun, India during 2009. Air Soil Water Res 6:37–45

Sharma N et al (2014) Atmospheric CO_2 variations in two contrasting environmental sites over India. Air Soil Water Res 7:61–68. https://doi.org/10.4137/aswr.s13987

Singh JS, Singh S, Raghubanshi AS, Singh S, Kashyap AK (1996) Methane flux from rice/wheat agroecosystem as affected by crop phenology, fertilization and water level. Plant Soil 183:323–327

Singh JS, Raghubanshi AS, Reddy VS, Singh S, Kashyap AK (1997) Methane flux from irrigated paddy and dryland rice fields, and from seasonally dry tropical forest and savanna soils of India. Soil Biol Biochem 30:135–139

Sreenivas G, Mahesh P, Subin J, Kanchana AL, Rao PVN, Dadhwal VK (2016) Influence of meteorology and interrelationship with greenhouse gases (CO_2 and CH_4) at a suburban site of India. Atmos Chem Phys 16:3953–3967

Sreenivas G, Mahesh P, Biswadip G, Suresh S, Rao PVN, Krishna Chaitanya M, Srinivasulu P (2019) Spatio-temporal distribution of CO_2 mixing ratio over Bhubaneswar, Varanasi and Jodhpur of India—airborne campaign, 2016. Atmos Environ. https://doi.org/10.1016/j.atmosenv.2019.01.010

Stocker TF et al (2013) IPCC, 2013: climate change 2013: the physical science basis. Contribution of working group I to the fifth assessment report of the Intergovernmental Panel on Climate Change

Thomson AJ, Giannopoulos G, Pretty J, Baggs EM, Richardson DJ (2012) Biological sources and sinks of nitrous oxide and strategies to mitigate emissions. Philos Trans R Soc B Biol Sci 367. https://doi.org/10.1098/rstb.2011.0415

Thompson RL, Lassaletta L, Patra PK et al (2019) Acceleration of global N_2O emissions seen from two decades of atmospheric inversion. Nat Clim Change 9:993–998. https://doi.org/10.1038/s41558-019-0613-7

Tiwari YK, Patra PK, Chevallier F, Francey RJ, Krummel PB, Allison CE, Revadekar JV, Chakraborty S, Langenfelds RL, Bhattacharya SK, Borole DV, Kumar KR, Steele LP (2011), CO_2 observations at Cape Rama, India for the period of 1993–2002: implications for constraining Indian emissions. Curr Sci 101 (12):1562–1568. ISSN: 0011-3891

Tiwari YK, Vellore RK, Kumar KR, van der Schoot M, Cho C-H (2014) Influence of monsoons on atmospheric CO_2 spatial variability and ground-based monitoring over India. Sci Total Environ 490:570–578. https://doi.org/10.1016/j.scitotenv.2014.05.045. ISSN: 0048-9697

Umezawa T, Niwa Y, Sawa Y, Machida T, Matsueda H (2016) Winter crop CO_2 uptake inferred from CONTRAIL measurements over Delhi, India. Geophys Res Lett 43:11859–11866. https://doi.org/10.1002/2016gl070939

Valsala V, Tiwari YK, Pillai P, Roxy M, Maksyutov S, Murtugudde R (2013) Intraseasonal variability of terrestrial biospheric CO_2 fluxes over India during summer monsoons. J Geophys Res Biogeosci 118:752–769

van Vuuren DP, Stehfest E, Gernaat D, Doelman J, van Berg M, Harmsen M, de Boer H-S, Bouwman LF, Daioglou V, Edelenbosch O, Girod B, Kram T, Lassaletta L, Lucas P, van Meijl H, Müller C, van Ruijven B, Tabeau A (2016) Energy, land-use and greenhouse gas emissions trajectories under a green growth paradigm. Glob Environ Change. https://doi.org/10.1016/j.gloenvcha.2016.05.008

Wassman R, Neue H, Ladha J, Aulakh M (2004) Mitigating greenhouse gas emissions from rice-wheat cropping systems in Asia. Environ Dev Sustain 6:65–90

Watham T, Kushwaha SP, Patel NR, Dadhwal VK (2014) Monitoring of carbon dioxide and water vapour exchange over a young mixed forest plantation using eddy covariance technique. Curr Sci 107:858–866

Watham T, Patel NR, Kushwaha SPS, Dadhwal VK, Kumar AS (2017) Evaluation of remote-sensing-based models of gross primary productivity over Indian sal forest using flux tower and MODIS satellite data. Int J Remote Sens 38:5069–5090

Watson RT, Noble I, Bolin B, Ravindranath NH, Verardo DJ, Dokken DJ (2000) Land use, land-use change, and forestry. A special report. Intergovernmental Panel on Climate Change, Washington. http://www.grida.no/climate/ipcc/land_use/024.htm

Coordinating Lead Authors

Suvarna Fadnavis, Indian Institute of Tropical Meteorology (IITM-MoES), Pune, India,
e-mail: suvarna@tropmet.res.in (corresponding author)
Anoop Sharad Mahajan, Indian Institute of Tropical Meteorology (IITM-MoES), Pune, India
Ayantika Dey Choudhury, Indian Institute of Tropical Meteorology (IITM-MoES), Pune, India

Lead Authors

Chaitri Roy, Indian Institute of Tropical Meteorology (IITM-MoES), Pune, India
Manmeet Singh, Indian Institute of Tropical Meteorology (IITM-MoES), Pune, India
Mriganka Shekhar Biswas, Indian Institute of Tropical Meteorology (IITM-MoES), Pune, India

Contributing Authors

G. Pandithurai, Indian Institute of Tropical Meteorology (IITM-MoES), Pune, India
Thara Prabhakaran, Indian Institute of Tropical Meteorology (IITM-MoES), Pune, India

Review Editors

Shyam Lal, Physical Research Laboratory, Ahmedabad, India
Chandra Venkatraman, Indian Institute of Technology Bombay, Mumbai, India
Dilip Ganguly, Indian Institute of Technology Delhi, New Delhi, India
Vinayak Sinha, Indian Institute of Science Education and Research, Mohali, India
M. M. Sarin, Physical Research Laboratory, Ahmedabad, India

Corresponding Author

Suvarna Fadnavis, Indian Institute of Tropical Meteorology (IITM-MoES), Pune, India,
e-mail: suvarna@tropmet.res.in

© The Author(s) 2020
R. Krishnan et al. (eds.), *Assessment of Climate Change over the Indian Region*,
https://doi.org/10.1007/978-981-15-4327-2_5

Key Messages

- Aerosol loading over India has substantially increased during the recent few decades. The annual mean 500 nm aerosol optical depth (AOD) from ground-based observations shows an overall increasing trend of $\sim 2\%$ year^{-1} during the last 30 years (high confidence). This trend in AOD is subject to seasonal variability. The rate of increase in AOD is significantly high during the dry months of December–March.

- The aerosol radiative forcing over India shows wide spatiotemporal variability resulting from the non-uniform distribution of aerosol burden over the region. Estimates of aerosol radiative forcing from measurements range from -49 to -31 W m^{-2} at the surface (high confidence) and from -15 to $+8$ W m^{-2} at the top-of-atmosphere (low confidence). The estimates at the top of the atmosphere are highly sensitive to the single scattering albedo values.

- The understanding of the aerosol indirect effect and aerosol impacts on precipitation has low confidence and needs to be addressed with process studies in different cloud systems and their environments.

- There is substantial spatiotemporal variability in the concentration of ozone (O$_3$) and its precursors over the Indian region. In general, there is an increasing trend in the ozone mixing ratios in the troposphere ($+0.7$ to $+0.9\%$ year^{-1} during 1979–2005, medium confidence) and a decreasing trend in the stratosphere (-0.05 to -0.4% year^{-1} during 1993–2015, medium confidence). Trends are driven by precursor gases emitted by anthropogenic activities.

- Over the Indian region, the estimates of radiative forcing (at the tropopause) due to tropospheric ozone increase since pre-industrial times vary between ~ 0.2 and 0.4 W m^{-2}.

5.1   Introduction

Aerosols and trace gases are essential drivers of climate change. They influence Earth's energy budget leading to climate change through various pathways. Their climatic impacts are eventually manifested as precipitation changes, increased evaporation, elevated temperatures, etc. Hence, information on their 'sources and sink,' 'physical and chemical processes,' and 'distribution' is important for an accurate prediction of the climate.

The Indian subcontinent is directly influenced by different aerosol species via changes in the insolation, atmospheric temperature structure, and alteration of the regional hydrological cycle. Along with absorption and scattering of

incoming solar radiation, aerosols interact with clouds modifying its radiative properties and precipitation efficiency (Box 5.1). The aerosol concentrations over the subcontinent are dominated by wind-driven desert dust, biomass burning, industrialization, agricultural activities, etc. Rapid growth in population, industrialization, and urbanization—over South, East, and Southeast Asia—has contributed to the significant rise in emissions producing different types of aerosol over the region. The associated increase in anthropogenic aerosol loading in recent decades (Satheesh et al. 2017) has led to increased reduction of surface insolation, contributing to solar dimming over the Indian landmass, affecting the energy balance at the surface (Ramanathan et al. 2005; Soni et al. 2012). The high aerosol burden has also been linked to changes in the hydrological cycle of the region (Box 5.3). The long-term decline in southwest monsoon precipitation has been associated with anthropogenic aerosol forcing over South Asia [(Krishnan et al. 2016), Box 5.3].

Box 5.1: How Aerosols Affect Regional Climate?
Atmospheric aerosols are tiny solid/liquid/mixed particles suspended in the air originating from natural or anthropogenic sources. With a typical lifetime of days to weeks in the troposphere and about a year in the stratosphere, aerosol size ranges from a few nanometers to several tens of micrometers. Aerosol particles influence the climate in different ways (Box 5.1, Fig. 5.1).

Fundamentally, aerosol particles absorb and scatter incoming solar radiation modifying the global and regional radiative budget. Non-absorbing aerosols like sulfate, nitrate, and sea spray scatter shortwave radiation back to space leading to a net cooling of the climate system while absorbing aerosols produce the opposite effect. Carbonaceous aerosols (black carbon, organic carbon) and mineral dust can absorb and scatter sunlight producing either warming or cooling effects determined by aerosol properties and environmental conditions. Absorbing aerosols also affect climate when present in surface snow by lowering surface albedo, yielding a positive radiative forcing, directly changing the melting of snow and ice. Although depending on the local emissions and transport processes, regionally, the anthropogenic aerosol radiative forcing can be either negative or positive; it is well established that globally, the radiative effect of anthropogenic aerosols produces cooling of the planet (IPCC 2013).

Additionally, aerosol particles act as cloud condensation nuclei (CCN) and ice nuclei (IN) and therefore have a significant impact on cloud properties

Fig. 5.1 A schematic showing the different sources of aerosols in the atmosphere and their effect on the radiative budget. Adapted from the figure provided by Brookhaven National Laboratory

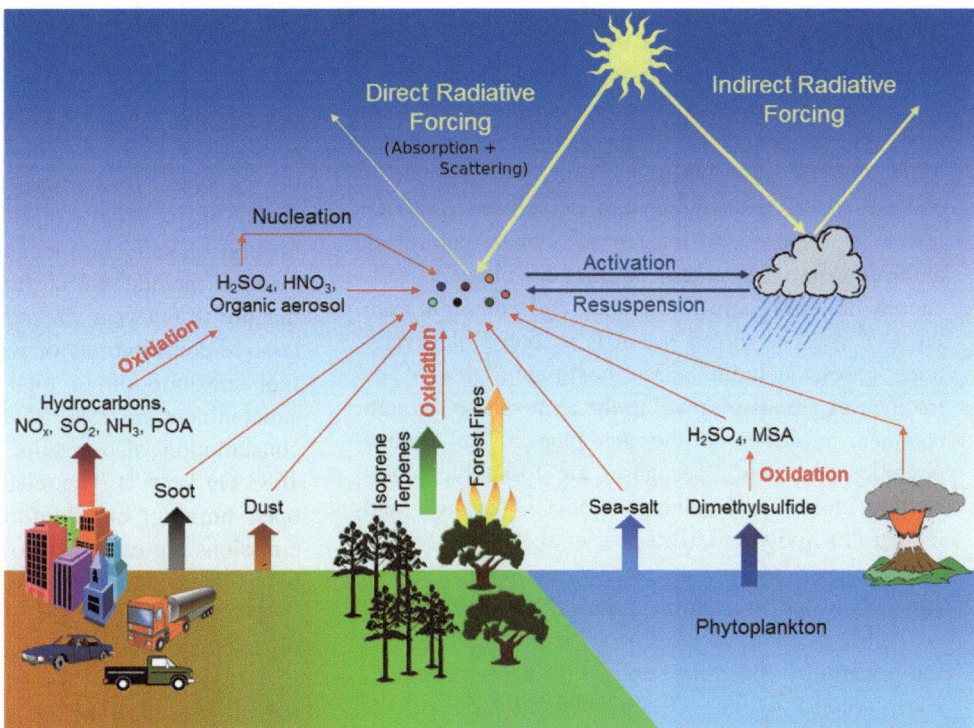

and precipitation. An increase in CCN forming aerosols in a cloudy region produces more, but smaller, cloud droplets reflecting more solar radiation to space leading to a cooling of the Earth's surface, known as the first indirect effect (cloud-albedo effect). Also, smaller droplets suppress collision coalescence requiring longer growth time to reach raindrop size, increasing the cloud albedo and enhancing the cooling effect, known as the cloud-lifetime effect or the second indirect effect. In contrast, in a saturated and buoyant environment, an increase in CCN forming aerosols can invigorate the cloud through microphysical processes; however, there is significant uncertainty in such impacts.

Furthermore, absorbing aerosols alter the air temperature causing an increase in lower level static stability inhibiting convection leading to a decrease in cloud cover, known as the semi-direct effect. However, the net effect of absorbing aerosols on precipitation depends on the vertical variation of the particles and background conditions.

Ozone, photo-chemically active trace gas, plays a crucial role in the climate system due to its implications on radiative processes and resulting dynamical changes. The majority of (ninety percent) the ozone in the atmosphere occurs in the stratosphere, where it is formed naturally by chemical reactions involving solar ultraviolet radiation (sunlight) and oxygen molecules (WMO 2019a). Remaining 10% of ozone

occurring in the troposphere is mainly produced from precursor gases (e.g., methane ($CH4$), nitrogen oxides (NO_x), volatile organic compounds (VOCs), carbon monoxide (CO), methane (CH_4)). The industrialization, vehicular emission, and other anthropogenic activities have accelerated the emission growth of ozone precursors leading to a continued rise in tropospheric ozone concentrations (Sinha et al. 2014). On the contrary, addition of these gases and ozone-depleting substances (ODSs) produces significant stratospheric ozone loss leading to chemical and dynamical changes in the troposphere and stratosphere. The measurements over India during the past two–three decades show rapid changes in ozone (O_3) mixing ratios in the troposphere and stratosphere. Considering the critical role of atmospheric ozone in the climate system, in this chapter, we document an assessment of emissions of ozone precursors, trends in ozone and related gases, the influence of transport processes on their distribution over the Indian region.

It is difficult to detangle the cause and effect of climate change. The temperature changes have significantly been affected by the atmospheric burden of aerosols and traces gases. The transport processes produce a significant impact on the redistribution of aerosols and trace gases (e.g., dust transport from West Asia in pre-monsoon season; monsoon convection-based lifting of aerosols and trace gases into the lower stratosphere, tropopause folding events in the winter/pre-monsoon season). The changes in their concentrations at the receptor region affect the temperature, radiative forcing, clouds, and aerosol-cloud interactions.

Box 5.2: What are Trace Gases?

The Earth's atmosphere consists of large amounts of nitrogen (78% by volume) and oxygen (21% by volume). The remaining 1% of the atmospheric gases are known as trace gases because they are present in small concentrations, mostly one part per billion (ppb) or lower. The sources of trace gases can be natural or anthropogenic. The natural sources are biogenic, volcanoes, lightning and forest fires, and emission from the Oceans. The global ocean is a source of several trace gases, including sulfur-containing gases. The trace gases are also formed in the atmosphere through chemical reactions in the gas phase. The anthropogenic sources of trace gases are fossil fuel combustion, fossil fuel mining, biomass burning, and industrial activity, etc. (Brasseur et al. 1999).

The most important trace gases found in the atmosphere are greenhouse gases. These trace gases are called greenhouse gases because they help to keep Earth warm by absorbing sunlight. In the troposphere, water vapor, ozone (O_3), carbon dioxide (CO_2), methane (CH_4), sulfur dioxide (SO_2), and nitrous oxide (N_2O) are the important greenhouse (trace) gases. The two most abundant greenhouse gases by volume are water vapor and CO_2 (Brasseur et al. 1999).

Ozone acts as a greenhouse gas in the troposphere, while in the stratosphere, filter out the incoming ultraviolet radiation coming from the Sun. Thus, it helps in protecting life on the Earth. The human-made processes have injected new trace gases into the atmosphere, for example, chlorofluorocarbons (CFCs), which damage the ozone layer in the stratosphere.

The increased burden of trace gases in the atmosphere leads to global warming and climate change. The focus of the current chapter is ozone and related trace gases, while other trace gases, e.g., CO_2, CH_4, and, N_2O, etc., are discussed in Chap. 4.

5.2 Aerosols

5.2.1 Emissions of Different Aerosol Species

Atmospheric aerosols originate from two distinct pathways—either by direct emission of primary aerosols into the atmosphere (e.g., dust, sea salt, OC, BC) or by the formation of secondary aerosols via atmospheric chemical reactions (e.g., sulfate, nitrate, ammonium, and SOA). While BC, along with sulfate, nitrate, and ammonium, has anthropogenic sources like incomplete combustion of biomass and fossil fuel, sea salt, dust, and primary biological aerosols are naturally produced in the atmosphere. Over the Indian subcontinent, the aerosol emission rates are 8.9 Tg year^{-1} for NMVOCs, 0.7 Tg year^{-1} for BC, 1.9 Tg year^{-1} for primary organic aerosol, 2.9 TgS year^{-1} for SO_2, 5.8 Tg year^{-1} for NH_3, and, 0.5 Tg year^{-1} for biomass burning aerosols (IPCC 2013).

The major source of BC emissions in India is the combustion of solid biofuel (172–340 Gg year^{-1}), while other sources include wood fuel (143 Gg year^{-1}), dried cattle manure (8 Gg year^{-1}), and crop waste (21 Gg year^{-1}). The relative contributions of fossil fuel, open burning, and biofuel consumption to total BC emissions over the Indian subcontinent are 25%, 35%, and 42%, respectively. Biofuel consumption also results in the emission of OC (583–1683 Gg year^{-1}). The relative contributions from fossil fuel, open burning, and biofuel consumption to the total OC emissions are estimated to be 13%, 43%, and 44%, respectively (Venkataraman et al. 2005).

The transport sector, the second-largest contributor to organic aerosols over India, shows an emission rate of 0.14 (0.1–0.3) Tg year^{-1} for BC and 0.07 (0.02–0.2) Tg year^{-1} for OC. For both the emissions, diesel vehicles were found to be the primary cause (92% to BC and 78% to OC). However, the combined emission rate of BC and OC from both the industry and transport sectors is 0.23 Tg year^{-1} and 0.15 Tg year^{-1}, respectively (Sadavarte and Venkataraman 2014). In the Indian rural sector, the emission estimate of organic and elemental carbon from biomass fuels over the IGP is 361.96 ± 170.18 Gg and 56.44 ± 29.06 Gg, respectively (Saud et al. 2012). The estimates of aerosols emissions from open burning (forest and crop waste) are 102–409 Gg year^{-1} for BC, 399–1529 Gg year^{-1} for OC, and 663–2303 Gg year^{-1} for organic matter. This overall contributes to about 25% of the total BC, OC/OM emissions (Venkataraman et al. 2006).

Residential sector SO_2 emissions have been estimated at 0.2 (0.08–0.4) Tg year^{-1} with major contribution from dung-based (56%) and coal-based stoves (19%). Emissions of SO_2 from agriculture were estimated at 0.09 (0.02–0.2) Tg year^{-1}. From the industry sector and transport sector, emission estimates of SO_2 were reported to be 7 (6.0–9.6) Tg year^{-1} and 0.08 (0.04–0.3) Tg year^{-1}, respectively, for the year 2015 (Pandey et al. 2014).

From 1996 to 2015, there has been a 30% increase in BC emissions due to increased emissions from informal industries, while OC emissions increased by only 4% (Pandey et al. 2014). Vehicular emissions of BC have increased by 112% during 1991–2001. From all the sources, the estimated BC emission for India is around 835.50 Gg for 1991 and 1343.78 Gg in 2001, indicating a growth of about 61% during the 1990s. During the same period, trends in SO_2 emissions increased by 32% (Pandey et al. 2014).

5.2.2 Long-Term Change in Aerosol Optical Depth

Long-term changes in aerosol loading over the Indian subcontinent have been estimated using ground-based single station measurements (Dani et al. 2012; Kaskaoutis et al. 2012) as well as multi-station long-term (>25 years for some stations) ground-based observational network database (Babu et al. 2013; Krishna Moorthy et al. 2013). Observations from ARFINET observatories spread over 35 locations across India reported an increase in annual mean AOD (0.008–0.02 year^{-1}) with a rate of 2.3% year^{-1} (of its value in 1985), while AOD trend in the last decade (2001–2011) alone showed a rapid increase of ~4% year^{-1} (Fig. 5.2) (Babu et al. 2013; Krishna Moorthy et al. 2013). However, the estimates of AOD trends vary from one location to another, with rural locations showing weak negative trends while industrial stations over peninsular India show high positive trends(Babu et al. 2013). AOD increasing trend of 2% per year is reported over Pune during the period 1998–2007 (Dani et al. 2012). Recent AERONET data over Pune from 2008 to 2017 also shows a positive trend in the range of 2.4–4% per year. The rate of increase of aerosol loading over the country is considerably high (0.0005–0.04 year^{-1}) during the dry winter months (December–March), while due to the contending effects of dust transportation and wet scavenging of aerosols by the monsoon precipitation, trends are weak or insignificant during the pre-monsoon and summer monsoon seasons (Babu et al. 2013). Furthermore, the long-term change in Angstrom wavelength exponent over India shows an increasing trend implying a relative buildup

of fine anthropogenic aerosols compared to coarser natural aerosols over the region (Dani et al. 2012; Satheesh et al. 2017). In contrast, recent ground observations have revealed a decreasing trend in BC concentrations over various locations in India (Ravi Kiran et al. 2018; Manoj et al. 2019; Sarkar et al. 2019). ARFINET observations (2007–2016) recorded a decreasing trend in BC mass concentrations over India, at a rate of ~242 ± 53 ng m^{-3} year^{-1} (Manoj et al. 2019). The negative trend in BC concentrations in the backdrop of rising BC emission trends (Sahu et al. 2008; Pandey et al. 2014) brings forth the uncertainty in BC estimates due to scattered observations over the source region (IGP) (Rana et al. 2019).

The trends in AOD over the Indian subcontinent have also been reported from satellite observations. The annual mean trend of AOD from MODIS observations for the period 2000–2014 shows an increase of ~40% over the Indian landmass (Srivastava 2017). MODIS regional trends show that the annual mean AODs have increased by >40% in major urban cities like Jaipur, Hyderabad, and Bengaluru during 2000–2009, while it has decreased (~10%) over high-altitude sites of Dehradun and Shimla (Ramachandran et al. 2012). Strong seasonal variability in AOD trends is also observed from satellite observations. During winter (DJF), a significantly increasing trend (1–2% year^{-1}, 0.02% year^{-1}) is observed over the subcontinent, especially IGP (Srivastava 2017). SeaWiFS (Sea-Viewing Wide Field-of-View Sensor) AOD shows an increasing trend of 0.0053 ± 0.0011 year^{-1} during DJF over 1998–2010 (Hsu et al. 2012). The post-monsoon/winter decadal trend from

Fig. 5.2 **a** Long-term 500 nm AOD trends derived from ARFINET station data. Different colors and symbols differentiate the stations and **b** long-term trend in regional mean aerosol optical depth at 500 nm. Adopted from Krishna Moorthy et al. (2013). © American Geophysical Union. Used with permission

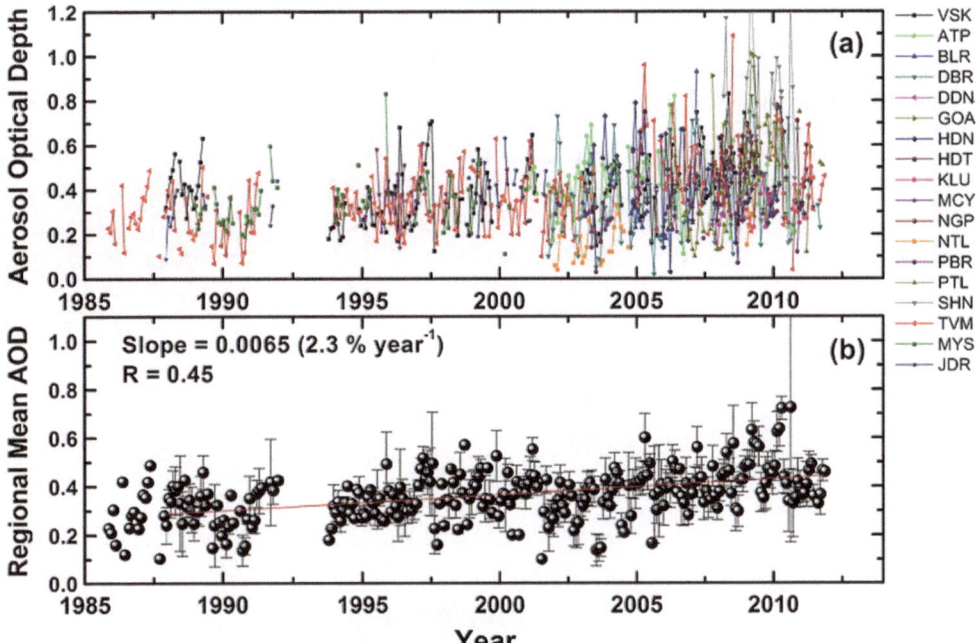

MISR (Multi-angle Imaging SpectroRadiometer) for the period (March 2000–February 2010) also shows an increasing trend in the range 0.01–0.04 year^{-1} over urban centers and densely populated rural areas (Dey and Di Girolamo 2011). On the contrary, a significant and widespread decreasing trend in AOD is observed in the pre-monsoon (MAM) period over certain parts of northern India from satellite measurements (MODIS, MISR, and OMI) (Pandey et al. 2017). However, the decrease in AOD is more prominent over northwest India (−0.01 to −0.02 year^{-1}), while a clear increasing trend (0.01–0.02 year^{-1}) is observed in the eastern part of the IGP. During the monsoon season, AOD trends do not exhibit any statistically significant signal but show slight (significant at 90%) positive values +0.003 to +0.0017 during the post-monsoon season (SON) over the IGP. During this period, strong positive trends (2–8%) are observed over northeast India. Over the oceanic regions, significant positive trends (2–6% year^{-1}) have been reported during DJF and MAM. A positive trend in AOD (>6% year^{-1}) was also observed during JJAS over the Northern Bay of Bengal.

Box 5.3: How Aerosols Impact Monsoon Precipitation?

The non-uniform distribution of aerosols in the atmosphere creates uneven atmospheric heating and surface cooling patterns, which drive changes in atmospheric circulation and regional rainfall. On longer timescales, the declining trend in ISM precipitation post-1950s has been linked to the rising anthropogenic aerosol burden over various regions (Ramanathan et al. 2005; Bollasina et al. 2011; Ganguly et al. 2012; Sanap and Pandithurai 2015; Sanap et al. 2015; Krishnan et al. 2016; Undorf et al. 2018). Local and remote aerosols alter the land–sea temperature contrast as well as the tropospheric temperature structure, both of which have a profound influence on the onset and sustenance of south Asian monsoon. An associated weakening of the monsoon overturning circulation (Fig. 5.3) due to anthropogenic aerosol-induced surface radiative changes results in suppression of ISM rainfall.

On shorter timescales, aerosols can either enhance or suppress monsoon convection depending on its properties and spatiotemporal variations. During pre-monsoon and early monsoon months, absorbing aerosols like locally emitted BC and soot from domestic and industrial sources, as well as transported dust from West Asia, accumulate over IGP and Tibetan plateau and contribute to invigoration of precipitation through the 'elevated heat pump' hypothesis (Lau and Kim 2006; Lau et al. 2006). Increased aerosol loading over the IGP and TP during pre-monsoon season due to transport associated with El Niño causes precipitation enhancements of ∼0.5–1.5 mm day^{-1} through an anomalous aerosol-induced warm core in the atmospheric column (Fadnavis et al. 2017). The heating induced relative strengthening of the cross-equatorial moisture flow reduces the severity of drought in El Niño years. On weekly timescale, atmospheric heating from accumulated dust aerosols over the Arabian Sea strengthens monsoon westerlies and helps in the

Fig. 5.3 Latitude-pressure section showing meridional overturning anomaly during the summer monsoon season. Adapted from Sanap et al. (2015). © Springer. Used with permission

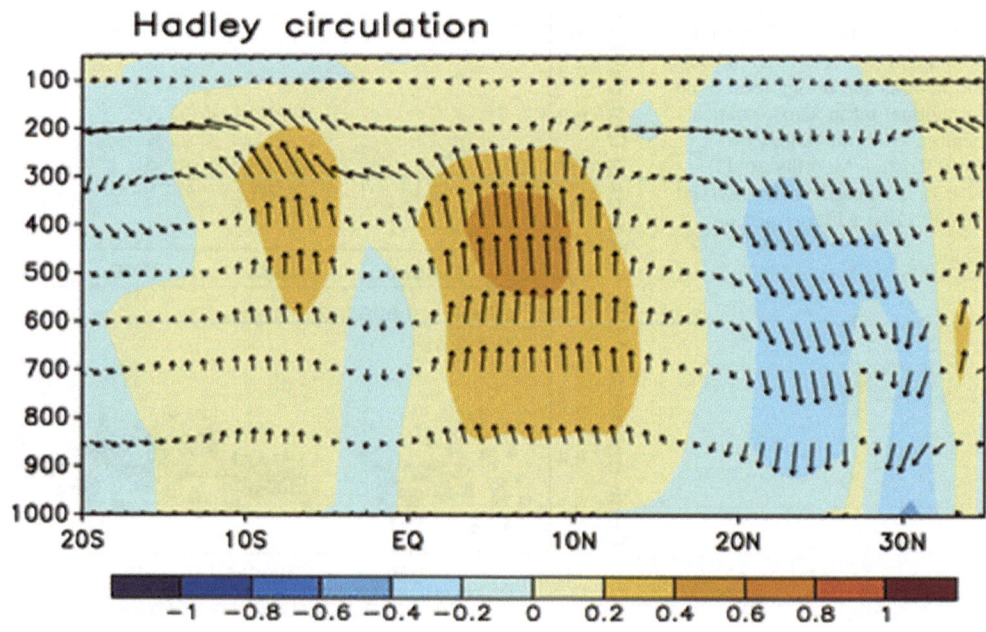

intensification of monsoon precipitation over central India (Vinoj et al. 2014). Additionally, aerosols can cause suppression of rainfall during monsoon breaks via atmospheric stabilization and increased moisture divergence (Dave et al. 2017).

5.2.3 Climate Model Simulations and Future Projections of Aerosol Properties

Quantifying the effects of anthropogenic aerosols on regional climate cannot be done based on observations alone. Modeling studies are needed to enumerate the aerosols–climate interactions and prediction of future climate change. Model intercomparison projects like CMIP and AEROCOM provide a multi-model platform for evaluating the capability of various climate models to simulate the observed variability of atmospheric aerosols. Evaluation of the present-day 550 nm AOD from several models participating in CMIP, AEROCOM with satellite observations (MODIS, MISR) reveals considerable bias (in the range $\sim \pm 0.3$) in aerosol optical properties estimations over the Indian subcontinent (Sanap et al. 2014; Pan et al. 2015; Misra et al. 2016). Very few models (HADGEM2-ES, HADGEM2-CC, IPSL-CM5A-MR) can capture the observed spatiotemporal distribution of aerosol loading over the subcontinent (Sanap et al. 2014; Misra et al. 2016). In the majority of the models, a negative anomaly in aerosol loading mainly occurs over IGP, western India, and the Arabian Sea and is attributed to significant underestimations in BC emissions and wind-driven dust transport. Also, the comparison of model-simulated extinction profiles with CALIOP observations shows that the bias in the models generally occurs in the lower tropospheric levels (below 2 km) and can be attributed to low emissions from agricultural waste burning and biofuel usage in the emission inventories

In the future, changing climate and changing emissions would result in changes in aerosol concentration and associated forcing. Future projections of aerosol emissions for

2015–2100 have been integrated into the nine different emission scenarios defined for CMIP6 based on new future pathways of societal growth, the Shared Socioeconomic Pathways (SSPs; Gidden et al. 2019). The yearly changes in aerosol optical properties for future scenarios derived using CMIP6 emission scenarios (Gidden et al. 2019) and MAC-SP parametrization show a pronounced decrease in global AOD by 2100 in all the scenarios excepting SSP3-70 and SSP4-60 (Fiedler et al. 2019). The time evolution of annual mean 550 nm AOD from nine different scenarios during the period 2015–2100 for the Indian landmass is plotted in Fig. 5.4. Most of the scenarios show increasing AOD during the initial period, with the maximum positive trend occurring in SSP5-85 and SSP5-34OS. Excepting SSP3-70, for the other scenarios, anthropogenic aerosol loading over India is projected to decline after 2030–2050 and reach levels much lower than the present aerosol level by the year 2100. The decrease in 550 nm AOD by 2100 for all the nine scenarios ranges between −66.5% and −0.63% with SSP1-19 producing the least aerosol forcing and SSP3-70 generating the maximum aerosol forcing over the Indian subcontinent. A projected decrease in dust aerosols over India by 2100 has also been reported causing changes in precipitation and soil moisture (Pu and Ginoux 2018). Climate projections of future emission scenarios show significant impacts on the north–south temperature gradient over India and Indian monsoon (Guo et al. 2015).

5.2.4 Aerosol Radiative Forcing

5.2.4.1 Radiative Forcing Due to All Aerosols
The annual global mean estimates of direct radiative forcing for aerosol speciation are (1) -0.4 ± 0.2 W m^{-2} for sulfates; (2) -0.09 ± 0.06 W m^{-2} for organic carbon; (3) $+0.40 \pm 0.40$ W m^{-2} for soot; (4) $+0.00 \pm 0.20$ W m^{-2}, -0.10 ± 0.2 W m^{-2}, and -0.11 ± 0.2 W m^{-2} for biomass burning, mineral dust, and nitrate, respectively (IPCC 2013). ACCMIP models show all-sky 1850 to 2000 global mean annual average total aerosol

Fig. 5.4 Time evolution of 550 nm AOD averaged over India for different scenarios

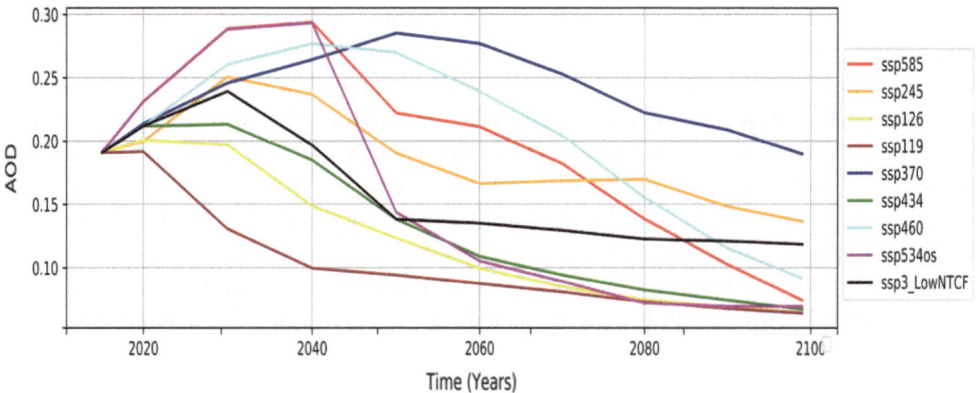

radiative forcing is -0.26 W m^{-2} (-0.06 to -0.49 W m^{-2}); however, screening based on model skill in capturing observed AOD yields the best estimate of -0.42 W m^{-2} (-0.33 to -0.50 W m^{-2}) (Shindell et al. 2013). The comparison of aerosol RF at the TOA (2005–1850) in CMIP5 models with the Modern Era Retrospective-analysis for Research and Analysis (MERRA) indicates that most of the CMIP5 models have underestimated aerosol RF over South Asia (Sanap et al. 2014). The ACCMIP model shows global mean pre-industrial to present-day aerosols direct RF (1850 and 2000) is -0.26 ± 0.14 W m^{-2} (-0.06 to -0.49 W m^{-2}), and top-of-atmosphere (TOA) is -1.2 ± 0.5 W m^{-2} although variability across models is large in many locations (Shindell et al. 2013).

Over the Indian landmass, there is a large spatiotemporal variation in both magnitude and sign ranging from -26 to $+14$ W m^{-2} at TOA and -63 to -2.8 W m^{-2} at the surface. A recent comprehensive analysis of aerosol direct radiative effect (DRE) by Nair et al. (2016) based on 27 ARFINET observatories, four AERONET stations, and four IMD stations estimated the seasonal variation of DRE over India. The regional mean DRE for TOA and surface is reported to be -8.6 ± 3 and -28.2 ± 12, -26 to $+14$ W m^{-2}, respectively, during winter and -6.8 ± 4, -33.7 ± 12–26 to $+14$ W m^{-2} during spring. The effective radiative forcing calculated by models indicates that over India, as high as 580 mW m^{-2} is because of residential biofuels. However, this is offset due to sulfate from power plants, which contribute about -30 mW m^{-2} (Streets et al. 2013).

In western India, observations of aerosol surface forcing over Ahmedabad, Gujarat, showed that surface forcing was found to be highest during post-monsoon (-63 ± 10 W m^{-2}), followed by dry (-54 ± 6 W m^{-2}) and lower values during pre-monsoon (-41.4 ± 5 W m^{-2}) and monsoon (-41 ± 11 W m^{-2}) seasons (Ganguly et al. 2005).

In the northeast of India, atmospheric aerosol radiative forcing estimated from the ARFINET observations over India (2010–2014) shows the highest radiative forcing in the pre-monsoon season ranging from 48.6 W m^{-2} in Agartala to 25.1 W m^{-2} in Imphal. Wintertime radiative forcing follows the pre-monsoon season at these locations. The heating rate resultant from this forcing is high at 1.2 K day^{-1} and 1.0 K day^{-1} over Shillong and Dibrugarh, respectively, in this season. However, Agartala experiences higher surface forcing (-56.5 W m^{-2}) and, consequently, larger heating of the atmosphere (1.6 K day^{-1}) in winter (Pathak et al. 2016). In the case of the top-of-atmosphere (TOA), radiative forcing is found to be negative during dry (-26 ± 3 W m^{-2}) and post-monsoon (-22 W m^{-2}), while positive values are obtained during monsoon (14 W m^{-2}) and pre-monsoon (8 W m^{-2}). Large differences between TOA and surface forcing during monsoon and pre-monsoon indicate a large

absorption of radiant energy (~ 50 W m^{-2}) within the atmosphere during these seasons. It can lead to heating rates as high as 5.6 K day^{-1} (Ganguly and Jayaraman 2006).

In north India, observations at New Delhi show a consistent increase in surface forcing, ranging from -39 W m^{-2} (March) to -99 W m^{-2} (June) and an increase in heating of the atmosphere from 27 W m^{-2} (March) to 123 W m^{-2} (June). Heating rates in the lower atmosphere (up to 5 km) are 0.6, 1.3, 2.1, and 2.5 K day^{-1} from March, April, May, and June 2006, respectively (Pandithurai et al. 2008). Observations at a semi-urban location, Hisar, showed an increase in the shortwave atmospheric forcing 16 W m^{-2} during clear periods to 49 W m^{-2} for foggy days. Longwave cooling of the atmosphere increased from about -2 W m^{-2} for clear conditions to about -3 W m^{-2} during foggy periods (Ramachandran et al. 2006).

Observations at Hyderabad displayed diurnally averaged values of direct shortwave radiative forcing in the range of -15 to -40 W m^{-2} at the surface, about 15% lower compared to that over the Bay of Bengal region and 22% higher than over the Arabian Sea. TOA forcing observed was in the range of $+0.7$ to -11 W m^{-2}, about 50% lower compared to both these regions. This results in a heating rate of nearly 0.8 K day^{-1} for the first 2 km in the atmosphere (Ganguly et al. 2005). Additional observations have also been made on cruises, with the INDOEX campaign in the 1990s, and showed that concentrations were relatively high throughout much of the cruise, even when the ship was at considerable distances from land. The northeast monsoonal low-level flow can transport sulfates, mineral dust, and other aerosols from the Indian subcontinent to the ITCZ within 6–7 days. These transports result in an increase in AOD at the equator by as much as 0.2 and a decrease in the solar radiative forcing at the sea surface by about 10–20 W m^{-2} (Krishnamurti et al. 1998).

The long-term historical surface temperature variations over the Indian subcontinent reveal an absorbing aerosol-induced statistically significant cooling of about 0.3 °C since the 1970s (Krishnan and Ramanathan 2002). In concurrence, the temporal and spatial variability in annually averaged global irradiance, diffuse irradiance, and bright sunshine duration over twelve stations of solar radiation network of India Meteorological Department (IMD) evaluated for the period 1971–2005. It showed a consistent decrease in the decadal mean all-sky global solar radiation at the surface for India, which was attributed to aerosols (Soni et al. 2012). The decadal mean of the global solar radiation for the decade was 221.5 W m^{-2} for 1976–1985. From 1986 to 1995, the observed global radiation decreased by 3.6 W m^{-2} and further by 9.5 W m^{-2} during the decade of 1996–2005 (Soni et al. 2012). The declining trend of all-sky global irradiance over India as a whole was 0.6 W m^{-2}

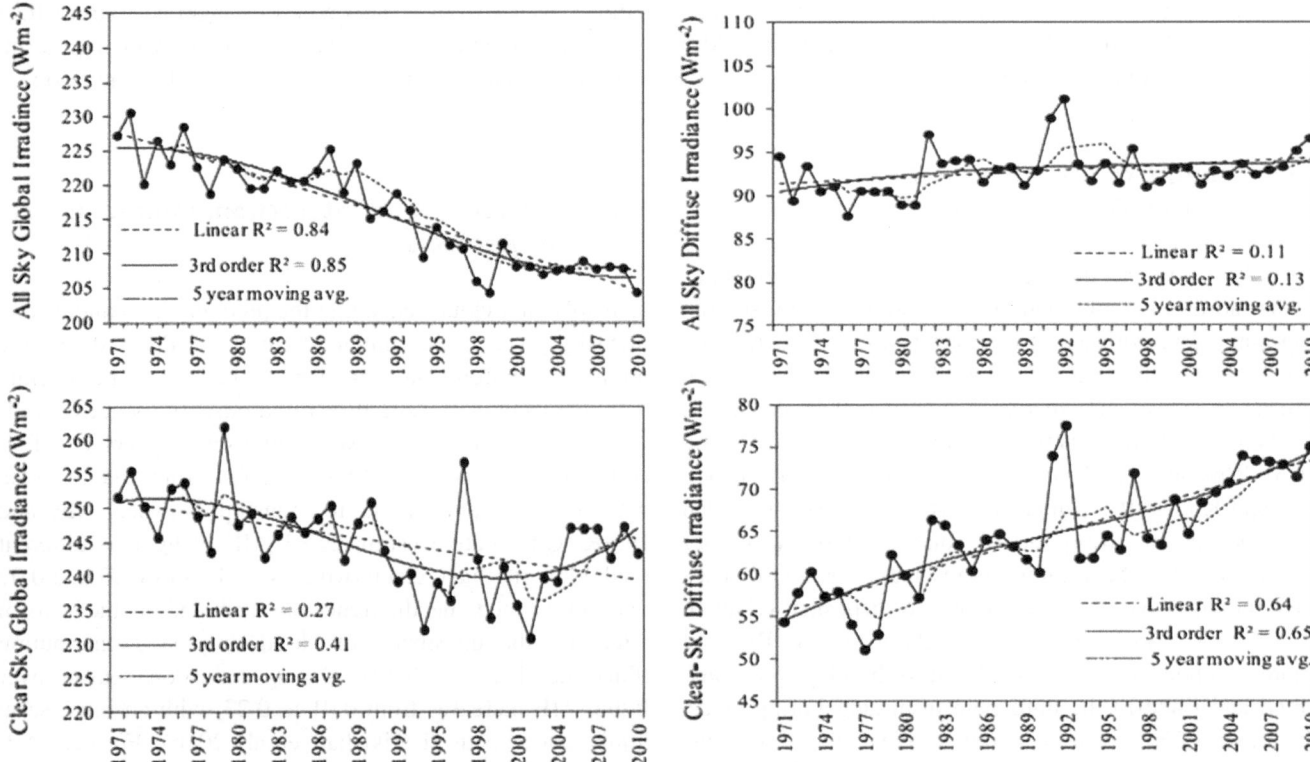

Fig. 5.5 (Left panels) Linear, third-order polynomial, and 5-year moving average fits to annual and seasonal time series of all-sky global irradiance averaged over all the twelve solar radiation stations and to clear-sky global irradiance averaged over eight stations. (Right panels) Linear, third-order polynomial, and 5-year moving average fits to annual and seasonal time series of all-sky diffuse irradiance averaged over all the twelve solar radiation stations and to clear-sky diffuse irradiance averaged over eight stations, adopted from (Soni et al. 2016). © Elsevier publications. Used with permission

year^{-1} during 1971–2000 and 0.2 W m^{-2} year^{-1} during 2001–2010 (Soni et al. 2016). This decrease in global irradiance is matched with an increase in the diffused radiation over the same period indicating an increase in the aerosol levels, as shown in Sect. 5.2.2 (Fig. 5.5).

5.2.4.2 Radiative Forcing Due to Different Species of Aerosols

Among the different species of aerosols, BC is the most important light-absorbing anthropogenic aerosol that contributes to atmospheric warming (Bond et al. 2013). During pre-monsoon season, high concentrations of BC are observed over northwest India and the IGP region. Radiative transfer calculations from observations suggest that from January to May, diurnal-averaged aerosol forcing at the surface is −33 W m^{-2}, and at the TOA above 100 km, it is observed to be +9 W m^{-2} (Badarinath and Madhavi Latha 2006). Similarly, large amounts of BC have been observed in Bangalore, both in absolute terms and fraction of total mass ($\sim$11%) and submicron mass ($\sim$23%). Estimated surface forcing is as high as −23 W m^{-2}, and TOA forcing is +5 W m^{-2} during relatively cleaner periods. The net atmospheric absorption translates to atmospheric heating of

0.8 K day^{-1} for cleaner periods and 1.5 K day^{-1} for less clean periods (Babu et al. 2002). It should be noted that a recent study indicates the reduction in BC over India, suggesting a decrease in BC caused heating of the atmosphere (Manoj et al. 2019).

Over the Indian region, the contribution to net cooling by sulfate aerosols is much larger than over other parts of the world (Verma et al. 2012). Estimates of monthly mean direct radiative forcing from sulfate aerosols using a coarse resolution model over India is high in December and January (−3.5 and −2.3 W m^{-2}), is moderate from February to April and November (−1.3 to −1.5 W m^{-2}) and low during May–October (−0.4 to −0.6 W m^{-2}) (Venkataraman et al. 1999). The sulfate aerosol radiative forcing over INDOEX (Indian Ocean Experiment) domain was found to be −1.2 W m^{-2} during INDOEX-FPP 1998 and −1.85 W m^{-2} during INDOEX-IFP 1999 (Verma et al. 2013). Aerosols originating from India, Africa, and West Asia lead to the reduction of total surface radiation by 40–60% (−3 to 8 W m^{-2}) over the Indian subcontinent and adjoining ocean (Verma et al. 2011). During the northeast winter monsoon, natural and anthropogenic aerosols reduce the solar flux reaching the surface by 25 W m^{-2}, leading to 10–15% less insolation at the surface. South Asia is the dominant contributor to sulfate aerosols over the INDOEX region and accounts for 60–70% of the AOD by sulfate (Reddy et al. 2004).

5.2.5 Impact of Absorbing Aerosols on Himalayan Snow/Ice Cover

Aerosols also impact the snow albedo in the high mountains in the north of India. During the pre-monsoon season, long-range transport and advection of desert dust over the Himalaya (Duchi et al. 2014) leads to its deposition on Himalayan snow, particularly over western Himalaya, and results in long-term reduction in snow albedo (Gautam et al. 2013). This snow darkening leads to accelerated snowmelt (Lau and Kim 2010). Similarly, dust above clean snow can also lead to the absorption of solar radiation at shorter wavelengths, but the warming is instantaneous and reduces when the dust layer is advected away or deposited (Gautam et al. 2013). Along with dust, light-absorbing BC and organic carbon aerosols from biomass burning get transported and lifted to the third pole from neighboring areas. The impact of BC on snow is similar to that of dust aerosols. On the deposition of BC on snow, it reduces surface albedo leading to increased absorption of shortwave radiation affecting the surface snowmelt.

Kopacz et al. (2011) found that emissions from northern India and central China contribute to the majority of BC in the Himalayas, although the precise source location varies with season. The Tibetan Plateau receives most BC from western and central China, as well as from India, Nepal, the Middle East, Pakistan, and other countries. The magnitude of contribution from each region also varies with season. The effect on this part of the cryosphere is highest in pre-monsoon and lowest during monsoon season (Gertler et al. 2016). Some studies (Xu et al. 2009) have analyzed ice cores for long-term trends of BC deposition on Himalayan glaciers. Xu et al. (2009) reported results from five ice core samples at various locations across Himalaya. Most of the locations show high concentrations in the 1950s–1960s and lower values in 1970–1980s. The authors attributed this temporal variation to the differential long-range transport of European emissions. The cryospheric soot concentrations over southern Himalaya show an increasing trend from the 1990s, which reflects an increase in present-day emissions (Xu et al. 2009). Jacobi et al. (2015) found a 4–5% reduction of albedo from BC over Himalaya. Kaspari et al. (2014) computed the change in albedo due to BC and dust over the Mera glacier and reported a 6–10% albedo reduction during the winter–spring season. The annual snow albedo surface forcing has been found in the range of 3–6 W m^{-2} (Jacobi et al. 2015). The radiative forcing in the snow-covered regions due to the BC-induced snow albedo effect can vary

from 5 to 15 W m^{-2}, an order of magnitude larger than radiative forcing due to the direct effect, and with significant seasonal variation in the northern Tibetan Plateau (Kopacz et al. 2011).

5.2.6 Aerosol-Cloud Interaction: Indirect Effect of Aerosols

Aerosols alter clouds changing the precipitation patterns and intensity over India. Initial attempts in the AIE in the monsoon environment were largely based on the satellite data. Panicker et al. (2010) reported positive (negative) AIE values during drought (excess monsoon) years for four consecutive years from 2001 to 2004 using satellite data demonstrating an inverse relationship between AIE and ISMR. Quantitative estimates of AIE using two different methodologies have given two close values of 0.13 and 0.07 over India, and the difference is attributed to the aerosol effect on the dispersion of cloud drop size distribution (Pandithurai et al. 2012). During monsoon season over India, AIE increases from 0.01 to 0.23, with an increase in liquid water path (Harikishan et al. 2016). Further, AIE derived from cloud drop number concentration and effective radius at different liquid water contents recorded at a high-altitude cloud physical laboratory in the Western Ghats provides a better understanding of the role of aerosol effect on cloud drop dispersion (Anil Kumar et al. 2016).

The Cloud-Aerosol Interaction and Precipitation Enhancement Experiment [CAIPEEX; (Kulkarni et al. 2012)] over India, mainly focusing on continental clouds, documented the precipitation process and associated aerosol-cloud interaction over the region. CAIPEEX in situ observations showed a significant increase in the cloud droplet number concentration with aerosol, as compared to the INDOEX study and several other reported results from around the world (Fig. 5.6). High aerosol loading into the atmosphere enhances the CCN, increasing the number concentration and decreasing the size of cloud droplets and a narrow droplet spectrum. This may further suppress collision coalescence and warm rain may form at an elevated layer (Konwar et al. 2012).

CAIPEEX observations revealed dilution in cumulus clouds over India (Nair et al. 2012) with pre-monsoon clouds in the high aerosol environment having more adiabatic cores than the monsoon clouds (Bera et al. 2019). Also, the clouds have a large amount of supercooled liquid water (>3 gm^{-3}) with mixed phase (Prabha et al. 2012). The cloud drop effective radius increases with the height from the cloud base (Prabha et al. 2012) and, the depth of warm rain was having a direct relationship with the subcloud aerosol. A decrease in the droplet spectral width of pre-monsoon cloud droplet sizes due to aerosol was reported by Prabha et al. (2012), and the

Fig. 5.6 Observations of cloud droplet number concentrations and associated aerosol number concentrations with data from INDOEX and CAIPEEX over the Indian continental region, error bars indicate spatial variability. Adapted from Prabha and Khain (2020). © John Wiley and Sons. Used with permission

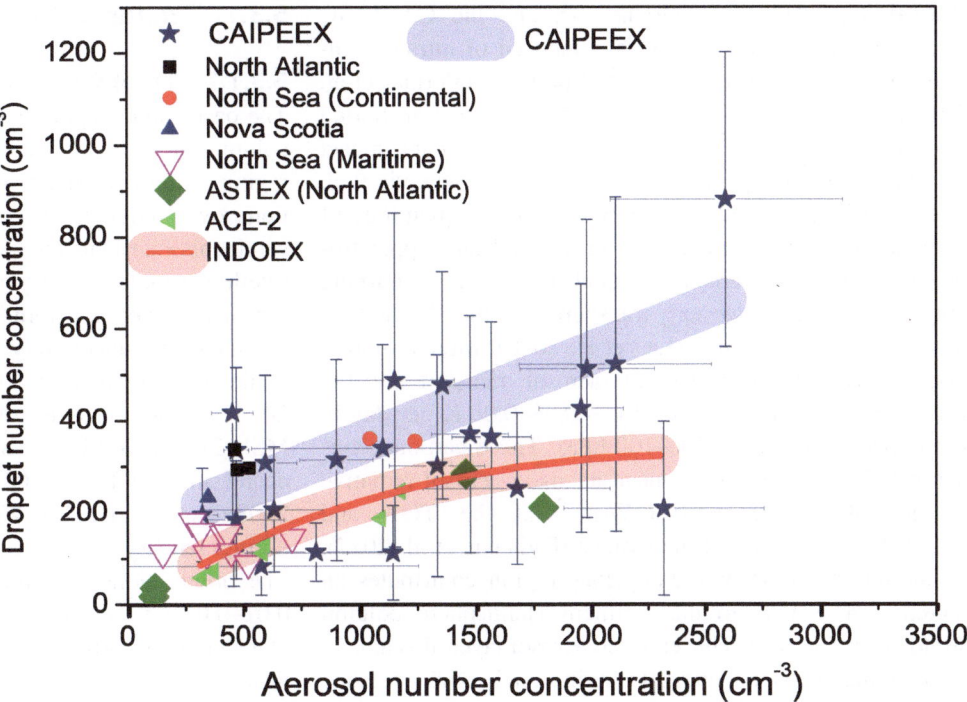

variation in spectral width was also largely variable with airmass characteristics (Bera et al. 2019). CAIPEEX observations have also documented both higher ice mass and number concentration in monsoon clouds, compared to pre-monsoon clouds with warm microphysics in the monsoon clouds determining the ice and mixed-phase microphysical properties whereas boundary layer moisture plays a key role in the initial developmental stages (Patade et al. 2014).

Documenting the aerosol and associated activation characteristics as cloud condensation nuclei or ice nuclei particles (INP) is essential for better characterization in numerical models. Observational data from Nainital shows that enhanced CCN concentrations coincide more with periods of aerosol absorption as compared to periods of aerosol scattering (Gogoi et al. 2015). Aerosol chemical composition, aerosol number size distribution, and CCN data show that the predictability of CCN improves when SOA component is considered in hygroscopicity estimates (Singla et al. 2017). Precipitation susceptibility estimates showed that clouds having medium liquid water content (0.6–0.8 mm) were highly affected due to aerosols (Leena et al. 2018). The vertical variation of aerosol as CCN is important, as the CCN spectral characteristics are significantly different near the surface and the cloud base (Varghese et al. 2016). Vertical distribution of aerosol types reveals a mixture of both biomass burning and dust aerosols (Padmakumari et al. 2013). The role of aged BC particle or bioaerosol acting as CCN or INP is yet to be investigated over the Indian region. Physical and chemical characterization of aerosol along with CCN and INP activity is required in future studies.

Numerical investigation of aerosol effect over the monsoon region shows that cloud microphysics processes are important for a break to active transition during monsoon season with higher concentrations of absorbing aerosols producing invigoration of convection strong moisture convergence and increased upper level heating (Hazra et al. 2013). Within the deep convective clouds during monsoon, an increase in soluble aerosol led to a marginal increase in precipitation attributing to enhanced updrafts in the warm phase and invigoration of mixed-phase cloud processes (Gayatri et al. 2017). Mixed-phase clouds contribute a significant part of monsoon clouds, which are least understood and need further focused process studies. A systematic approach aerosol impact on the cloud system effects needs to be investigated, and the regional impacts on precipitation through redistribution of clouds need to be understood.

5.2.7 Impact of Convective Transport of Aerosols

During the monsoon season, deep convection transports boundary layer aerosols from Asia to the UTLS (Fadnavis et al. 2013, 2017). These aerosols form a layer near the tropopause (13–18 km) known as the Asian Tropopause Aerosol Layer 'ATAL' (Vernier et al. 2009). Development of the ATAL is associated with convective transport of aerosols from the lower atmosphere to the UTLS (Fadnavis et al. 2013; Vernier et al. 2015). Observations from satellites

(CALIPSO, SAGE-II), balloonsonde, and the CAREBIC aircraft reveal that the ATAL is composed of nitrates, sulfate, BC, organic aerosols, and dust particles (Vernier et al. 2018). Studies indicate that these aerosols are transported into the lower stratosphere and produce a significant impact on stratospheric temperature and circulations (Fadnavis et al. 2017). MERRA-2 data also shows abundant quantities of carbonaceous aerosols and dust in the mid and upper troposphere over India, arising from enhanced biomass burning emissions as well as westerly transport from the Middle East deserts during May–June (Lau et al. 2018). Model simulations indicate that carbonaceous aerosol transport into the UTLS enhances heating rates by ~ 0.03–0.08 K per day in the upper troposphere (300–100 hPa). These carbonaceous aerosols induce a seasonal mean anomaly aerosol radiative forcing of $\sim + 0.37 \pm 0.26$ W m^{-2} at the TOA and -4.74 ± 1.42 W m^{-2} at the surface (Fadnavis et al. 2017). Asian summer monsoon anticyclone region contributes an increase of $\sim 15\%$ to the Northern Hemisphere column stratospheric aerosol. This elevated aerosol layer also aids in aggravating monsoon droughts during an El Niño episode (Fadnavis et al. 2019).

5.3　Trace Gases

Ozone variations in the troposphere and stratosphere play a key role in maintaining the Earth's radiative budget and climate change (Logan et al. 2012); it is important to know its assessment. In this section, we provide an assessment of its trends from past literature. Trend estimates in total ozone column, tropospheric ozone column, and surface measurements are reported from in situ observations, satellites remote sensing, and model simulations.

5.3.1　Trends in Ozone

5.3.1.1　Total Ozone Column

The past studies report estimates of trends in the total ozone column at various stations widespread over the Indian region. Although, there are limited ozone monitoring stations over India, trend estimates from ground-based measurements like Dobson spectrophotometer (DU year^{-1}) and satellite remote sensing (% year^{-1}) are consistent. Measurements over north India (20–35° N) show statistically significant (σ uncertainty level) negative (declining) trends while they are positive (increasing) over south India (8–20° N). For example, Multi-Sensor Reanalysis (MSR/MSR-2) and TOMS observations over north India show a decreasing trend of -0.08 to -0.15% year^{-1} during 1979–2008 (Tandon and Attri 2011) and -0.03 to -0.11% year^{-1} during January 1979–December 2012 (Sahu et al. 2014). The observations in south India show a positive trend of 0.01–0.03% year^{-1} during January 1979–December 2012. Dobson spectro-photometer tropospheric column ozone measurements also show a decreasing trend -0.01 DU year^{-1} at Varanasi, (in north India), and positive trend +0.14 DU year^{-1} at Kodaikanal, (in south India) during 1957 and 2015 (Pathakoti et al. 2018). The above studies indicate that amplitude of trend varies with location and time period of measurement, and the trend estimates have medium confidence.

In the global context, the ozone assessment report shows that total ozone has been stable since about 1996 in response to emission control of ozone-depleting substances (ODSs) (Chehade et al. 2014; Zvyagintsev et al. 2015). Future trends in total column ozone over the globe and tropics (25° S–25° N) are tabulated in Table 5.1 (Cionni et al. 2011; Eyring et al. 2013).

Table 5.1 Ozone trend over the Indian region and CMIP5 annual mean future trends over tropics

Ozone trends	Indian region	Tropics 25° S–25° N from CMIP5 annual mean future trends (2090s–2010s) in the RCPs (Cionni et al. 2011)
Total column ozone	North India: -0.03 to -0.11% year^{-1} South India: $+0.01$ to $+0.03\%$ year^{-1} (1979–2012) (Sahu et al. 2017)	RCP 2.6: -2 DU (-1%) RCP 4.5: 0 DU (0%) RCP 6.0: 0 DU (0%) RCP 8.5: 7 DU (4%)
Troposphere	Tropospheric column: 0.3 ± 2.6 to $2.7 \pm 2.3\%$, year^{-1} (Saraf and Beig 2004)	RCP 2.6: -4 DU (-17%) RCP 4.5: -2DU (-10%) RCP 6.0: -2DU (-10%) RCP 8.5: $+5$ DU (18%)
Stratosphere	0.27 ± 0.67 to $1.3 \pm 0.65\%$ year^{-1} (~ 200–50 mb) (1993–2005) -0.45 ± 0.8 to $-0.57 \pm 0.62\%$ year^{-1} (30–10 mb) (1993–2005) (Fadnavis et al. 2014a)	RCP 2.6: 2 DU (1%) RCP 4.5: 2DU (1%) RCP 6.0: 2DU (1%) RCP 8.5: 2DU (1%)

5.3.1.2 Tropospheric Ozone

Although a majority of the ozone is concentrated in the stratosphere, tropospheric ozone plays a vital role in atmospheric chemistry, determining the oxidative capacity of the atmosphere through the production of the hydroxyl radical (OH), and can also act as a pollutant affecting human health and crop productivity. Efforts have been taken toward estimating trends in surface and tropospheric column ozone. The estimated amplitude of trend varies with season and location (urban and rural). The annual mean trends in tropospheric column ozone derived by integrating the vertical profiles of ozonesonde data (in situ observations on balloon platforms) over Delhi, Pune, and Trivandrum (1972–2001) show an increasing trend of $2.7 \pm 2.3\%$, $0.9 \pm 1.8\%$, and $0.3 \pm 2.6\%$ year^{-1}, respectively. These values of trends are in close agreement with that obtained from TOMS data (Saraf and Beig 2004). Nimbus-7 and Earth Probe satellite—Total Ozone Mapping Spectrometer (TOMS) data for the period of 1979–2005 show positive trends 0.7–0.9% year^{-1} over South Asia. Also, the trends estimated from the MOZART model are in agreement with observations over the Bay of Bengal region (0.4 ± 0.29–$0.6 \pm 0.43\%$ year^{-1}, Beig and Singh 2007). The regressed tropospheric ozone residual (TOR) data shows an annual trend of $\sim 0.4 \pm 0.25$ $1\sigma\%$ per year over the northeastern Gangetic region (Lal et al. 2012). Similar estimates of trends in ozone at various stations in India are reported (Lal et al. 2013, 2014, 2017).

The multi-model ensemble-mean from CMIP5 historical simulations (2009–2000) shows an increase in tropospheric column ozone ~ 25–35 DU (decadal mean) over the Indian region. Future projections (2090–2100) are tabulated in Table 5.1. ACCMIP multi-model simulations for relative changes of tropospheric ozone between 2000 and 2030 (2100) for the different RCPs show decrease of -5% (-22%) in RCP 2.6, 3% (-8%) in RCP 4.5, 0% (-9%) and increase in 15 RCP 6.0, and 5% (15%) in RCP 8.5 (Young et al. 2013). However, there are large uncertainties in model simulations over the Indian region due to uncertainties in emission inventories, model parameterization, chemistry representation, etc. (Fadnavis et al. 2015).

Surface ozone observations have shown an increasing trend at various locations around India, which are attributed to increasing anthropogenic activity. Naja and Lal (1996) reported increasing ozone by 14.7 ppbv during 1954–1955 and 25.3 ppbv during 1991–1993, which results in linearly increasing trends of 1.45% year^{-1}. The seasonal trends also show a significant increase, e.g., winter $\sim 1.91\%$ year^{-1} and summer 0.86% year^{-1} at Ahmedabad (23° N, 75.6° E). At southern peninsula station, Thiruvananthapuram (8.542° N, 76.858° E), surface ozone measurements were obtained during 1973–1975, 1983–1985, 1997–1998, and 2004–2014 (Nair et al. 2018). These measurements show a slow increase

of ~ 0.1 ppb year^{-1} during 1973 to 1997 and faster growth of 0.4 ppb year^{-1} afterward till 2009 after which it showed a steady-state till 2012 followed by a minor decrease. The above studies show that trends in surface and tropospheric column ozone are positive but have low confidence.

Profiles of ozone show variations in the vertical structure of ozone. From ozonesonde observations, Saraf and Beig, (2004) reported long-term trends in ozone at Trivandrum, Pune, and Delhi. The observed trend at Delhi was increasing between 1.5% year^{-1} and 7.3% year^{-1} during 1972–2001, but negative trends with low confidence (statically insignificant) ~ -0.5 to -3% year^{-1} were observed at Pune and Trivandrum. Seasonal trends in tropospheric ozone are positive, particularly around 500 hPa and 200–300 hPa during months of January–March. The role of biomass burning and stratosphere–troposphere exchange is evident in these two layers. There is also a prominent increasing trend in ozone near the tropopause (~ 100 hPa) during June and July (monsoon month) (Fadnavis et al. 2014a). The GEOS-Chem model simulations also show a positive trend of 0.19 ± 0.07 ppb year^{-1} (p-value < 0.01) (an annual mean) in the lower troposphere between 1990 and 2010 (Lu et al. 2018).

5.3.1.3 Ozone Trends in the Upper Troposphere and Stratosphere

It is important to understand ozone variability and trends in the UTLS since a small amount of variation of ozone in the UTLS has a large impact on radiative forcing and climate change (Forster and Shine 1997). Fadnavis et al. (2014a) reported an increasing ozone trend between 0.6 ± 0.65 and 2.35 ± 1.3 year^{-1} in the upper troposphere and in the lower stratosphere from multiple satellite data sets and model simulations. The estimated trends are slightly positive up to 30 hPa and then negative between 30 and 10 hPa. Seasonal mean trends vary between -0.04 ± 0.3 and $3.48 \pm 2\%$ year^{-1} (low confidence). In the stratosphere (20–50 km), ozone shows a decreasing trend of (medium confidence) $\sim -0.4 \pm 0.1\%$ year^{-1} near 16–20 km while trends values are positive near 24–30 km (0.05 ± 0.04 to $0.1 \pm 0.9\%$ year^{-1}) (low confidence) during 1993–2015 (Raj et al. 2018).

A decrease in abundance of ozone-depleting substances (ODSs) under the compliance of the Montreal Protocol was the start of the recovery of stratospheric ozone. Since the atmospheric burden of ozone-depleting substances is declining, changes in CO_2, N_2O, and CH_4 will have an increasing influence on the ozone layer (WMO 2019b). Ozone layer changes in the latter half of this century will be complicated, with projected increase or decrease in different regions. Eyring et al. (2013) reported the evolution of stratospheric ozone over the CMIP5 historical period (1960 to 2005) and the sensitivity of ozone to future GHGs (2006–

2100) for the four different RCPs (2.6, 4.5, 6.0, and 8.5). The simulations with the 1980 baseline-adjusted stratospheric column ozone (time series from 1960 to 2100) over the tropics ($25°$ S–$25°$ N) show a decrease in tropical lower stratospheric ozone (100–30 hPa) and increase in the upper stratosphere (Cionni et al. 2011). A summary of annual mean trends is ozone in the troposphere and stratosphere discussed above from observations, and CMIP5 multi-models future projections are listed in Table 5.1.

5.3.2 Emissions of Ozone Precursors

One of the largest uncertainties in modeling studies is emission inventories. Global and regional emission inventories carry large uncertainties, especially in regions where observational data are sparse. In this section, we provide a brief overview of emissions of ozone precursors, e.g., NO_x, CO, and NMVOCs, over India. Jena et al. (2015) reported total NO_x flux $\sim$1.5, 2.1, 2.4, 1.9, 1.7, and 1.4 Tg N year^{-1} over India from six different inventories. Thermal power plants contribute 30% of the total NO_x emissions in India (Garg et al. 2006). The total surface NO_2 emissions in India are $\sim$3.5 Tg year^{-1} in 1991 and $\sim$4.3 Tg year^{-1} in 2001 (Beig and Brasseur 2006). The total NO_x emissions in 2005 amount to $\sim$1.9 Tg N year^{-1} (Ghude et al. 2013b). The growth in oil and coal consumption resulted in a growth rate of 3.8% $\pm$ 2.2% year^{-1} between 2003 and 2011 for anthropogenic NO_x (Ghude et al. 2013a). This growth rate is comparable with the estimate made by EDGAR (V4.2; 4.2% year^{-1}), GAINS (3.6% year^{-1}), or (Garg et al. 2006) (4.4% year^{-1}) emission inventories. Sadavarte and Venkataraman, (2014) reported estimates of NO_x emissions $\sim$5.6 (1.7–15.9) Tg year^{-1} in 2015.

CO emissions show annual growth rate of 1.1% during 1985–2005 (Garg et al. 2006). The annual growth rate of CO from the transport sector is $\sim$8.8% during 2001–2013 (Singh et al. 2017). The total CO emissions from India were 59.3 Tg year^{-1} in 1991 and 69.4 Tg year^{-1} in 2001 (Beig and Brasseur 2006). The CO emission from wheat straw burning in 2000 was 541 $\pm$ 387 Gg year^{-1} (Sahai et al. 2007). Venkataraman et al. (2006) estimated $\sim$13–81 Gg year^{-1} of CO from biomass burning during 1995–2000. In 2000, CO emissions in India (63.3 Tg) were $\sim$23% of Asia (279 Tg) (Streets et al. 2003).

Biogenic emissions are the largest natural source ($\sim$90%) of volatile organic compounds (VOCs) in the atmosphere (Guenther et al. 2006). The annual emissions of VOCs in India from anthropogenic and biomass burning sources were $\sim$10.8 Tg and 2.2 Tg, respectively, in 2000 (Streets et al. 2003). Total anthropogenic emissions of non-methane volatile organic compounds (NMVOCs) were

9.81 Tg in 2010 (Sharma et al. 2015). The majority of NMVOCs emissions (60%) originated from residential combustion of biomass for cooking. Venkataraman et al. (2006) estimated NMVOC emission $\sim$2.04–7.41 Tg year^{-1} from biomass burning during 1995–2000 over India.

5.3.3 Trends of Tropospheric NO_x, CO, NMVOCs, and PAN

Trends in some of the ozone precursors (NO_x, CO, VOCs) are reported over the Indian region. Satellite observations from GOME, GOME-2a, OMI, and SCIAMACHY during 2002–2011 show a trend of 2.20 $\pm$ 0.73% year^{-1} in NO_2 volume mixing ratios over India (Mahajan et al. 2015). While NO_2 volume mixing ratios from 1996 to 2006 showed an increasing trend of 1.65 $\pm$ 0.52% year^{-1} in over India. The industrial regions of Mumbai and Delhi show increasing trends of 2.1 $\pm$ 1.1 and 2.4 $\pm$ 1.2% year^{-1}, respectively (Ghude et al. 2008). CO observations from MOPITT (Measurements of Pollution in the Troposphere) satellite during 2000–2014 show contrasting trends in the lower and upper troposphere. Estimated trends in lower-troposphere and columnar CO are negative −2.0 to −3.4 ppb year^{-1} (−1.1 to −2.0% year^{-1}) and positive 1.4–2.4 ppb year^{-1} (1.8–3.2% year^{-1}) in the upper troposphere (Girach et al. 2017). AIRS/AMSU satellite (2003–2012) shows a 2% increase in tropospheric CO concentration over the Indian region (Ul-Haq et al. 2015) (Fig. 5.7a). Emission estimates based on technology also show increasing trends $\sim$19 Tg year^{-1}, e.g. (Sadavarte and Venkataraman 2014). Peroxyacetyl nitrate (PAN) is formed in biomass burning plumes. It is a secondary pollutant produced through the oxidation of VOCs and NO_X released from anthropogenic and biogenic sources. Recent satellite observations show an increasing trend in PAN $\sim$0.1 $\pm$ 0.05 to 2.7 $\pm$ 0.8 ppt year^{-1} during 2005–2012 in the UTLS over Asia (Fadnavis et al. 2015) (Fig. 5.7a). A significant increase in amounts of VOCs and air pollutants is observed (May 2012) in the Indo-Gangetic Plain (IGP). These observations show extremely high levels of both VOCs and the primary air pollutants in the evening and early morning hours in May 2012 (Fig. 5.7b). These increasing levels of VOCs may be contributing to postive trends in PAN in the UTLS. The observed trends in NO_x and PAN have high confidence, while CO and VOCs have low confidence. Ozone and its precursor gases, $PM_{2.5}$, PM_{10}, are being monitored since 2010 at various Indian stations by System of Air Quality Forecasting and Research (SAFAR) which is developed by the Indian Institute of Tropical Meteorology. Long-term observations from SAFAR will be helpful in obtaining future trends in ozone and its precursors over the India region.

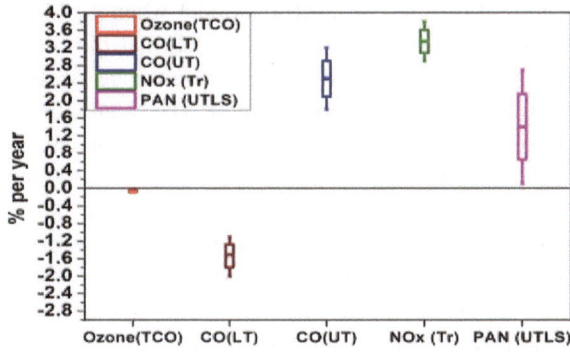 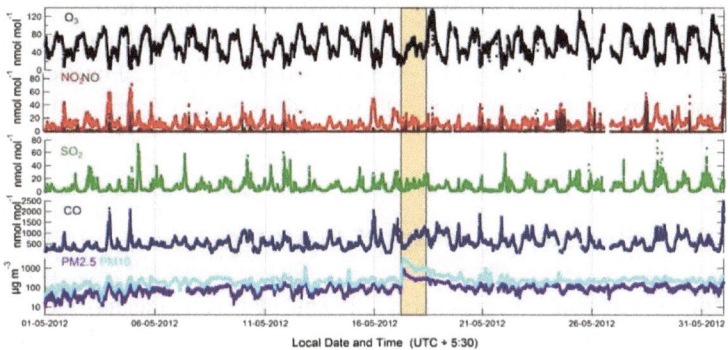

Fig. 5.7 **a** Trends in trace gases over the Indian region. These trends are adopted from Beig and Singh 2007; Fadnavis et al. 2014a; Mahajan et al. 2015; Girach et al. 2017; Sahu et al. 2017. **b** Time series of the one-minute data in May 2012 for the mixing ratios of ozone (top panel), and NO_2 and NO (second panel), SO_2 (third panel), CO (fourth panel), and mass concentrations of $PM_{2.5}$ and PM_{10} (bottom panel) adopted from (Sinha et al. 2014), Fig 5.7b. © Copernicus publications. Used with permission

5.3.4 Variations in Ozone and NO$_x$ Due to Lightning

The composition of trace gases in the troposphere is influenced by lightning in addition to anthropogenic emissions. It is estimated that lightning contributes to about 10% of the global annual NO source (Schumann and Huntrieser 2007). It can contribute up to 90% variation of NO_x at the altitudes of 5–15 km. Variation of ozone in the middle/upper troposphere due to lightning may change ozone heating rates and may have an impact on Asian monsoon circulation (Roy et al. 2017).

Over the Asian region, lightning contributes $\sim$40% to NO_x and 20% to ozone production in the middle and upper troposphere during the monsoon season (Fadnavis et al. 2014a). Previous studies (Bharali et al. 2015) have reported an increase in the O_3 mixing ratio $\sim$18 ppbv during pre-monsoon and $\sim$12 ppbv during summer associated with the lightning activity over Dibrugarh (27.4° N, 94.9° E) in northern India and over Hyderabad (17.44° N, 78.30° E) (a station in southern peninsular India) (Venkanna et al. 2016). Kavitha et al. (2018) reported an enhancement in NO_x (5.2–8.7 ppbv) and an associated reduction in surface O_3 mixing ratio (9.9–18.8 ppbv) during pre-monsoon and monsoon seasons due to lightning activity.

5.3.5 Radiative Forcing due to Ozone and Precursor Gases

The radiative forcing (RF) due to changes in tropospheric and stratospheric ozone is the third-largest GHGs contributor to RF since pre-industrial times. According to the (IPCC 2013), the total increase in global radiative forcing due to changes in ozone is +0.35 (0.15–0.55) W m^{-2} (high confidence), with radiative forcing due to tropospheric ozone +0.40 (0.20–0.60) W m^{-2} (high confidence) and due to stratospheric ozone −0.05 (−0.15 to +0.05) W m^{-2} (high confidence). ACCMIP tropospheric ozone future projections (2100–1850) show global mean annual average anthropogenic forcing $\sim$0.14 ± 0.12 W m^{-2} in RCP 2.6, 0.23 ± 0.15 W m^{-2} in RCP 4.5, 0.25 ± 0.09 W m^{-2} in RCP 6.0, and 0.55 ± 0.30 W m^{-2} in RCP 8.5.

According to the CMIP5 estimates, the tropospheric ozone radiative forcing from the 1850s to the 2000s is +0.23 W m^{-2}, lower than the IPCC estimate (IPCC 2013). The lower value is mainly due to (i) a smaller increase in biomass burning emissions; (ii) a larger influence of stratospheric ozone depletion on upper tropospheric ozone at high southern latitudes; and possibly (iii) a larger influence of clouds (which act to reduce the net forcing). Over the same period, decreases in stratospheric ozone, mainly at high latitudes, produce an RF of −0.08 W m^{-2}, which is more negative than the IPCC but is within the stated range of −0.15 to +0.05 W m^{-2} (Cionni et al. 2011; Eyring et al. 2013).

Estimates over India suggest that the radiative forcing has changed in the range between 0.2 and 0.4 W m^{-2} since pre-industrial times (Chalita et al. 1996). The radiative forcing effect from tropospheric ozone is regional due to its short lifetime. The model simulations with 10% reductions in the precursor's emission over India resulted in a decrease of $\sim$0.59 m W m^{-2} (Naik et al. 2005).

For the other trace gases mentioned in this chapter, the resultant effect on radiative forcing is not direct. Gases such as CO, NO_x, and VOCs are precursors of ozone and hence have an indirect impact on radiative forcing. Additionally, gases such as sulfur dioxide (SO_2) and NO_x also contribute to the formation of sulfate and nitrate aerosols, which can have a net cooling effect on the atmosphere. Globally, the contribution from CO and NMVOCs toward ozone radiative forcing is estimated to be about +0.2 (−0.18 to +0.9) W m^{-2}

and 0.1 (−0.06 to +0.14) W m^{-2}. For NO$_x$, due to its role in nitrate aerosol formation, the best estimate is a resultant negative forcing of −0.15 (−0.34 to +0.02) W m^{-2}. For sulfur dioxide, the estimate is −0.41(−0.62 to −0.21) W m^{-2} (IPCC 2013).

5.4　Influence of Transport Processes

Monsoon sustains a remarkably efficient cleansing mechanism in which contaminants are rapidly oxidized and deposited to Earth's surface. However, some pollutants are lifted above the monsoon clouds due to deep convection and are chemically processed in a reactive reservoir before being redistributed globally, including to the stratosphere. Numbers of studies based on satellite remote sensing indicate the transport of CO, H$_2$O, PAN, Hydrogen cyanide (HCN), CH$_4$, NO$_x$, etc., from the Asian boundary layer to the UTLS (Fadnavis et al. 2013, 2015, 2017). Enhancement of trace gases in the UTLS during the monsoon season alters local heating rates and radiative balance. Transported NO$_x$ and associated ozone variations in the UTLS enhances the ozone heating rates by ∼1–1.4 K day^{-1} in the upper troposphere (400–200 hPa) and radiative forcing ∼16.3 m W m^{-2} over the Indian region. There is a positive impact of ozone heating rates and radiative forcing on the Indian monsoon circulation (Roy et al. 2017). There is a convective injection of polluted water vapor from the Asian region into the UTLS, which is then dispersed into the global stratosphere by the large-scale upward motion (Fu et al. 2006). The H$_2$O feedback amplifies the radiative forcing of anthropogenic greenhouse gases by a factor of ∼2.

In the lower troposphere, the transport of trace gases occurs to and from India with seasonal variations in the wind. The seasonal variation in most trace gases shows a dip during the monsoon season due to efficient wet scavenging by precipitation and the transport of clean marine air. Integrated Campaign for Aerosols, gases, and Radiation Budget (ICARB) measurements during the pre-monsoon season show elevated levels of CO (∼100 ppb) over the Bay of Bengal and the Arabian Sea. These studies reveal that high amounts of marine CO are attributed to transport from the Indian subcontinent (Aneesh et al. 2008). Satellite observations also show high values of CO (130–160 ppb) and ozone (120–130 ppb) at 825 hPa near the location of cyclones occurring in the Bay of Bengal and Arabian Sea (Fadnavis et al. 2011).

The in situ observations are unable to explain the different atmospheric processes accountable for high pollution events. Therefore, chemistry transport models are valuable for providing a large-scale view of the regional impact of these gases and are useful for the interpretation of observations on local to global scale (Yarragunta et al. 2017). The simulated ozone concentrations from the MOZART4 model when evaluated against ground-based observations revealed that the model captures the seasonal cycle of ozone amounts but overestimates the values of ozone concentration. The magnitude of observed ozone is in the range of 7–60 ppbv, whereas the quantity of simulated ozone is ∼27–53 ppbv (Yarragunta et al. 2018). Lower tropospheric ozone over India during 2006–2010 as observed from OMI showed the highest concentrations (54.1 ppbv) in the pre-summer monsoon season (May) and the lowest concentrations (40.5 ppbv) in the summer monsoon season (August). Analyses from the GEOS-Chem model showed that the onset of the summer monsoon brings ozone-unfavorable meteorological conditions which all lead to substantial decreases in the lower tropospheric ozone burden (Lu et al. 2018). The influence of springtime (MAM) biomass burning in central India, the Indo-Gangetic region and the Bay of Bengal, on regional ozone distribution has been evaluated using a regional chemical transport model (WRF-Chem), and the Fire Inventory from NCAR (FINNv1). These simulations demonstrated that the springtime fire emissions have a significant impact on the ozone in this region (Jena et al. 2015).

5.4.1　Influence of Stratosphere to Troposphere Transport

Transport associated with tropopause folding produces a significant variation in ozone, humidity, and temperature. MLS and AIRS satellites show intrusion events of ozone-rich dry stratospheric air over northern India and the Tibetan Plateau region occurring every winter and pre-monsoon season. It enhances ozone amounts by ∼100–200 ppmv in the UTLS (300–100 hPa) (Fadnavis et al. 2010). Tropopause folding in the subtropical westerly jet during the monsoon seasons sheds eddies into the deep troposphere (∼700 hPa) which are a carrier of ozone-rich cold and dry air. These eddies spread stratospheric air in the upper troposphere, increasing the static stability of the troposphere (Fig. 5.8). These stratospheric dry air intrusions are associated with monsoon breaks and are evident in observations during 1979–2007 (Fadnavis and Chattopadhyay 2017).

The stratospheric folding tends to occur on the northwestern side of the upper-level anticyclone resulting in intensified subsidence and reduces extreme rainfall upstream of the fold, while it enhances the precipitation at downstream of the fold. A typical pattern of suppression of extreme rainfall upstream and promotion downstream of the fold persists for about 1–2 days. Rossby wave breaking over West Asia inhibits deep monsoonal convection and thereby leading to a dry spell over India (1998–2010) (Samanta et al. 2016).

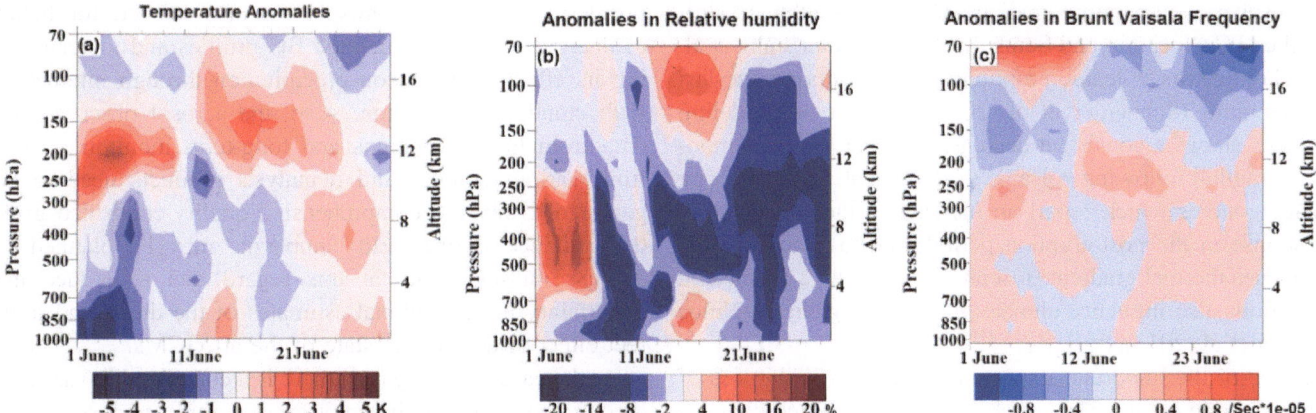

Fig. 5.8 Time-pressure cross section of anomalies in **a** temperature (K) averaged over 30–50° N, 75–110° E, **b** RH (%) averaged over 25–40° N, 60–75° E, **c** square of Brunt–Väisälä frequency (per sec*1E-5) averaged over 30–50° N, 75–110° E. Adapted from Fadnavis and Chattopadhyay (2017). © American Meteorological Society. Used with permission

Distribution of trace gases in the UTLS is also affected by the stratospheric Brewer–Dobson circulation (Brewer 1949; Dobson 1956). The Asian summer monsoon is an important pathway for the transport of Asian tropospheric constituents into the stratosphere (Fadnavis et al. 2013, 2017). The HALOE aircraft observations of N_2O, CO, and O_3 indicate a significant increase in the impact of the South Asian tropospheric pollutants on the extratropical lower stratosphere (Müller et al. 2016). Inter-annual variations of stratospheric N_2O, CFC-11 (CCl3F), and CFC-12 (CCl2F2) are modulated by the BDC. Satellite observations show that the transport of water vapor and HCN from the South and Southeast Asia occurs into the lower stratosphere by the monsoon convection and is then re-circulated by the Brewer–Dobson circulations. Thus, Asian trace gases and aerosols affect the chemical composition of the extratropical stratosphere (Fadnavis et al. 2013).

5.4.2 Influence of Transport Associated with Quasi-biennial Oscillation

The phenomenon of the equatorial quasi-biennial oscillation (QBO) is known to produce a significant impact on dynamics and chemistry over the tropical region. Studies indicate that the secondary meridional circulation induced by QBO produces a double peak structure in the stratosphere at the equator with maximum amplitude in the temperature and ozone at two pressure levels 30 and 9 hPa and a node at 14 hPa. Phase structure reveals that the temperature QBO descends faster than the ozone QBO (Fadnavis et al. 2008).

Past studies indicate that cyclones are modulated by the phases of the QBO (Fadnavis et al. 2011, 2014b). In post-monsoon season, during the east phase, cyclones move westward/northwestward while during the westerly phase, they move northward/northeastward. During pre-monsoon

season, cyclones move northward/northeastward irrespective of phases of QBO. The possible interaction between the stratospheric QBO and cyclone is explained from the variation of winds, geopotential height, tropopause pressure, OLR, and SST (Fadnavis et al. 2011). QBO shows an influence on Indian Summer Monsoon Rainfall (ISMR). The ISMR is stronger during the west phase of QBO and weak during the east phase (Rai and Dimri 2017). QBO also influences the stratospheric aerosol layer. Satellite observations show that QBO modulates the vertical extent of the stratospheric aerosol layer in the tropics by up to 6 km, or ∼35% of its mean vertical extent between 100 and 7 hPa (about 16–33 km) (Hommel et al. 2015).

5.5 Impact of Volcanic Eruptions

Volcanoes inject huge amounts of aerosols and trace gases in the upper troposphere and stratosphere, thereby drive the natural mode of climate variability through alteration of radiative forcing (Robock 2015). A volcanic eruption in the vicinity of India, e.g., Mt. Nabro during 11–13 June 2011, injected a large amount of water vapor, and SO2 (1.3–2.0 Tg) in the upper troposphere and lower stratosphere over India. The aerosols injected into the stratosphere traveled large distances and thickened the stratospheric aerosol layer. The global lidar networks (EARLINET, MPLNET, and NDACC) and satellite (Cloud-Aerosol Lidar and Infrared Pathfinder Satellite Observation, (CALIPSO)) show that Mt. Nabro has increased stratospheric volcanic AOD by 0.003–0.04 (global mean) and heating by ∼0.3 K day^{-1} between 16 and 17 km altitude (Fairlie et al. 2014). The aerosol surge causes tropospheric cooling and stratospheric warming by scattering and reflecting incoming solar radiation (von Glasow et al. 2009). Large volcanoes modulate the Inter-Tropical Convergence Zone via changes in the hemispheric temperature gradient. There is a southward shift in the ITCZ location and South Asian Monsoon (Sinha et al. 2011) after the volcanic eruptions occurred during the last millennium (Schneider et al. 2014). A host of modeling studies shows a consistent decrease in Asian summer monsoon rainfall following volcanic (Zambri et al. 2017), barring a few studies which report an increase in the precipitation response in the post-eruption period, due to change in the land–sea thermal gradient (Joseph and Zeng 2011). Volcanic eruptions also influence climate; that is, it triggers El Niños (Ohba et al. 2013).

One of the important impacts of volcanoes is the loss of stratospheric ozone. The ozone loss and associated changes in photolysis rates affect the tropospheric/stratospheric temperature (cooling/warming) (Santer et al. 2003). The stratospheric ozone loss is linked with chemical reactions occurring over aerosol surfaces. There is a reduction of nitrogen oxides and chlorine activation, which leads to an increase in Equivalent Effective Stratospheric Chlorine (EESC) (Tie and Brasseur 1995; Tabazadeh et al. 2002). A drastic decrease in the stratospheric ozone over Antarctica due to a series of volcanic eruptions has been proposed to lead to large-scale changes in atmospheric dynamics resulting in massive de-glaciations in the past (McConnell et al. 2017). Previous work also indicates that volcanic eruptions can serve as a source of potential predictability (Gaddis 2013) by having links with the tropical precipitation via modulations of stratospheric ozone.

5.6 Knowledge Gaps

Aerosol-cloud-precipitation-meteorology interaction is one of the most challenging scientific issues requiring intensive observational and modeling with focused research from the climate science community. The complexity in the aerosol-cloud interaction arises from variations in dominant phase changes and microphysical and dynamical processes associated with different types of clouds. Concurrent measurements of aerosol size distribution, composition, cloud properties, microphysical parameters as well as the development of physical process scale studies based on observations over a varying space and times-scales, and translating them to climate models are essential to gain a good understanding on the role of aerosols in modifying weather and climate over India.

In addition, accurate representation of the absorbing aerosol hotspots, particularly BC and dust, is crucial to comprehend its impact on regional climate. Uncertainty in the measurement of single scattering albedo, the parameter determining the absorptive nature of aerosols, limits the correct quantification of the sign of TOA radiative forcing at regional scales. More in situ measurements of vertical profiles of absorbing aerosols are also needed for better evaluation of model-simulated BC profiles and understanding its effect on monsoon precipitation through interaction with clouds and radiation. Also, the aerosol observational data from field campaigns and long-term monitoring sites from various sources and reanalysis products need to be gathered to make a comprehensive quality-controlled gridded product. Future field campaigns may be planned to address missing links in this regard and to reduce the uncertainty in the regional estimates of the direct and indirect effect of aerosols in state of the art GCMs.

In the case of trace gases, there is a considerable variation among the emission inventories of ozone precursors and related trace gases. Dedicated modeling and observational efforts are needed to improve ozone emission inventories over the Indian region. Model simulations show seasonal transport of chemical species over the Indian Ocean which affects the air–sea interaction and convective processes. However, the models show significant biases over the Oceans. There is a need to improve chemical processes and parameterization in the model to reduce the biases. Finally, there are limited studies quantifying the radiative impact of trace gases and associated climate change over the Indian region, and further modeling and observational studies in this direction are required.

5.7 Summary

The regional assessment of long-term in situ and remotely sensed observations over India shows a significant increase in aerosol loading over the subcontinent accompanied by robust seasonal variations. The trend in AOD is $\sim 2\%$ year^{-1} (high confidence) during the last 30. The temporal build-up of aerosols is significantly high in the dry winter months, while changes are smaller in the pre-monsoon and monsoon season. This change has been attributed to rise in fine mode particles due to rapid growth in anthropogenic activities over the region in recent decades. CMIP5 multi-model simulations also capture the large increase in AOD over the Indian region between 1980 and 2000 with considerable bias in the three-dimensional heterogeneous distribution of different aerosol species.

There is a large seasonal as well as spatiotemporal variability in the aerosol radiative forcing. In general, the estimates of aerosol radiative forcing from measurements range from -49 to -31 W m^{-2} at the surface (high confidence), and -15 to $+8$ W m^{-2} at top-of-atmosphere (low confidence). The positive forcing at TOA is linked with the absorptive nature of the aerosols over the Indian region. Aerosols produce a declining trend of all-sky global irradiance over India. During 1986–1995, the observed global radiation decreased by 3.6 W m^{-2} and further by 9.5 W m^{-2}

during the decade of 1996–2005 (Soni et al. 2012). The declining trend of all-sky global irradiance over India as a whole was 0.6 W m^{-2} year^{-1} during 1971–2000 and 0.2 W m^{-2} year^{-1} during 2001–2010 (Soni et al. 2016). This decrease in global irradiance is matched with an increase in the diffused radiation over the same period indicating an increase in the aerosol levels.

Efforts were taken to understand aerosol-cloud interaction over the Indian region. The Cloud-Aerosol Interaction and Precipitation Enhancement Experiment [CAIPEEX; (Kulkarni et al. 2012)] has documented important processes associated with aerosol-cloud interaction over the Indian region. There is a significant increase in the cloud droplet number concentration with an increase in aerosols (Kulkarni et al. 2012). Very high aerosol loading causes narrowing of the droplet spectrum, collision coalescence is suppressed, and warm rain forms at an elevated layer (Konwar et al. 2012). During high aerosol loading conditions, clouds have a large amount of super-cooled liquid water (>3 gm^{-3}) with the dominant mixed-phase (Prabha et al. 2012). Mixed-phase clouds contribute a significant part of monsoon clouds, which are not understood completely and need further focused process studies. Aerosols acting as CCN, and INP and their variability over the Indian region need further observations and can be used for the models or fine-tune the parameterization schemes.

Long-term observations of ozone (total column, vertical profiles, and surface measurements) and its precursors (CO, NO$_x$, VOCs) have been studied to estimate linear trends over the Indian region. Tropospheric ozone trends show spatiotemporal variations. Trend estimates vary with time due to changes in the emission of precursors gases. In general, ozone observations show increasing trends in the troposphere (0.7–0.9% year^{-1} during 1979–2005) (high confidence) and decreasing trends in the stratosphere (-0.05 ± 0.04 to -0.4 ± 0.1% year^{-1} during 1993–2015) (medium confidence). The reported ozone observations over the Indian regions are of different time periods. However, the System of Air Quality Forecasting and Research (SAFAR) developed by the Indian Institute of Tropical Meteorology is monitoring ozone and its precursors, since 2010. These long-term observations will be helpful in obtaining future ozone trends over the India region. The CMIP5 multi-model future projections (the 2090s–2010s) over the tropics (25° S–25° N) show that the annual mean ozone trend is decreasing in the troposphere (except RCP8.5) and increasing in the stratosphere (Cionni et al. 2011). Seasonal trends in the troposphere and stratosphere, both, are influenced by emissions and, seasonal stratopheric intrusions, etc. The reported ozone trends have low confidence. Long-range transport processes (e.g., seasonal variations, transport between extra-tropics and tropics, stratosphere and troposphere, etc.) produce a significant variation in the loading of tropospheric ozone leading to large changes in radiative forcing and dynamics. The model simulations show that increased tropospheric ozone since pre-industrial times has imposed ozone radiative forcing (at the tropopause) ~ 0.2–0.4 W m^{-2} over the Indian region (Chalita et al. 1996).

References

Aneesh VR, Mohankumar G, Sampath S (2008) Spatial distribution of atmospheric carbon monoxide over Bay of Bengal and Arabian Sea: measurements during pre-monsoon period of 2006. J Earth Syst Sci 117:449–455. https://doi.org/10.1007/s12040-008-0044-8

Anil Kumar V, Pandithurai G, Parambil Leena P et al (2016) Investigation of aerosol indirect effects on monsoon clouds using ground-based measurements over a high-altitude site in Western Ghats. Atmos Chem Phys 16:8423–8430. https://doi.org/10.5194/acp-16-8423-2016

Babu SS, Satheesh SK, Moorthy KK (2002) Aerosol radiative forcing due to enhanced black carbon at an urban site in India. Geophys Res Lett 29:27-1–27-4. https://doi.org/10.1029/2002GL015826

Babu SS, Manoj MR, Moorthy KK et al (2013) Trends in aerosol optical depth over Indian region: potential causes and impact indicators. J Geophys Res Atmos 118:11794–11806. https://doi.org/10.1002/2013JD020507

Badarinath KVS, Madhavi Latha K (2006) Direct radiative forcing from black carbon aerosols over urban environment. Adv Sp Res 37:2183–2188. https://doi.org/10.1016/j.asr.2005.10.034

Beig G, Brasseur GP (2006) Influence of anthropogenic emissions on tropospheric ozone and its precursors over the Indian tropical region during a monsoon. Geophys Res Lett 33:1–5. https://doi.org/10.1029/2005GL024949

Beig G, Singh V (2007) Trends in tropical tropospheric column ozone from satellite data and MOZART model. Geophys Res Lett 34:1–5. https://doi.org/10.1029/2007GL030460

Bera S, Prabha TV, Malap N et al (2019) Thermodynamics and Microphysics Relation During CAIPEEX-I. Pure Appl Geophys 176:371–388. https://doi.org/10.1007/s00024-018-1942-6

Bharali C, Pathak B, Bhuyan PK (2015) Spring and summer night-time high ozone episodes in the upper Brahmaputra valley of North East India and their association with lightning. Atmos Environ 109:234–250. https://doi.org/10.1016/j.atmosenv.2015.03.035

Bollasina MA, Ming Y, Ramaswamy V (2011) Anthropogenic aerosols and the summer monsoon. Science 80(334):502–505

Bond TC, Doherty SJ, Fahey DW et al (2013) Bounding the role of black carbon in the climate system: a scientific assessment. J Geophys Res Atmos 118:5380–5552. https://doi.org/10.1002/jgrd.50171

Brasseur GP, Orlando JJ, Tyndall GS, Atmospheric Research (U.S.) NC (1999) Atmospheric chemistry and global change, topics in environmental chemistry. Oxford University Press, New York

Brewer AW (1949) Evidence for a world circulation provided by the measurements of helium and water vapour distribution in the stratosphere. Q J R Meteorol Soc 75:351–363. https://doi.org/10.1002/qj.49707532603

Chalita S, Hauglustaine DA, Le Treut H, Müller JF (1996) Radiative forcing due to increased tropospheric ozone concentrations. Atmos Environ 30:1641–1646. https://doi.org/10.1016/1352-2310(95)00431-9

Chehade W, Weber M, Burrows JP (2014) Total ozone trends and variability during 1979–2012 from merged data sets of various satellites. Atmos Chem Phys 14:7059–7074. https://doi.org/10.5194/acp-14-7059-2014

Cionni I, Eyring V, Lamarque JF et al (2011) Ozone database in support of CMIP5 simulations: results and corresponding radiative forcing. Atmos Chem Phys 11:11267–11292. https://doi.org/10.5194/acp-11-11267-2011

Dani KK, Ernest Raj P, Devara PCS et al (2012) Long-term trends and variability in measured multi-spectral aerosol optical depth over a tropical urban station in India. Int J Climatol 32:153–160. https://doi.org/10.1002/joc.2250

Dave P, Bhushan M, Venkataraman C (2017) Aerosols cause intraseasonal short-term suppression of Indian monsoon rainfall. Sci Rep 7:17347. https://doi.org/10.1038/s41598-017-17599-1

Dey S, Di Girolamo L (2011) A decade of change in aerosol properties over the Indian subcontinent. Geophys Res Lett 38:1–5. https://doi.org/10.1029/2011GL048153

Dobson GMB (1956) Origin and distribution of the polyatomic molecules in the atmosphere. Proc R Soc London Ser A Math Phys Sci 236:187–193. https://doi.org/10.1098/rspa.1956.0127

Duchi R, Cristofanelli P, Marinoni A et al (2014) Synoptic-scale dust transport events in the southern Himalaya. Aeolian Res 13:51–57. https://doi.org/10.1016/j.aeolia.2014.03.008

Eyring V, Arblaster JM, Cionni I et al (2013) Long-term ozone changes and associated climate impacts in CMIP5 simulations. J Geophys Res Atmos 118:5029–5060. https://doi.org/10.1002/jgrd.50316

Fadnavis S, Chattopadhyay R (2017) Linkages of subtropical stratospheric intraseasonal intrusions with Indian summer monsoon deficit rainfall. J Clim 30:5083–5095. https://doi.org/10.1175/JCLI-D-16-0463.1

Fadnavis S, Beig G, Polade SD (2008) Features of ozone quasi-biennial oscillation in the vertical structure of tropics and subtropics. Meteorol Atmos Phys 99:221–231. https://doi.org/10.1007/s00703-007-0270-7

Fadnavis S, Chakraborty T, Beig G (2010) Seasonal stratospheric intrusion of ozone in the upper troposphere over India. Ann Geophys 28:2149–2159. https://doi.org/10.5194/angeo-28-2149-2010

Fadnavis S, Chakraborty T, Ghude SD et al (2011) Modulation of cyclone tracks in the Bay of Bengal by QBO. J Atmos Solar-Terr Phys 73:1868–1875. https://doi.org/10.1016/j.jastp.2011.04.014

Fadnavis S, Semeniuk K, Pozzoli L et al (2013) Transport of aerosols into the UTLS and their impact on the Asian monsoon region as seen in a global model simulation. Atmos Chem Phys 13:8771–8786. https://doi.org/10.5194/acp-13-8771-2013

Fadnavis S, Dhomse S, Ghude S et al (2014a) Ozone trends in the vertical structure of upper troposphere and lower stratosphere over the Indian monsoon region. Int J Environ Sci Technol 11:529–542. https://doi.org/10.1007/s13762-013-0258-4

Fadnavis S, Ernest Raj P, Buchunde P, Goswami BN (2014b) In search of influence of stratospheric quasi-biennial oscillation on tropical cyclones tracks over the Bay of Bengal region. Int J Climatol 34:567–580. https://doi.org/10.1002/joc.3706

Fadnavis S, Semeniuk K, Schultz MG et al (2015) Transport pathways of peroxyacetyl nitrate in the upper troposphere and lower stratosphere from different monsoon systems during the summer monsoon season. Atmos Chem Phys 15:11477–11499. https://doi.org/10.5194/acp-15-11477-2015

Fadnavis S, Kalita G, Ravi Kumar K et al (2017) Potential impact of carbonaceous aerosol on the upper troposphere and lower stratosphere (UTLS) and precipitation during Asian summer monsoon in a global model simulation. Atmos Chem Phys 17:11637–11654. https://doi.org/10.5194/acp-17-11637-2017

Fadnavis S, Sabin TP, Roy C et al (2019) Elevated aerosol layer over South Asia worsens the Indian droughts. Sci Rep 9:1–11. https://doi.org/10.1038/s41598-019-46704-9

Fairlie TD, Vernier JP, Natarajan M, Bedka KM (2014) Dispersion of the Nabro volcanic plume and its relation to the Asian summer monsoon. Atmos Chem Phys 14:7045–7057. https://doi.org/10.5194/acp-14-7045-2014

Fiedler S, Stevens B, Gidden M et al (2019) First forcing estimates from the future CMIP6 scenarios of anthropogenic aerosol optical properties and an associated Twomey effect. Geosci Model Dev 12:989–1007. https://doi.org/10.5194/gmd-12-989-2019

Forster PM, Shine KP (1997) Radiative forcing and temperature trends from stratospheric ozone changes. J Geophys Res Atmos 102:10841–10855. https://doi.org/10.1029/96jd03510

Fu R, Hu Y, Wright JS et al (2006) Short circuit of water vapor and polluted air to the global stratosphere by convective transport over the Tibetan Plateau. Proc Natl Acad Sci 103:5664–5669. https://doi.org/10.1073/pnas.0601584103

Gaddis AL (2013) Evaluating predictability in the community earth system model in response to the eruption of Mount Pinatubo. University of Tennessee

Ganguly D, Jayaraman A (2006) Physical and optical properties of aerosols over an urban location in western India: implications for shortwave radiative forcing. J Geophys Res 111:D24207. https://doi.org/10.1029/2006JD007393

Ganguly D, Gadhavi H, Jayaraman A et al (2005) Single scattering albedo of aerosols over the central India: implications for the regional aerosol radiative forcing. Geophys Res Lett 32:1–4. https://doi.org/10.1029/2005GL023903

Ganguly D, Rasch PJ, Wang H, Yoon JH (2012) Climate response of the South Asian monsoon system to anthropogenic aerosols. J Geophys Res Atmos 117:1–20. https://doi.org/10.1029/2012JD017508

Garg A, Shukla PR, Kapshe M (2006) The sectoral trends of multigas emissions inventory of India. Atmos Environ 40:4608–4620. https://doi.org/10.1016/j.atmosenv.2006.03.045

Gautam R, Hsu NC, Lau WKM, Yasunari TJ (2013) Satellite observations of desert dust-induced Himalayan snow darkening. Geophys Res Lett 40:988–993. https://doi.org/10.1002/grl.50226

Gayatri K, Patade S, Prabha TV (2017) Aerosol-cloud interaction in deep convective clouds over the Indian Peninsula using spectral (bin) microphysics. J Atmos Sci 74:3145–3166. https://doi.org/10.1175/JAS-D-17-0034.1

Gertler CG, Puppala SP, Panday A et al (2016) Black carbon and the Himalayan cryosphere: a review. Atmos Environ 125:404–417. https://doi.org/10.1016/j.atmosenv.2015.08.078

Ghude SD, Fadnavis S, Beig G et al (2008) Detection of surface emission hot spots, trends, and seasonal cycle from satellite-retrieved NO_2 over India. J Geophys Res Atmos 113:1–13. https://doi.org/10.1029/2007JD009615

Ghude SD, Kulkarni SH, Jena C et al (2013a) Application of satellite observations for identifying regions of dominant sources of nitrogen oxides over the Indian subcontinent. J Geophys Res Atmos 118:1075–1089. https://doi.org/10.1029/2012JD017811

Ghude SD, Pfister GG, Jena C et al (2013b) Satellite constraints of nitrogen oxide (NO_x) emissions from India based on OMI observations and WRF-Chem simulations. Geophys Res Lett 40:423–428. https://doi.org/10.1029/2012GL053926

Gidden MJ, Riahi K, Smith SJ et al (2019) Global emissions pathways under different socioeconomic scenarios for use in CMIP6: a dataset of harmonized emissions trajectories through the end of the century. Geosci Model Dev 12:1443–1475. https://doi.org/10.5194/gmd-12-1443-2019

Girach IA, Ojha N, Nair PR et al (2017) Variations in O_3, CO, and CH_4 over the Bay of Bengal during the summer monsoon season: shipborne measurements and model simulations. Atmos Chem Phys 17:257–275. https://doi.org/10.5194/acp-17-257-2017

Gogoi MM, Babu SS, Jayachandran V et al (2015) Optical properties and CCN activity of aerosols in a high-altitude Himalayan environment: results from RAWEX-GVAX. J Geophys Res 120:2453–2469. https://doi.org/10.1002/2014JD022966

Guenther A, Karl T, Harley P et al (2006) Estimates of global terrestrial isoprene emissions using MEGAN (model of emissions of gases and aerosols from nature). Atmos Chem Phys 6:3181–3210. https://doi.org/10.5194/acpd-6-107-2006

Guo L, Turner AG, Highwood EJ (2015) Impacts of 20th century aerosol emissions on the South Asian monsoon in the CMIP5 models. Atmos Chem Phys 15:6367–6378. https://doi.org/10.5194/acp-15-6367-2015

Harikishan G, Padmakumari B, Maheskumar RS et al (2016) Aerosol indirect effects from ground-based retrievals over the rain shadow region in Indian subcontinent. J Geophys Res 121:1–14. https://doi.org/10.1002/2015JD024577

Hazra A, Goswami BN, Chen JP (2013) Role of interactions between aerosol radiative effect, dynamics, and cloud microphysics on transitions of monsoon intraseasonal oscillations. J Atmos Sci 70:2073–2087. https://doi.org/10.1175/JAS-D-12-0179.1

Hommel R, Timmreck C, Giorgetta MA, Graf HF (2015) Quasi-biennial oscillation of the tropical stratospheric aerosol layer. Atmos Chem Phys 15:5557–5584. https://doi.org/10.5194/acp-15-5557-2015

Hsu NC, Gautam R, Sayer AM et al (2012) Global and regional trends of aerosol optical depth over land and ocean using SeaWiFS measurements from 1997 to 2010. Atmos Chem Phys 12:8037–8053. https://doi.org/10.5194/acp-12-8037-2012

IPCC (2013) Fifth assessment report of the intergovernmental panel on climate change (IPCC)

Jacobi HW, Lim S, Ménégoz M et al (2015) Black carbon in snow in the upper Himalayan Khumbu Valley, Nepal: observations and modeling of the impact on snow albedo, melting, and radiative forcing. Cryosphere 9:1685–1699. https://doi.org/10.5194/tc-9-1685-2015

Jena C, Ghude SD, Beig G et al (2015) Inter-comparison of different NOx emission inventories and associated variation in simulated surface ozone in Indian region. Atmos Environ 117:61–73. https://doi.org/10.1016/j.atmosenv.2015.06.057

Joseph R, Zeng N (2011) Seasonally modulated tropical drought induced by volcanic aerosol. J Clim 24:2045–2060. https://doi.org/10.1175/2009JCLI3170.1

Kaskaoutis DG, Singh RP, Gautam R et al (2012) Variability and trends of aerosol properties over Kanpur, northern India using AERONET data (2001–10). Environ Res Lett 7(2):024003. https://doi.org/10.1088/1748-9326/7/2/024003

Kaspari S, Painter TH, Gysel M et al (2014) Seasonal and elevational variations of black carbon and dust in snow and ice in the Solu-Khumbu, Nepal and estimated radiative forcings. Atmos Chem Phys 14:8089–8103. https://doi.org/10.5194/acp-14-8089-2014

Kavitha MP, Nair R, Renju R (2018) Thunderstorm induced changes in near-surface O3, NOx and CH4 and associated boundary layer meteorology over a tropical coastal station. J Atmos Solar Terr Phys 179. https://doi.org/10.1016/j.jastp.2018.08.008

Konwar M, Maheskumar RS, Kulkarni JR et al (2012) Aerosol control on depth of warm rain in convective clouds. J Geophys Res 117:1–10. https://doi.org/10.1029/2012JD017585

Kopacz M, Mauzerall DL, Wang J et al (2011) Origin and radiative forcing of black carbon transported to the Himalayas and Tibetan Plateau. Atmos Chem Phys 11:2837–2852. https://doi.org/10.5194/acp-11-2837-2011

Krishna Moorthy K, Suresh Babu S, Manoj MR, Satheesh SK (2013) Buildup of aerosols over the Indian Region. Geophys Res Lett 40:1011–1014. https://doi.org/10.1002/grl.50165

Krishnamurti TN, Jha B, Prospero J et al (1998) Aerosol and pollutant transport and their impact on radiative forcing over the tropical Indian Ocean during the January–February 1996 pre-INDOEX cruise. Tellus, Ser B Chem Phys Meteorol 50:521–542. https://doi.org/10.3402/tellusb.v50i5.16235

Krishnan R, Ramanathan V (2002) Evidence of surface cooling from absorbing aerosols. Geophys Res Lett 29:54-1–54-4. https://doi.org/10.1029/2002gl014687

Krishnan R, Sabin TP, Vellore R et al (2016) Deciphering the desiccation trend of the South Asian monsoon hydroclimate in a warming world. Clim Dyn 47:1007–1027. https://doi.org/10.1007/s00382-015-2886-5

Kulkarni JR, Maheskumar RS, Morwal SB et al (2012) The cloud aerosol interaction and precipitation enhancement experiment (CAIPEEX): overview and preliminary results. Curr Sci 102:413–425

Lal DM, Ghude SD, Patil SD et al (2012) Tropospheric ozone and aerosol long-term trends over the Indo-Gangetic Plain (IGP), India. Atmos Res 116:82–92. https://doi.org/10.1016/j.atmosres.2012.02.014

Lal S, Venkataramani S, Srivastava S et al (2013) Transport effects on the vertical distribution of tropospheric ozone over the tropical marine regions surrounding India. J Geophys Res Atmos 118:1513–1524. https://doi.org/10.1002/jgrd.50180

Lal S, Venkataramani S, Chandra N et al (2014) Transport effects on the vertical distribution of tropospheric ozone over western India. J Geophys Res 119:n/a. https://doi.org/10.1002/2014JD021854

Lal S, Peshin SK, Naja M, Venkataramani S (2017) Variability of ozone and related trace gases over India. In: Rajeevan MN, Nayak S (eds) Observed climate variability and change over the Indian region. Springer geology. Springer, Singapore

Lau KM, Kim KM (2006) Observational relationships between aerosol and Asian monsoon rainfall, and circulation. Geophys Res Lett 33:1–5. https://doi.org/10.1029/2006GL027546

Lau WKM, Kim KM (2010) Fingerprinting the impacts of aerosols on long-term trends of the Indian summer monsoon regional rainfall. Geophys Res Lett 37:1–5. https://doi.org/10.1029/2010GL043255

Lau KM, Kim MK, Kim KM (2006) Asian summer monsoon anomalies induced by aerosol direct forcing: the role of the Tibetan Plateau. Clim Dyn 26:855–864. https://doi.org/10.1007/s00382-006-0114-z

Lau WKM, Yuan C, Li Z (2018) Origin, maintenance and variability of the Asian tropopause aerosol layer (ATAL): the roles of monsoon dynamics. Sci Rep 8:1–14. https://doi.org/10.1038/s41598-018-22267-z

Leena PP, Anilkumar V, Sravanthi N et al (2018) On the precipitation susceptibility of monsoon clouds to aerosols using high-altitude ground-based observations over Western Ghats, India. Atmos Environ 185:128–136. https://doi.org/10.1016/j.atmosenv.2018.05.001

Logan JA, Staehelin J, Megretskaia IA et al (2012) Changes in ozone over Europe: analysis of ozone measurements from sondes, regular aircraft (MOZAIC) and alpine surface sites. J Geophys Res Atmos 117:1–23. https://doi.org/10.1029/2011JD016952

Lu X, Zhang L, Liu X et al (2018) Lower tropospheric ozone over India and its linkage to the South Asian monsoon. Atmos Chem Phys 18:3101–3118. https://doi.org/10.5194/acp-18-3101-2018

Mahajan AS, De Smedt I, Biswas MS et al (2015) Inter-annual variations in satellite observations of nitrogen dioxide and formaldehyde over India. Atmos Environ 116:194–201. https://doi.org/10.1016/j.atmosenv.2015.06.004

Manoj MR, Satheesh SK, Moorthy KK et al (2019) Decreasing trend in black carbon aerosols over the Indian region. Geophys Res Lett 46:2903–2910. https://doi.org/10.1029/2018GL081666

McConnell JR, Burke A, Dunbar NW et al (2017) Synchronous volcanic eruptions and abrupt climate change ∼17.7 ka plausibly linked by stratospheric ozone depletion. Proc Natl Acad Sci 114:10035–10040. https://doi.org/10.1073/pnas.1705595114

Misra A, Kanawade VP, Tripathi SN (2016) Quantitative assessment of AOD from 17 CMIP5 models based on satellite-derived AOD over India. Ann Geophys 34:657–671. https://doi.org/10.5194/angeo-34-657-2016

Müller S, Hoor P, Bozem H et al (2016) Impact of the Asian monsoon on the extratropical lower stratosphere: trace gas observations during TACTS over Europe 2012. Atmos Chem Phys 16:10573–10589. https://doi.org/10.5194/acp-16-10573-2016

Naik V, Mauzerall D, Horowitz L et al (2005) Net radiative forcing due to changes in regional emissions of tropospheric ozone precursors. J Geophys Res Atmos 110:1–14. https://doi.org/10.1029/2005JD005908

Nair S, Sanjay J, Pandithurai G et al (2012) On the parameterization of cloud droplet effective radius using CAIPEEX aircraft observations for warm clouds in India. Atmos Res 108:104–114. https://doi.org/10.1016/j.atmosres.2012.02.002

Nair VS, Babu SS, Manoj MR et al (2016) Direct radiative effects of aerosols over South Asia from observations and modeling. Clim Dyn 49:1411–1428. https://doi.org/10.1007/s00382-016-3384-0

Nair PR, Ajayakumar RS, David LM et al (2018) Decadal changes in surface ozone at the tropical station Thiruvananthapuram (8.542° N, 76.858° E), India: effects of anthropogenic activities and meteorological variability. Environ Sci Pollut Res 25:14827–14843. https://doi.org/10.1007/s11356-018-1695-x

Naja M, Lal S (1996) Changes in surface ozone amount and its diurnal and seasonal patterns, from 1954-55 to 1991-93, measured at Ahmedabad (23 N), India. Geophys Res Lett 23:81–84

Ohba M, Shiogama H, Yokohata T, Watanabe M (2013) Impact of strong tropical volcanic eruptions on ENSO simulated in a coupled gcm. J Clim 26:5169–5182. https://doi.org/10.1175/JCLI-D-12-00471.1

Padmakumari B, Maheskumar RS, Harikishan G et al (2013) In situ measurements of aerosol vertical and spatial distributions over continental India during the major drought year 2009. Atmos Environ 80:107–121. https://doi.org/10.1016/j.atmosenv.2013.07.064

Pan X, Chin M, Gautam R et al (2015) A multi-model evaluation of aerosols over South Asia: common problems and possible causes. Atmos Chem Phys 15:5903–5928. https://doi.org/10.5194/acp-15-5903-2015

Pandey A, Sadavarte P, Rao AB, Venkataraman C (2014) Trends in multi-pollutant emissions from a technology-linked inventory for India: II. Residential, agricultural and informal industry sectors. Atmos Environ 99:341–352. https://doi.org/10.1016/j.atmosenv.2014.09.080

Pandey SK, Vinoj V, Landu K, Babu SS (2017) Declining pre-monsoon dust loading over South Asia: signature of a changing regional climate. Sci Rep 7:1–10. https://doi.org/10.1038/s41598-017-16338-w

Pandithurai G, Dipu S, Dani KK et al (2008) Aerosol radiative forcing during dust events over New Delhi, India. J Geophys Res Atmos 113:1–13. https://doi.org/10.1029/2008JD009804

Pandithurai G, Dipu S, Prabha T V et al (2012) Aerosol effect on droplet spectral dispersion in warm continental cumuli. J Geophys Res Atmos 117. https://doi.org/10.1029/2011JD016532

Panicker AS, Pandithurai G, Dipu S (2010) Aerosol indirect effect during successive contrasting monsoon seasons over Indian subcontinent using MODIS data. Atmos Environ 44:1937–1943. https://doi.org/10.1016/j.atmosenv.2010.02.015

Patade S, Nagare B, Wagh S et al (2014) Deposition ice nuclei observations over the Indian region during CAIPEEX. Atmos Res 149:300–314. https://doi.org/10.1016/j.atmosres.2014.07.001

Pathak B, Subba T, Dahutia P et al (2016) Aerosol characteristics in north-east India using ARFINET spectral optical depth measurements. Atmos Environ 125:461–473. https://doi.org/10.1016/j.atmosenv.2015.07.038

Pathakoti M, Asuri LK, Venkata MD et al (2018) Assessment of total columnar ozone climatological trends over the Indian sub-continent. Int J Remote Sens 39:3963–3982. https://doi.org/10.1080/01431161.2018.1452066

Prabha TV and Khain A (2020) Water vapor and pollutants, aerosol–cloud interactions. In: Patricia A. Maurice (ed) Encyclopedia of water: science, technology and society. https://doi.org/10.1002/9781119300762.wsts0093

Prabha TV, Patade S, Pandithurai G et al (2012) Spectral width of premonsoon and monsoon clouds over Indo-Gangetic valley. J Geophys Res Atmos 117:1–15. https://doi.org/10.1029/2011JD016837

Pu B, Ginoux P (2018) How reliable are CMIP5 models in simulating dust optical depth? Atmos Chem Phys 18:12491–12510. https://doi.org/10.5194/acp-18-12491-2018

Rai P, Dimri AP (2017) Effect of changing tropical easterly jet, low level jet and quasi-biennial oscillation phases on Indian summer monsoon. Atmos Sci Lett 18:52–59. https://doi.org/10.1002/asl.723

Raj STA, Venkat Ratnam M, Narayana Rao D, Krishna Murthy BV (2018) Long-term trends in stratospheric ozone, temperature, and water vapor over the Indian region. Ann Geophys 36:149–165. https://doi.org/10.5194/angeo-36-149-2018

Ramachandran S, Rengarajan R, Jayaraman A et al (2006) Aerosol radiative forcing during clear, hazy, and foggy conditions over a continental polluted location in north India. J Geophys Res Atmos 111:1–12. https://doi.org/10.1029/2006JD007142

Ramachandran S, Kedia S, Srivastava R (2012) Aerosol optical depth trends over different regions of India. Atmos Environ 49:338–347. https://doi.org/10.1016/j.atmosenv.2011.11.017

Ramanathan V, Chung C, Kim D et al (2005) Atmospheric brown clouds: impacts on South Asian climate and hydrological cycle. Proc Natl Acad Sci U S A 102:5326–5333. https://doi.org/10.1073/pnas.0500656102

Rana A, Jia S, Sarkar S (2019) Black carbon aerosol in India: a comprehensive review of current status and future prospects. Atmos Res 218:207–230. https://doi.org/10.1016/j.atmosres.2018.12.002

Ravi Kiran V, Talukdar S, Venkat Ratnam M, Jayaraman A (2018) Long-term observations of black carbon aerosol over a rural location in southern peninsular India: role of dynamics and meteorology. Atmos Environ 189:264–274. https://doi.org/10.1016/j.atmosenv.2018.06.020

Reddy MS, Boucher O, Venkataraman C et al (2004) General circulation model estimates of aerosol transport and radiative forcing during the Indian Ocean Experiment. J Geophys Res Atmos 109:1–15. https://doi.org/10.1029/2004JD004557

Robock A (2015) Important research questions on volcanic eruptions and climate. PAGES Mag 23. https://doi.org/10.22498/pages.23.2.68

Roy C, Fadnavis S, Müller R et al (2017) Influence of enhanced Asian NO_x emissions on ozone in the upper troposphere and lower stratosphere in chemistry-climate model simulations. Atmos Chem Phys 17:1297–1311. https://doi.org/10.5194/acp-17-1297-2017

Sadavarte P, Venkataraman C (2014) Trends in multi-pollutant emissions from a technology-linked inventory for India: I. Industry and transport sectors. Atmos Environ 99:353–364. https://doi.org/10.1016/j.atmosenv.2014.09.081

Sahai S, Sharma C, Singh DP et al (2007) A study for development of emission factors for trace gases and carbonaceous particulate species from in situ burning of wheat straw in agricultural fields in India. Atmos Environ 41:9173–9186. https://doi.org/10.1016/j.atmosenv.2007.07.054

Sahu SK, Beig G, Sharma C (2008) Decadal growth of black carbon emissions in India. Geophys Res Lett 35:1–5. https://doi.org/10.1029/2007GL032333

Sahu LK, Sheel V, Kajino M et al (2014) Seasonal and interannual variability of tropospheric ozone over an urban site in India: a study based on MOZAIC and CCM vertical profiles over Hyderabad. J Geophys Res Atmos 119:3615–3641. https://doi.org/10.1002/2013JD021215

Sahu BS, Tandon A, Attri AK (2017) Roles of ozone depleting substances and solar activity in observed long-term trends in total ozone column over Indian region. Int J Remote Sens 38:5091–5105. https://doi.org/10.1080/01431161.2017.1333654

Samanta D, Dash MK, Goswami BN, Pandey PC (2016) Extratropical anticyclonic Rossby wave breaking and Indian summer monsoon failure. Clim Dyn 46:1547–1562. https://doi.org/10.1007/s00382-015-2661-7

Sanap SD, Pandithurai G (2015) The effect of absorbing aerosols on Indian monsoon circulation and rainfall: a review. Atmos Res 164–165:318–327. https://doi.org/10.1016/j.atmosres.2015.06.002

Sanap SD, Ayantika DC, Pandithurai G, Niranjan K (2014) Assessment of the aerosol distribution over Indian subcontinent using CMIP5 models. Atmos Environ, 123–137 (in press)

Sanap SD, Pandithurai G, Manoj MG (2015) On the response of Indian summer monsoon to aerosol forcing in CMIP5 model simulations. Clim Dyn 45:2949–2961. https://doi.org/10.1007/s00382-015-2516-2

Santer BD, Wehner MF, Wigley TML et al (2003) Contributions of anthropogenic and natural forcing to recent tropopause height changes. Science 80-(301):479–483. https://doi.org/10.1126/science.1084123

Saraf N, Beig G (2004) Long-term trends in tropospheric ozone over the Indian tropical region. Geophys Res Lett 31:n/a. https://doi.org/10.1029/2003GL018516

Sarkar C, Roy A, Chatterjee A et al (2019) Factors controlling the long-term (2009–2015) trend of $PM_{2.5}$ and black carbon aerosols at eastern Himalaya, India. Sci Total Environ 656:280–296. https://doi.org/10.1016/j.scitotenv.2018.11.367

Satheesh S, Babu S, Padmakumari B et al (2017) Variability of atmospheric aerosols over India, pp 221–248

Saud T, Gautam R, Mandal TK et al (2012) Emission estimates of organic and elemental carbon from household biomass fuel used over the Indo-Gangetic Plain (IGP), India. Atmos Environ 61:212–220. https://doi.org/10.1016/j.atmosenv.2012.07.030

Schneider T, Bischoff T, Haug GH (2014) Migrations and dynamics of the intertropical convergence zone. Nature 513:45–53. https://doi.org/10.1038/nature13636

Schumann U, Huntrieser H (2007) The global lightning-induced nitrogen oxides source. Atmos Chem Phys 7:3823–3907. https://doi.org/10.5194/acp-7-3823-2007

Sharma S, Goel A, Gupta D et al (2015) Emission inventory of non-methane volatile organic compounds from anthropogenic sources in India. Atmos Environ 102:209–219. https://doi.org/10.1016/j.atmosenv.2014.11.070

Shindell DT, Lamarque JF, Schulz M et al (2013) Radiative forcing in the ACCMIP historical and future climate simulations. Atmos Chem Phys 13:2939–2974. https://doi.org/10.5194/acp-13-2939-2013

Singh R, Sharma C, Agrawal M (2017) Emission inventory of trace gases from road transport in India. Transp Res Part D Transp Environ 52:64–72. https://doi.org/10.1016/j.trd.2017.02.011

Singla V, Mukherjee S, Safai PD et al (2017) Role of organic aerosols in CCN activation and closure over a rural background site in Western Ghats, India. Atmos Environ 158:148–159. https://doi.org/10.1016/j.atmosenv.2017.03.037

Sinha A, Berkelhammer M, Stott L et al (2011) The leading mode of Indian Summer Monsoon precipitation variability during the last millennium. Geophys Res Lett 38:2–6. https://doi.org/10.1029/2011GL047713

Sinha V, Kumar V, Sarkar C (2014) Chemical composition of pre-monsoon air in the Indo-Gangetic Plain measured using a new air quality facility and PTR-MS: high surface ozone and strong influence of biomass burning. Atmos Chem Phys 14:5921–5941. https://doi.org/10.5194/acp-14-5921-2014

Soni VK, Pandithurai G, Pai DS (2012) Evaluation of long-term changes of solar radiation in India. Int J Climatol 32:540–551. https://doi.org/10.1002/joc.2294

Soni VK, Pandithurai G, Pai DS (2016) Is there a transition of solar radiation from dimming to brightening over India? Atmos Res 169:209–224. https://doi.org/10.1016/j.atmosres.2015.10.010

Srivastava R (2017) Trends in aerosol optical properties over South Asia. Int J Climatol 37:371–380. https://doi.org/10.1002/joc.4710

Streets DG, Bond TC, Carmichael GR et al (2003) An inventory of gaseous and primary aerosol emissions in Asia in the year 2000. J Geophys Res Atmos 108. https://doi.org/10.1029/2002jd003093

Streets DG, Shindell DT, Lu Z, Faluvegi G (2013) Radiative forcing due to major aerosol emitting sectors in China and India. Geophys Res Lett 40:4409–4414. https://doi.org/10.1002/grl.50805

Tabazadeh A, Drdla K, Schoeberl MR et al (2002) Arctic ozone hole in a cold volcanic stratosphere. Proc Natl Acad Sci U S A 99:2609–2612. https://doi.org/10.1073/pnas.052518199

Tandon A, Attri AK (2011) Trends in total ozone column over India: 1979–2008. Atmos Environ 45:1648–1654. https://doi.org/10.1016/j.atmosenv.2011.01.008

Tie XX, Brasseur G (1995) The response of stratospheric ozone to volcanic eruptions: sensitivity to atmospheric chlorine loading. Geophys Res Lett 22:3035–3038. https://doi.org/10.1029/95GL03057

Ul-Haq Z, Rana AD, Ali M et al (2015) Carbon monoxide (CO) emissions and its tropospheric variability over Pakistan using satellite-sensed data. Adv Sp Res 56:583–595. https://doi.org/10.1016/j.asr.2015.04.026

Undorf S, Polson D, Bollasina MA et al (2018) Detectable impact of local and remote anthropogenic aerosols on the 20th century changes of West African and South Asian Monsoon precipitation. J Geophys Res Atmos 123:4871–4889. https://doi.org/10.1029/2017JD027711

Varghese M, Prabha TV, Malap N et al (2016) Airborne and ground based CCN spectral characteristics: inferences from CAIPEEX—2011. Atmos Environ 125:324–336. https://doi.org/10.1016/j.atmosenv.2015.06.041

Venkanna R, Nikhil GN, Sinha PR et al (2016) Role of lightning phenomenon over surface O_3 and NO_x at a semi-arid tropical site Hyderabad, India: inter-comparison with satellite retrievals. Theor Appl Climatol 125:691–701. https://doi.org/10.1007/s00704-015-1538-3

Venkataraman C, Chandramouli B, Patwardhan A (1999) Anthropogenic sulphate aerosol from India: estimates of burden and direct radiative forcing. Atmos Environ 33:3225–3235. https://doi.org/10.1016/S1352-2310(98)00140-X

Venkataraman C, Habib G, Eiguren-Fernandez A et al (2005) Residential biofuels in South Asia: carbonaceous aerosol emissions and climate impacts. Science 80-(307):1454–1456. https://doi.org/10.1126/science.1104359

Venkataraman C, Habib G, Kadamba D et al (2006) Emissions from open biomass burning in India: integrating the inventory approach with high-resolution Moderate Resolution Imaging Spectroradiometer (MODIS) active-fire and land cover data. Global Biogeochem Cycles 20:1–12. https://doi.org/10.1029/2005GB002547

Verma S, Venkataraman C, Boucher O (2011) Attribution of aerosol radiative forcing over India during the winter monsoon to emissions from source categories and geographical regions. Atmos Environ 45:4398–4407. https://doi.org/10.1016/j.atmosenv.2011.05.048

Verma S, Boucher O, Shekar Reddy M et al (2012) Tropospheric distribution of sulphate aerosols mass and number concentration during INDOEX-IFP and its transport over the Indian Ocean: a GCM study. Atmos Chem Phys 12:6185–6196. https://doi.org/10.5194/acp-12-6185-2012

Verma S, Boucher O, Upadhyaya HC, Sharma OP (2013) Variations in sulphate aerosols concentration during winter monsoon season for two consecutive years using a general circulation model. Atmosfera 26:359–367. https://doi.org/10.1016/S0187-6236(13)71082-8

Vernier JP et al (2009) Tropical stratospheric aerosol layer from CALIPSO lidar observations. J Geophys Res 114, D00H10. https://doi.org/10.1029/2009JD011946

Vernier JP, Fairlie TD, Natarajan M et al (2015) Increase in upper tropospheric and lower stratospheric aerosol levels and its potential connection with Asian pollution. J Geophys Res 120:1608–1619. https://doi.org/10.1002/2014JD022372

Vernier JP, Fairlie TD, Deshler T et al (2018) BATAL: the balloon measurement campaigns of the Asian tropopause aerosol layer. Bull Am Meteorol Soc 99:955–973. https://doi.org/10.1175/BAMS-D-17-0014.1

Vinoj V, Rasch PJ, Wang H et al (2014) Short-term modulation of Indian summer monsoon rainfall by West Asian dust. Nat Geosci 7:308–313. https://doi.org/10.1038/ngeo2107

von Glasow R, Bobrowski N, Kern C (2009) The effects of volcanic eruptions on atmospheric chemistry. Chem Geol 263:131–142. https://doi.org/10.1016/j.chemgeo.2008.08.020

WMO (2019a) Scientific assessment of ozone depletion: 2014 global ozone research and monitoring project—report no. 55

WMO (2019b) Scientific assessment of ozone depletion: 2018 world meteorological organization global ozone research and monitoring project-report no. 58. World Meteorological Organization United Nations Environment Programme National Oceanic and Atmospheric Administration

Xu B, Cao J, Hansen J et al (2009) Black soot and the survival of Tibetan glaciers. Proc Natl Acad Sci 106:22114–22118. https://doi.org/10.1073/pnas.0910444106

Yarragunta Y, Srivastava S, Mitra D (2017) Validation of lower tropospheric carbon monoxide inferred from MOZART model simulation over India. Atmos Res 184:35–47. https://doi.org/10.1016/j.atmosres.2016.09.010

Yarragunta Y, Srivastava S, Mitra D, Chandola HC (2018) Seasonal and spatial variability of ozone inferred from global chemistry transport model simulations over India. ISPRS Ann Photogramm Remote Sens Spat Inf Sci IV 5:375–381. https://doi.org/10.5194/isprs-annals-IV-5-375-2018

Young PJ, Archibald AT, Bowman KW et al (2013) Pre-industrial to end 21st century projections of tropospheric ozone from the Atmospheric Chemistry and Climate Model Intercomparison Project (ACCMIP). Atmos Chem Phys 13:2063–2090. https://doi.org/10.5194/acp-13-2063-2013

Zambri B, LeGrande AN, Robock A, Slawinska J (2017) Northern Hemisphere winter warming and summer monsoon reduction after volcanic eruptions over the last millennium. J Geophys Res 122:7971–7989. https://doi.org/10.1002/2017JD026728

Zvyagintsev AM, Vargin PN, Peshin S (2015) Total ozone variations and trends during the period 1979–2014. Atmos Ocean Opt 28:575–584. https://doi.org/10.1134/S1024856015060196

Droughts and Floods

Coordinating Lead Authors

Milind Mujumdar, Indian Institute of Tropical Meteorology (IITM-MoES), Pune, India,
e-mail: mujum@tropmet.res.in (corresponding author)
Preethi Bhaskar, Indian Institute of Tropical Meteorology (IITM-MoES), Pune, India
M. V. S. Ramarao, Indian Institute of Tropical Meteorology (IITM-MoES), Pune, India

Lead Authors

Umakanth Uppara, Indian Institute of Tropical Meteorology (IITM-MoES), Pune, India
Mangesh Goswami, Indian Institute of Tropical Meteorology (IITM-MoES), Pune, India
Hemant Borgaonkar, Indian Institute of Tropical Meteorology (IITM-MoES), Pune, India
Supriyo Chakraborty, Indian Institute of Tropical Meteorology (IITM-MoES), Pune, India
Somaru Ram, Indian Institute of Tropical Meteorology, Pune (IITM-MoES), India

Review Editors

Vimal Mishra, Indian Institute of Technology Gandhinagar, Palaj, Gandhinagar, India
M. Rajeevan, Ministry of Earth Sciences (MoES), Government of India, New Delhi, India
Dev Niyogi, Purdue University, West Lafayette, IN and University of Texas at Austin, Austin, TX, USA

Corresponding Author

Milind Mujumdar, Indian Institute of Tropical Meteorology (IITM-MoES), Pune, India,
e-mail: mujum@tropmet.res.in

© The Author(s) 2020
R. Krishnan et al. (eds.), *Assessment of Climate Change over the Indian Region*,
https://doi.org/10.1007/978-981-15-4327-2_6

Key Messages

- The frequency and spatial extent of droughts over India have increased significantly during 1951–2015. An increase in drought severity is observed mainly over the central parts of India, including parts of Indo-Gangetic Plains (*high confidence*). These changes are consistent with the observed decline in the mean summer monsoon rainfall.
- Increased frequency of localized heavy rainfall on sub-daily and daily timescales has enhanced flood risk over India (*high confidence*). Increased frequency and impacts of floods are also on the rise in urban areas.
- Climate model projections indicate an increase in frequency, spatial extent and severity of droughts over India during the twenty-first century (*medium confidence*), while flood propensity is projected to increase over the major Himalayan river basins (e.g. Indus, Ganga and Brahmaputra) (*high confidence*).

6.1 Introduction

Hydroclimatic extremes such as droughts and floods are inherent aspects of the monsoonal landscape. Droughts over India are typically associated with prolonged periods of abnormally low monsoon rainfall that can last over a season or longer and extend over large spatial scales across the country (Sikka 1999). The slow evolutionary nature of monsoon droughts and enhanced surface dryness exert significant impacts on water availability, agriculture and socio-economic activities over India (Bhalme and Mooley 1980; Swaminathan 1987; Sikka 1999; Gadgil and Gadgil 2006; Asoka et al. 2017; Pai et al. 2017). Compared to droughts, floods typically occur over smaller locales in association with heavy precipitation and stream flows on shorter timescales (Dhar and Nandargi 2003; Kale 2003, 2012; Mishra et al. 2012a; Sharma et al. 2018). Every year, nearly 8 million hectares of the land area is affected by floods over India (Ray et al. 2019). Droughts and floods across India are known to have complex linkages with the space-time distribution of monsoon rainfall and socio-economic demand (Sikka 1999; see Chap. 3 for details).

Observations for the recent decades, from post-1950, clearly show a significant rising trend in frequency and intensity of both heavy rain events as well as consecutive dry days (CDD). These trends are particularly notable over central parts of the Indian subcontinent during the south-west (SW) monsoon and southern peninsular India

during the north-east (NE) monsoon (see Chap. 3 for details). The observed rainfall data indicates that there have been 22 monsoon droughts since 1901 (Fig. 6.1a). Interestingly, studies have shown that drought, as well as flood frequency, have increased since the 1950s. India experienced an increase in intensity and percentage of area affected by moderate droughts along with frequent occurrence of multi-year droughts during recent decades (Niranjan Kumar et al. 2013; Mallya et al. 2016). In this chapter, an assessment based on observational evidences from instrumental, palaeoclimatic records and likely future changes from climate model projections on droughts and floods across India is presented.

6.2 Observed Variability of Droughts

Droughts are broadly categorized into four major classes: (1) meteorological drought, as a deficit in precipitation; (2) hydrological drought, as a deficit in streamflow, groundwater level or water storage; (3) agricultural drought, as a deficit in soil moisture; and (4) socio-economic drought, incorporating water supply and demand (Wilhite and Glantz 1985; Anderson et al. 2011). All these four categories of droughts usually initiate with a deficiency in precipitation. Some of the prominent drought indices for the categorization of meteorological droughts in India are summarized in Table 6.1. Out of these indices, standardized precipitation evapotranspiration index (SPEI) has been used for analysing drought trends and variability over India (Mallya et al. 2016). The SPEI has also been used for evaluating reanalysis products during drought monsoon years (Shah and Mishra 2014); for drought monitoring (Aadhar and Mishra 2017), and adopted by the India Meteorology Department (IMD) for the operational purpose (http://imdpune.gov.in/hydrology/hydrg_index.html). As SPEI index is considered better suited to explore the effects of warming temperatures on droughts (Table 6.1; also Box 6.1), the present chapter uses SPEI for assessing the variability of droughts over India.

Box 6.1: Details of SPEI drought indicator
SPEI was computed at horizontal grid spacing of 0.5° longitude x 0.5° latitude, using monthly rainfall (0.25° x 0.25°) from IMD and potential evapotranspiration (PET; 0.5° x 0.5°) from the Climate Research Unit (CRU) for the period 1901-2016, with respect to the base period 1951–2000. PET was calculated from a variant of the Penman–Monteith formula (Sheffield et al. 2012) recommended by the United Nations Food

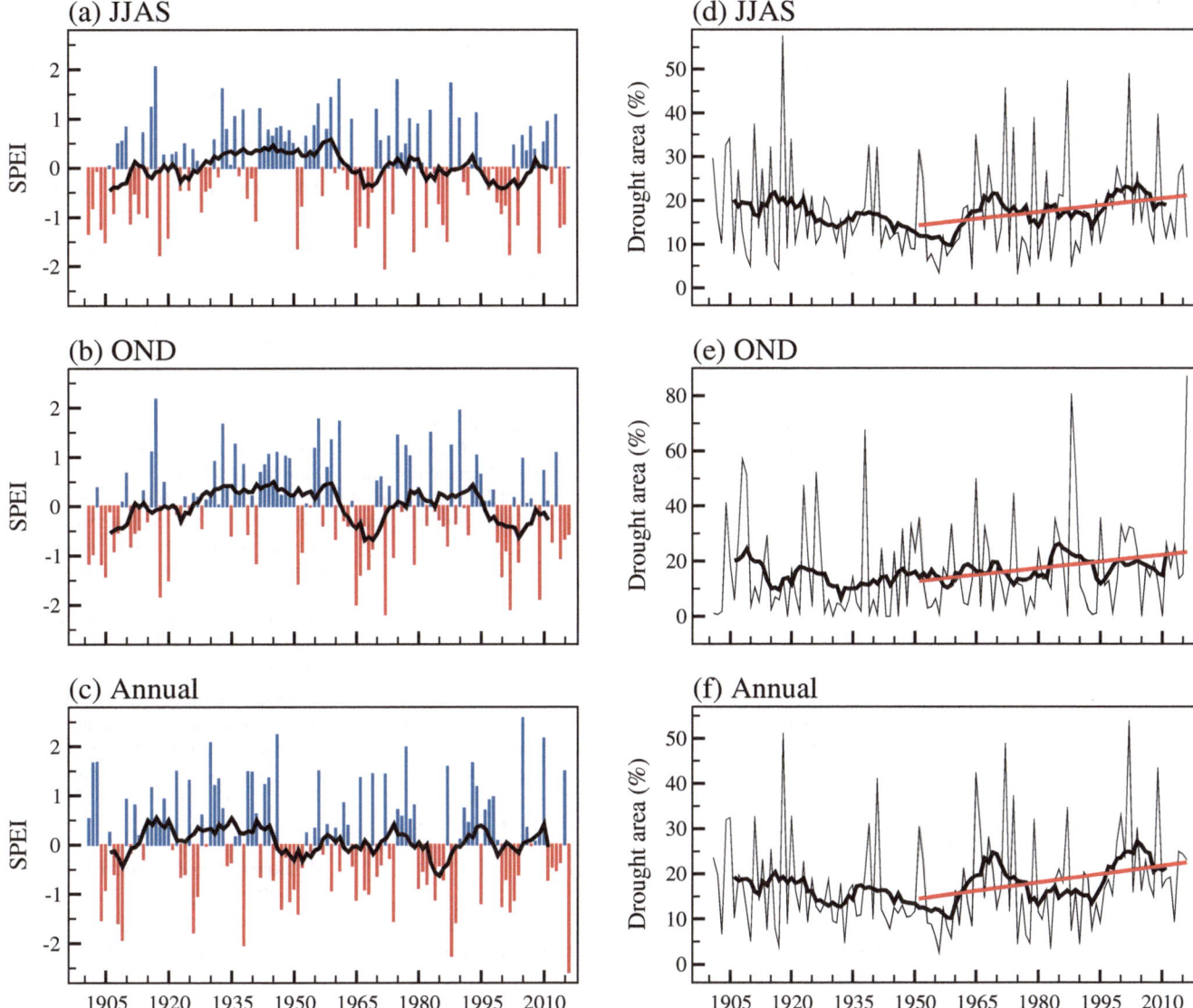

Fig. 6.1 Time series of **a–c** SPEI and **d–f** percentage area affected by drought (SPEI < −1) for **a, d** SW monsoon, **b, e** NE monsoon and **c, f** annual scale, during 1901–2016. SPEI in **a–c** is computed for **a** all India during JJAS, **b** NE monsoon region for OND, and **c** all India for the entire year, with respect to the base period 1951–2000. Black lines indicate 11-year smoothed time series. Blue (red) bars in **a–c** denote wet (dry) years. Red lines in (**d–f**) indicate a linear trend for the period 1951–2016

and Agriculture Organization (FAO; http://www.fao.org/docrep/x0490e/x0490e06.htm). The Penman-Monteith formulation is based on physical principles of energy balance over a wet surface and is considered superior to empirically based formulations, which usually consider the effects of temperature and/or radiation only (see Ramarao et al. 2019 and references therein). The SPEI is a normalized index and can be used to infer both wet (positive SPEI) and dry (negative SPEI) conditions over the region of interest. Although the theoretical limits are (-∞, +∞), SPEI value normally ranges from -2.5 to +2.5. An index of +2 and above indicates extremely wet; (1.5 to 1.99) very wet; (1.0 to 1.49) moderately wet; (0.99 to -0.99) near normal; (-1.0 to -1.49) moderately dry; (-1.5 to -1.99) severely dry; (-2.0 or less) extremely dry. For the analysis presented in this chapter, SPEI is computed for 4 month, 3 month and 12 month timescales spanning the JJAS (SPEI-SW), OND (SPEI-NE) and Annual from January to December (SPEI-ANN) to represent SW, NE monsoons, and annual scale respectively. The SPEI-SW and SPEI-ANN are computed for the Indian subcontinent, whereas SPEI-NE is computed for the southern peninsular India, the region under the

Table 6.1 Various meteorological drought indices

Index	Computation	Strength and weakness
Percent of normal precipitation (PNP)	Actual precipitation divided by normal precipitation—typically a 30 year mean and multiplied by 100 (%)	**Strength**: Simple measurement, very effective in a single region or a single season, can be calculated for a variety of timescales **Weakness**: Biased by the aridity of the region, cannot compare with different locations, cannot identify the specific impact of drought
Palmer drought severity index (PDSI)	Computed from precipitation and temperature (Palmer 1965; Dai et al. 2004)	**Strength**: Widely used for drought characterization **Weakness**: Lags the detection of drought over several months due to its dependency on soil moisture, which is simplified to one value in each climate zone
Standardized precipitation index (SPI)	SPI is defined based on the cumulative probability of a given rainfall event. It is derived from the transformation of fitted gamma distribution of historical rainfall to a standard normal distribution (Mckee et al. 1993)	**Strength**: Not biased by aridity, better than PNP and PDSI. It can be computed for different timescales. Considers multi-scalar nature of droughts. Allows comparison of drought severity at two or more locations, regardless of climatic conditions **Weakness**: Only precipitation is used and does not consider other crucial variables, e.g. temperature
Standardized precipitation evapotranspiration index (SPEI)	SPEI uses accumulations of precipitation minus potential evapotranspiration (PET) and thereby accounts for changes in both supply and demand in moisture variability over the region of interest (Vicente-Serrano et al. 2010)	**Strength**: Similar to SPI. Includes the effect of temperature via evaporative demand. More suited to explore impacts of warming temperatures on the occurrence of droughts. A more extensive range of applications than SPI **Weakness**: Sensitive to PET computation

Several other drought indices have been developed based on different indicator variables such as soil moisture, run-off and evapotranspiration (Karl and Karl 1983; Mo 2008; Shukla and Wood 2008; Hao and AghaKouchak 2013)

Table 6.2 List of SW monsoon droughts from 1901 to 2015. Years in bold letters represent severe droughts

Period	Drought years	Total number of droughts (per decade)
1901–1930	1901, 1904, **1905**, 1911, **1918**, 1920	6 (2)
1931–1960	1941, **1951**	2 (0.7)
1961–1990	**1965**, 1966, 1968, **1972**, **1979**, 1982, 1986, 1987	8 (2.7)
1991–2015	**2002**, 2004, **2009**, 2014, 2015	5 (1.9)

influence of NE monsoon. The NE monsoon region comprises of 5 meteorological sub-divisions over the southern peninsular India, namely, coastal Andhra Pradesh, Rayalaseema, South interior Karnataka, Kerala and Tamil Nadu.

For both SW and NE monsoons, the time series of SPEI shows considerable interannual and multidecadal variations with a slight negative trend (Fig. 6.1a, b), corresponding to the respective monsoon rainfall variations. The declining trend in SPEI time series is indicative of an increase in the intensity of droughts. The annual scale SPEI time series is shown in Fig. 6.1c. The variability in the frequency of SW monsoon droughts during different epochs can be noted in Table 6.2. The drought frequency for the period 1901–2016 revealed 21, 19 and 18 cases of moderate to extreme droughts (SPEI ≤ -1) for the SW, NE monsoons and annual timescale, respectively, with almost 2 droughts per decade on an average. The number of wet monsoon years (SPEI ≥ 1) is found to be 16, 14 and 19 for the SW, NE

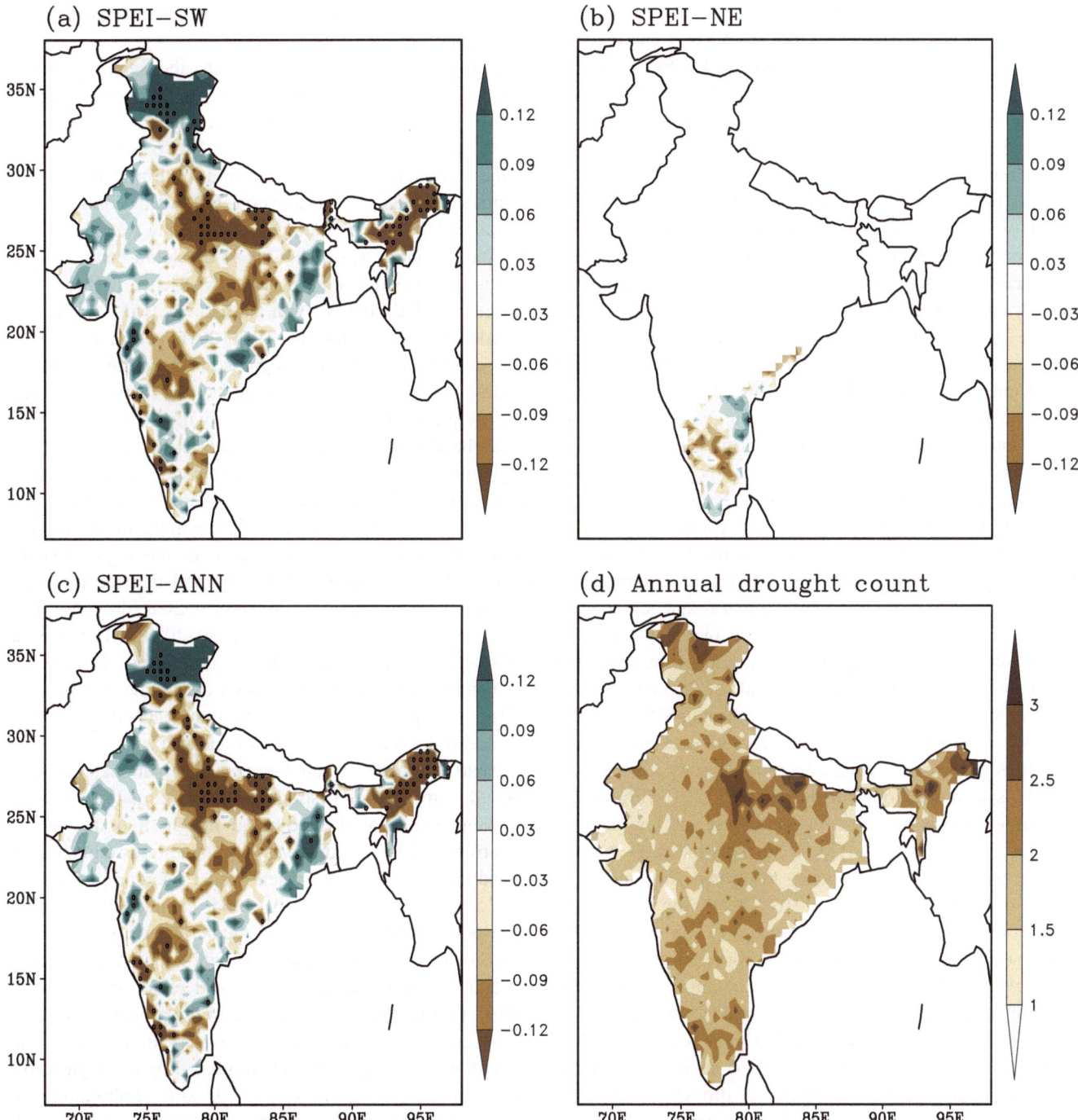

Fig. 6.2 Spatial pattern of trends (decade^{-1}) in **a** SPEI-SW for JJAS, **b** SPEI-NE for OND and **c** SPEI-ANN for annual, during 1951–2016. Regions with statistically significant (at 95% confidence level) trends are hatched. **d** Frequency of annual droughts (SPEI-ANN $\leq$ −1.0) per decade, from 1951 to 2016

monsoon seasons and annual timescale, respectively, with about 1–2 wet monsoon years per decade. The second half of data period (1951–2016) has witnessed frequent droughts with 14, 11 and 12 cases, compared to the previous period 1901–1950 (7, 8 and 6), for the SW, NE monsoon seasons and annual timescale, respectively.

The number of severe to extreme drought cases (SPEI $\leq$ −1.5) for 1951–2016 period is 6, 4 and 5 for the SW, NE monsoon seasons, and annual timescales respectively, compared to 2, 5 and 1 during 1901–1950 (Fig. 6.1a–c). Additionally, an increasing trend in drought area is observed for the entire period of analysis (1901–2016), with a statistically significant trend (at 95% confidence level) for JJAS season and annual timescale for 1951–2016 (Fig. 6.1 d–f). The period 1951–2016 also witnessed 1.2%, 1.2% and 1.3% increase in dry area per decade for SW, NE monsoon seasons and annual timescale, respectively. It is interesting to note that the drying trends are slightly higher for annual scale droughts. The analysis thus shows that the period 1951–2016 witnessed an increase in frequency and areal extent of droughts. Consistent with this, previous studies also reported an increase in frequency, duration as well as the intensity of the monsoon droughts for the post-1960 period compared to the pre-1960 period (Mallya et al. 2016; Mishra et al. 2016). Further, a relative enhancement of moderate to severe drought frequency has occurred during the recent epoch of 1977–2010 compared to 1945–1977 (Niranjan Kumar et al. 2013). Interestingly, an increase in the episodes of two consecutive years with deficient monsoon has also occurred during the post-1960 period (Fig. 6.1a; Niranjan Kumar et al. 2013). Studies highlight an increasing trend in dry areas (Niranjan Kumar et al. 2013; Mallya et al. 2016). The similar conclusions reached by different studies using different datasets and approaches provides a "high confidence" finding that frequency, as well as percentage area under drought, have increased over the Indian subcontinent during the second half of the twentieth century when compared to the first half of the century.

Significant drying trend (negative values in SPEI), during the SW monsoon season, was observed over the humid regions of Central India, and over some regions of north-east as well as west coast of India during 1951–2016 (Fig. 6.2a). A wetting trend is noticed over north-west and few parts of southern peninsular India (Fig. 6.2a). This indicates that the humid regions exhibit a tendency towards drying and more intense droughts during 1951–2016. This drying tendency is seen prominently during recent decades (Yang et al. 2019). Long-term (1901–2002) multiple data sources and methods also revealed that droughts are becoming much more regional in recent decades and depict a general migration from west to east and over the Indo-Gangetic plain (Mallya et al. 2016). This study also identified an increase in the duration, severity and spatial extent of droughts during the recent decades, highlighting the Indo-Gangetic plain, parts of coastal south India and central Maharashtra as regions that are becoming increasingly vulnerable to droughts. Strong drying over the central and the north Indian regions (Fig. 6.2a) has also been revealed from other observational studies using rainfall observations (Krishnan et al. 2013; Preethi et al. 2017a) and various drought indices (Pai et al. 2011; Niranjan Kumar et al. 2013; Damberg and Agha-Kouchak 2014; Yang et al. 2019). It is to be noted that these regions are also accompanied by an increase in aridity (Ramarao et al. 2019; Yang et al. 2019). As a result, the conclusion regarding, the drying and potential for increasing drought propensity over central and northern India, is a high confidence finding.

During the NE monsoon season, the spatial trends in SPEI depict an increase in drought intensity over the majority of region (Fig. 6.2b). A similar pattern as that of SPEI-SW is seen for the entire year (Fig. 6.2c) probably due to the dominance of rainfall contribution from SW monsoon compared to that of NE monsoon. It is worth noting that the regions which witnessed significant drying trend, e.g. Central India, Kerala, some regions of the south peninsula, and north-eastern parts of India, also experience higher annual frequency of droughts, with more than two droughts per decade on average for the 1951–2016 period (Fig. 6.2d), thus confirming that these regions are becoming more vulnerable to droughts during recent decades (high confidence). The frequent and intense droughts will likely pose significant challenges for food and water security in India by depleting soil moisture and groundwater storages (Asoka et al. 2017). Soil moisture droughts hamper crop production in India, where the majority of the population depends on agriculture and leads to famines over the region (Mishra et al. 2019). Past studies have reported that the frequency and areal extent of soil moisture-based droughts have increased substantially during 1980–2008 (Mishra et al. 2014), and hence, efforts are being made to provide forecasts of standardized soil moisture index over India (Mishra et al. 2018; https://sites.google.com/iitgn.ac.in/expforecastland surfaceproducts/erf-forecasted-sri-and-ssi).

Apart from the aforementioned observational studies, a limited number of investigations using climate models are available that provide additional insight into the drought occurrence and variability. Among the various climate models participated in the Coupled Model Intercomparison Project 5 (CMIP5), very few could capture the observed monsoon rainfall variability, particularly the frequent occurrence of droughts and spatial variability of rainfall during drought years in the recent historical period (Preethi et al. 2019). Further, a marked increase in the propensity of monsoon droughts similar to the observations during the post-1950s is reasonably well simulated by the high resolution (horizontal grid size ∼35 km) Laboratoire de Météorologie Dynamique (LMDZ4) global model with telescopic zooming over the South Asia region (Krishnan et al. 2016). It is reported that the SPEI index at 12-month and 24-month timescales in historical simulation (with both natural and anthropogenic forcings) exhibits an increase in the frequency and intensity of droughts during 1951–2005, which is possibly attributed to the influence of anthropogenic forcing on the weakening monsoon circulation and rainfall over the India subcontinent (Krishnan et al. 2016). It is important to note that the climate models have a large bias in simulating monsoon rainfall and its variability on different timescales (Turner and Annamalai 2012; Chaturvedi et al. 2012;

Rajeevan et al. 2012; Jayasankar et al. 2015; Preethi et al. 2010, 2017b). Large uncertainties are also found in reconstructing agricultural drought events for the period 1951–2015 based on simulated soil moisture from three different land surface models (Mishra et al. 2018). These uncertainties are mainly due to differences in model parameterizations and hence the study highlighted the importance of considering the multi-model ensemble for real-time monitoring and prediction of soil moisture drought over India.

6.2.1 Drought Mechanism

General features associated with SW monsoon droughts are weaker meridional pressure gradient, a larger northward seasonal shift of the monsoon trough, more break days, reduction in the frequency of depressions and shorter westward extent of depression tracks (Mooley 1976; Parthasarathy et al. 1987; Raman and Rao 1981; Sikka 1999). Droughts during the SW monsoon are, in general, significantly related to external forcings such as sea surface temperature (SST) variations in the tropical oceans, particularly with the warm phase of El Niño–Southern Oscillation (ENSO; Sikka 1980, 1999; Pant and Parthasarathy 1981; Pai

et al. 2011, 2017; Mishra et al. 2012b; Preethi et al. 2017a and the references therein) events in the eastern equatorial Pacific, central Pacific El Niño (Kumar et al. 2006) or El Niño Modoki events (Ashok et al. 2007) and also the negative Indian Ocean Dipole (IOD) events (Saji et al. 1999; Ashok et al. 2001). Apart from the tropical teleconnections, impacts on SW monsoon droughts from extra-tropics are evident from negative phase of the North Atlantic Oscillation (NAO; Goswami et al. 2006) on interannual timescale, negative phase of Atlantic Multidecadal Oscillation (AMO; Goswami et al. 2006) and positive phase of Pacific Decadal Oscillation (PDO; Krishnan and Sugi 2003) on multidecadal timescales. On the other hand, NE monsoon droughts are associated with a negative phase of ENSO (La Niña) and negative phase of IOD (Kripalani and Kumar 2004). In addition to the tropical influence, extratropical influence is evident as a relationship between the positive phase of the NAO and NE monsoon drought (Balachandran et al. 2006). Further details can be obtained from Box 3.2 in Chap. 3. It is to be noted that these teleconnections exhibit a secular variation, with epochs of strong and weak relationship with SW as well as NE monsoon rainfall (Kripalani and Kulkarni 1997; Kumar et al. 1999; Pankaj Kumar et al. 2007; Yadav 2012; Rajeevan et al. 2012).

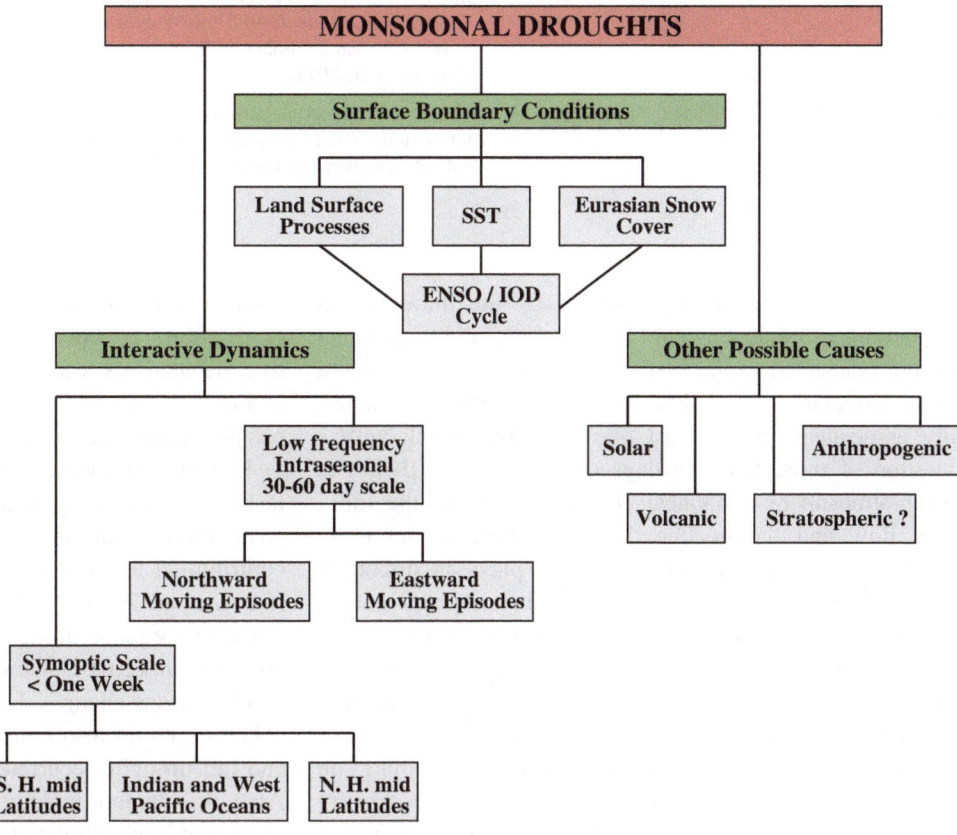

Fig. 6.3 Schematic diagram representing the interactive mechanisms leading to droughts (This Schematic is an adaptation of Fig. 4.4 in Joint COLA/CARE Technical Report No.2, July 1999 Monsoon Drought in India by D. R. Sikka, and is used with permission of the Center for Ocean-Land-Atmosphere Studies.)

Table 6.3 List of recent major droughts over India

Year	Region affected	Cause
1987	Central and North India	Warmer SSTs over equatorial Indian and Pacific oceans related to ENSO and IOD phases were unfavourable and lead to suppressed rainfall over India (Krishnamurti et al. 1989)
2000	North-west and Central India	Enhanced convective activity associated with the warmer equatorial and southern tropical Indian Ocean SSTs induced anomalous subsidence over the Indian subcontinent and thereby weakened the monsoon Hadley cell which ultimately decreased the rainfall. The warmer SSTs also led to a higher probability of occurrence of dry spells and prolonged break monsoon conditions over the subcontinent (Krishnan et al. 2003)
2002	North-west parts of India	The anomalous atmospheric convective activity over north-west and north-central Pacific associated with moderate El Niño conditions induced subsidence and rainfall deficiency over the Indian landmass (Mujumdar et al. 2007). A slower 30–60 days mode dominated the season and led to deficit monsoon rainfall (Kripalani et al. 2004). Prevailing circulation features over mid-latitudes of Eurasia and the south Indian ocean, the negative phase of SOI, warmer SST over South china sea and El Niño conditions have favoured the monsoon drought (Sikka 2003)
2008	Central India	The abnormal SST warming in southern tropical Indian Ocean due to the combined influence of a warming trend in the tropical Indian Ocean and warming associated with the IOD, resulted in enhancement of convection in the south-west tropical Indian Ocean and forced anticyclonic circulation anomalies over the Bay of Bengal and Central India, leading to suppressed rainfall over this region (Rao et al. 2010)
2009	Most of the country except north and south interior Karnataka	The unfavourable phases of the two important modes, viz., El Niño and the equatorial Indian Ocean Oscillation (EQUINOO) along with the reversal of the SST gradient between the Bay of Bengal and eastern equatorial Indian Ocean, played a critical role in the rainfall deficit over the Bay of Bengal and the Indian region (Francis and Gadgil 2010). Also, monsoon break conditions extended by the incursion of western Asian desert dry air towards Central India (Krishnamurti et al. 2010) and by the westward propagating convectively coupled planetary-scale equatorial Rossby waves (Neena et al. 2011) leading to a seasonal deficit in rainfall. Thus, weak cross-equatorial flow, monsoon systems not moving in land, penetration of mid-latitude upper tropospheric westerlies and the circulation associated with the Walker and Hadley circulation, with descending motion over the Indian landmass, collectively resulted in less moisture supply, leading to a drought (Preethi et al. 2011)
2015	Indo-Gangetic plains and western peninsular India	Enhanced convective activity associated with the pronounced meridional sea surface temperature (SST) gradient across the central-eastern Pacific ocean induce large-scale subsidence over the monsoon region (Mujumdar et al. 2017a)

Also, internal variability induced by the intraseasonal oscillations of monsoon could lead to seasonal droughts that are not connected to the known external forcing (Goswami 1998; Kripalani et al. 2004). Monsoon droughts are generally associated with at least one very long break with a duration of more than ten days (Joseph et al. 2010). The ocean-atmosphere dynamical coupling, between the monsoon flow and thermocline depth on intraseasonal timescales, in the equatorial Indian Ocean, plays an important role in forcing extended monsoon breaks and causes droughts over the Indian subcontinent (Krishnan et al. 2006). Complex thermodynamical interactions among equatorial Indo-Pacific and off-equatorial northern Indian Ocean (between $10° \text{N}—25° \text{N}$) convective systems on intraseasonal timescale as well as the ocean-atmosphere coupling on interannual timescale can also trigger the occurrence of very long breaks (Joseph et al. 2010). Moreover, the initiation of extended breaks resulting in drought conditions could be influenced by the extratropical systems as well (Krishnan et al. 2009). A schematic diagram representing the interactive mechanisms leading to large-scale droughts is provided in Fig. 6.3. Major SW monsoon droughts along with their possible causes are also listed, in Table 6.3.

In recent decades, warming of the Indian Ocean, at a faster rate than the global oceans (Roxy et al. 2014) could have implications on the variability of rainfall over India, contributing to the declining trend of SW monsoon rainfall (Roxy et al. 2014; Preethi et al. 2017a) and aiding occurrence of frequent droughts (Niranjan Kumar et al. 2013). Niyogi et al. (2010) have shown using observational analysis that landscape changes due to agricultural intensification and irrigation could also contribute to declining monsoon rains and aid drought occurrences particularly in northern India. Also, the increase in anthropogenic aerosol emissions might have contributed to the observed decline in monsoon rainfall (see Chap. 3 and Box 5.3 of Chap. 5). In this context, it is important to have an estimation of likely future changes in rainfall, particularly droughts, under warming scenario. Additionally, quantitative information on rainfall and related atmospheric and oceanic parameters prior to the period of recorded meteorological data is also essential for understanding and possibly mitigating the effects of projected climate change. Hence, a look into the palaeoclimatic records has also been made, in the following section, for understanding the variability of monsoon droughts in the past.

6.2.2 Palaeoclimatic Evidences

Evidences from proxy records indicate that past monsoonal variations were dominated by decadal- to millennial-scale variability and long-term trends (Kelkar 2006; Sinha et al. 2018; Band et al. 2018 and references therein). Reconstruction of SW monsoon variability based on stalagmite oxygen isotope ratios from Central India indicates a gradual decrease in monsoon during the beginning of the mid-Holocene from 8.5 to 7.3 ka BP (Before Present: 1950 AD), followed by a steady increase in monsoon intensity between 6.3 and 5.6 ka BP. This overall trend of monsoon during the mid-Holocene is punctuated by abrupt megadrought events spanning 70–100 years. During the past 1500 years, centennial-scale climate oscillations include the Medieval Warm Period (MWP) during 900–1300 AD—with relatively stronger monsoon, and the Little Ice Age (LIA) during 1400–1850 AD—with the relatively weaker monsoon (e.g. Sinha et al. 2007, 2011a, b; Goswami et al. 2015; Kathayat et al. 2017 and references therein). Severe drought, in India, lasting decades occurred during fourteenth and mid-fifteenth centuries in LIA. Nearly every major famine, including the devastating Durga Devi famine during 1396–1409 AD, coincides with a period of reduced monsoon rainfall, reconstructed from $\delta^{18}O$ of speleothems collected from Central India (Sinha et al. 2007). A possible influence of ENSO is suggested for the Indian monsoon variability during the mid-Holocene, MWP and LIA (Mann et al. 2009; Band et al. 2018; Tejavath et al. 2019).

Indian monsoon drought history for past 500 years and its association with El Nino was derived by Borgaonkar et al. (2010) from 523-year (1481–2003 AD) tree-ring chronology from Kerala, south India (Fig. 6.4a). This chronology exhibits a significant positive relationship with the observed SW monsoon rainfall for the instrumental period (1871–2003 AD; Fig. 6.4b, c). LIA with weaker monsoon is also evident (Fig. 6.4a: 1600–1700 AD). Higher frequency of low tree growth occurrences (Fig. 6.4b) was observed in years of monsoon droughts (Fig. 6.4c), these events are associated with El Niño since the late eighteenth century. Prior to that, many low tree growth years were detected during known El

Niño events, probably related to deficient Indian monsoon rainfall (Fig. 6.4a; Borgaonkar et al. 2010). It is noteworthy, however, that the mid-eighteenth century is a time where drought is indicated in northern Thailand (Buckley et al. 2007) and northern Vietnam (Sano et al. 2009), suggestive of a weakened monsoon in the late eighteenth century. Most of these periods, including those prior to the mid-eighteenth century, have also been reported to have widespread droughts in India (Pant et al. 1993). Aforementioned studies thus indicate a strong influence of Indo-Pacific SSTs on past monsoon droughts at decadal to millennial timescales.

6.3 Observed Variability of Floods

Floods, as compared to droughts, have regional characteristics and are typically confined to shorter timescales ranging from several hours to days. Floods are classified into different types such as riverine (extreme rainfall for longer periods), flash (heavy rainfall in cities or steep slopes), urban (lack of drainage), coastal (storm surge) and pluvial (rainfall over a flat surface) flooding. Regions prone to frequent floods mainly include river basins, hilly, coastal areas and in some instances, cities. In India, different types of floods frequently occur primarily during the SW monsoon season, the major rainy season. In addition, south peninsular India experiences floods during the NE monsoon season (Dhar and Nandargi 2000, 2003). The majority of floods in India are closely associated with heavy rainfall events, and not all of these heavy rain events translate into floods. Apart from the rainfall extremes, flood occurrences are linked to other factors such as antecedent soil moisture, storm duration, snowmelt, drainage basin conditions, urbanization, dams and reservoirs, and also proximity to the coast (Rosenzweig et al. 2010; Mishra et al. 2012a; Sharma et al. 2018). In addition, several other factors, such as infrastructure, siltation of rivers, deforestation, and backwater effect, can accelerate the impacts of floods.

In India, the spatial variation of floods mostly follows the monsoon intraseasonal oscillations. For example, Central India experiences the majority of flood events during the active monsoon phase, primarily due to heavy rainfall received from monsoon disturbances (Dhar and Nandargi 2003; Kale 2003, 2012; Ranade et al. 2007; Sontakke et al. 2008). On the contrary, regions near the foothills of the Himalayas typically experience floods during the break monsoon condition, due to heavy rainfall associated with the movement of monsoon trough towards the foothills, orographic uplift of moist monsoon flows and also due to tropical and mid-latitude interactions (Dhar and Nandargi 2000; Krishnan et al. 2000, 2009; Vellore et al. 2014). Occasionally, low pressure systems and western disturbances interact to give rise to heavy rains and floods (Sikka

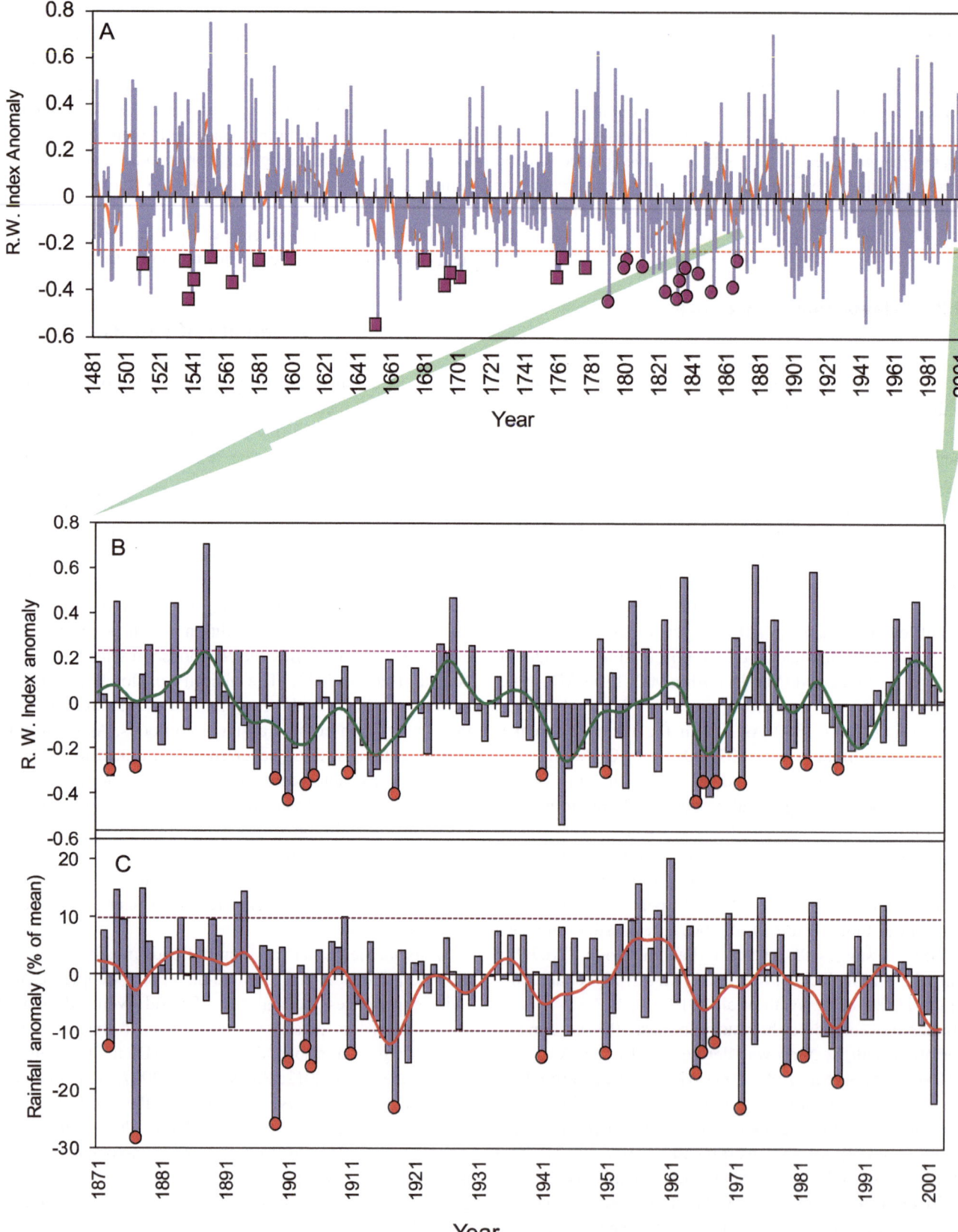

Fig. 6.4 Correspondence between tree-ring width index anomaly of Kerala tree-ring chronology (KTRC) and SW monsoon rainfall. **a** KTRC anomaly for the period 1481–2003. Anomalies in (**b**) KTRC and (**c**) SW monsoon rainfall for the instrumental period of 1871–2003. Smooth line denotes 10-year cubic spline fit. Dashed lines in the figures indicate "mean ± std.dev." limits. In Figure (**a**), circles indicate low growth events during the years of deficient rainfall (droughts) associated with El Niño and Squares are low growth associated with El Niño years. Circles in Figure (**b**) are low growth years and have one to one correspondence with deficient ISM rainfall (drought) years associated with El Niño, shown as circle in Figure (**c**)

Fig. 6.5 a Schematic diagram representing various types of floods and causative interactive mechanisms. **b** Time series of the frequency of severe flood events over India, during 1985–2019 based on flood database of the Dartmouth Flood Observatory (http://www.dartmouth.edu/ ~floods/Archives/index.html). The red line in (**b**) indicates linear trend

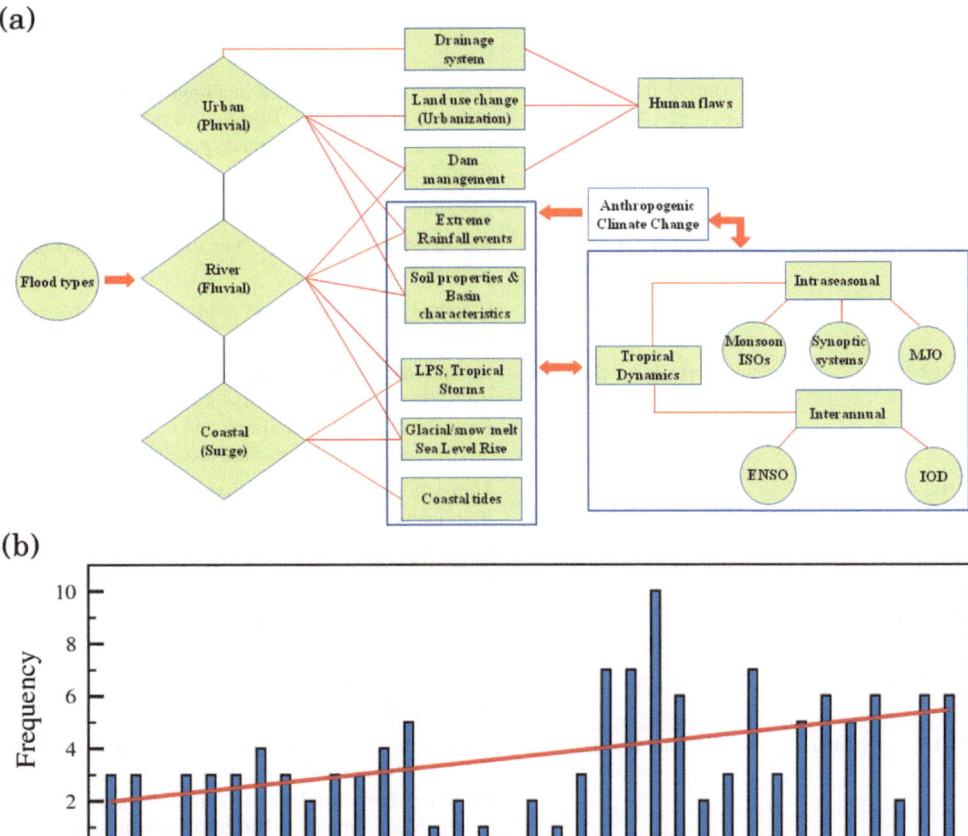

et al. 2015). Low pressure systems during the monsoon season or active monsoon conditions or monsoon breaks are the root cause of extreme floods in the South Asian rivers (Ramaswamy 1962; Dhar and Nandargi 2003). Further details on heavy rainfall occurrences and monsoon disturbances are given in Chaps. 3 and 7, respectively. In addition to the intraseasonal variability, floods exhibit variations on interannual to multidecadal timescales, in association with the flood producing extreme rainfall events. The variations in floods are reported to be also linked to the large-scale climatic drivers such as ENSO, NAO, AMO, PDO (Chowdhury 2003; Mirza 2003; Ward et al. 2016; Najibi and Devineni 2018). In particular, flood duration appears to be more sensitive to ENSO conditions in current climate. Long duration floods mainly occur during El Niño and La Niña years, compared to neutral years (Ward et al. 2016), while the influence of ENSO on flood frequency is not so strong as that on the flood duration. A schematic diagram representing the various types of floods and causative interactive mechanisms is provided in Fig. 6.5a.

Under changing climate, an intensification of the global water cycle could accelerate the risk of floods and exposure to flooding on a global scale (Milly et al. 2002; Dentener et al. 2006; Trenberth 2011; Schiermeier 2011; Hirabayashi et al.

2013). Observations for the period 1985–2015 reported an increase in frequency as well as long duration floods over the globe with a fourfold increase in frequency of floods in tropics after 2000 (Najibi and Devineni 2018). The increasing trend in extreme rainfall over the Indian subcontinent, in spite of the weakening of monsoon circulation observed during the post-1950s, also hints towards an increase in flood risk in a warming environment (Rajeevan et al. 2008; Guhathakurta et al. 2011; Roxy et al. 2017). A noticeable increase in the flood events has also occurred over the Indian subcontinent. In particular, urban and river floods (discussed in detail in the following subsections) have increased considerably along with the increasing trend in heavy rainfall events. A brief description of the major flood events that occurred over the Indian subcontinent since 2000 can be found in Table 6.4. The analysis of severe flood events using the flood database of Dartmouth Flood Observatory indicates a statistically significant increasing trend (1 flood event per decade) in the frequency of severe flood events over India during the period 1985–2019 (Fig. 6.5b). The severity of the flood events is calculated following the formulation used by the Dartmouth Flood Observatory. Studies have shown that extreme floods over South Asia cluster during excess monsoons and these extremes are rising post-1950s in the river basins across India

Table 6.4 List of recent major flood events over India

Year	Region	Cause
2005 (July)	Mumbai flood	Heavy downpour resulted in a huge rainfall as much as 994 mm of rain fell in just 24 h and 684 mm in only 12 h. Resulted in massive flooding of the Mithi river. The impact was further amplified by the inadequate drainage and sewage resulting in massive flooding (Gupta and Nikam 2013)
2007 (August)	Bihar flood	Extremely heavy and long-term rainfall flooded various rivers in Bihar and Uttar Pradesh
2008 (August)	Bihar flood	The flooding of the Kosi river valley in the northern Bihar due to breaking of Kosi embankment
2012 (June)	Brahmaputra floods	Extremely heavy monsoon rainfall resulted in over-flowing of Brahmaputra river and its tributaries
2013 (June)	North India floods (Uttarakhand)	Notable natural disaster in Uttarakhand. Continuous heavy monsoon rainfall followed by landslides in the hills led to flash flooding (Vellore et al. 2016). This region has recently experienced frequent flooding and landslides (e.g. flash floods in 2010; floods and landslides in 2011; and Himalayan flash floods in 2012)
2013 (July)	Brahmaputra floods	Similar to the 2012 event, extremely heavy monsoon rainfall resulted in over-flowing of Brahmaputra river and its tributaries
2014 (September)	Kashmir floods	Continuous rainfall for more than three days resulted in floods and landslides in Jammu and Kashmir after the Jhelum river reached above the dangerous level
2015 (June and August)	Assam floods	Extremely heavy monsoon rainfall resulted in the bursting of Brahmaputra river and its tributaries causing landslides in the region
2015 (July)	Gujarat flood	Monsoon deep depression over the Arabian Sea caused intense rainfall and flooding across the coast of Gujarat
2015 (November)	Chennai floods	The transition of low pressure into a deep depression after crossing the coast resulted in very rainfall. It is likely due to blocking of clouds by the Eastern Ghats which led continuous rainfall and produced massive urban flooding. (Assessment AR 2016; Van Oldenborgh et al. 2016). Also, rampant urban development could have played a vital role
2016 (July)	Assam floods	Extremely heavy monsoon rainfall mostly during monsoon breaks resulted in the bursting of Brahmaputra river and its tributaries
2017 (June and July)	North-east India floods	Extremely heavy monsoon rainfall mostly during monsoon breaks resulted in the bursting of Brahmaputra river and its tributaries
2017 (July)	Bihar flood	The torrential rain in the Nepal region resulted in a sudden increase in the discharge in all the eight rivers in Bihar, which led to massive flooding
2017 (July)	West Bengal floods	Week-long continuous rainfall due to cyclone Komen during monsoon resulted in dangerous floods in West Bengal and Jharkhand
2017 (July)	Gujarat flood	Simultaneous occurrence of rainfall due to low pressure systems from Arabian sea as well as Bay of Bengal. It is also likely that the heavy inflow into dams Dharoj and Dantiwada resulted in the massive flooding
2017 (August)	Mumbai flood	Massive Mumbai flood after 2005. The high tide and the extreme rainfall (468 mm in 12 h) along with inadequate drainage and sewage resulted in massive flooding
2018 (August)	Kerala floods	The unusual rainfall during the monsoon season has resulted in massive flooding. Other reasons are a sudden discharge of water from the reservoir, land-use changes and landslides (Mishra and Shah 2018)
2019 (July, August and September)	Widespread over Indian regions	A series of devastating floods over areas of several states (such as Maharashtra, Karnataka, Kerala, Gujarat, Rajasthan, Andhra Pradesh, Orissa, Uttarakhand, Madhya-Pradesh, Bihar, Uttar Pradesh, West Bengal, Assam and Punjab) due to persisting monsoonal deluges with excessive rain rates, stream flow and run-off during peak monsoon months extending into September (Global Disaster Alert and Coordination System, GDACS www.gdacs.org, https://erccportal.jrc.ec.europa.eu/ and http://floodlist.com/tag/india)

(Kale 2012; Nandargi and Shelar 2018; Mirza 2011; Ali et al. 2019). Increase in extreme rainfall events (Goswami et al. 2006; Rajeevan et al. 2008; Guhathakurta et al. 2011), rate of intensification of cyclones into severe cyclones (Niyas et al. 2009; Kishtawal et al. 2012; Chap. 8 on Extreme storms) and prolonged breaks (Ramesh Kumar et al. 2009; Chap. 3 on

Precipitation changes in India) are suggested to be the possible reasons for the intensification of river floods during post-1950 period, in addition to the anthropogenic induced changes in the catchment and river hydrology (Kale 2012). The increasing trend in floods is also possibly attributed to long-term climate variability (Ward et al. 2016; Najibi and Devineni 2018).

6.3.1 Urban/Coastal Floods

In general, urban areas are prone to river or flash flooding. Additionally, the major factors for urban floods include the effect of anthropogenic geographical alterations, inadequate drainage and storm water management system as well as high structural inhomogeneity due to intense land-use changes in proportion to increased urban population, and also the increasing population (Carvalho et al. 2002; Shepherd 2005; Goswami et al. 2010; Yang et al. 2015; Liu and Niyogi 2019). Under global warming, the observed increasing trend in heavy rainfall events has resulted in more frequent and intense flash floods over urban areas (Kishtawal et al. 2010; Guhathakurta et al. 2011; Mishra and Lilhare 2016). It is also reported that the regions which are not traditionally prone to floods experience severe inundation due to downpour and cloud burst during recent decades.

The major urban flood events of India have occurred in Mumbai (2005, 2014, 2017), Bangalore (2005, 2007, 2015), Chennai (2002, 2004, 2005, 2006, 2007, 2015), Ahmadabad (2017) and Kolkata (2007, 2017). It is to be noticed that three major metropolitan Indian cities experienced severe flooding in the same year 2005, i.e. Mumbai in July 2005, Bangalore and Chennai in October and December 2005, respectively (Guhathakurta et al. 2011). In the case of Mumbai flood of 2005, apart from about 944 mm rainfall recorded in 24-h, the intrusion of sea water into the city resulted in mass inundation due to the complex drainage system (Gupta and Nikam 2013). Studies also suggest that Mumbai region is highly vulnerable to climate change due to sea-level rise, storm surge and extreme precipitation (Hallegatte et al. 2010). The coastal city of Kolkata is also prone to flooding due to extreme rainfall activities associated with tropical cyclones. The subsidence of land in this region combined with high-tide results in heavy flooding and is a major problem for the river-side dwellers of the city which could be exacerbated in this era of climate change (Dasgupta et al. 2013). The Chennai city is more prone to tropical disturbances, and cyclones, which often leads to flooding of major rivers and clogging of drainage systems (Boyaj et al. 2018). A major flood event occurred in December 2015 was reported as one of the most disastrous floods in the history of

the region (Assessment AR 2016, vandenborgh et al. 2016). In spite of numerous flood occurrences, there is a knowledge gap in assessing the impact of climate change on flooding over urban areas. However, floods in Mumbai and Kolkata are attributed to the impact of climate shifts, urbanization, sea-level rise and other regional factors.

6.3.2 River Floods

The major river basins of South Asia such as the Brahmaputra, Ganga, Meghna, Narmada, Godavari and Mahanadi are mainly driven by the SW and NE monsoons apart from snow and glacier melt for Himalayan rivers (Mirza 2011). River basin scale flooding is generally due to the occurrence of extreme rainfall as well as variations in the factors associated with the basin catchment characteristics (e.g. Mishra and Lilhare 2016). Several intense floods were recorded in all the large river basins in South Asia during the second half of the twentieth century, such as the 1968 flood in the Tapi river, the 1970 flooding of the Narmada, the 1978 and 1987 floods on the Ganga, the 1956 and 1986 floods in the Indus river, the 1979 flood of the Luni river, the 1982 flooding of the Mahanadi river, the 1986 flooding of the Godavari river, the 1988 and 1998 floods of the Brahmaputra and the catastrophic flood of 2010 along the Indus basin (Kale 2012). River basins located in Central India, i.e. Ganga, Narmada-Tapi and Godavari, exhibit a significant increasing trend in the area covered by heavy rainfall episodes, during the monsoon season for the period 1951–2014 (Deshpande et al. 2016), which has lead to increased flooding over these basins. The increase in flood events over the Ganga-Brahmaputra basin is compounded by subsidence of land (Higgins et al. 2014) as well as glacier and water from snowmelt feeding into these rivers (Lutz et al. 2014). Hence, in a global warming scenario, melting of glaciers and snow could get accelerated and could lead to larger flood risks in the Himalayan rivers.

River floods are found to have a close association with ENSO events. A strong connection between rainfall over the Ganga-Brahmaputra-Meghna basin and the Southern Oscillation Index (SOI) was identified (Chowdhury 2003), with less than normal rainfall during the negative phase of the SOI (El Niño) whereas severe flooding due to significant increase in rainfall during positive SOI (La Niña). Moreover, major floods over the basin have occurred during La Niña years and also La Niña years co-occurring with negative IOD events (Pervez and Henebry 2015). The extreme floods across the Brahmaputra river in 1988 (Bhattacharyya and Bora 1997), 1998 (Dhar and Nandargi 2000), 2012, 2016, 2017; over the Narmada river in 1970, 2012, 2016 and over the Indus river in 1956, 1973, 1976, 2010 (Houze et al. 2011;

Webster et al. 2011; Mujumdar et al. 2012; Priya et al. 2015) have also occurred during La Niña years. Thus indicating that, in addition to the regional factors, remote forcing also has a strong influence on the flood occurrences in the Indian river basins.

In general, the increasing trend in the heavy rainfall events is found to be the major factor for the rising trend in flood occurrences in India. However, with the limited observational flood records, it is difficult to ascertain whether the increasing trend in floods is attributed to natural climate variability or to anthropogenically driven climate change. In this context, an assessment of palaeoclimatic records from the Indian subcontinent can provide crucial information on the natural variations in floods during the pre-instrumental era and the same is provided in the next section.

6.3.3 Palaeoclimatic Evidences

Palaeoclimate records from Indian peninsular rivers have indicated the occurrence of floods in the ancient period as well (Kale and Baker 2006; Kale 2012). Moreover, considerable variations in the frequency and magnitude of large floods during the last two millennia are observed in some of the western, central and south Indian rivers such as Luni, Narmada, Tapi, Godavari, Krishna, Pennar and Kaveri. The Late Holocene period witnessed clustering of large floods whereas extreme floods were absent during the late MWP and LIA (Kale and Baker, 2006; Kale 2012). This suggests a close association of century-scale variations in river floods with the variations in monsoon rainfall across the Indian subcontinent. However, a comparison of the Late Holocene floods with the post-1950 floods over palaeoflood sites in the Indian peninsular rivers indicates that the recent flood events are more intense than those during the past (Kale and Baker 2006; Kale 2012).

6.4 Future Projections

India has witnessed an increase in the frequency of droughts and floods during the past few decades. Notably, the humid regions of the central parts of India have become drought-prone regions. Also, the flood risk has increased over the east coast, West Bengal, eastern Uttar Pradesh, Gujarat and Konkan region, as well as a majority of urban areas such as Mumbai, Kolkata and Chennai (Guhathakurta et al. 2011). Given the adverse impacts of droughts and floods on food and water security in India, it is imperative to understand the future changes in drought and flood characteristics projected to develop suitable adaptation and mitigation policies.

6.4.1 Droughts

Climate model projections indicate an increase in monsoon rainfall, however, the models also show a large inter-model spread leading to uncertainty (Turner and Annamalai 2012; Chaturvedi et al. 2012; Jayasankar et al. 2015). Apart from this, a probable increase in the severity and frequency of both strong and weak monsoon as indicated by strong interannual variability in future climate is suggested by a reliable set of CMIP5 models, identified based on their ability to simulate monsoon variability in the current climate (Menon et al. 2013; Sharmila et al. 2015; Jayasankar et al. 2015). Along with this, an increase in consecutive dry days has also been projected for the future (see Chap. 3 for details). However, drought severity and frequency in the future warming climate remain largely unexplored over India and are considered in a limited number of studies. Hence, to bring out characteristics of future droughts, additional analysis is undertaken using six dynamically downscaled simulations using the regional climate model RegCM4 for historical, RCP 4.5 and RCP 8.5 scenarios till the end of the twenty-first century. These simulations are available from CORDEX South Asia experiments, and details are provided in Box 2.3, Table 2.6 (see list of IITM-RegCM4).

Similar to the observations, discussed in Sect. 6.2 (see Box 6.1), the SPEI drought index is computed for 4-month, 3-month and 12-month timescales spanning the JJAS (SPEI-SW), OND (SPEI-NE) seasons and annual from January to December (SPEI-ANN), for 1951–2099 with respect to the base period 1976–2005. Monthly rainfall and PET computed using the Penman-Monteith formula, from the six downscaled historical, RCP 4.5 and RCP 8.5 experiments, are used to derive SPEI. Consistent with the observation (Fig. 6.1a–c), the ensemble mean of CORDEX simulations (Fig. 6.6a–c) depicts a weak negative trend in SPEI for both the monsoon seasons and annual timescale. The large spread among the different members indicates low skill in simulating the rainfall variability over the Indian subcontinent, as mentioned earlier. The future projections, however, depict a spread larger than the historical period for both the scenarios RCP 4.5 and RCP 8.5, for all the seasons (Fig. 6.6a–c). The spread is seen more notably, especially for the SW monsoon season (Fig. 6.6a) and for RCP 8.5 scenario (Fig. 6.6a–c). In spite of the large spread, the ensemble mean projected a weak declining trend till 2070 for all the time series in both the scenarios. A stronger decreasing trend is projected for the post-2070 period by the high emission RCP 8.5 scenario compared to the medium emission scenario of RCP 4.5. Also, a weak increasing trend in drought area is simulated for the historical period, compared to that of observations (Fig. 6.1d–f). Similar to drought intensity (Fig. 6.6a–c), large spread among the

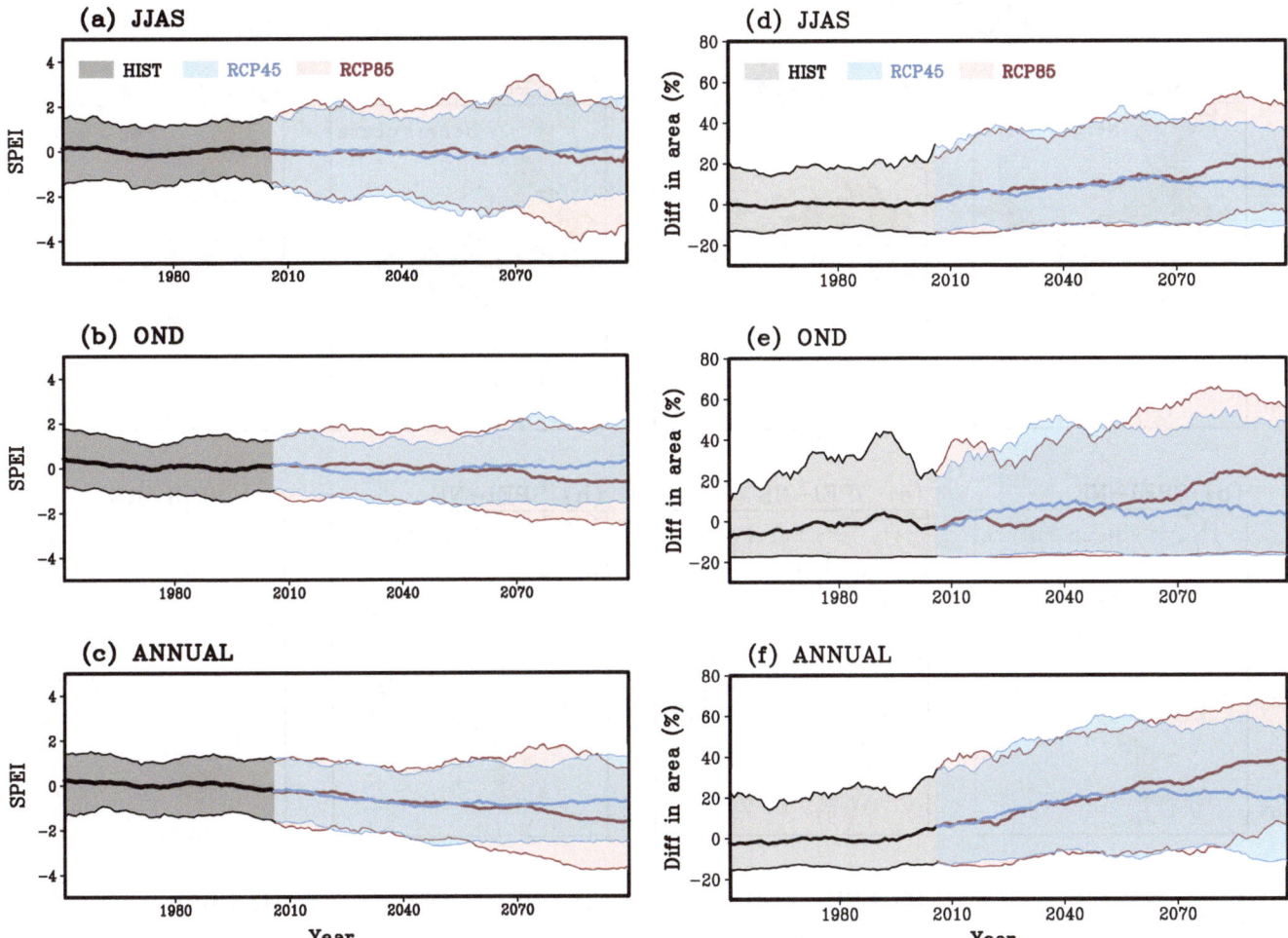

Fig. 6.6 Time series of **a–c** SPEI and **d–f** difference in percent area affected by drought during 1951–2099 (relative to 1976–2005), from six CORDEX South Asia downscaled regional climate simulations for **a, d** SW monsoon **b, e** NE monsoon and **c, f** annual scale. The historical simulations (grey) and the downscaled projections are shown for RCP4.5 (blue) and RCP8.5 (red) scenarios for the multi-RCM ensemble mean (solid lines) and the minimum to the maximum range of the individual RCMs (shading)

ensemble members is noted in all the timescales (Fig. 6.6d–f). Compared to the historical period, the spread is larger in the future projections, indicating the difficulty in estimating the drought characteristics for the future. For 2006–2099, the dry area is projected to increase by 0.84%, 0.35% and 1.64% per decade under RCP 4.5 scenario for SW, NE monsoons and annual timescale, respectively. A significant (95% confidence level) increase of 1.87, 3.22 and 3.81% per decade are projected for the respective seasons under RCP 8.5 scenario. It is interesting to note that the projected trends in dry areas are slightly higher for annual scale droughts. Under the RCP 4.5 scenario, the drought area is projected to reduce slightly after the 2070s until the end of the century.

Under RCP 4.5 scenario, the ensemble mean frequency of SW monsoon droughts is projected to increase by 1–2 events per decade over central and northern parts of India with more than two droughts over eastern parts of India in both near (2040–2069) and far (2070–2099) future, against the

reference period of 1976–2005. On the other hand, a decrease of 1–2 drought events per decade is projected over southern peninsular India under RCP 4.5 scenario (Fig. 6.7a, d). Under RCP 8.5, more frequent occurrence of moderate and severe SW monsoon droughts (>2 events per decade) is projected over Gangetic plains, north-west and central parts of India while a reduction in drought frequency is projected over south peninsular India (Fig. 6.7g, j). For NE monsoon, the drought frequency is projected to increase by 1–2 events under RCP 4.5 scenario (Fig. 6.7b, e), whereas an increase by more than two events per decade in the far future is indicated under RCP 8.5 scenario (Fig. 6.7h, k). An increase of more than three annual scale drought events per decade is suggested for north-west India during the twenty-first century under RCP 4.5 scenario (Fig. 6.7c, f). The increase in drought frequency (>3 events per decade) for the near future (Fig. 6.7i) is larger under RCP 8.5 scenario in comparison with RCP 4.5. Moreover, the area with more than three

RCP 4.5　　　　　　　RCP 8.5

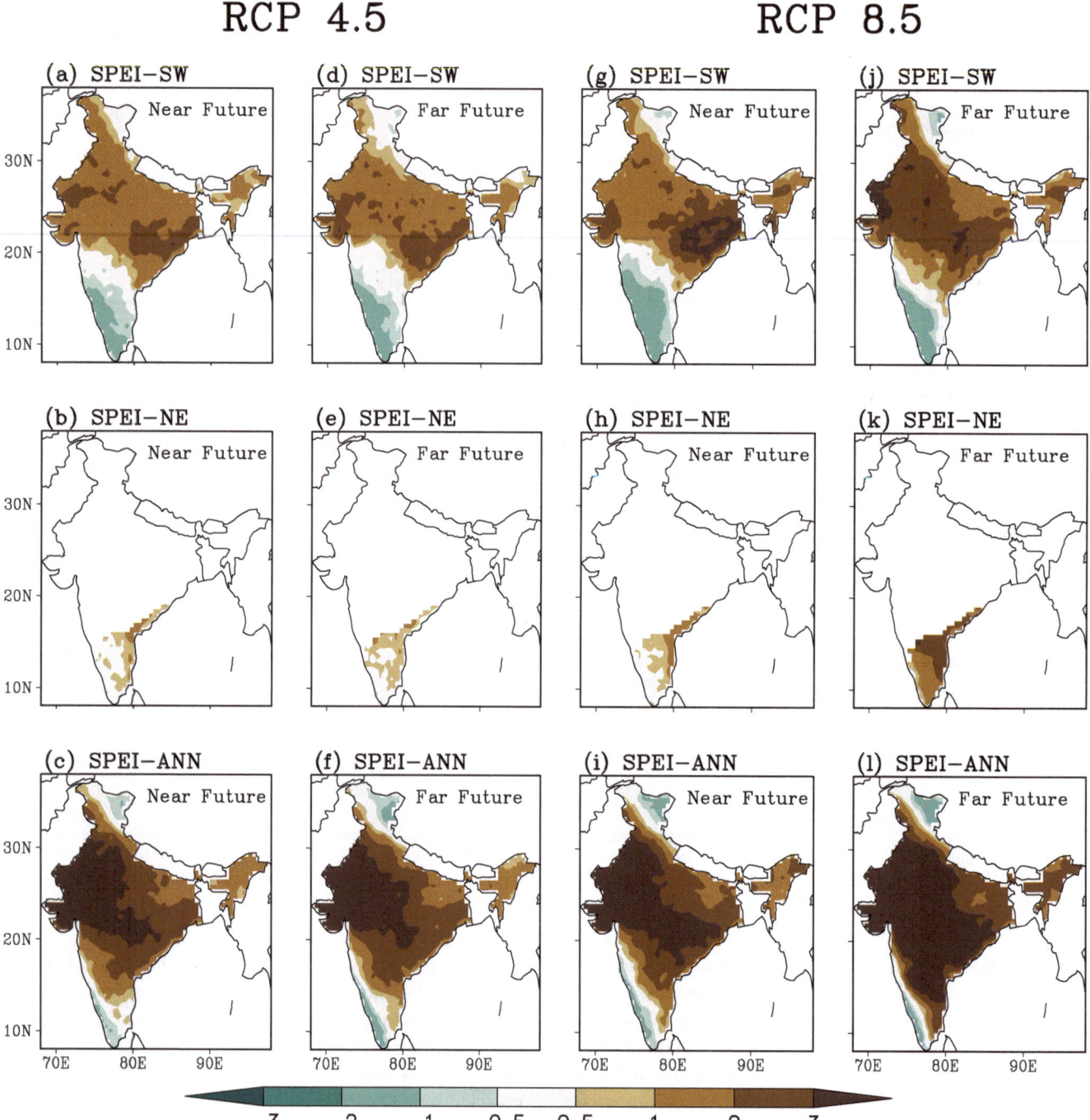

Fig. 6.7 Changes in frequency (per decade) of moderate and severe droughts (SPEI ≤ −1) for **a, d, g, j** JJAS, **b, e, h, k** OND and **c, f, i, l** annual timescale, projected by multi-model ensemble of six CORDEX simulations for **a–c, g–i** near future (2040–2069) and **d–f, j–l** far future (2070–2099) from **a–f** RCP4.5 and **g–l** RCP8.5 scenarios, with respect to the historical period (1976–2005)

drought events per decade is projected to expand across most of the regions in India by the end of the twenty-first century under RCP 8.5 scenario (Fig. 6.7l). Thus, more frequent droughts are projected under RCP 8.5 in comparison with RCP 4.5 for all the timescales.

Similar results are obtained from various climate model studies. A probable increase in drought intensity over the

Indian region towards the end of the twenty-first century is suggested under the RCP 4.5 scenario (Krishnan et al. 2016). SPEI index derived from 5 CMIP5 climate models indicates a possibility of more frequent occurrences of severe droughts by the end of the twenty-first century under both RCP 4.5 and RCP 8.5 scenarios. The similar results are reproduced using CORDEX South Asia experiments (Spinoni et al.

2020). The area affected by severe drought is also projected to increase by 150% with warming under RCP 8.5 scenario (Aadhar and Mishra 2018). Another study using a subset of 9 CMIP5 model simulations also suggests a high likelihood of above moderate drought conditions along with a significant rising trend in a drought area and an increase in the average drought length in a warming climate under RCP 4.5 and RCP 8.5 scenarios (Bisht et al. 2019). Using multiple drought indices like PNP, SPI and percentage area of droughts, two CMIP5 models, which adequately simulate frequent droughts during recent decades, have projected frequent occurrence of droughts during near and mid future, with a pronounced intensification over Central India, dynamically consistent with the modulation of the monsoon trough under RCP 4.5 scenario (Preethi et al. 2019).

On the other hand, few studies have revealed contradicting drought projections. For example, a study using drought projections based on SPI shows a decrease in the drought frequency in the twenty-first century (Aadhar and Mishra 2018). A global analysis of the CMIP5 projections of a drought hazard index based on precipitation in a warming climate (Carrão et al. 2018) evaluated that although drought has been reported in the agriculture dominated parts of India at least once in every 3 years during the past five decades, the CMIP5 ensemble mean for the present time period is found to be less consistent with the observed drought hazard index over subtropical western India. Further, this study concluded that although the clear signals of wetting are found in the CMIP5 simulations for the core monsoon zone in South Asia-east India, the projected future changes in drought hazard are neither robust nor significant for this region. Also the SPEI analysis, based on CORDEX simulations, suggests that the projected future change in drought frequency is not robust over India (Spinoni et al. 2020).

It is interesting to note that recent climate modelling studies suggest a possible frequent occurrence of El Niño events in the future (Cai et al. 2014; Azad and Rajeevan 2016) with a stable inverse relationship between El Niño and monsoon rainfall (Azad and Rajeevan 2016). This is in turn indicative of the persistent influence of El Niño events on the Indian monsoon droughts in the future. Moreover, in a warming climate, the rise in atmospheric water demand (or PET) can lead to depletion of soil moisture and prolonged drought conditions (Scheff and Frierson 2014; Ramarao et al. 2015; Krishnan et al. 2016). In summary, the above studies using regional as well as global models indicate that there is a high likelihood of an increase in the frequency, intensity and area under drought conditions even in a wetter and warmer future climate scenario. However, the large spread in the model simulations and implementation of different drought indices introduces uncertainties in analysis which eventually brings down confidence in future projections of droughts. In spite of that, an increase in droughts in the future can pose a severe threat to the availability of regional water resources in India and highlight the need for better adaptation and water management strategies.

6.4.2 Floods

Many studies have projected a possible increase in extreme precipitation events in a warming environment (see Chaps. 3, 7, and 8 for further details), which could likely increase the flood risk over the Indian subcontinent. Analysis of precipitation extremes under 1.5 and 2.0 °C global warming levels (GWL), committed under the "Paris Agreement", suggested a rise in the short duration rainfall extremes and associated flood risk over urban areas of India (Ali and Mishra 2018). The increase in temperature over Indus-Ganga-Brahmaputra river basins, which are highly sensitive to climate change, is projected to be in the range of 1.4–2.6 °C (2.0–3.4 °C) under 1.5 °C (2 °C) GWL. A further amplified warming is projected under RCP 4.5 and RCP 8.5 scenarios, possibly leading to severe impacts on streamflow and water availability over these river basins (Lutz et al. 2019). Due to the proximity of Indus-Ganga-Brahmaputra river basins to the foothills of the Himalayas, the run-off is projected to increase primarily by an increase in precipitation and accelerated meltwater in a warming environment, at least until 2050 (Lutz et al. 2014). Other major river basins of India also suggest an increase in run-off in the future, with the most significant change over the Meghna basin, indicating a high probability for flood occurrences (Mirza et al. 2003; Mirza 2011; Masood et al. 2015).

The projected changes in the frequency of extreme flooding events of 1-day, 3-day and 5-day duration for the periods 2020–2059 and 2060–2099 estimated based on the 20-year return period streamflow values with respect to the historical base period (1966–2005) are provided in Fig. 6.8 (modified from Ali et al. 2019). A higher increase in 1-day flood events is projected for the far future than that of the near future under RCP 8.5 scenario (Fig. 6.8a). The highest increase is located over the Brahmaputra basin as well as the river basins in the central parts of the Indian subcontinent, while the least increase is seen over the Indus basin. It can also be noticed that the projected increase in multi-day (3 and 5 days) flood events is more compared to one-day events across all the river basins under both RCP 2.6 and RCP 8.5 scenarios (Fig. 6.8). The increase in the frequency of all the flood events of different duration is more in the high emission scenario of RCP 8.5 compared to low emission scenarios of RCP 2.6. In another study, a rise in flood frequency, with respect to the magnitude of floods of 100-year return periods in the historical simulation, is projected over the majority of the Indian subcontinent in the twenty-first century under the RCP8.5 scenario by CMIP5

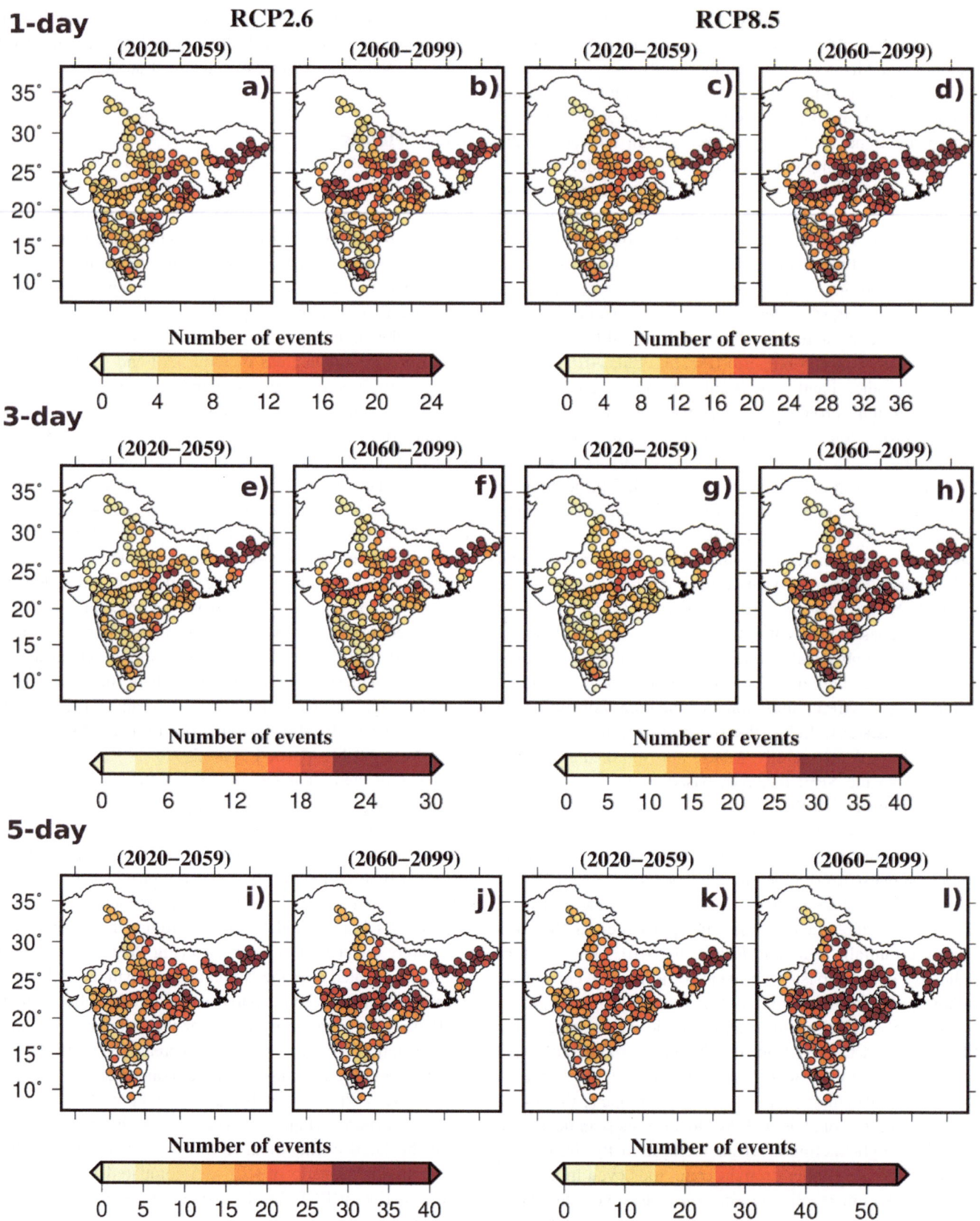

Fig. 6.8 Changes in frequency of **a–d** 1-day, **e–h** 3-day and **i–l** 5-day duration extreme flood events, projected for **a, e, i; c, g, k** near future 2020–2059 and **b, f, j; d, h, l** far future 2060–2099, exceeding 20-year return level based on the historic period 1966–2005, as derived from the ensemble mean of five GCMs for **a, b; e, f; i, j** RCP2.6 and **c, d; g, h; k, l** RCP8.5 scenario (Modified from Ali et al. 2019)

(a)

COSMOS-INDIA network

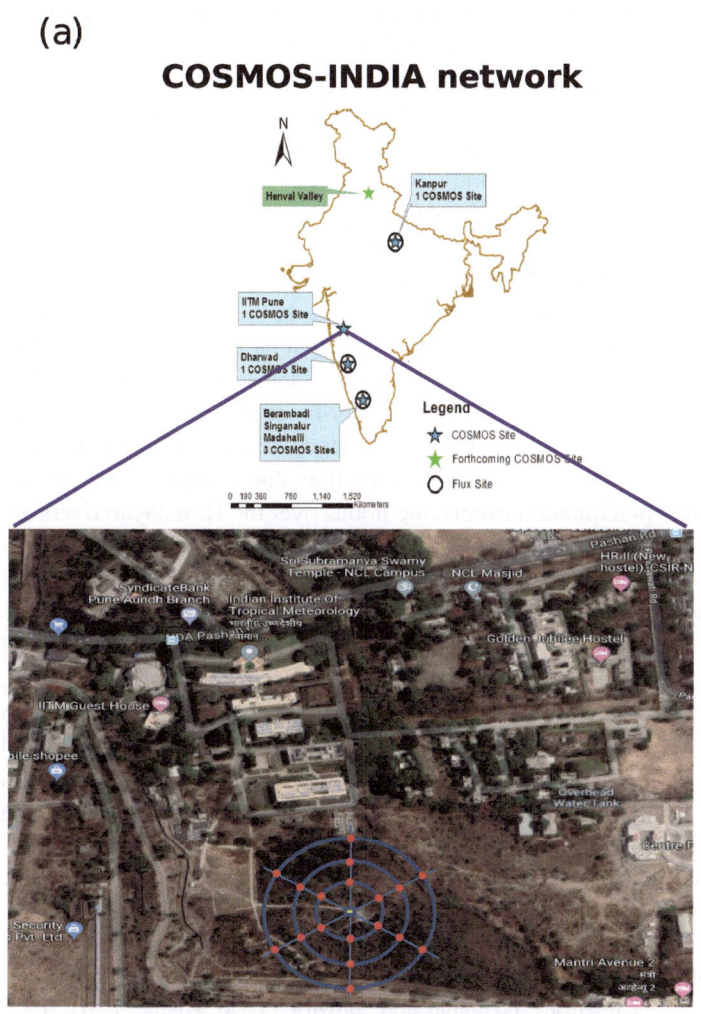

(b)

Fig. 6.9 **a** COSMOS-INDIA network, **b** COSMOS-IITM site. Notation A, B, C, D and E are used to indicate different hydro-meteorological sensor installed at COSMOS-IITM site Pune. A

—COSMOS Probe; B—Data logger; $C_{1,2,3}$—Multiweather component sensor; at 2 heights 10 and 20 m; D—Net radiometer E—Eddy covariance system

models. The southern peninsular India, Ganges, and Brahmaputra basins are projected to experience floods of similar magnitude at a higher frequency (<15 years) in the twenty-first century, with high consistency among the models (Hirabayashi et al. 2013). The majority of studies projected an increasing flood risk for Indus-Ganga-Brahmaputra river basins (Higgins et al. 2014; Shrestha et al. 2015; Kay et al. 2015; Wijngaard et al. 2017; Lutz et al. 2019) and these basins are considered as hotspots in a changing climate (De Souza et al. 2015). The increased flood risk in terms of frequency and duration of flood events can exert profound impacts on food production, water resources and management. Additionally, the human-induced influences such as land-use changes, irrigation, mismanagement of dams and reservoirs can aggravate the magnitude and the frequency of flood events in the future.

6.5 Knowledge Gaps

This section highlights some of the knowledge gaps that would be of relevance for future studies on drought and flood assessments.

1. Lack of dense observational networks for essential climate variables like soil moisture, surface and sub-surface energy, water fluxes, stream flow, etc. limits our scientific understanding of the complex multiscale (spatial and temporal) interactions taking place in the climate system. To better understand the processes involved in the variations of intensity and duration of floods and droughts over India in a warming climate, novel observational datasets are highly required. For example, network of the

newly developed neutron scattering method, used in non-invasive Cosmic ray soil moisture monitoring system (COSMOS), could potentially help scale gap between the conventional point scale, remote sensing techniques and model simulations of surface soil moisture (see Fig. 6.9 and Mujumdar et al. 2017b).

2. Attribution of anthropogenically induced climate change to the variability of drought and floods in historical as well as future projections remains a challenging issue and an open problem for further scientific research.

3. Model uncertainties in reproducing the observed variability of droughts and floods, as well as the spread among the models, also hamper our confidence in assessing future changes. Thus efforts are needed for reducing the model uncertainties.

4. Assessing the impact of increasing urbanization, as well as agricultural intensification on the hydroclimatic extremes of heavy rains/floods and droughts, continues to be a challenge for the Indian monsoon region, and additional multiscale assessments are critically needed.

6.6 Summary

A detailed assessment of the long-term variability of droughts and floods in the current as well as future climate is presented in this chapter, in view to support a better framing of climate mitigation and adaptation strategies in India. Indian subcontinent witnessed a decline in monsoon rainfall along with frequent occurrences of droughts and flood events in the past few decades, in association with the changes in regional and remote forcings. Besides, many studies projected a probable increase in these hydroclimatic extreme events in a warming environment.

The analysis of SPEI over India for the period 1901–2016 identified more droughts (~ 2 per decade) compared to wet (1–2 per decade) monsoon years. For the post-1950 period, a high frequency of droughts along with an expansion of dry area at a rate of 1.2, 1.2 and 1.3% per decade is observed in SW, NE monsoon seasons and annual timescale, respectively. In the humid regions of the country, particularly the parts of Central India, Indo-Gangetic plains, south peninsula and north-east India experienced significant drying trend with more intense droughts during SW monsoon season. On the other hand, the east coast and southern tip of India show slight wetting trend during the NE monsoon season. In recent decades (post-1950 period), droughts have been more frequent (>2 droughts per decade on average) over Central India, Kerala, some regions of the south peninsula, and north-eastern parts of India, making these regions more vulnerable. These results are consistent among various studies (Pai et al. 2011, 2017; Niranjan Kumar et al. 2013;

Damberg and AghaKouchak 2014; Mallya et al. 2016; Krishnan et al. 2016; Mishra et al. 2016; Preethi et al. 2019; Yang et al. 2019). Thus, it is assessed with high confidence that the frequency and spatial extent of droughts over the country have increased significantly along with an increase in intensity, mainly confining to the central parts including the Indo-Gangetic plains of India, during 1951–2016. These changes are observed in association with the decline in monsoon rainfall, which is likely due to an increase in anthropogenic aerosol emissions in the northern hemisphere, regional land-use changes as well as warming of the Indian Ocean. During this period, an increasing trend in floods is also reported over the majority of the Indian river basins associated with the rise in heavy rainfall episodes. In addition to the enhanced stream flow due to increase in extreme precipitation events, the floods over the Himalayan rivers are compounded by subsidence of land as well as glacier and snowmelt water feeding into these rivers. The observed increasing trend in heavy rainfall events combined with the intense land-use changes has resulted in more frequent and intense flash floods over urban areas, like Mumbai, Chennai, Bangalore, Kolkata, etc. (Guhathakurta et al. 2011). Though there is high confidence in the rising trend in extreme rainfall events and the associated flood risk over India, its attribution of climate change remains a challenging issue and an open problem for further scientific research.

Future projections of regional as well as global climate models indicate a high likelihood of an increase in frequency, intensity and area under drought conditions over India, with medium confidence due to large spread in model projections (Aadhar and Mishra 2018; Bisht et al. 2019; Preethi et al. 2019). Though the climate models project an enhanced mean monsoon rainfall, the projected increase in droughts could be due to the larger interannual variability of rainfall and the increase in atmospheric water vapour demand (potential evapotranspiration) over the country (Menon et al. 2013; Scheff and Frierson 2014; Jayasankar et al. 2015; Sharmila et al. 2015; Krishnan et al. 2016). Moreover, climate model projections also indicate frequent El Niño events in the Pacific Ocean with a stable inverse relation with the monsoon, which could also result in more number of monsoon droughts in future (Cai et al. 2014; Azad and Rajeevan 2016). The climate projections for India also indicate an increase in frequency of urban and river floods, under different levels of warming, 1.5 and 2.0 °C, as well as for different emission scenarios in association with an expected rise in heavy rainfall occurrences (Hirabayashi et al. 2013; Ali and Mishra 2018; Lutz et al. 2019). However, larger changes in flood frequency are projected in the high emission scenario of RCP 8.5. Flood frequency and associated risk are projected to increase over the major river basins of India, with a higher risk for the Indus-Ganges-Brahmaputra river basins in a warming climate (Lutz et al. 2014). The enhanced flood risk is likely due to increasing stream flow and run-off associated with the projected increase in frequency of extreme rainfall events over the major Indian river basins and is compounded by glacier and snowmelt over the Indus-Ganges-Brahmaputra. The projected enhanced droughts and flood risk over India highlight the potential need for a better adaptation and mitigation strategies.

References

Aadhar S, Mishra V (2017) High-resolution near real-time drought monitoring in South Asia. Sci Data 4:170145. https://doi.org/10.1038/sdata.2017.145

Aadhar S, Mishra V (2018) Impact of climate change on drought frequency over India. Book Climate Change and Water Resources in India Publisher: Ministry of Environment, Forest and Climate Change (MoEF&CC), Government of India

Ali H, Mishra V (2018) Increase in subdaily precipitation extremes in india under 1.5 and 2.0 °C warming worlds. Geophys Res Lett 45 (14):6972–6982

Ali H, Modi P, Mishra V (2019) Increased flood risk in Indian sub-continent under the warming climate. Weather Clim Extrems 25:100212. https://doi.org/10.1016/J.WACE.2019.100212

Anderson MC, Hain C, Wardlow B, Agustin P, John RM, William PK (2011) Evaluation of drought indices based on thermal remote sensing of evapotranspiration over the continental United States. J Clim 24:2025–2044. https://doi.org/10.1175/2010JCLI3812.1

Ashok K, Guan Z, Yamagata T (2001) Impact of the Indian Ocean dipole on the relationship between the Indian monsoon rainfall and ENSO. Geophys Res Lett 28:4499–4502. https://doi.org/10.1029/2001GL013294

Ashok K, Behera SK, Rao SA, Weng Hengyi, Yamagata Toshio (2007) El Niño Modoki and its possible teleconnection. J Geophys Res 112:C11007. https://doi.org/10.1029/2006JC003798

Asoka A, Gleeson T, Wada Y, Mishra V (2017) Relative contribution of monsoon precipitation and pumping to changes in groundwater storage in India. Nat Geosci 10:109–117. https://doi.org/10.1038/ngeo2869

Azad S, Rajeevan M (2016) Possible shift in the ENSO-Indian monsoon rainfall relationship under future global warming. Sci Rep 6:20145. https://doi.org/10.1038/srep20145

Balachandran S, Asokan R, Sridharan S (2006) Global surface temperature in relation to northeast monsoon rainfall over Tamil Nadu. J Earth Syst Sci 115:349–362. https://doi.org/10.1007/BF02702047

Band S, Yadava MG, Lone MA, Shen C-C, Sree K, Ramesh R (2018) High-resolution mid-Holocene Indian summer monsoon recorded in a stalagmite from the Kotumsar Cave, Central India. Quat Int 479:19–24. https://doi.org/10.1016/J.QUAINT.2018.01.026

Bhalme HN, Mooley DA (1980) Large-scale droughts/floods and monsoon circulation. Mon Weather Rev 108:1197–1211. https://doi.org/10.1175/1520-0493(1980)108%3c1197:LSDAMC%3e2.0.CO;2

Bhattacharyya NN, Bora AK (1997) Floods of the Brahmaputra River in India. Water Int 22:222–229. https://doi.org/10.1080/02508069708686709

Bisht DS, Sridhar V, Mishra A, Chatterjee C, Raghuwanshi NS (2019) Drought characterization over India under projected climate scenario. Int J Climatol 39:1889–1911. https://doi.org/10.1002/joc.5922

Borgaonkar HP, Sikder AB, Ram S, Pant GB (2010) El Niño and related monsoon drought signals in 523-year-long ring width records of teak (Tectona grandis L.f.) trees from south India. Palaeogeogr Palaeo climatol Palaeoecol 285:74–84. https://doi.org/10.1016/J.PALAEO.2009.10.026

Boyaj A, Ashok K, Ghosh S, Devanand A, Dandu G (2018) The Chennai extreme rainfall event in 2015: The Bay of Bengal connection. Clim Dyn 50:2867–2879. https://doi.org/10.1007/s00382-017-3778-7

Buckley BM, Palakit K, Duangsathaporn K, Sanguantham P, Prasomsin P (2007) Decadal scale droughts over northwestern Thailand over the past 448 years: links to the tropical Pacific and Indian Ocean sectors. Clim Dyn 29:63–71. https://doi.org/10.1007/s00382-007-0225-1

Cai W, Santoso A, Wang G, Weller E, Wu L, Ashok K, Masumoto Y, Yamagata T (2014) Increased frequency of extreme Indian Ocean Dipole events due to greenhouse warming. Nature 510:254–258. https://doi.org/10.1038/nature13327

Carrão H, Naumann G, Barbosa P (2018) Global projections of drought hazard in a warming climate: a prime for disaster risk management. Clim Dyn 50:2137–2155. https://doi.org/10.1007/s00382-017-3740-8

Carvalho LMV, Jones C, Liebmann B (2002) Extreme precipitation events in Southeastern South America and large-scale convective patterns in the South Atlantic Convergence Zone. J Clim 15:2377–2394. https://doi.org/10.1175/1520-0442(2002)015%3c2377:EPEISS%3e2.0.CO;2

Chaturvedi RK, Joshi J, Jayaraman M, Bala G, Ravindranath NH (2012) Multi-model climate change projections for India under representative concentration pathways. Curr Sci 103:791–802

Chennai floods 2015: a satellite and field based assessment study, Decision Support Centre (DSC) Disaster Management Support Division (DMSD) National Remote Sensing Centre (NRSC), ISRO Balanagar, Hyderabad-500037

Chowdhury MR (2003) The El Nino-Southern Oscillation (ENSO) and seasonal flooding? Bangladesh. Theor Appl Climatol 76:105–124. https://doi.org/10.1007/s00704-003-0001-z

Dai A, Trenberth KE, Qian T (2004) A global dataset of palmer drought severity index for 1870–2002: relationship with soil moisture and effects of surface warming. J Hydrometeorol 5:1117–1130. https://doi.org/10.1175/JHM-386.1

Damberg L, AghaKouchak A (2014) Global trends and patterns of drought from space. Theor Appl Climatol 117:441–448. https://doi.org/10.1007/s00704-013-1019-5

Dasgupta S, Gosain AK, Rao S, Roy S, Sarraf M (2013) A megacity in a changing climate: the case of Kolkata. Clim Change 116:747–766. https://doi.org/10.1007/s10584-012-0516-3

De Souza K, Kituyi E, Harvey B, Leone M, Murali KS, Ford JD (2015) Vulnerability to climate change in three hot spots in Africa and Asia: key issues for policy-relevant adaptation and resilience-building research. Reg Environ Change 15:747–753. https://doi.org/10.1007/s10113-015-0755-8

Dentener F, Stevenson D, Ellingsen K, van Noije T, Schultz M, Amann M, Atherton C, Bell N, Bergmann D, Bey I, Bouwman L, Butler T, Cofala J, Collins B, Drevet J, Doherty R, Eickhout B, Eskes H, Fiore A, Gauss M, Hauglustaine D, Horowitz L, Isaksen ISA, Josse B, Lawrence M, Krol M, Lamarque JF, Montanaro V, Müller JF, Peuch VH, Pitari G, Pyle J, Rast S, Rodriguez J, Sanderson M, Savage NH, Shindell D, Strahan S, Szopa S, Sudo K, Van Dingenen R, Wild O, Zeng G (2006) The global atmospheric environment for the next generation. Environ Sci Technol 40(11):3586–3594. https://doi.org/10.1021/ES0523845

Deshpande NR, Kothawale DR, Kulkarni A (2016) Changes in climate extremes over major river basins of India. Int J Climatol 36:4548–4559. https://doi.org/10.1002/joc.4651

Dhar ON, Nandargi S (2000) A study of floods in the Brahmaputra basin in India. Int J Climatol 20:771–781. https://doi.org/10.1002/1097-0088(20000615)20:7%3c771:AID-JOC518%3e3.0.CO;2-Z

Dhar ON, Nandargi S (2003) Hydrometeorological aspects of floods in India. Nat Hazards. https://doi.org/10.1023/A:1021199714487

Francis PA, Gadgil S (2010) Towards understanding the unusual Indian monsoon in 2009. J Earth Syst Sci 119:397–415. https://doi.org/10.1007/s12040-010-0033-6

Gadgil S, Gadgil S (2006) The Indian monsoon, GDP and agriculture. Econ Polit Weekly 4887–4895

Goswami BN (1998) Interannual variations of Indian summer monsoon in a GCM: external conditions versus internal feedbacks. J Clim 11:501–522. https://doi.org/10.1175/1520-0442(1998)011%3c0501:IVOISM%3e2.0.CO;2

Goswami BN, Venugopal V, Sengupta D et al (2006) Increasing trend of extreme rain events over india in a warming environment. Science 314:1442–1445. https://doi.org/10.1126/science.1132027

Goswami P, Shivappa H, Goud BS (2010) Impact of urbanization on tropical mesoscale events: investigation of three heavy rainfall events. Meteorol Zeitschrift 19:385–397. https://doi.org/10.1127/0941-2948/2010/0468

Goswami BN, Kripalani RH, Borgaonkar HP, Preethi B (2015) Multi-decadal variability in Indian summer monsoon rainfall using proxy data. In: Chang CP, Ghil M, Latif M, Wallace J (eds) Climate change multidecadal and beyond. World Scientific Publishing Co., Pvt. Ltd., Singapore, pp 327–345

Guhathakurta P, Sreejith OP, Menon PA (2011) Impact of climate change on extreme rainfall events and flood risk in India. J Earth Syst Sci 120:359–373. https://doi.org/10.1007/s12040-011-0082-5

Gupta K, Nikam V (2013) A methodology for rapid inundation mapping for a megacity with sparse data: case of Mumbai, India. IAHS-AISH Publication 357:385–391

Hallegatte S, Ranger N, Bhattacharya S, Bachu M, Priya S, Dhore K, Rafique F, Mathur P, Naville N, Henriet F, Patwardhan A, Narayanan K, Ghosh S, Karmakar S, Patnaik U, Abhayankar A, Pohit S, Corfee-Morlot J, Herweijer C (2010) Flood risks climate change impacts and adaptation benefits in Mumbai. OECD Environ. In: Working paper. https://doi.org/10.1787/5km4hv6wb434-en

Hao Z, AghaKouchak A (2013) Multivariate standardized drought index: a parametric multi-index model. Adv Water Resour 57 (2013):12–18

Higgins SA, Overeem I, Steckler MS, Syvitski JPM, Seeber L, Akhter SH (2014) InSAR measurements of compaction and subsidence in the Ganges-Brahmaputra Delta Bangladesh. J Geophys Res Earth Surf 119:1768–1781. https://doi.org/10.1002/2014JF003117

Hirabayashi Y, Mahendran R, Koirala S, Konoshima L, Yamazaki D, Watanabe S, Kim H, Kanae S (2013) Global flood risk under climate change. Nat Clim Chang 3:816–821. https://doi.org/10.1038/nclimate1911

Houze RA, Rasmussen KL, Medina S, Brodzik SR, Romatschke U (2011) Anomalous atmospheric events leading to the summer 2010 floods in Pakistan. Bull Am Meteorol Soc 92:291–298. https://doi.org/10.1175/2010BAMS3173.1

Jayasankar CB, Surendran S, Rajendran K (2015) Robust signals of future projections of Indian summer monsoon rainfall by IPCC AR5 climate models: Role of seasonal cycle and interannual variability. Geophys Res Lett 42:3513–3520. https://doi.org/10.1002/2015GL063659

Joseph S, Sahai AK, Goswami BN (2010) Boreal summer intraseasonal oscillations and seasonal Indian monsoon prediction in DEMETER coupled models. Clim Dyn 35:651–667. https://doi.org/10.1007/s00382-009-0635-3

Kale VS (2003) Geomorphic effects of monsoon floods on Indian rivers. Nat Hazards 28:65–84. https://doi.org/10.1023/A:1021121815395

Kale V (2012) On the link between extreme floods and excess monsoon epochs in South Asia. Clim Dyn 39:1107–1122. https://doi.org/10.1007/s00382-011-1251-6

Kale VS, Baker VR (2006) An extraordinary period of low-magnitude floods coinciding with the Little Ice Age: Palaeoflood evidence from central and western India. J GeolSoc India 68:477–483

Karl TR, Karl TR (1983) Some spatial characteristics of drought duration in the United States. J Clim Appl Meteorol 22:1356–1366. https://doi.org/10.1175/1520-0450(1983)022%3c1356:SSCODD%3e2.0.CO;2

Kathayat G, Cheng H, Sinha A, Yi L, Li X, Zhang H, Li H, Ning Y, Edwards RL (2017) The Indian monsoon variability and civilization changes in the Indian subcontinent. Sci Adv 3:1–9. https://doi.org/10.1126/sciadv.1701296

Kay S, Caesar J, Wolf J, Bricheno L, Nicholls RJ, Saiful Islam AKM, Haque A, Pardaens A, Lowe JA (2015) Modelling the increased frequency of extreme sea levels in the Ganges-Brahmaputra-Meghna delta due to sea level rise and other effects of climate change. Environ Sci Process Impacts 17(7):1311–1322. https://doi.org/10.1039/C4EM00683F

Kelkar RR (2006) The Indian monsoon as a component of the climate system during the Holocene. J Geol Soc India 68(3):347–352

Kishtawal CM, Niyogi D, Tewari M, Pielke RA Sr, Shepherd JM (2010) Urbanization signature in the observed heavy rainfall climatology over India. Int J Climatol 30:1908–1916. https://doi.org/10.1002/joc.2044

Kishtawal CM, Jaiswal N, Singh R, Niyogi D (2012) Tropical cyclone intensification trends during satellite era (1986–2010). Geophys Res Lett 39(10)

Kripalani RH, Kulkarni A (1997) Climatic impact of El Niño/La Niña on the Indian monsoon: a new perspective. Weather 52:39–46. https://doi.org/10.1002/j.1477-8696.1997.tb06267.x

Kripalani RH, Kumar P (2004) Northeast monsoon rainfall variability over south peninsular India vis-à-vis the Indian Ocean dipole mode. Int J Climatol 24:1267–1282. https://doi.org/10.1002/joc.1071

Kripalani RH, Kulkarni A, Sabade SS, Revadekar JV, Patwardhan SK, Kulkarni JR (2004) Intra-seasonal oscillations during monsoon 2002 and 2003. Curr Sci 87:327–331

Krishnamurti TN, Bedi HS, Subramaniam M et al (1989) The summer monsoon of 1987. J Clim 2:321–340. https://doi.org/10.1175/1520-0442(1989)002%3c0321:TSMO%3e2.0.CO;2

Krishnamurti TN, Thomas A, Simon A, Kumar V (2010) Desert air incursions, an overlooked aspect, for the dry spells of the Indian summer monsoon. J Atmos Sci 67:3423–3441. https://doi.org/10.1175/2010JAS3440.1

Krishnan R, Sugi M (2003) Pacific decadal oscillation and variability of the Indian summer monsoon rainfall. Clim Dyn 21:233–242. https://doi.org/10.1007/s00382-003-0330-8

Krishnan R, Zhang C, Sugi M (2000) Dynamics of breaks in the Indian summer monsoon. J Atmos Sci 57:1354–1372. https://doi.org/10.1175/1520-0469(2000)057%3c1354:DOBITI%3e2.0.CO;2

Krishnan R, Mujumdar M, Vaidya V, Ramesh KV, Satyan V (2003) The abnormal Indian summer monsoon of 2000. J Clim 16:1177–1194. https://doi.org/10.1175/1520-0442(2003)16%3c1177:TAISMO%3e2.0.CO;2

Krishnan R, Ramesh KV, Samala BK, Meyers G, Slingo JM, Fennessy MJ (2006) Indian Ocean-monsoon coupled interactions and impending monsoon droughts. Geophys Res Lett 33:L08711. https://doi.org/10.1029/2006GL025811

Krishnan R, Kumar V, Sugi M, Yoshimura J (2009) Internal feedbacks from monsoon midlatitude interactions during droughts in the Indian summer monsoon. J Atmos Sci 66:553–578. https://doi.org/10.1175/2008JAS2723.1

Krishnan R, Sabin TP, Ayantika DC, Kitoh A, Sugi M, Murakami H, Turner AG, Slingo JM, Rajendran K (2013) Will the South Asian monsoon overturning circulation stabilize any further? Clim Dyn 40:187–211. https://doi.org/10.1007/s00382-012-1317-0

Krishnan R, Sabin TP, Vellore R, Mujumdar M, Sanjay J, Goswami BN, Hourdin F, Dufresne JL, Terray P (2016) Deciphering the desiccation trend of the South Asian monsoon hydroclimate in a warming world. Clim Dyn 47:1007–1027. https://doi.org/10.1007/s00382-015-2886-5

Kumar KK, Rajagopalan B, Cane MA (1999) On the weakening relationship between the Indian monsoon and ENSO. Science 284:2156–2159. https://doi.org/10.1126/science.284.5423.2156

Kumar KK, Rajagopalan B, Hoerling M, Bates G, Cane M (2006) Unraveling the mystery of Indian monsoon failure during El Nino. Science 80(314):115–119. https://doi.org/10.1126/science.1131152

Kumar P, Rupa Kumar K, Rajeevan M, Sahai AK (2007) On the recent strengthening of the relationship between ENSO and northeast monsoon rainfall over South Asia. ClimDyn 28:649–660. https://doi.org/10.1007/s00382-006-0210-0

Liu J, Niyogi D (2019) Meta-analysis of urbanization impact on rainfall modification. Sci Rep 9(1):7301

Lutz AF, Immerzeel WW, Shrestha AB, Bierkens MFP (2014) Consistent increase in High Asia's runoff due to increasing glacier melt and precipitation. Nat Clim Change 4:587–592. https://doi.org/10.1038/nclimate2237

Lutz AF, TerMaat HW, Wijngaard RR, Biemans H, Syed A, Shrestha AB, Wester P, Immerzeel WW (2019) South Asian river basins in a 1.5 °C warmer world. Reg Environ Change 19:833–847. https://doi.org/10.1007/s10113-018-1433-4

Mallya G, Mishra V, Niyogi D, Tripathi S, Govindaraju RS (2016) Trends and variability of droughts over the Indian monsoon region. Weather Clim Extremes 12:43–68. https://doi.org/10.1016/j.wace.2016.01.002

Mann ME, Zhang Z, Rutherford S, Bradley RS, Hughes MK, Shindell D, Ammann C, Faluvegi G, Ni F (2009) Global signatures and dynamical origins of the little ice age and medieval climate anomaly. Science (80):326 1256–1260. https://doi.org/10.1126/science.1177303

Masood M, Yeh PJF, Hanasaki N, Takeuchi K (2015) Model study of the impacts of future climate change on the hydrology of Ganges-Brahmaputra-Meghna basin. Hydrol Earth Syst Sci 19:747–770. https://doi.org/10.5194/hess-19-747-2015

McKee TB, Nolan J, Kleist J (1993) The relationship of drought frequency and duration to time scales. Prepr Eighth Conf Appl Climatol Am Meteor Soc

Menon A, Levermann A, Schewe J, Lehmann J, Frieler K (2013) Consistent increase in Indian monsoon rainfall and its variability across CMIP-5 models. Earth Syst Dyn 4:287–300. https://doi.org/10.5194/esd-4-287-2013

Milly PCD, Wetherald RT, Dunne KA, Delworth TL (2002) Increasing risk of great floods in a changing climate. Nature 415:514–517. https://doi.org/10.1038/415514a

Mirza MMQ (2003) Three recent extreme floods in Bangladesh: a hydro-meteorological analysis. Nat Hazards 28:35–64. https://doi.org/10.1023/A:1021169731325

Mirza MMQ (2011) Climate change, flooding in South Asia and implications. Reg Environ Chang 11:95–107. https://doi.org/10.1007/s10113-010-0184-7

Mirza MMQ, Warrick RA, Ericksen NJ (2003) The implications of climate change on floods of the Ganges, Brahmaputra and Meghna rivers in Bangladesh. Clim Change. https://doi.org/10.1023/a:1022825915791

Mishra V, Lilhare R (2016) Hydrologic sensitivity of Indian sub-continental river basins to climate change. Glob Planet Change 139:78–96. https://doi.org/10.1016/j.gloplacha.2016.01.003

Mishra V, Shah HL (2018) Hydroclimatological perspective of the Kerala flood of 2018. J GeolSoc India 92:645–650. https://doi.org/10.1007/s12594-018-1079-3

Mishra V, Dominguez F, Lettenmaier DP (2012a) Urban precipitation extremes: how reliable are regional climate models? Geophys Res Lett 39:1–8. https://doi.org/10.1029/2011GL050658

Mishra V, Smoliak BV, Lettenmaier DP, Wallace JM (2012b) A prominent pattern of year-to-year variability in Indian Summer Monsoon Rainfall. Proc Natl Acad Sci USA 109:7213–7217. https://doi.org/10.1073/pnas.1119150109

Mishra V, Kumar D, Ganguly AR, Sanjay J, Mujumdar M, Krishnan R, Shah RD (2014) Reliability of regional and global climate models to simulate precipitation extremes over India. J Geophys Res Atmos. 119:9301–9323. https://doi.org/10.1002/2014JD021636

Mishra V, Aadhar S, Asoka A, Pai S, Kumar R (2016) On the frequency of the 2015 monsoon season drought in the Indo-Gangetic Plain. Geophys Res Lett 43:12102–12112. https://doi.org/10.1002/2016GL071407

Mishra V, Shah R, Azhar S, Shah H, Modi P, Kumar R (2018) Reconstruction of droughts in India using multiple land-surface models (1951–2015). Hydrol Earth Syst Sci 22:2269–2284. https://doi.org/10.5194/hess-22-2269-2018

Mishra V, Tiwari AD, Aadhar S et al (2019) Drought and Famine in India, 1870–2016. Geophys Res Lett 46:2075–2083. https://doi.org/10.1029/2018gl081477

Mooley DA (1976) Worst summer monsoon failures over the Asiatic monsoon area. Proc Ind Natl Sci Acad 42:34–43

Mujumdar M, Kumar V, Krishnan R (2007) The Indian summer monsoon drought of 2002 and its linkage with tropical convective activity over northwest Pacific. Clim Dyn 28:743–758. https://doi.org/10.1007/s00382-006-0208-7

Mujumdar M, Preethi B, Sabin TP, Ashok K, Saeed S, Pai DS, Krishnan R (2012) The Asian summer monsoon response to the La Niña event of 2010. Meteorol Appl 19:216–225. https://doi.org/10.1002/met.130

Mujumdar M, Sooraj KP, Krishnan R, Preethi B, Joshi MK, Varikoden H, Singh BB, Rajeevan M (2017a) Anomalous convective activity over sub-tropical east Pacific during 2015 and associated boreal summer monsoon teleconnections. Clim Dyn 48:4081–4091. https://doi.org/10.1007/s00382-016-3321-2

Mujumdar M, Goswami M, Ganeshi N, Sabade SS, Morrison R, Muddu S, Krishnan R, Ball L, Cooper H, Evans J, Jenkins A (2017b) The field scale soil moisture analysis using COSMOS-India network to explore water resource quantity and quality for water supply, agriculture and aquaculture over the Indian regions. India UK Water Centre (IUKWC) Workshop: Enhancing Freshwater Monitoring through Earth Observation, Stirling, UK, 19–21 June 2017 (http://nora.nerc.ac.uk/id/eprint/517252/)

Najibi N, Devineni N (2018) Recent trends in the frequency and duration of global floods. Earth Syst Dyn 9:757–783. https://doi.org/10.5194/esd-9-757-2018

Nandargi SS, Shelar A (2018) Rainfall and flood studies of the Ganga River Basin in India. Ann Geogr Stud 1:34–50

Neena JM, Suhas E, Goswami BN (2011) Leading role of internal dynamics in the 2009 Indian summer monsoon drought. J Geophys Res 116:D13103. https://doi.org/10.1029/2010JD015328

Niranjan Kumar K, Rajeevan M, Pai DS, Srivastava AK, Preethi B (2013) On the observed variability of monsoon droughts over India. Weather Clim Extremes 1:42–50. https://doi.org/10.1016/J.WACE.2013.07.006

Niyas NT, Srivastava AK, Hatwar HR (2009) Variability and trend in the cyclonic storms over north indian ocean. India Meteorological Department, MET Monograph No. Cyclone Warning—3/2009

Niyogi D, Kishtawal C, Tripathi S, Govindaraju RS (2010) Observational evidence that agricultural intensification and land use change may be reducing the Indian summer monsoon rainfall. Water Resour Res 46(3)

Pai DS, Sridhar L, Guhathakurta P, Hatwar HR (2011) District-wide drought climatology of the southwest monsoon season over India based on standardized precipitation index (SPI). Nat Hazards 59:1797–1813. https://doi.org/10.1007/s11069-011-9867-8

Pai DS, Guhathakurta P, Kulkarni A, Rajeevan MN (2017) Variability of meteorological droughts over India. 73–87

Palmer W (1965) Meteorological drought. Research paper no. 45, U.S. Department of Commerce Weather Bureau

Pant GB, Parthasarathy SB (1981) Some aspects of an association between the southern oscillation and indian summer monsoon. Arch Meteorol Geophys Bio Climatol Ser B 29:245–252. https://doi.org/10.1007/BF02263246

Pant GB, Rupa Kumar K, Sontakke NA, Borgaonkar HP (1993) Climate variability over India on century and longer time scales. In: Keshavamurty RN, Joshi PC (eds) Adv Trop Meteorol. Tata McGraw Hill Pub. Ltd., New Delhi, pp 71–84

Parthasarathy B, Sontakke NA, Monot AA, Kothawale DR (1987) Droughts/floods in the summer monsoon season over different meteorological subdivisions of India for the period 1871–1984. J Climatol 7:57–70. https://doi.org/10.1002/joc.3370070106

Pervez MS, Henebry GM (2015) Spatial and seasonal responses of precipitation in the Ganges and Brahmaputra river basins to ENSO and Indian Ocean dipole modes: implications for flooding and drought. Nat Hazards Earth Syst Sci 15:147–162. https://doi.org/10.5194/nhess-15-147-2015

Preethi B, Kripalani RH, Kumar K (2010) Indian summer monsoon rainfall variability in global coupled ocean-atmospheric models. Clim Dyn 35:1521–1539. https://doi.org/10.1007/s00382-009-0657-x

Preethi B, Revadekar JV, Munot AA (2011) Extremes in summer monsoon precipitation over India during 2001–2009 using CPC high-resolution data. Int J Remote Sens 32:717–735. https://doi.org/10.1080/01431161.2010.517795

Preethi B, Mujumdar M, Kripalani RH, Prabhu A, Krishnan R (2017a) Recent trends and tele-connections among South and East Asian summer monsoons in a warming environment. Clim Dyn 48:2489–2505. https://doi.org/10.1007/s00382-016-3218-0

Preethi B, Mujumdar M, Prabhu A, Kripalani R (2017b) Variability and teleconnections of South and East Asian summer monsoons in present and future projections of CMIP5 climate models. Asia-Pacific J Atmos Sci 53:305–325. https://doi.org/10.1007/s13143-017-0034-3

Preethi B, Ramya R, Patwardhan SK, Mujumdar M, Kripalani RH (2019) Variability of Indian summer monsoon droughts in CMIP5 climate models. Clim Dyn 53:1937–1962. https://doi.org/10.1007/s00382-019-04752-x

Priya P, Mujumdar M, Sabin TP, Terray P, Krishnan R (2015) Impacts of Indo-Pacific Sea surface temperature anomalies on the summer monsoon circulation and heavy precipitation over northwest India-Pakistan region during 2010. J Clim 28:3714–3730. https://doi.org/10.1175/JCLI-D-14-00595.1

Rajeevan M, Bhate J, Jaswal AK (2008) Analysis of variability and trends of extreme rainfall events over India using 104 years of gridded daily rainfall data. Geophys Res Lett 35:L18707. https://doi.org/10.1029/2008GL035143

Rajeevan M, Unnikrishnan CK, Preethi B (2012) Evaluation of the ENSEMBLES multi-model seasonal forecasts of Indian summer monsoon variability. Clim Dyn 38:2257–2274. https://doi.org/10.1007/s00382-011-1061-x

Raman CRV, Rao YP (1981) Blocking highs over Asia and monsoon droughts over India. Nature 289:271–273. https://doi.org/10.1038/289271a0

Ramarao MVS, Krishnan R, Sanjay J, Sabin TP (2015) Understanding land surface response to changing South Asian monsoon in a warming climate. Earth Syst Dyn 6(2):569–582. https://doi.org/10.5194/esd-6-569-2015

Ramarao MVS, Sanjay J, Krishnan R, Mujumdar M, Bazaz A, Revi A (2019) On observed aridity changes over the semiarid regions of India in a warming climate. Theor Appl Climatol 136:693–702. https://doi.org/10.1007/s00704-018-2513-6

Ramaswamy C (1962) Breaks in the Indian summer monsoon as a phenomenon of interaction between the easterly and the sub-tropical westerly jet streams. Tellus 14:337–349. https://doi.org/10.1111/j.2153-3490.1962.tb01346.x

Ramesh Kumar MR, Krishnan R, Sankar S, Unnikrishnan AS, Pai DS (2009) Increasing trend of break-monsoon conditions over india-role of ocean-atmosphere processes in the Indian ocean. IEEE Geosci Remote Sens Lett 6:332–336. https://doi.org/10.1109/LGRS.2009.201336

Ranade AA, Singh N, Singh H, Sontakke N (2007) Characteristics of hydrological wet season over different river basins of India. Indian Institute of Tropical Meteorology, IITM research report RR-119, ISSN 0252-1075

Rao SA, Chaudhari HS, Pokhrel S, Goswami BN (2010) Unusual central Indian drought of summer monsoon 2008: role of southern tropical Indian Ocean warming. J Clim 23:5163–5174. https://doi.org/10.1175/2010JCLI3257.1

Ray K, Pandey P, Pandey C, Dimri AP, Kishore K (2019) On the recent floods in India. Current science, vol 117, no 2

Rosenzweig C, Solecki W, Hammer SA, Mehrotra S (2010) Cities lead the way in climate-change action. Nature 467(7318):909–911. https://doi.org/10.1038/467909a

Roxy MK, Ritika K, Terray P, Masson S (2014) The curious case of Indian Ocean warming. J Clim 27:8501–8509. https://doi.org/10.1175/JCLI-D-14-00471.1

Roxy MK, Ghosh S, Pathak A, Athulya R, Mujumdar M, Murtugudde R, Terray P, Rajeevan M (2017) A threefold rise in widespread extreme rain events over central India. Nat Commun 8:708. https://doi.org/10.1038/s41467-017-00744-9

Saji NH, Goswami BN, Vinayachandran PN, Yamagata T (1999) A dipole mode in the tropical Indian Ocean. Nature 401:360–363. https://doi.org/10.1038/43854

Sano M, Buckley BM, Sweda T (2009) Tree-ring based hydroclimate reconstruction over northern Vietnam from Fokieniahodginsii: eighteenth century mega-drought and tropical Pacific influence. Clim Dyn 33:331–340. https://doi.org/10.1007/s00382-008-0454-y

Scheff J, Frierson DMW (2014) Scaling potential evapotranspiration with greenhouse warming. J Clim 27:1539–1558. https://doi.org/10.1175/JCLI-D-13-00233.1

Schiermeier Q (2011) Increased flood risk linked to global warming. Nature 470:316. https://doi.org/10.1038/470316a

Shah R, Mishra V (2014) Evaluation of the reanalysis products for the monsoon season droughts in India. J Hydrometeorol 15:1575–1591. https://doi.org/10.1175/jhm-d-13-0103.1

Sharma A, Wasko C, Lettenmaier DP (2018) If precipitation extremes are increasing, why aren't floods? Water Resour Res 54:8545–8551. https://doi.org/10.1029/2018WR023749

Sharmila S, Joseph S, Sahai AK, Abhilash S, Chattopadhyay R (2015) Future projection of Indian summer monsoon variability under climate change scenario: an assessment from CMIP5 climate models. Glob Planet Change 124:62–78. https://doi.org/10.1016/j.gloplacha.2014.11.004

Sheffield J, Livneh B, Wood EF (2012) Representation of terrestrial hydrology and large-scale drought of the continental United States from the North American regional reanalysis. J Hydrometeorol 13:856–876. https://doi.org/10.1175/JHM-D-11-065.1

Shepherd JM (2005) A review of current investigations of urban-induced rainfall and recommendations for the future. Earth Interact 9:1–27. https://doi.org/10.1175/EI156.1

Shrestha A, Agrawal N, Alfthan B, Bajracharya S, Maréchal J, van Oort B, (2015) The Himalayan climate and water atlas. 0493(1983) 111<1830:TSOALR>2.0.CO;2

Shukla S, Wood AW (2008) Use of a standardized runoff index for characterizing hydrologic drought. Geophys Res Lett 35:L02405. https://doi.org/10.1029/2007GL032487

Sikka DR (1980) Some aspects of the large scale fluctuations of summer monsoon rainfall over India in relation to fluctuations in the planetary and regional scale circulation parameters. Proc Ind Acad Sci Earth Planet Sci 89:179–195. https://doi.org/10.1007/BF02913749

Sikka DR (1999) Monsoon drought in India. Joint COLA/CARE technical report No. 2. Center for Ocean–Land–Atmosphere Studies (COLA), Center for the Application of Research on the Environment (CARE), 243 pp

Sikka DR (2003) Evaluation of monitoring and forecasting of summer monsoon over India and a review of monsoon drought of 2002. Proc Natl Acad Sci India Sect A 69:479–504

Sikka DR, Ray K, Chakravarthy K, Bhan SC, Tyagi A (2015) Heavy rainfall in the Kedarnath valley of Uttarakhand during the advancing monsoon phase in June 2013. Curr Sci 109:353–361

Sinha A, Cannariato KG, Stott LD, Cheng H, Edwards RL, Yadava MG, Ramesh R, Singh IB, (2007) A 900-year (600–1500 A.D.) record of the Indian summer monsoon precipitation from the core monsoon zone of India. Geophys Res Lett 34. https://doi.org/10.1029/2007GL030431

Sinha A, Berkelhammer M, Stott L, Mudelsee M, Cheng H, Biswas J (2011a) The leading mode of Indian Summer Monsoon precipitation variability during the last millennium. Geophys Res Lett 38: L15703. https://doi.org/10.1029/2011GL047713

Sinha A, Stott L, Berkelhammer M, Cheng H, Edwards RL, Buckley B, Aldenderfer M, Mudelsee M (2011b) A global context for megadroughts in monsoon Asia during the past millennium. Quat Sci Rev 30:47–62. https://doi.org/10.1016/j.quascirev.2010.10.005

Sinha N, Gandhi N, Chakraborty S, Krishnan R, Yadava M, Ramesh R (2018) Abrupt climate change at ∼2800 yr BP evidenced by a stalagmite record from peninsular India. Holocene 28:1720–1730. https://doi.org/10.1177/0959683618788647

Sontakke NA, Singh N, Singh HN (2008) Instrumental period rainfall series of the Indian region (AD 1813—2005): revised reconstruction, update and analysis. Holocene 18:1055–1066. https://doi.org/10.1177/0959683608095576

Spinoni et al (2020) Future global meteorological drought hot spots: a study based on CORDEX data. J Clim 33:3635–3661. https://doi.org/10.1175/JCLI-D-19-0084.1

Swaminathan MS (1987) The Indian experience. Monsoons JS, Fein, Stephens PL (eds) Wiley, New York, pp 121–132

Tejavath CT, Ashok K, Chakraborty S, Ramesh R (2019) A PMIP3 narrative of modulation of ENSO teleconnections to the Indian summer monsoon by background changes in the Last Millennium. Clim Dyn 53:3445–3461. https://doi.org/10.1007/s00382-019-04718-z

Trenberth K (2011) Changes in precipitation with climate change. Clim Res 47:123–138. https://doi.org/10.3354/cr00953

Turner AG, Annamalai H (2012) Climate change and the South Asian summer monsoon. Nat Clim Chang 2:587–595. https://doi.org/10.1038/nclimate1495

van Oldenborgh GJ, Otto FEL, Haustein K, AchutaRao K (2016) The heavy precipitation event of December 2015 in Chennai, India [in Explaining extreme events of 2015 from a climate perspective], Bulletin of the American Meteorological Society 97(12):S87–S91. http://www.ametsoc.net/eee/2015/17_india_precip.pdf

Vellore RK, Krishnan R, Pendharkar J, Choudhury AD, Sabin TP (2014) On the anomalous precipitation enhancement over the Himalayan foothills during monsoon breaks. Clim Dyn 43:2009–2031. https://doi.org/10.1007/s00382-013-2024-1

Vellore RK, Kaplan ML, Krishnan R., Lewis JM, Sabade S, Deshpande N, Singh BB, Madhura RK, Rama Rao MVS (2016) Monsoon-extratropical circulation interactions in Himalayan extreme rainfall. Climate Dynamics 46:3517–3546. https://doi.org/10.1007/s00382-015-2784-x

Vicente-Serrano SM, Beguería S, López-Moreno JI (2010) A multiscalar drought index sensitive to global warming: the standardized precipitation evapotranspiration index. J Clim 23:1696–1718. https://doi.org/10.1175/2009JCLI2909.1

Ward PJ, Kummu M, Lall U (2016) Flood frequencies and durations and their response to El Niño Southern oscillation: global analysis. J Hydrol 539:358–378. https://doi.org/10.1016/j.jhydrol.2016.05.045

Webster PJ, Toma VE, Kim H-M (2011) Were the 2010 Pakistan floods predictable? Geophys Res Lett 38:n/a-n/a. https://doi.org/10.1029/2010gl046346

Wijngaard RR, Lutz AF, Nepal S, Khanal S, Pradhananga S, Shrestha AB, Immerzeel WW (2017) Future changes in hydro-climatic extremes in the Upper Indus Ganges and Brahmaputra River basins. PLoS ONE 12:e0190224. https://doi.org/10.1371/journal.pone.0190224

Wilhite DA, Glantz MH (1985) Understanding: the drought phenomenon: the role of definitions. Water Int 10:111–120. https://doi.org/10.1080/02508068508686328

Yadav RK (2012) Why is ENSO influencing Indian northeast monsoon in the recent decades? Int J Climatol 32:n/a-n/a. https://doi.org/10.1002/joc.2430

Yang L, Tian F, Niyogi D (2015) A need to revisit hydrologic responses to urbanization by incorporating the feedback on spatial rainfall patterns. Urban Clim 12:128–140

Yang T, Ding J, Liu D, Wang X, Wang T, Yang T, Ding J, Liu D, Wang X, Wang T (2019) Combined use of multiple drought indices for global assessment of dry gets drier and wet gets wetter paradigm. J Clim 32:737–748. https://doi.org/10.1175/JCLI-D-18-0261.1

Coordinating Lead Authors

Savita Patwardhan, Indian Institute of Tropical Meteorology (IITM-MoES), Pune, India
K. P. Sooraj, Indian Institute of Tropical Meteorology (IITM-MoES), Pune, India,
e-mail: sooraj@tropmet.res.in (corresponding author)

Lead Authors

Hamza Varikoden, Indian Institute of Tropical Meteorology (IITM-MoES), Pune, India
S. Vishnu, Indian Institute of Tropical Meteorology (IITM-MoES), Pune, India
K. Koteswararao, Indian Institute of Tropical Meteorology (IITM-MoES), Pune, India
M. V. S. Ramarao, Indian Institute of Tropical Meteorology (IITM-MoES), Pune, India

Review Editor

D. R. Pattanaik, India Meteorological Department (IMD-MoES), New Delhi, India

Corresponding Author

K. P. Sooraj, Indian Institute of Tropical Meteorology (IITM-MoES), Pune, India,
e-mail: sooraj@tropmet.res.in

© The Author(s) 2020
R. Krishnan et al. (eds.), *Assessment of Climate Change over the Indian Region*,
https://doi.org/10.1007/978-981-15-4327-2_7

Key Messages

- While the frequency of summer monsoon lows has significantly increased, the frequency of monsoon depressions has declined during 1951–2015 (high confidence)
- A significant rising trend in the amplitude of wintertime western disturbances is observed during 1951–2015 (medium confidence).
- Climate models project a decline in the frequency of monsoon low-pressure systems (LPS) by the end of twenty-first century (medium confidence).
- Climate models also project a poleward shift in monsoon LPS activity by the end of the twenty-first century which is likely to enhance heavy rainfall occurrences over northern India (medium confidence).

7.1 Introduction

The weather over the Indian subcontinent is distinctly influenced by various synoptic-scale weather systems, viz., monsoon lows, monsoon depressions, mid-tropospheric cyclones, tropical cyclones and western disturbances (Sikka 2006). These synoptic-scale weather systems have horizontal dimensions varying from 100 to 2000 km and temporal dimensions ranging from a few to several days (Wallace and Hobbs 2006; Ding and Sikka 2006). They bring floods, snowstorms, and avalanches to Indian landmass, subsequently modulating the annual mean Indian rainfall (Ajayamohan et al. 2010; Revadekar et al. 2016; Sikka 2006; Hunt et al. 2016a, b). Since Indian agriculture depends on seasonal rains, the inter-annual variability in the synoptic-scale weather systems plays an instrumental role in the socio-economic fabric of the country (e.g. Ajayamohan et al. 2010; Hunt et al. 2016a, b). This chapter provides an assessment of observed and future changes in synoptic weather systems during boreal summer (i.e., monsoon lows and depressions) and winter to spring season (i.e., western disturbances).

7.1.1 Monsoon Low-Pressure Systems

The synoptic-scale weather systems formed during the Indian summer monsoon season (June to September, JJAS)

have varying intensities, collectively referred as the low-pressure systems (LPS; Mooley 1973; Sikka 1977; Saha et al. 1981; Mooley and Shukla 1989; Krishnamurthy and Ajayamohan 2010; Praveen et al. 2015). The India Meteorological Department (IMD) classifies the LPS based on their intensity and their characterization is described viz., (i) Low, a weaker system with wind speed less than 8.75 m s^{-1} and a closed isobar in the surface pressure chart in the radius of 3° from the center, (ii) Monsoon Depressions having wind speeds between 8.75 and 17 m s^{-1} and more than two closed isobars with an interval of 2 hPa in the radius of 3° from the center, and (iii) Cyclonic storms having wind speed more than 17 m s^{-1} and more than four closed isobars at 2 hPa interval on the surface pressure chart (see the summary in Table 7.1).

Summer monsoon LPS mostly comprises lows and depressions, with a very few cyclonic storms. The majority of LPS originate from the head of the Bay of Bengal (BoB, 76%) (Godbole 1977; Sikka 1977, 2006; Saha et al. 1981) and move northwestward and/or westward towards the Indian subcontinent with an average speed of $1.4–2.8 \text{ m s}^{-1}$ (Sikka 1980). A few systems also form over the Indian landmass (15%) and the Arabian Sea (9%) and move towards the Indian subcontinent (Sikka 2006). Most of the LPS forming within the Indian monsoon trough region are generally cyclonic systems with weaker intensity as compared to tropical cyclones (Mooley 1973; Godbole 1977; Sikka 1977).

The most efficient rain-producing systems are the LPS forming over BoB and moving along the monsoon trough region in northwesterly/westerly direction and they also regulate the seasonal monsoon rains over the Indian landmass (Raghavan 1967; Krishnamurti et al. 1975; Saha et al. 1981; Yoon and Chen 2005; Sikka 2006; Vishnu et al. 2016). Among the various LPS, monsoon depressions are usually associated with widespread to heavy rainfall over the central part of India and their contribution to seasonal rainfall is as high as 45% (Krishnamurti et al. 1975; Saha et al. 1981; Yoon and Chen 2005; Pai et al. 2014, 2015). The west coast of India also receives a significant amount of rainfall during the occurrence of depressions over the BoB (Krishnamurthy and Ajayamohan 2010; Vishnu et al. 2016). In contrast, observations indicate rainfall reduction over the northeast part of India and the southern Peninsula during the passage of monsoon depression (Raghavan 1967). Monsoon lows, unlike depressions, are not often associated with

Table 7.1 Classification of monsoon LPS following IMD

LPS	Closed isobars	Wind speeds
Low	1	$<8.75 \text{ m s}^{-1}$
Monsoon depression	>2	$8.75–17 \text{ m s}^{-1}$
Cyclonic storms	>4	$>17 \text{ m s}^{-1}$

extreme rainfall events. But they can bring substantial rains to the Indian landmass, and monsoon lows contribute to about 40% of monsoon seasonal rains over the central Indian landmass (Hurley and Boos 2015).

Generally, the development and intensification of LPS have associations with warm sea surface temperatures (SSTs), and environmental factors, such as the presence of low level (850 hPa) cyclonic vorticity, high mid-tropospheric (500 hPa) humidity and strong vertical wind shear (difference in the zonal winds between 850 and 200 hPa) (Sikka 1977). Further, other large-scale synoptic environments which favor the LPS genesis also includes the following: (i) upper-tropospheric easterly waves, (ii) westward-moving residual low of tropical cyclones from the Western Tropical Pacific–South China Sea (WTP-SCS) region and (iii) slow descent of mid-tropospheric cyclonic circulations (Sikka 2006). While in all other northern hemispheric basins the cyclone activity peaks in July–August, the strong vertical wind shear during summer monsoon season generally restricts the LPS activity over the Arabian Sea and the BoB to further intensify into tropical cyclones (Gray 1968; Sikka 1977; Ding and Sikka 2006; see Chap. 8). Accordingly, intense systems such as Cyclonic Storms and/or Severe Cyclonic Storms (commonly referred hereafter as simply Cyclonic storms throughout the text) very rarely form in the summer monsoon season (e.g. Sikka 2006).

The spatio-temporal variations in monsoon rainfall are often associated with the genesis and movement of the LPS, and the associated rainfall distribution over its domain of influence. According to the pioneering study by Eliot (1884), the heaviest rainfall occurs in the southern quadrant of monsoon depressions over the head BoB in the formative stage, and in the southwest quadrant during its west/west-northwest translation. Monsoon depressions typically produce heavy rainfall amounts of 30–60 cm day^{-1} within the 200–300 km radius located in the southwestern sector of depressions (Sikka 2006).

In addition to LPS, there is another distinct class of summer monsoon (JJAS) synoptic systems known as mid-tropospheric cyclones (MTCs) which are quasi-stationary cold-core systems associated with the strongest cyclonic vorticity between 700 and 500 hPa levels (Miller and Keshavamurthy 1968; Krishnamurti and Hawkins 1970; Carr 1977; Mak 1983; Choudhury et al. 2018). Further, MTCs show strong midlevel convergence, with anomalous temperature field exhibiting cold (warm) signatures below (above) 500 hPa. MTCs seen over the Arabian Sea have received special attention in recent times, as they often produce flood-producing rainfall situations over the western states of India (Maharashtra and Gujarat) during JJAS. Choudhury et al. (2018) showed that some of the heaviest 3-day rain accumulations over the western Indian regions (e.g. south Gujarat and adjoining areas) during 1998–2007 co-occurred with MTC signatures. For example,

the MTC occurrence during the 24 June–3 July 2005 period was associated with record 3-day rainfall accumulations of 700 mm at 72.7° E, 20.87° N located just north of Mumbai on 28 June 2005. A few other cases include: the MTC event during 9–20th July 2018 produced heavy rainfall over Saurashtra, Kutch, Gujarat, and interior Maharashtra. The extreme rainfall events over Mumbai that occurred on 29th June, 1st July, and 5th September 2019 (24-h rainfall accumulations exceeding 200 mm, as recorded at the Santa Cruz observatory in Mumbai; Indian Daily Weather Report, IMD) have co-occurred with MTCs seen over north Konkan and adjoining south Gujarat region. Choudhury et al. (2018) also showed that the formation of heavily precipitating MTCs over western India has linkage to stratiform heating structure within the northward propagating organized monsoon convection on sub-seasonal timescales. There are, however, very limited studies on MTCs and ascertaining its association with extreme rainfall events over western India (Miller and Keshavamurthy 1968; Krishnamurti and Hawkins 1970; Carr 1977; Choudhury et al. 2018), and so far no studies have documented the future projections in the MTCs. Hence for JJAS period, we mainly focus on the present and future changes in LPS characteristics.

7.1.2 Western Disturbances

During boreal winter and early spring season (December to April; DJFMA), high-pressure conditions are prevalent over north India and the associated weather is usually clear skies and dry. The conditions of cloudy, dense fog, snow, and light to heavy precipitation also occur intermittently during this season by the eastward passage of synoptic-scale weather disturbances, known as 'western disturbances (WDs)', originating from the Mediterranean (Pisharoty and Desai 1956; Mooley 1957; Singh and Kumar 1977; Kalsi 1980; Kalsi and Halder 1992; De et al. 2005; Schiemann et al. 2009; Madhura et al. 2015; Cannon et al. 2015, 2016; Dimri 2007, 2008, Dimri et al. 2015, Dimri and Chevuturi 2016; Krishnan et al. 2019; Hunt et al. 2018a, b). IMD defined WDs as follows: a cyclonic circulation/trough in the mid and lower tropospheric levels or as a low-pressure area on the surface, which occurs in middle latitude westerlies and originates over the Mediterranean Sea, Caspian Sea, and Black Sea and moves eastwards across north India (http://imd.gov.in/section/nhac/wxfaq.pdf). The WDs are basically synoptic-scale perturbations embedded in subtropical westerly jet stream (STJs) at upper levels, and also latitudinal positioning of these STJs has a greater influence on the frequency of WDs (Hunt et al. 2018a).

The WDs are modulated by the tropical air mass and the Himalayas. Accordingly, the WDs are preceded by warm and moist air mass of tropical origin and succeeded by the cold and dry air mass of extra-tropical character (Mooley 1957). So the interaction between the tropics and mid-latitude systems is manifested in WDs with associated extensive cloudiness in the mid and high levels (Kalsi 1980; Kalsi and Halder 1992; Dimri 2007).

In association with WD passages, the Karakoram, Hindu Kush Mountain Ranges and also the northern part of India oftentimes experience extreme winter precipitation and flooding conditions, and the snowfall from WDs is the major precipitation input for the Himalayan Rivers (Pisharoty and Desai 1956; Mooley 1957; Rangachary and Bandyopadhyay 1987; Lang and Barros 2004; Hunt et al. 2018c; Roy and Roy Bhowmik 2005; Kotal et al. 2014; Dimri et al. 2015). The wintertime precipitation from the WDs, a non-monsoonal type of precipitation (Krishnan et al. 2019), contributes significantly by about 30% to the annual mean precipitation over the north Indian region (e.g. Dimri 2013a, Dimri 2013b).

On an average, 4–6 intense WDs are observed during the DJFMA (Pisharoty and Desai 1956; Rao and Srinivasan 1969; Chattopadhyay 1970; Dhar et al. 1984; Rangachary and Bandyopadhyay 1987; Mohanty et al. 1998; Hatwar et al. 2005; Dimri et al. 2015; Cannon et al. 2016; Hunt et al. 2018b). The life cycle of WDs typically ranges between 2 and 4 days, and WDs are relatively rapidly moving weather systems with zonal speeds of about 8–10° longitude/day (about 10–12 m s^{-1}) (Datta and Gupta 1967; Rao and Srinivasan 1969). The periodicity of WDs ranges from 4 to 12 days as noted by various studies (Krishnan et al. 2019; Rao and Rao 1971; Chattopadhyay 1970).

7.2 Observed Variability and Future Projections

7.2.1 Monsoon LPS

LPS plays a significant role in the Indian summer monsoon seasonal total rainfall. Hence, it is of paramount importance to understand their statistics on frequency, duration, etc. Table 7.2 shows the seasonal mean statistics in the frequency of lows, depressions, and LPS for two time periods (1901–2015 and 1951–2015). Note that the statistics is prepared without distinguishing them based on their origin (i.e., irrespective of land or sea). The LPS frequency shown in Table 7.2 includes the total number of summertime synoptic systems (i.e., lows, depressions, and cyclonic storms). The data sources for depressions are from the cyclone eAtlas archived by the IMD (for 1901–2015; http://www.rmcchennaieatlas.tn.nic.in). The data for lows are from published documentations from Mooley and Shukla (1987) for the period 1901–1983, from Sikka (2006) for the period 1984–2002 and from the Journal of Mausam published by Indian Meteorological Society for the latest period (i.e., since 2003).

Consistent with the statistical inferences from previous studies (Godbole 1977; Mooley and Shukla 1987), Table 7.2 also shows that LPS is generally dominated by monsoon lows and depressions as there are only a few intense cyclonic storms during JJAS. The long-term (1901–2015) seasonal mean frequency of monsoon lows is about 7 per season, while it increases to 8 per season during the 1951–2015 period. In contrast, the monsoon depression shows a slight decrease in its frequency during 1951–2015 relative to 1901–2015.

Table 7.2 further shows that the variability in lows and depressions tends to remain the same irrespective of the data period (i.e., for 1901–2015 and 1951–2015, respectively). The mean of LPS days constitutes about 45% of the total number of days in JJAS (i.e., on average, LPS is observed for 59 out of 122 days in the season; as obtained by Krishnamurthy and Ajayamohan 2010, for the period of 1901–2003).

A time series, from 1901 to 2015, of LPS forming (in addition to lows and depressions, frequency of cyclonic storms are also included in the figure) over BoB, Arabian Sea and also on land during JJAS is shown in Fig. 7.1. There is no significant change in trend in the frequency of LPS for

Table 7.2 Statistics of summer monsoon LPS frequency for two time periods (1901–2015 and 1951–2015)

Data period	Lows			Depressions			LPS		
	Mean	Standard deviation	Trend per decade (*p*-value)	Mean	Standard deviation	Trend per decade (*p*-value)	Mean	Standard deviation	Trend per decade (*p*-value)
1901–2015	6.8	3.46	0.38* (0.0001)	4.8	2.37	−0.11 (0.07)	13.0	2.30	0.09 (0.12)
1951–2015	7.7	3.68	1.01* (0.00001)	4.6	2.42	−0.69* (0.00001)	13.3	2.40	0.16 (0.28)

Significant trends at 5% level of significance, as estimated using the *F*-test, are marked with an asterisk (*) and the corresponding *p*-values indicated in parentheses

Note that the frequency of LPS includes all summer synoptic systems (i.e., lows, depressions, cyclonic storms, and severe cyclonic storms) originated from BoB, Arabian Sea and land

the 1901–2015 period (0.09 decade^{-1}; see Fig. 7.1a and Table 7.2) as well as for the post-1950 period (0.16 decade^{-1}). While the frequency of LPS does not change significantly in the last 100 years, there is a rise in the duration of LPS (Jadhav and Munot 2009). The long-term (1901–2015) trend in the frequency of lows shows a significant increase (+0.38 decade^{-1}, see Fig. 7.1b and Table 7.2), with a pronounced rise (+1.01 decade^{-1}) during the post-1950 period. In contrast, trends in the frequency of depressions during the 1901–2015 period show a decline of −0.11 decade^{-1} (Fig. 7.1c and Table 7.2), which gets prominently significant (−0.69 decade^{-1}) during 1951–2015. This decreasing (increasing) trend in depressions (lows) may also suggest compensating effect as evidenced in the insignificant trends in LPS frequency (Table 7.2). Similar to depressions, the frequency of cyclonic storms (see Fig. 7.1d) also shows a decreasing trend which is significant in both periods, i.e., −0.17 decade^{-1} for 1901–2015 and −0.15 decade^{-1} for 1951–2015. The long-term trends in the LPS frequency has

been investigated in several earlier studies (Rajeevan et al. 2000; Patwardhan and Bhalme 2001; Rajendra Kumar and Dash 2001; Mandke and Bhide 2003; Jadhav and Munot 2009; Ajayamohan et al. 2010; Prajeesh et al. 2013; Vishnu et al. 2016; Mohapatra et al. 2017). Though the data periods of these studies are different, major conclusions are, moreover, the same. However, the trends in the frequency of lows and depressions are not significant during the recent three decades (1986–2015).

Also, note that periods associated with frequent lows apparently coincide with periods of fewer depressions which clearly suggest that the frequency of lows and depressions also exhibits significant inter-decadal variations (Fig. 7.1b, c). Earlier studies showed that the frequency of lows and depressions displays an epochal behavior on inter-decadal time scale. For example, Rajendra Kumar and Dash (2001) examined the inter-decadal variations in lows and depressions based on long-term observations (i.e., for 110 years, 1889–1998) and to understand their relationship with the

Fig. 7.1 Low-pressure systems (LPS) forming over the Indian region (Bay of Bengal, Arabian Sea, and Indian land) during the Summer monsoon season (JJAS) for the period 1901–2015. The data sources for depressions are from the cyclone eAtlas archived by the IMD (http://www.rmcchennaieatlas.tn.nic.in), and data sources for lows include published documentations (Mooley and Shukla 1987; Sikka 2006) and from the Journal of Mausam published by the Indian Meteorological Society. In **a** LPS, i.e., total frequency of lows, depressions and cyclonic storms, **b** lows, **c** depressions and **d** cyclonic storms (including both cyclonic storms as well as severe cyclonic storms)

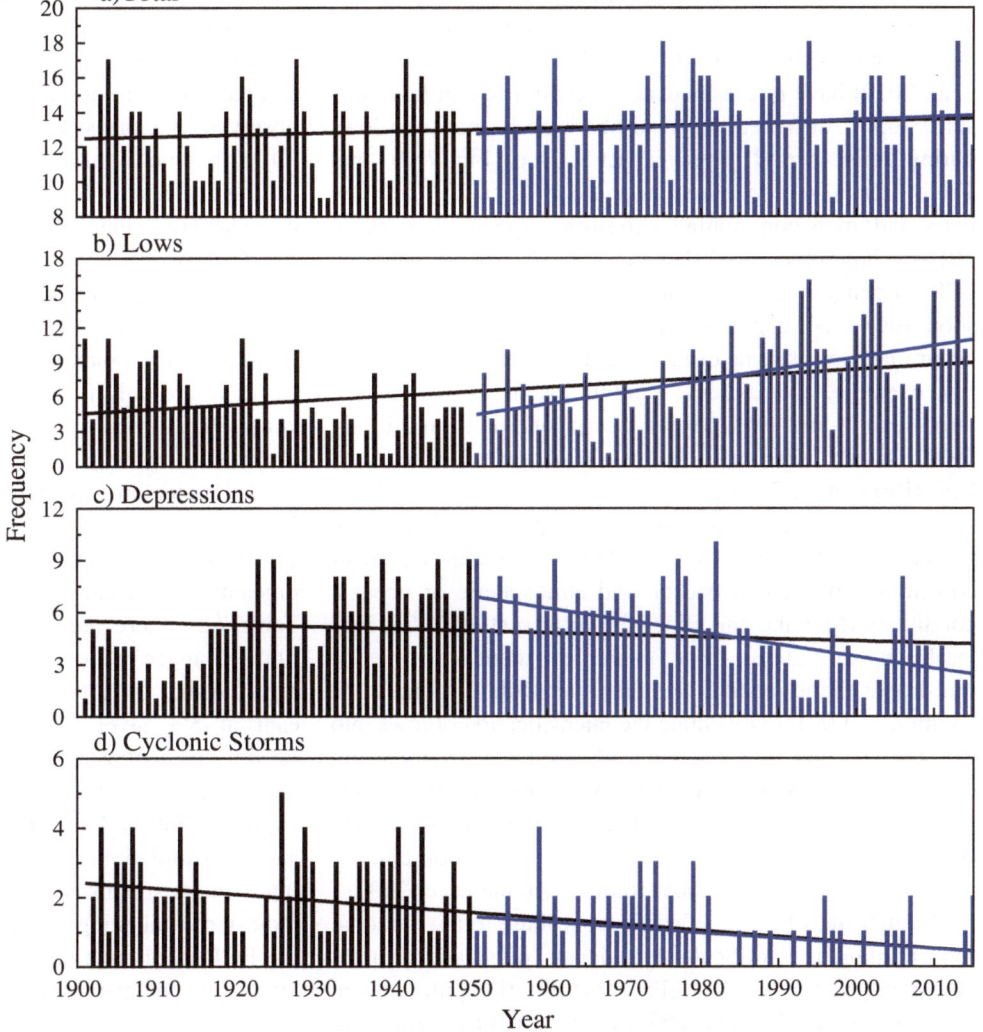

Indian summer Monson rainfall over a 30-year time period (i.e., 30-year periodicity of above normal or below normal epochs of the Indian summer monsoon rainfall). This study showed that intense LPS (i.e., except lows) are more (less) seen during the epochs of above (below) normal Indian summer monsoon rainfall. A recent study by Vishnu et al. (2018) examined the inter-decadal aspects of LPS, and they showed that number of monsoon depressions (stronger LPS) over BoB has out-of-phase relationship with the Pacific Decadal Oscillation (PDO). The PDO induced warming in the Western Equatorial Indian Ocean decreases the moisture advection into the BoB, thereby reducing the relative humidity and suppresses the monsoon depression activity. This is in contrast to PDO induced cooling (i.e., over Western Equatorial Indian Ocean) which increases the moisture advection into the BoB, thereby enhancing the monsoon depression activity.

LPS is generally found to be associated with heaviest rain intensities (Sikka 2006). Despite the decreasing trend seen in the occurrence of stronger LPS (monsoon depressions, as noted above), the frequency of monsoon rainfall extremes (i.e., heavy rainfall events, rainfall ≥ 100 mm day^{-1}, and very heavy rainfall events, rainfall ≥ 150 mm day^{-1}; as defined in Goswami et al. 2006; Roxy et al. 2017; Nikumbh et al. 2019) have increased over the central Indian landmass since 1950. An increasing trend observed in monsoon lows during this period (weaker LPS, see Fig. 7.1b and Table 7.2), also implies an in-phase relationship between lows and monsoon rainfall extremes (Ajayamohan et al. 2010). Nikumbh et al. (2019) also noticed that the monsoon LPS, in general, (i.e., without distinguishing between monsoon lows and depressions) is conducive for increasing occurrences of extreme events over the central part of India. For example, the extreme rainfall events which caused large-scale floods over central Indian landmass on 24 July 1989, 18 July 2000 and 7 August 2007, are associated with LPS (Roxy et al. 2017).

The contrasting trends in lows and depressions imply that the intensification from lows to depressions may be rather constrained by certain background atmospheric or oceanic conditions (Mandke and Bhide 2003; Rao et al. 2004; Prajeesh et al. 2013). For example, reduction in the mid-tropospheric relative humidity over BoB is found to be an important factor preventing the intensification of lows into depressions, and thus the reduced frequency of monsoon depressions (Prajeesh et al. 2013; Vishnu et al. 2016). This was also attributed to the weakening of the low-level jet, consistent with the weakening of summer monsoon circulation (Joseph and Simon 2005; Ramesh Kumar et al. 2009).

Though the LPSs significantly contribute to the seasonal total rainfall, it is difficult to designate flood and drought monsoon years in terms of LPS variability (i.e., inter-annual variability of LPS; Sikka 2006) alone. Krishnamurthy and

Ajayamohan (2010) have shown that the LPS contribution (to the total seasonal monsoon rainfall) remains invariant during the periods of monsoon floods or droughts, even though LPS frequency is slightly seen higher during flood years. However, they have shown that the track of LPS shows a marked difference between flood and drought years. The LPS reach up to northwest India during the flood years, while they are confined to central India during the drought years.

On the large-scale modes of variability influencing LPS, Hunt et al. (2016a) inferred a significant relationship between El Niño–Southern Oscillation (ENSO) and LPS activities (particularly for monsoon depressions). Their study indicated that there are more monsoon depressions during El Niño years (approximately 16% more) than La Niña years. This study differs from the investigation of Krishnamurthy and Ajayamohan (2010) which suggests that there is no significant relationship. This may be due to the consideration of total LPS in their study, instead of only monsoon depressions. There are few other studies that focused on the association of LPS activity with the Indian Ocean Dipole (IOD). Krishnan et al. (2011) reported that positive IOD is favorable for long-lived LPS. They found an approximate 12% increased lifetime of LPS during the positive IOD as compared to the normal years. Hunt et al. (2016a), however, observed that the state of IOD (i.e., positive and negative IODs) has no significant impact on depressions. Thus far, contrasting results from different studies imply that there is no clear consistency to assert the association of LPS with ENSO/IOD.

Given the prominent dependency of the Indian summer monsoon seasonal rainfall on LPS, it is important to understand the potential impact of climate change on LPS; yet there are only few studies in this direction. Patwardhan et al. (2012), with a focus on stronger LPS, showed that the frequency (intensity) of LPS may reduce (increase) by about 9% (11%) towards the end of the twenty-first century (under SRES-A2 scenario). They focused mainly on stronger LPS, except lows. Although observational evidence portray significantly increasing long-term trends (Fig. 7.1b and Table 7.2), there are no studies to diagnose the potential future changes in monsoon lows so far. Sandeep et al. (2018) reported that there would be about 45% reduction (significant at 5% level) in the LPS activity during the late twenty-first century (2071–2095) following RCP8.5 scenario (i.e., stronger warming climate scenario) from the High-Resolution Atmospheric Model (HiRAM) simulations, and the simulations from the fifth phase of Coupled Model Intercomparison Project (CMIP5; Taylor et al. 2011) also indicated weakening of LPS activity (over central India) in the RCP8.5 simulation. They used a combined measure of frequency, intensity, and duration of LPS to determine the LPS activity. The HiRAM projections also showed a

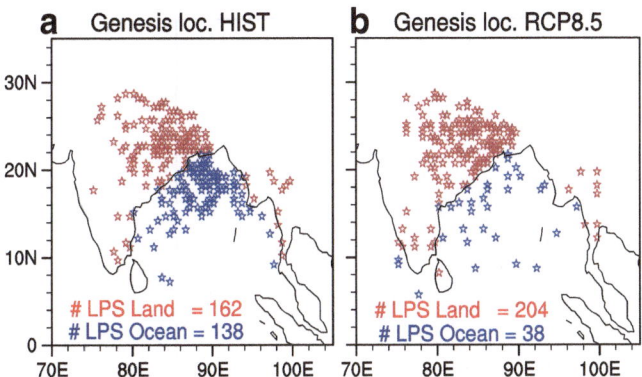

Fig. 7.2 Genesis locations of LPS formed during the monsoon season (June–September) from **a** HIST and **b** RCP8.5 simulations from High-Resolution Atmospheric Model (HiRAM). The HIST refers to historical simulation which includes both natural and anthropogenic forcing. The RCP8.5 is the simulation following the Representative Concentration Pathway 8.5 (RCP8.5) scenario. The red (blue) color indicates the genesis location overland (ocean). Adapted by permission from Sandeep et al. (2018)

poleward shift in the distribution of LPS genesis with a reduction by about 60% from the oceanic regions, and a rise by about 10% over the continental regions (see Fig. 7.2 for more details). The poleward shift in LPS activity is further stated to have wider implications and societal impacts, as it may possibly dry up central India as well as increase the frequency of extreme rainfall events over northern India. On the other hand, the future projection results from another recent study using CMIP5 models do not suggest a significant change in frequency and trajectory of the monsoon depressions during the twenty-first century (using RCP8.5 scenario, Rastogi et al. 2018). The contrasting inferences from different model projections may be attributed to the differences in experimental designs and methods of analysis.

Though climate models show a decline in LPS activity under the global warming scenario, there is medium confidence in the projected changes in LPS frequency. In this context, it is noteworthy to mention here that there is a big challenge in detecting the LPS from the model simulations, as LPS has weaker structure compared to other tropical storms (e.g. Cohen and Boos 2014; Hurley and Boos 2015; Praveen et al. 2015). Praveen et al. (2015) showed that only a very few CMIP5 models capture the observed characteristic of LPS. Moreover, the CMIP5 models usually being coarser in resolution show poor representation of LPS structure raising concerns on the reliability of future projections in LPS (Sandeep et al. 2018). The aforesaid clearly suggest an inherent uncertainty of GCMs to simulate and represent the observed and future characteristics of LPS (such as frequency, track, variability, trends, etc.). This clearly warrants careful evaluation of the model's ability to capture the observed LPS activity and its characteristics,

along with continued efforts to find better modeling and identification strategies for LPS.

7.2.2 WDs

Observational studies have reported a significant warming trend in the winter and annual temperatures over the WH (Kothawale and Rupa Kumar 2005; Bhutiyani et al. 2007), and there is, however, less spatially coherent trend in the non-monsoon precipitation observed over this region (Madhura et al. 2015). The estimates from contemporary studies of Cannon et al. (2015) and Madhura et al. (2015) show a rising trend (significant at 95% confidence level) in the frequency of WD activity and in the associated localized heavy precipitation over the WH region. Madhura et al. (2015) further attributed it to anomalous warming of the Tibetan Plateau and associated mid-to-upper-level meridional temperature gradients over the sub-tropics and mid-latitudes, i.e., pronounced mid-tropospheric warming in recent decades over the west-central Asia increases the baroclinic instability of the mean westerly winds. These changes tend to favor increased variability of WDs (Puranik and Karekar 2009; Raju et al. 2011) and also increase the tendency of extreme precipitation events over WH. Krishnan et al. (2019) also highlighted the significant rising trend of WD activity and precipitation extremes over the WH through the use of daily filtered geopotential height anomalies at 200 hPa averaged over the WH region (Fig. 7.3; see also Fig. 11.8b). The reader is referred to Fig. 11.8b for the corresponding changes in precipitation over the WH region. Using a global variable-grid climate model simulations (with telescopic zooming over the South Asian region; see Fig. 7.4 for more details) for the period 1900–2005, they further attributed that wintertime regional changes over WH come from human activity.

There are some observational studies showing either no significant trends or decreasing trends in the frequency of WDs (Das et al. 2002; Shekhar et al. 2010; Kumar et al. 2015). Shekhar et al. (2010) suggested a decreasing amount of snowfall in boreal winter (using data for the period 1984–2008), but with no appreciable and consistent trend in the occurrence of WDs. Kumar et al. (2015) identified (based on the data for the period 1977–2007) a decreasing trend in total precipitation over Himachal Pradesh with significant (at 95% confidence level) reduction in the frequency of WDs. Note that these studies used a shorter period for detecting the trends as compared to studies by Madhura et al. (2015) and Krishnan et al. (2019) which also suggests that the results may be sensitive to the analysis period. In addition to the climate change associated changes in WDs, WD activity can also be modulated by large-scale modes of variability such

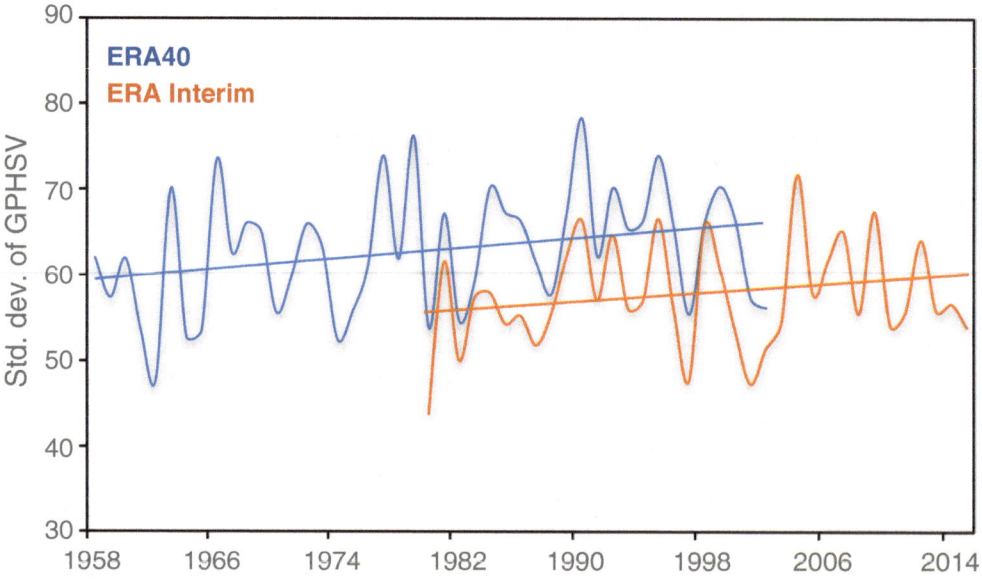

Fig. 7.3 Time-series of standard deviations of daily filtered (4–15 days band-pass) index (in gpm units) computed for every DJFMA season using 200 hPa geopotential anomalies averaged over the region 58° E–62° E and 32° N–36° N, from ERA-40 (1958–2002) and ERA-Interim (1979–2015) datasets. Adapted by permission from Krishnan et al. (2019)

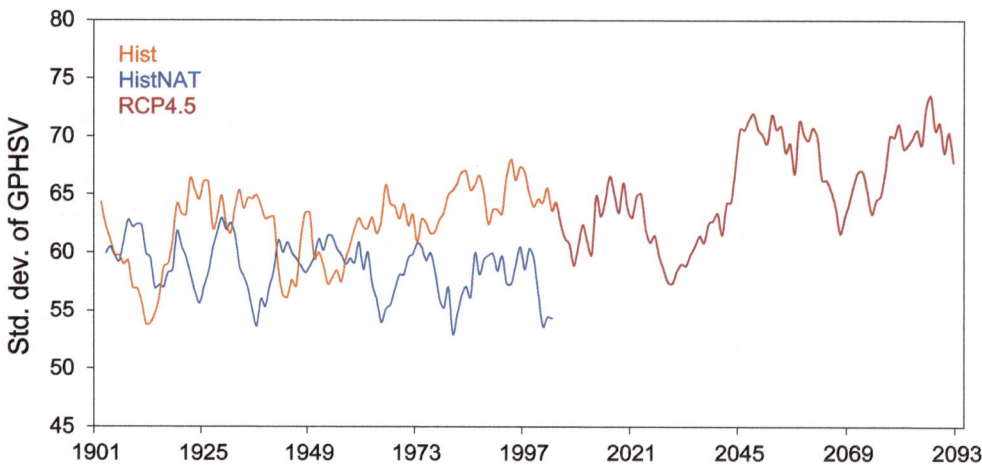

Fig. 7.4 Time-series of standard deviation of daily filtered (4–15 days band-pass) index (in gpm units) computed for every DJFMA season using upper-level geopotential anomalies at 200 hPa averaged over the region bounded by 58° E–62° E and 32° N–36° N, from the HIST (orange), HISTNAT (blue) and RCP4.5 (red) experiments. A 5-year moving average has been applied on the time-series. The first two experiments (HIST and HISTNAT) are for the twentieth-century period 1900–2005. The HIST experiment includes both natural and anthropogenic forcing, whereas the HISTNAT includes only natural forcing. The third experiment is performed in continuation with HIST into the twenty-first century period 2006–2095, following the Representative Concentration Pathway 4.5 (RCP4.5) scenario. Adapted by permission from Krishnan et al. (2019)

as Madden Julian Oscillation, ENSO, Arctic Oscillation, North Atlantic Oscillation and PDO (for more details, see Sect. 11.2).

There are only a few studies that examined the changes in WD activity (e.g. frequency) in a changing climate. For example, Ridley et al. (2013) have investigated the future projection of WD frequencies and the associated winter snowfall using two simulations of regional climate model, HadRM3 (i.e., HadRM3-H and HadRM3-E). HadRM3-H projected an increased occurrence of WDs and an increase in total winter snowfall by 2100, whereas HadRM3-E did not indicate any significant future change in snowfall or occurrences of WDs, suggesting that their RCM result was sensitive to the boundary forcing from GCM. Krishnan et al. (2019) showed a projected increase in the trend of WDs and precipitation extremes over WH using RCP4.5 scenario (i.e., a warming climate scenario; see Figs. 7.4 and 11.8d for more details) and attributed the rising trend to strong surface warming over eastern Tibetan Plateau compared to that over western side thus creating enhanced zonal gradient across Tibetan Plateau. That is, the warming trend tends to alter the background mean circulation to favor enhancements of the WDs and the orographic precipitation over the WH (see Fig. 11.8d for the precipitation changes). On the contrary, Hunt et al. (2019) found a significant 10% reduction in the annual frequency of WDs at the end of the century in future projections (i.e., in a warming climate scenario, RCP8.5) and attributed the changes in WD frequency to the projected widening and weakening of the STJ. Their study also suggests that as a consequence of this falling WD activity, the winter precipitation in north India will decrease at the end of the century.

7.3 Knowledge Gaps

Most of the studies on climate model projections focused mainly on stronger LPS (monsoon depressions) and not on monsoon lows, which remains a knowledge gap, thus requiring more studies in this direction. Further, there is no clear understanding of the various factors causing changes in LPS characteristics (i.e., both observed and projected changes), thus calling for more process-oriented studies in this direction. Secondly, the monsoon LPS associated rainfall contribution is not well represented in climate models, despite their prominent contribution to the monsoon seasonal rainfall. More studies to fully comprehend the genesis mechanisms (i.e., originating mechanism of LPS) of LPS may help in its improved representation in climate models. Finally, it is a challenge to detect the synoptic systems (i.e., LPS and WDs) from reanalysis and model simulations. Particularly, detecting monsoon LPS, having a weaker structure compared to the other tropical storms, poses a serious problem. Further, the number of synoptic systems detected in the reanalysis dataset, and climate model simulations are quite sensitive to the algorithm used for its detection and tracking. The discrepancies due to the tracking algorithm subsequently generate uncertainty in the climate model projections of synoptic systems. Furthermore, the model constraints to represent

the small-scale process associated with synoptic systems (e.g. coarse resolution in CMIP5 models lead to the poor representation of LPS) raises concerns on the reliability of its future projections. In view of this, continued efforts are required to find better modeling and identification strategies for the synoptic systems.

7.4 Summary

This chapter documents a review on the current understanding of observed variability and future changes in synoptic systems that form during boreal summer and winter–spring [i.e., monsoon low-pressure systems (LPS) and western disturbances] seasons. Generally, majority of LPS originates from the head BoB migrates northwestward into the Indian subcontinent producing heavy rainfall (30–50 cm day^{-1}), thereby significantly contributing to the total seasonal (June–September) rainfall in India. On interannual time scales, rainfall contribution from LPS (to the total seasonal monsoon rainfall) remains invariant during flood or drought years, even though the LPS frequency is slightly higher during flood years. The frequency of LPS also shows significant inter-decadal variations with more (less) monsoon depressions in the epochs of above (below) normal Indian summer monsoon rainfall. On the large-scale modes of variability influencing LPS, there is no consistency between the investigations to infer the associations of monsoon LPS with ENSO or IOD. However, studies show an out-of-phase relationship between the monsoon depressions and PDO. There is no detectable long-term trend in the observed frequency of total LPS during both periods of our analysis (1901–2015 and 1951–2015). The observed frequency of monsoon depressions shows a decreasing long-term trend (1901–2015), but the decline is more significant during the 1951–2015 period. In contrast, monsoon lows show a significant increasing trend during the 1901–2015 period. There are only few studies that are currently available to understand the potential impact of climate change on LPS. Climate model projections generally show a weakening of LPS activity and a poleward shift of the genesis locations of LPS at the end of the twenty-first century. The poleward shift in the LPS activity is expected to have profound societal impacts, as it may possibly dry up central India as well as increase the frequency of heavy rainfall events over northern India.

The WDs are the synoptic weather systems that propagate eastward from the Mediterranean region towards south Asia during boreal winter, impacting the northern parts of India and the WH region. There is a significant rise in the observed WD activity during the 1951–2015 period. Yet, there is no clear consensus from the climate model projections for potential changes in the occurrence of WDs under global warming.

References

Ajayamohan RS, Merryfield WJ, Kharin VV (2010) Increasing trend of synoptic activity and its relationship with extreme rain events over central India. J Clim 23:1004–1013

Bhutiyani MR, Kale VS, Pawar NJ (2007) Long-term trends in maximum, minimum and mean annual air temperatures across the northwestern Himalaya during the 20th century. Clim Change 85:159–177

Cannon F, Carvalho LMV, Jones C, Bookhagen B (2015) Multi-annual variations in winter westerly disturbance activity affecting the Himalaya. Clim Dyn 44:441–455

Cannon F, Carvalho LMV, Jones C, Norris J (2016) Winter westerly disturbance dynamics and precipitation in the western Himalaya and Karakoram: a wave-tracking approach. Theor Appl Clim 125:27–44

Carr FH (1977) Mid-tropospheric cyclones of the summer monsoon. Pure Appl Geophys (PAGEOPH) 115:1383–1412. https://doi.org/10.1007/BF00874415

Chattopadhyay J (1970) Power spectrum analysis of atmospheric ozone content over north India. Pure Appl Geophys 83(1):111–119

Choudhury AD, Krishnan R, Ramarao MVS, Ramesh KV, Manmeet S, Brian Mapes (2018) A phenomenological paradigm for mid-tropospheric cyclogenesis in the Indian Summer Monsoon. J Atmos Sci https://journals.ametsoc.org/doi/abs/10.1175/JAS-D-17-0356.1

Cohen NY, Boos WR (2014) Has the number of Indian summer monsoon depressions decreased over the last 30 years? Geophys Res Lett 41(22):7846–7853

Das MR, Mukhopadhyay RK, Dandekar MM, Kshirsagar SR (2002) Pre-monsoon western disturbances in relation to monsoon rainfall, its advancement over NW India and their trends. Curr Sci 82:1320–1321

Datta RK, Gupta MG (1967) Synoptic study of the formation and movement of western depressions. Indian J Meteorol Geophys 18(1):45–50

De US, Dube RK, Rao GP (2005) Extreme weather events over India in the last 100 years. J Indian Geophys Union 9(3):173–187

Dhar ON, Kulkarni AK, Sangam RB (1984) Some aspects of winter and monsoon rainfall distribution over the Garhwal-Kumaon Himalaya: a brief appraisal. Himal Res Dev 2:10–19

Dimri AP (2006) Surface and upper air fields during extreme winter precipitation over the western Himalayas. Pure appl Geophys 163:1679–1698

Dimri AP (2007) The transport of mass, heat and moisture over Western Himalayas during winter season. Theor Appl Climatol 90:49–63

Dimri AP (2008) Diagnostic studies of an active western disturbance over western Himalaya. Mausam 59(2):227–246

Dimri AP (2013a) Interannual variability of Indian winter monsoon over the western Himalaya. Glob Planet Chang 106:39–50

Dimri AP (2013b) Intraseasonal oscillation associated with the Indian winter monsoon. J Geophys Res 118:1–10

Dimri AP, Chevuturi A (2016) Western disturbances—an Indian meteorological perspective. Springer, Berlin, 146 pp

Dimri AP, Niyogi D, Barros AP, Ridley J, Mohanty UC, Yasunari T, Sikka DR (2015) Western disturbances: a review. Rev Geophys 53:225–246. https://doi.org/10.1002/2014RG000460

Ding Y, Sikka DR (2006) Synoptic systems and weather. In: Wang B (ed) The Asian monsoon, pp 132–201

Eliot J (1884) Account of southwest monsoon storms generated in the Bay of Bengal during 1877–1881. Mem Ind Met Dept 2:217–448

Godbole RV (1977) The composite structure of the monsoon depression. Tellus 29:25–40

Goswami BN, Venugopal V, Sengupta D, Madhusoodanan MS, Xavier PK (2006) Increasing trend of extreme rain events over India in a warming environment. Science 314:1442–1445

Gray WM (1968) Global view of the origin of tropical disturbances and storms. Mon Wea Rev 96:669–700

Hatwar HR, Yadav BP, Rama Rao YV (2005) Prediction of western disturbances and associated weather over Western Himalayas. Curr Sci 88:913–920

Hunt KM, Turner AG, Inness PM, Parker DE, Levine RC (2016a) On the structure and dynamics of Indian monsoon depressions. Mon Wea Rev 144:3391–3416

Hunt KMR, Turner AG, Parker DE (2016b) The spatiotemporal structure of precipitation in Indian monsoon depressions. Q J R Meteorol Soc 142:3195–3210

Hunt KM, Curio J, Turner AG, Schiemann R (2018a) Subtropical westerly jet influence on occurrence of western disturbances and Tibetan Plateau vortices. Geophys Res Lett 45:8629–8636. https://doi.org/10.1029/2018GL077734

Hunt KMR, Turner AG, Shaffrey LC (2018b) The evolution, seasonality, and impacts of western disturbances. Q J R Meteorol Soc 144(710):278–290. https://doi.org/10.1002/qj.3200

Hunt KMR, Turner AG, Shaffrey LC (2018c) Extreme daily rainfall in Pakistan and north India: scale-interactions, mechanisms, and precursors. Mon Wea Rev 146(4):1005–1022

Hunt KM, Turner AG, Shaffrey LC (2019) Falling trend of western disturbances in future climate simulations. J Clim 32:5037–5051

Hurley JV, Boos WR (2015) A global climatology of monsoon low-pressure systems Quart J Roy Meteor Soc 141(689):1049–1064

Jadhav SK, Munot AA (2009) Warming SST of bay of Bengal and decrease in formation of cyclonic disturbances over the Indian region during southwest monsoon season. Theor Appl Climatol 96:327–336. https://doi.org/10.1007/s00704-008-0043-3

Joseph PV, Simon A (2005) Weakening trend of the southwest monsoon current through peninsular India from 1950 to the present. Curr Sci 89:687–694

Kalsi SR (1980) On some aspects of interaction between middle latitude westerlies and monsoon circulation. Mausam 31(2):305–308

Kalsi SR, Halder SR (1992) Satellite observations of interaction between tropics and mid-latitudes. Mausam 43:59–64

Kotal SD, Roy SS, Roy Bhowmik SK (2014) Catastrophic heavy rainfall episode over Uttarakhand during 16–18 June 2013–observational aspects. Curr Sci 107:234–245

Kothawale DR, Rupa Kumar K (2005) On the recent changes in surface temperature trends over India. Geophys Res Lett 32:L18714. https://doi.org/10.1029/2005GL023528

Krishnamurthy V, Ajayamohan RS (2010) Composite structure of monsoon low pressure systems and its relation to Indian rainfall. J Clim 23(16):4285–4305

Krishnamurti TN, Hawkins RS (1970) Mid-tropospheric cyclones of the southwest monsoon. J Appl Meteorol 9:442–458

Krishnamurti TN, Kanamitsu M, Godbole R, Chang CB, Carr F, Chow JH (1975) Study of a monsoon depression (I): synoptic structure. J Meteorol Soc Jpn 53:227–240

Krishnan R, Ayantika D, Kumar V, Pokhrel S (2011) The long-lived monsoon depressions of 2006 and their linkage with the Indian

Ocean Dipole. Int J Climatol 31:1334–1352. https://doi.org/10.1002/joc.2156

Krishnan R, Sabin TP, Madhura RK, Vellore RK, Mujumdar M, Sanjay J, Nayak S, Rajeevan M (2019) Non-monsoonal precipitation response over the Western Himalayas to climate change. Clim Dyn 52:4091–4109. https://doi.org/10.1007/s00382-018-4357-2

Kumar N, Yadav BP, Gahlot S, Singh M (2015) Winter frequency of western disturbances and precipitation indices over Himachal Pradesh, India: 1977–2007. Atmosfera 28(1):63–70

Lang TJ, Barros AP (2004) Winter storms in the central Himalayas. J Meteorol Soc Jpn 82:829–844

Madhura RK, Krishnan R, Revadekar JV, Mujumdar M, Goswami BN (2015) Changes in western disturbances over the western Himalayas in a warming environment. Clim Dyn 44:1157–1168

Mak MK (1983) A moist baroclinic model for monsoonal mid-tropospheric cyclogenesis. J Atmos Sci 40:1154–1162

Mandke S, Bhide UV (2003) A study of decreasing storm frequency over Bay of Bengal. Ind J Meteorol Geophys 7:53–58

Miller FR, Keshavamurthy RN (1968) Structure of an Arabian Sea summer monsoon system. Meteorological monographs no. 1. East–West Center Press, 94 pp

Mohanty UC, Madan OP, Rao PLS, Raju PVS (1998) Meteorological fields associated with western disturbances in relation to glacier basins of western Himalayas during winter season. Center for Atmospheric Sciences, Indian Institute of Technology Technical report, New Delhi

Mohapatra M, Srivastava AK, Balachandran S, Geetha B (2017) Inter-annual variation and trends in Tropical Cyclones and Monsoon Depressions over the North Indian Ocean. In: Rajeevan M, Nayak S (Eds) Observed Climate Variability and Change over the Indian Region. Springer Geology. Springer, Singapore

Mooley DA (1957) The role of western disturbances in the prediction of weather over India during different seasons. Ind J Meteorol Geophys 8:253–260

Mooley DA (1973) Some aspects of Indian monsoon depression and associated rainfall. Mon Wea Rev 101:271–280

Mooley DA, Shukla J (1987) Characteristics of the westward-moving summer monsoon low pressure systems over the Indian region and their relationship with the monsoon rainfall. University of Maryland Center for Ocean-Land-Atmosphere Interactions Report, p 47

Mooley DA, Shukla J (1989) Main features of the westward-moving low pressure systems which form over the Indian region during the summer monsoon season and their relation to the monsoon rainfall. Mausam 40:137–152

Nikumbh AC, Chakraborty A, Bhat GS (2019) Recent spatial aggregation tendency of rainfall extremes over India. Sci Rep. https://doi.org/10.1038/s41598-019-46719-2

Pai DS, Sridhar L, Rajeevan M, Sreejith O, Satbhai N, Mukhopadhyay B (2014) Development of a new high spatial resolution (0.25 × 0.25) long period (1901–2010) daily gridded rainfall data set over India and its comparison with existing data sets over the region. Mausam 65:1–18

Pai DS, Sridhar L, Badwaik MR, Rajeevan M (2015) Analysis of the daily rainfall events over India using a new long period (1901–2010) high resolution (0.25° × 0.25°) gridded rainfall data set. Climate Dynamics 45 (3–4):755–776

Patwardhan S, Bhalme HN (2001) A study of cyclonic disturbances over India and the adjacent ocean. Int J Climatol 21:527–534

Patwardhan S, Kulkarni A, Kumar K (2012) Impact of global warming on cyclonic disturbances over south Asian region. J Earth Syst Sci 121(1):203–210

Pisharoty PR, Desai BN (1956) Western disturbances and Indian weather. Ind J Meteorol Geophys 8:333–338

Prajeesh AG, Ashok K, Rao DVB (2013) Falling monsoon depression frequency: a gray-sikka conditions perspective. Scientific Reports 3:2989. https://doi.org/10.1038/srep02989

Praveen V, Sandeep S, Ajayamohan RS (2015) On the relationship between mean monsoon precipitation and low pressure systems in climate model simulations. J Clim 28:5305–5324

Puranik DM, Karekar RN (2009) Western disturbances seen with AMSU-B and infrared sensors. J Earth Syst Sci 118(1):27–39

Raghavan K (1967) Influence of tropical storms on monsoon rainfall in India. Weather 22:250–256

Rajeevan M, De US, Prasad RK (2000) Decadal variation of sea surface temperatures, cloudiness and monsoon depressions in the north Indian Ocean. Curr Sci 79:283–285

Rajendra Kumar J, Dash S (2001) Interdecadal variations of characteristics of monsoon disturbances and their epochal relationships with rainfall and other tropical features. Int J Climatol 21:759–771

Raju PVS, Bhatla R, Mohanty UC (2011) A study on certain aspects of kinetic energy associated with western disturbances over northwest India. Atmósfera 24:375–384

Ramesh Kumar MR, Krishnan R, Syam S, Unnikrishnan AS, Pai DS (2009) Increasing trend of 'break-monsoon' conditions over India—role of ocean–atmosphere processes in the Indian Ocean. Geosci Rem Sens Lett IEEE 6(2):332–336

Rangachary N, Bandyopadhyay BK (1987) An analysis of the synoptic weather pattern associated with extensive avalanching in Western Himalaya. Int Assoc Hydrol Sci Publ 162:311–316

Rao VB, Rao ST (1971) A theoretical and synoptic study of western disturbances. Pure appl Geophys 90:193–208

Rao YP, Srinivasan V (1969) Forecasting manual, Part II Discussion of typical weather situation: winter western disturbances and their associated features. India Meteorological Department, FMU Report No. III-1

Rao B, Rao D, Rao VB (2004) Decreasing trend in the strength of tropical easterly jet during the Asian summer monsoon season and the temporal variation of tropical cyclonic systems over bay of Bengal. Geophys Res Lett 31. https://doi.org/10.1029/2004gl019817

Rastogi D, Ashfaq M, Leung LR, Ghosh S, Saha A, Hodges K, Evans K (2018) Characteristics of Bay of Bengal monsoon depressions in the 21st century. Geophys Res Lett 45(13):6637–6645

Revadekar JV, Varikoden H, Preethi B, Mujumdar M (2016) Precipitation extremes during Indian summer monsoon: role of cyclonic disturbances. Nat Hazards 81(3):1611–1625

Ridley J, Wiltshire A, Mathison C (2013) More frequent occurrence of westerly disturbances in Karakoram up to 2100. Sci Total Environ. https://doi.org/10.1016/j.scitotenv.2013.03.074

Roxy MK, Ghosh S, Pathak A, Athulya R, Mujumdar M, Murtugudde R, Terray P, Rajeevan M (2017) A threefold rise in widespread extreme rain events over central India. Nat Commun 8 (1):708

Roy SS, Roy Bhowmik SK (2005) Analysis of thermodynamics of the atmosphere over north west India during the passage of a western disturbance as revealed by model analysis field. Curr Sci 88:947–951

Saha K, Sanders F, Shukla J (1981) Westward propagating predecessors of monsoon depressions. Mon Wea Rev 109:330–343

Sandeep S, Ajayamohan RS, Boos WR, Sabin TP, Praveen V (2018) Decline and poleward shift in Indian summer monsoon synoptic activity in a warming climate. Proc Natl Acad Sci 115(11):2681–2686

Schiemann R, Lüthi D, Schär C (2009) Seasonality and interannual variability of the westerly jet in the Tibetan Plateau region. J Clim 22:2940–2957. https://doi.org/10.1175/2008JCLI2625.1

Shekhar MS, Chand H, Kumar S, Srinivasan K, Ganju A (2010) Climate-change studies in the western Himalaya. Ann Glaciol 51 (54):105–112

Sikka DR (1977) Some aspects of the life history, structure and movement of monsoon depressions. Pure appl Geophys 115:1501–1529

Sikka DR (1980) Some aspects of the large scale fluctuations of summer monsoon rainfall over India in relation to fluctuations in the planetary and regional scale circulation parameters. Proc Ind Acad Sci (Earth Planet Sci) 89:179–195

Sikka DR (2006) A study on the monsoon low pressure systems over the Indian region and their relationship with drought and excess monsoon seasonal rainfall. Center for Ocean-Land-Atmosphere Studies Technical Report 217, p 61

Singh MS, Kumar S (1977) Study of a western disturbance. Mausam 28:233–242

Taylor KE, Stouffer RJ, Meehl GA (2011) An overview of CMIP5 and the experiment design. Bull Am Meteorol Soc 93:485–498

Vishnu S, Francis PA, Shenoi SSC, Ramakrishna SSVS (2016) On the decreasing trend of the number of monsoon depressions in the Bay of Bengal. Environ Res Lett 11(1):014011

Vishnu S, Francis PA, Shenoi SSC, Ramakrishna SSVS (2018) On the relationship between the Pacific Decadal Oscillation and the Monsoon Depressions over the Bay of Bengal. Atmos Sci Lett. https://doi.org/10.1002/asl.825

Wallace J, Hobbs P (2006) Atmospheric Science: an introductory survey (2nd ed). Elsevier Academic Press, Amsterdam, p 504

Yoon JH, Chen TC (2005) Water vapor budget of the Indian monsoon depression. Tellus A 57:770–782

Extreme Storms

Coordinating Lead Authors

Ramesh K. Vellore, Indian Institute of Tropical Meteorology (IITM-MoES), Pune, India,
e-mail: rameshv@tropmet.res.in (corresponding author)
Nayana Deshpande, Indian Institute of Tropical Meteorology (IITM-MoES), Pune, India

Lead Authors

P. Priya, Indian Institute of Tropical Meteorology (IITM-MoES), Pune, India
Bhupendra B. Singh, Indian Institute of Tropical Meteorology (IITM-MoES), Pune, India
Jagat Bisht, Indian Institute of Tropical Meteorology (IITM-MoES), Pune, India; Japan Agency for Marine Earth Science Technology (JAMSTEC), Yokohama, Japan

Review Editor

Subimal Ghosh, Indian Institute of Technology Bombay, Mumbai, India

Corresponding Author

Ramesh K. Vellore, Indian Institute of Tropical Meteorology (IITM-MoES), Pune, India,
e-mail: rameshv@tropmet.res.in

© The Author(s) 2020
R. Krishnan et al. (eds.), *Assessment of Climate Change over the Indian Region*,
https://doi.org/10.1007/978-981-15-4327-2_8

Key Messages

- Long-term observations (1951–2018) indicate a significant reduction in annual frequency of tropical cyclones (TCs) in the North Indian Ocean (NIO) basin [−0.23 per decade over the entire NIO; −0.26 per decade over the Bay of Bengal]. A significant rise [+0.86 per decade] in the frequency of post-monsoon (October–December) season very severe cyclonic storms (VSCS) is observed in the NIO during the past two decades (2000–2018) (high confidence).

- Observations indicate that frequency of extremely severe cyclonic storms (ESCS) over the Arabian Sea has increased during the post-monsoon seasons of 1998–2018 (high confidence). There is medium confidence in attributing this observed increase to human-induced SST warming.

- Analyses from the observations show a decline in number of thunderstorm days (1981–2010 relative to 1950–1980) by 34% over the Indian region, while there is a rise in short-span high-intensity rain occurrences (mini-cloudbursts) along the west coast of India (5 per decade) and along the foothills of western Himalayas (1 per decade) during the 1969–2015 period (high confidence).

- Climate model simulations project a rise in TC intensity (medium confidence) and TC precipitation intensity (medium-to-high confidence) in the NIO basin.

8.1 Introduction

High-impact weather phenomena associated with cyclonic storms (synoptic-scale weather disturbances that last for a few days), thunderstorms (occur on less than a day time scale), and short-lived cloudbursts (a time scale of few hours) that can produce intense rainfall amounts are generally categorized as severe or extreme weather events in the Indian weather chronology. The extreme weather events over the Indian region have profound socio-economic implications (e.g., De et al. 2005). The North Indian Ocean (NIO) rim countries (India, Bangladesh, Myanmar, Sri Lanka, Oman; countries within the Equator region—30° N; 50–100° E) comprising large coastal areas are severely affected by tropical cyclones (TCs) every year (see Singh et al. 2016; Ramsay 2017; Mohapatra et al. 2014, 2017). For example, the year 2018 witnessed four very severe TCs over this region during the pre-monsoon (March–May) and post-monsoon (October–December) seasons of India [Mekunu and Luban over the Arabian Sea (AS) in May and October 2018, Titli and Gaja in October

and November 2018 over the Bay of Bengal (BOB); Source: Annual cyclone review report, India Meteorological Department (IMD); see also Table 8.2].

Convective storms such as thunderstorms in general are considered as hazard to aviation, and also cause severe loss to life, agriculture, and property. The eastern and north-eastern states of India (West Bengal, Bihar, Assam, Chhattisgarh, Jharkhand, and Orissa) and adjoining regions in Bangladesh experience violent thunderstorms known as "Nor'wester" during the pre-monsoon season (Mukhopadhyay et al. 2005; Ghosh et al. 2008; Tyagi et al. 2012). Rainfall occurrences of unprecedented intensity have also been witnessed in the recent decades during the Indian summer monsoon (ISM; June–September) season. For example, heavy downpour that caused calamitous floods and heavy casualties in Mumbai during July 2005 (Bohra et al. 2006), in Leh of the trans-Himalayan region during August 2010 (Thayyen et al. 2013; Rasmussen and Houze 2012), in the northern states of Uttarakhand and Jammu and Kashmir during June 2013 and September 2014 (Lotus 2015; Ranalkar et al. 2016; Vellore et al. 2016, 2019; Priya et al. 2017), and in the southern state of Kerala during August 2018 (Mishra and Shah 2018) are to name a few. Various synoptic-scale signatures have been recognized in connection with these extreme rain situations, viz localized convective instabilities, large-scale organized monsoon activity and anomalous extratropical circulation, and their interactions with the monsoon circulation across the Himalayas (e.g., Vellore et al. 2014, 2016; see also Krishnan et al. 2019). The aforesaid extreme storm phenomena are significant threat to lives, property, and agricultural yields and cause huge revenue losses for the Indian subcontinent every year. It is also noteworthy to mention that extreme rain situations over the Indian subcontinent exhibit significant variations on a regional scale (Guhathakurta et al. 2011). A list of extreme rain events occurred in the recent times that resulted in calamitous flood situations over the Indian region is documented in Table 6.4. Long-term observations show significant rising trends in the frequency and magnitude of extreme rain occurrences over the ISM core rain-fed regions of central India during the later half of the twentieth century (Goswami et al. 2006) in an unequivocally warming climate (IPCC 2007, 2014; see also Krishnan et al. 2016; Singh et al. 2019).

Nonetheless, a clear deciphering of anthropogenic climate change manifestations on extreme rain or storm occurrences broadly remains elusive. Greater challenges are with detection and attribution of trends in high-impact rain events due to representation constraints of finer scale physical processes in the state-of-the-art climate modeling systems, as opposed to challenges in understanding of changes in large-scale environment in a warming climate (Mukherjee et al. 2018;

Table 8.1 Classification of cyclonic disturbances observed in the NIO region and definition of cloudbursts referenced in this study

Classification of cyclonic disturbance	Maximum sustained surface wind speed (MSW)		Number of closed isobars (2 hPa interval)	Dvorak intensity (category)
	Knots	km h^{-1}		
Low-pressure area	<17	<31	1	1
Depression (D)	17–27	31–50	2	1.5
Deep depression (DD)	28–33	51–62	3	2
Cyclonic storm (CS)	34–47	63–88	4–7	2.5–3
Severe cyclonic storm (SCS)	48–63	89–117	8–10	3.5
Very severe cyclonic storm (VSCS)	64–89	118–165	11–25	4–5
Extremely severe cyclonic storm (ESCS)	90–119	166–221	26–39	5–6
Super cyclonic storm (SuCS)	≥ 120	≥ 222	40 or more	6.5
Classification of cloudbursts	Surface rainfall amount			
Cloudburst	>100 mm in an hour			
Mini-cloudburst	>50 mm in two consecutive hours			

Source IMD (2003), Dvorak (1984), Velden et al. (2006), Deshpande et al. (2018)
The categories CS, SCS, VSCS, ESCS, SuCS described in the table are generally referenced as TCs in this chapter

Knutson et al. 2019a, b). Furthermore, projections of synoptic-scale TCs, mesoscale convective storms or thunderstorms (e.g., Tyagi et al. 2012), and very short-lived cloudburst occurrences (Deshpande et al. 2018) into the future from modeling framework are an ambitiously demanding task due to stringent model constraints in reproducing their occurrences on localized scale settings. Notwithstanding these challenges, considerable progress has been generally realized in the understanding of changes in TC activity over the global ocean basins (Walsh et al. 2016; Knutson et al. 2019a, b). The understanding of changes in regional-scale extreme or convective storm phenomena over the Indian subcontinent is still rudimentary. In this chapter, we present an assessment of occurrences and changes in extreme storm categories suitable for the Indian subcontinent.

8.2 Synoptic-Scale Extreme Storms

TCs are synoptic-scale, warm-core, non-frontal low-pressure systems embedded in a weakly sheared environment over tropical warm waters characterized by organized convection within a closed cyclonic circulation (see Gray 1968; Anthes 1982; IMD 2003). The NIO basin typically contributes about 7–10% of world's TCs, and these TCs are further considered to be the most deadly ones in the world (e.g., Mohapatra et al. 2014; Sahoo and Bhaskaran 2016; Gupta et al. 2018). The India Meteorological Department (IMD) classifies NIO TC activity into different storm categories based on the

intensity of maximum sustained surface wind speeds (MSW), and a classification of storm categories is given in Table 8.1. The gusty winds from the categories severe cyclonic storm (SCS) and above have larger devastating potential to life and property.

The category 4 and above TCs observed in the NIO region from the year 2000 during the cyclone seasons are given in Table 8.2. Various studies in the past have documented the long-term trends in frequency and intensity of NIO TCs (e.g., Raghavendra 1973; Mooley 1980; Mooley and Mohile 1984; Singh and Khan 1999; Raghavan and Rajesh 2003; Singh 2007; Niyas et al. 2009; Sikka 2006; Mohanty et al. 2010, 2012; Mohapatra et al. 2012, 2014; Gupta et al. 2018). In continuation with the past literatures, an update of historical and projected changes in TC intensity and frequency is documented in the following.

8.2.1 Historical Changes

Based on long-term (1891–2018) TC datasets, there were 1433 synoptic-scale cyclonic disturbances observed in the NIO basin on annual scale (Source: Cyclone eAtlas, IMD; http://www.rmcchennaieatlas.tn.nic.in). Of which 55%, 24%, and 21% of these disturbances were characterized as D, CS, and SCS (see Table 8.1 for the classification), respectively. The TC contribution from BOB [AS] is about 80% [20%] in the NIO basin on annual scale. Also, a major percentage of cyclonic disturbances evolve to SCS and

Table 8.2 Extreme category (category 4 and above; see Table 8.1) pre- and post-monsoon TCs observed over the Indian seas during the period 2000–2018

Year	Category—basin—month—name of the storm
2000	VSCS—BOB—Nov—BB 05; VSCS—BOB—Dec—BB 06
2001	ESCS—AS—May—ARB 01
2003	VSCS—BOB—May—BB 01
2004	VSCS—BOB—May—BB 02
2006	VSCS—BOB—Apr—Mala
2007	VSCS—BOB—Nov—Sidr
2008	ESCS—BOB—Apr—Nargis
2010	ESCS—BOB—Oct—Giri
2011	VSCS—BOB—Dec—Thane
2013	ESCS—BOB—Oct—Phailin; VSCS—BOB—Nov—Lehar; VSCS—BOB—Dec—Madi
2014	ESCS—BOB—Oct—Hudhud; ESCS—AS—Oct—Nilofar
2015	ESCS—AS—Oct—Chapala; ESCS—AS—Nov—Megh
2016	VSCS—BOB—Dec—Vardah
2017	VSCS—BOB—Nov—Ockhi
2018	ESCS—AS—May—Mekunu; VSCS—AS—Oct—Luban; VSCS—BOB—Oct—Titli

Source IMD

Table 8.3 Number of CS and SCS observed in the NIO basin during pre-monsoon (March–May; MAM) and post-monsoon (October–December; OND) seasons for the period 1891–2018

Storm category	Ocean basin	Pre-monsoon (MAM)	Post-monsoon (OND)	Annual
CS	NIO	47 (14%)	149 (44%)	339
	BOB	37 (13%)	124 (43%)	286
	AS	10 (19%)	25 (47%)	53
SCS	NIO	82 (27%)	171 (56%)	307
	BOB	58 (25%)	141 (60%)	234
	AS	24 (33%)	30 (41%)	73

Source Cyclone eAtlas IMD
The numbers during pre- and post-monsoon seasons with respect to annual frequency are shown in percent

above categories during the pre-monsoon (March–May) and post-monsoon (October–December) seasons. Table 8.3 shows the number of synoptic-scale cyclonic storms observed in the NIO basin during the pre- and post-monsoon seasons. Note that, 56% [27%] of annually observed storms of severe category is seen in the NIO basin during the post-monsoon [pre-monsoon] season. On annual scale, 286 [53] CS and 234 [73] SCS were observed from the BOB [AS] during the 1891–2018 period.

Figure 8.1 shows the observed monthly frequency of CS and SCS for the period 1891–2018. The TC activity in the NIO basin exhibits a bimodal distribution with larger number of occurrences during the pre-monsoon and post-monsoon seasons (see also Haggag et al. 2010; Mohapatra et al. 2017). Notice that large number of CS formation is observed during the months of October and November in BOB and AS. A maximum of 72 [20] SCS events is observed during November [May] in BOB [AS] (see also Singh et al. 2016). The larger number of TCs in

BOB is possibly due to prevalence of upper-ocean warm stratification and higher sea-surface temperatures (SSTs) as compared to AS. The TC frequency in BOB constantly poses serious threat to eastern coastal Indian states Orissa, Andhra Pradesh, and Tamil Nadu. Mishra (2014) indicates that uneven surface temperature rise in the coastal regions during the recent decades also appears to have a bearing on increased cyclone vulnerability in these states. On average, about 5–6 TCs are generally observed in BOB and AS every year, of which 2–3 reach severe stages (Mohapatra et al. 2014). Based on long-period records, Tyagi et al. (2010) also indicated that more than 60% of BOB TCs endure landfall in various parts of the Indian east coast, 30% experience recurvature and landfall over Bangladesh and Myanmar, while 10% generally dissipate over the oceanic regions.

Large vertical wind shear during the ISM season generally prevents the formation of TCs. However, the ISM month of August during the 1891–2018 period witnessed a

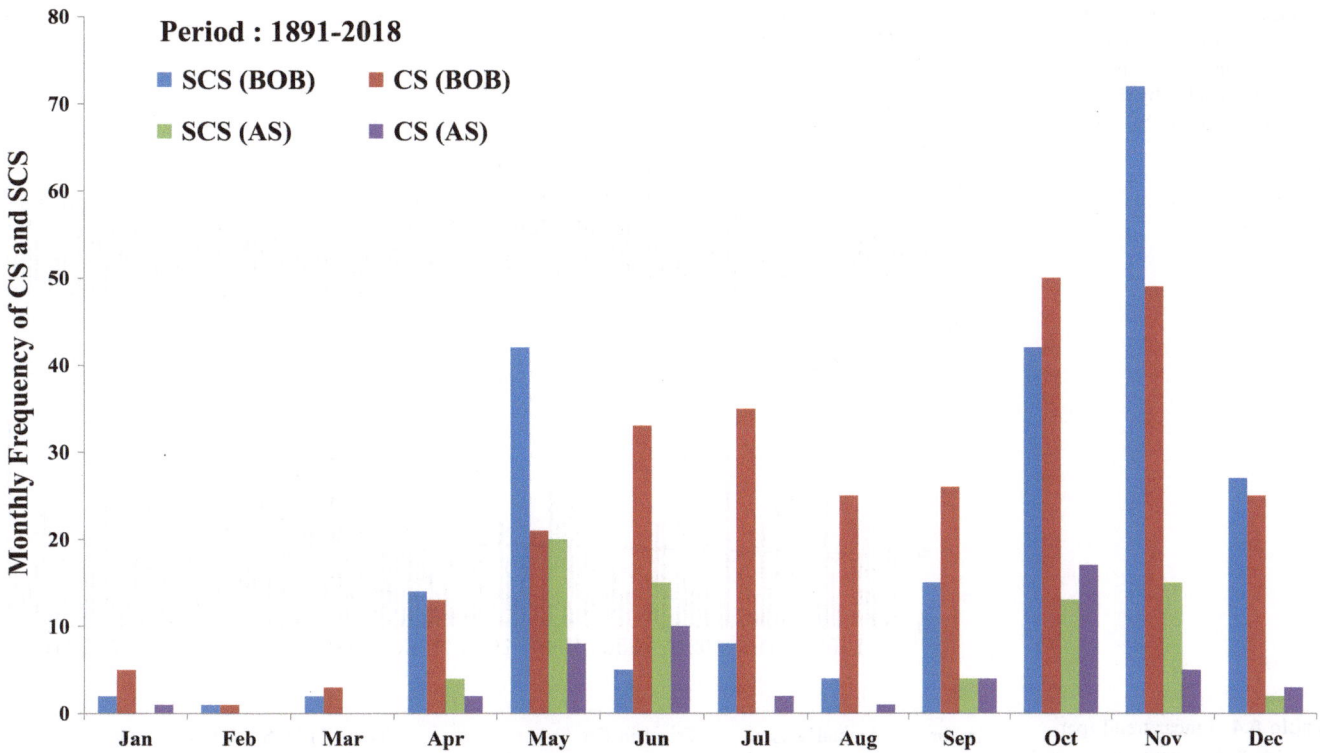

Fig. 8.1 Observed monthly frequency of CS and SCS in the NIO basin during the 1891–2018 period. *Source* Cyclone eAtlas IMD; http://www. rmcchennaicatlas.tn.nic.in. BOB = Bay of Bengal, AS = Arabian Sea

maximum number of cyclonic disturbances in the BOB region which evolved into depression (D) category, i.e., 74 in June, 106 in July, 152 in August, and 125 in September, while the month of June witnessed a maximum number of 28 depressions in the AS region (For more details, see also Chap. 7; see also Singh et al. 2016; see also Prajeesh et al. 2013; Hunt et al. 2016).

A comparison between pre-1950 and the post-1950 (commonly referred to as the warming era) periods indicates that there is a rise in number of SCS from 94 to 140 (i.e., 49% rise) in the BOB region, and from 29 to 44 (i.e., 52% rise) in the AS region on annual scale. On seasonal scale, there is a rise in number of SCS from 93 to 160 (72% rise) in the NIO basin—of which there is a marked increase of +105% [+21%] observed during the post-monsoon [pre-monsoon] season. A consistent rise in number of SCS in the BOB region is noted irrespective of pre- or post-monsoon cyclone seasons, i.e., +100% [+42%] rise in the number of SCS in the aforesaid comparison periods during the post-monsoon [pre-monsoon] seasons of 1891–2018. On the contrary, there is a 15% reduction in the number of SCS in the AS region during the pre-monsoon season, while there is a 130% rise during the post-monsoon season (see also Mohanty et al. 2012; Singh et al. 2016 who also considered long-period TC records in their

assessments). The mean life period of SCS is estimated to be about 4 days, 6 days, and more for the extreme TC categories during the cyclone seasons. In particular, severe storm categories (VSCS and SuCS; see Table 8.1) show longer life period over AS [BOB] during pre-monsoon [post-monsoon] seasons (Kumar et al. 2017).

Figure 8.2 and Table 8.4 show the frequency distribution of TCs observed in the NIO region and trends in the frequency of CS and SCS, respectively, for the period 1891–2018. One can clearly see inter-annual and inter-decadal variability in the frequency distribution (Fig. 8.4). There is a significant decline in the annual frequency of NIO TCs, i.e., −0.18 per decade (1891–2018), and −0.23 per decade (1951–2018) (see also Singh et al. 2000, 2001; Singh 2007; Mohapatra et al. 2014, 2017). While there is a significant upward trend (+0.07 per decade) in SCS observed in the NIO region during the post-monsoon seasons of 1891–2018, there is a significant decline in the annual frequency of CS and SCS in the BOB region during the 1951–2018 period with trend values of −0.26 per decade and −0.15 per decade for CS and SCS, respectively. There is also an upward trend of CS and SCS observed in the AS region both on annual and seasonal scales during the 1951–2018 period, with larger trend values (+0.02 per decade; <90% confidence level) during the post-monsoon season. Further, these long-term

Fig. 8.2 Observed annual frequency of **a** CS and above **b** SCS in the NIO. Linear trend lines are indicated by dashed lines —black (1891–2018), blue (1951–2018), red (2000–2018). Also, 10-year running mean is shown by a solid-green line. *Source* IMD

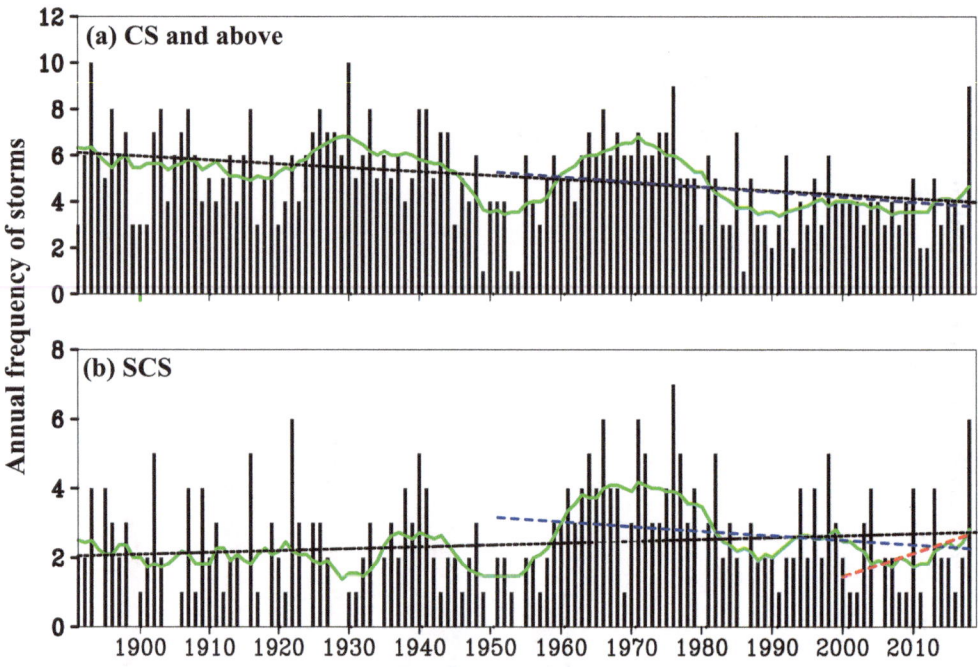

Table 8.4 Linear trend (per decade) in frequency of CS and SCS observed in the NIO region based on the 1951–2018 and 1891–2018 periods

Basin	Annual/season	Trend in CS		Trend in SCS (VSCS)	
		1951–2018	1891–2018	1951–2018 (2000–2018)	1891–2018
NIO	Annual	**−0.231**	**−0.180**	−0.140 (+0.590)	+0.050
	MAM	−0.034	−0.002	−0.041 (−0.230)	−0.004
	OND	−0.065	+0.011	−0.072 **(+0.860)**	**+0.070**
BOB	Annual	**−0.264**	**−0.190**	**−0.154**	+0.036
	MAM	−0.035	−0.001	−0.040	+0.005
	OND	−0.100	+0.002	−0.085	**+0.060**
AS	Annual	+0.045	+0.020	+0.021	+0.020
	MAM	+0.002	−0.000	−0.001	−0.010
	OND	+0.023	−0.001	+0.013	+0.020

Source Cyclone Atlas; IMD. Trends in VSCS occurrences (for the period 2000–2018) are shown in brackets Confidence level above 90% is only shown in bold here

datasets also show a rising trend in TC frequency in the BOB region particularly during the months of November and May which is consistent with the earlier studies of Singh et al. (2000, 2007).

Further, based on IMD long-period datasets, Singh et al. (2000) identified a rise in intensification rates in the transformation of tropical disturbances into TCs and also a significant rise in SCS and VSCS in the NIO basin (see also Mohanty et al. 2012). Figure 8.3 shows the probabilities of TC intensification observed in the NIO basin, where one can notice that intensification probability from tropical depression D to CS [CS to SCS] is about 45% [48%] on annual scale. Also, there is a 21% [32%] probability of intensification from depression to extreme storm categories in the BOB [AS] region on annual scale. On seasonal scale, there

are higher probabilities of transformation observed during pre- and post-monsoon seasons, i.e., probability of transformation from D to CS is 72% and from CS to SCS is 64% during the pre-monsoon season, and 58% for D to CS and 53% for CS to SCS during the post-monsoon season. There is only a 27% probability for the transformations from D to CS and from CS to SCS during the ISM season. Although the number of storms in the AS region is relatively smaller than observed in the BOB region, the probabilities of transformation from [D to CS] CS to SCS are [33%] 27% larger in AS compared to BOB. While the translational speeds are larger for VSCS and SuCS categories (Chinchole and Mohapatra 2017; see also Kossin 2018), the intensification rates in the NIO basin were found to be rather pronounced in November during the post-monsoon season and

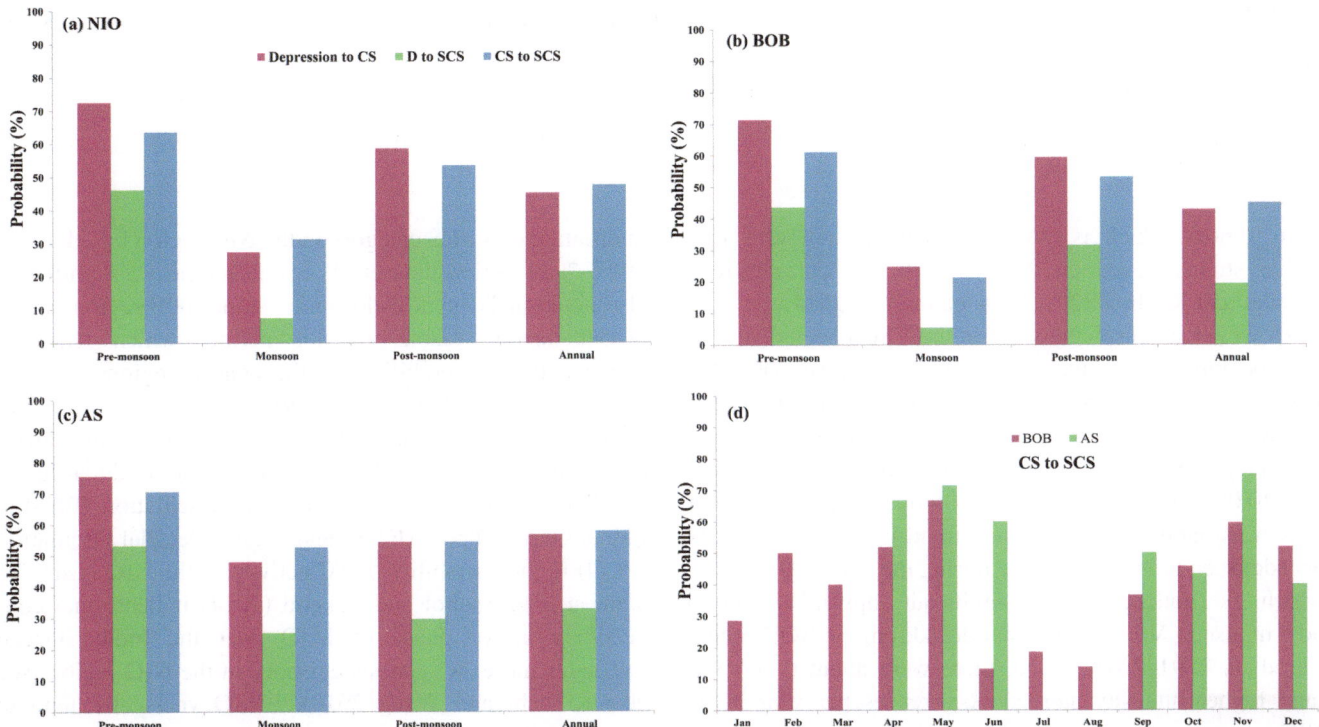

Fig. 8.3 Probability of TC intensification (expressed in %) by seasons for the **a** NIO, **b** BOB, **c** AS, **d** by months in the BOB and AS region diagnosed for the period 1891–2018. *Source* Cyclone eAtlas IMD

Fig. 8.4 **a** Annual and **b**, **c** seasonal VSCS occurrences in the NIO assessed for the period 1965–2018. Linear trend lines (black: 1965–2018, red: 2000–2018) are indicated by dashed lines. *Source* Mohapatra et al. (2014; IMD)

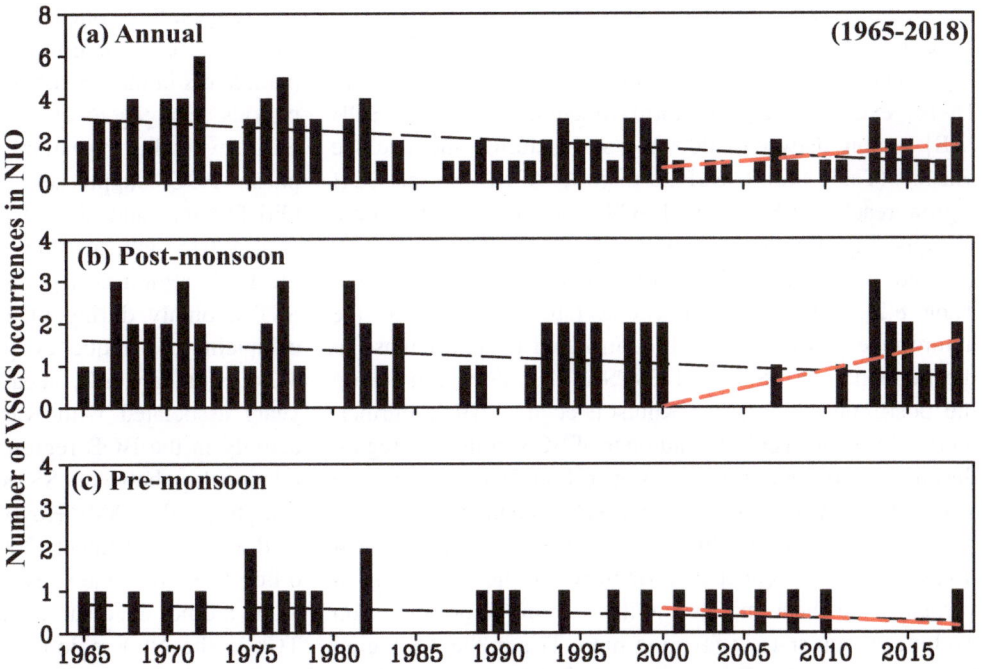

in May during pre-monsoon season (Fig. 8.3d). A few studies also indicate that the trends in TC intensity over the Atlantic Basin are more significant as compared to NIO which has been ascribed to inadequate observational records from the NIO region (Elsner et al. 2008; see also Walsh et al. 2016).

Figure 8.4 shows the observed number of VSCS occurrences from the NIO basin both on annual and seasonal scales based on the 1965–2018 period (see also Fig. 3 of Mohapatra et al. 2014). There were a total of 105 VSCS occurrences, of which 60% of them occurred during the post-monsoon season. There is a significant decline in the number of VSCS

Table 8.5 Number of VSCS
observed during the pre-monsoon
and post-monsoon months in the
BOB and AS regions for the
period 1981–2018

Basin	Pre-monsoon season	Post-monsoon season
AS	6	5
BOB	9	28

Source Balaji et al. (2018); Annual cyclone review reports, IMD

from 1960s—which is consistent with the analyses from earlier studies based on the long-term TC datasets (e.g., Mandal and Krishna 2009; Mohapatra et al. 2014).

However, a conspicuous rise in the VSCS is evident from the beginning of the twenty-first century (Fig. 8.4; Table 8.4). A linear trend analysis shows that the frequency of VSCS category storms significantly decreased at the rate of 0.41 per decade on annual scale, and on seasonal scale 0.17 [0.08] per decade during post-monsoon [pre-monsoon] season. Notably, there is a rising trend in VSCS (+0.59 per decade) seen in the NIO region during the 2000–2018 period which is dominated by significant upward trend in post-monsoon VSCS (+0.86 per decade) in the BOB region. Based on 1891–2018 period, there were about two VSCS (and higher intensity storms) occurrences per year in the NIO region on average—with predominance during the post-monsoon season.

Table 8.5 shows the number of VSCS occurrences observed in the BOB and AS regions during the pre-monsoon and post-monsoon seasons of 1981–2018. Note that on 40% of the occasions, the SCS in BOB transforms into VSCS during these cyclone seasons. There were 15 [6] extreme category storms documented in the BOB [AS] region between 2000 and 2018 during the cyclone seasons (Table 8.2). Clearly, a majority of SCS in the BOB region reaches VSCS and ESCS categories during these seasons. In particular, 6 out of 11 SCS in the AS region reached VSCS and ESCS (category 4 and above; see Table 8.2). There is also a detectable signal from the anthropogenic forcing which contributed to the increase in the frequency of VSCS and ESCS in the AS region during the post-monsoon season (Knutson et al. 2019a). Further, there is fivefold rise in the number of SCS in the AS region during the pre-monsoon season (2000–2018 relative to 1981–1999) while there is no notable change during the post-monsoon season. This is also consistent with the recent investigations of epochal variability in the reduction in vertical wind shear and also an increase in the pre-monsoon cyclone season span in favoring more TCs in the AS region (Wang et al. 2012; Rajeevan et al. 2013; Deo and Ganer 2014). In a recent study, Balaji et al. (2018) also indicated that there is a 34% rise in the frequency and also a 37% rise

in duration of VSCS category in the NIO region based on the 1997–2014 period dataset. A significant eastward shift was also noted in TC genesis locations in the BOB region during post-monsoon seasons of this period—which tends to enhance the vulnerability for the coastal regions of Bangladesh (see also Rao et al. 2019).

Various studies indicate the possible role of atmospheric and oceanic variability elements such as the Madden–Julian Oscillation (MJO), El Nino Southern Oscillation (ENSO), Indian Ocean Dipole (IOD), and Pacific Decadal Oscillation (PDO) in the variability of TC activity in the NIO basin. To name a few, Tsuboi and Takemi (2014) indicate that convectively active phase of MJO over the Indian Ocean facilitates more TC genesis episodes in the NIO region (see also Kikuchi and Wang 2010). ENSO years also tend to favor smaller number of post-monsoon cyclonic storms in the NIO region (Singh and Rout 1999). Sumesh and Kumar (2013) indicate that TC activity tends to be more [less] over the AS [BOB] region during the El Nino-Modoki instances, and suggest that concurrent occurrences of positive IOD and El Nino can significantly modulate the cyclogenesis parameters in the AS region as compared to El Nino-Modoki periods. Haggag et al. (2010) indicate that cold phase of PDO favors TC formation in the NIO region while the warm phase PDO suppresses the TC formation. In contrast, Girishkumar and Ravichandran (2012) and Girishkumar et al. (2015) indicate that accumulated cyclone energy from the BOB region is negatively correlated with the Niño 3.4 SST anomaly during October–December months, thereby enhancing the frequency, genesis, location, and intensity of TCs. In other words, negative IOD events and the La Niña years associated with warm phase PDO favor more TC activity in the BOB region. A few studies proposed a close relationship between SSTs and frequency of intense TCs (Singh et al. 2000; Hoyos et al. 2006). However, some studies (e.g., Pattanaik 2005; Sebastian and Behera 2015) differ from this view by suggesting that changes in SST alone are not adequate enough to establish the variability of TCs in the NIO region on different time scales, and these studies further emphasized on better understanding of large-scale atmospheric circulation and their links to TC variability in the NIO basin.

8.2.2 Climate Change Implications and Projected Changes

One of the serious concerns of climate change is the impacts of SST warming on frequency, intensity, and duration of TCs which remain elusive particularly for the tropical oceans (e.g., IPCC 2007, 2014; Elsner and Kocher 2000; Pielke 2005; Emanuel 2005; Anthes et al. 2006; Elsner et al. 2008; Xie et al. 2010; Knutson et al. 1998, 2010a, b; Ramesh Kumar and Sankar 2010). The NIO region witnessed a rapid rise in SSTs from 1950s by about 0.6 °C as compared to other tropical ocean basins (see Mohanty et al. 2012). Some of the earlier studies implied that changing SST trends and rising intense TCs in the NIO region may have consequences from anthropogenic climate change (e.g., Knutson et al. 2006; Elsner et al. 2008; Knutson et al. 2019b). Knutson et al. (2010b) also suggested that climate change signal to changes in SST and associated TC activity might emerge sooner in the Indian Ocean as compared to other ocean basins. Another potential concern in the NIO region is that TC intensities particularly in the AS region exhibit an unprecedented rise in the recent years (see Table 8.2). High-resolution global climate model experiments indicate that anthropogenic global warming has increased the probability of extremely severe cyclonic systems during the post-monsoon season in the AS region (Murakami et al. 2017). Although the investigations concerned with TC changes in the AS region are limited at this time, some recent studies suggest that increasing anthropogenic emissions of black carbon and sulfate can play a role in reducing the vertical wind shear so as to favor more intense TC activity in the AS region (Evan et al. 2011, 2012; Wang et al. 2012).

Many climate modeling investigations generally suggest that increasing or doubling of carbon dioxide (CO_2) may enhance the frequency of severe/most intense cyclonic storms (e.g., Knutson et al. 2001; Knutson and Tuleya 2004; Webster et al. 2005; Oouchi et al. 2006; Bengtsson et al. 2007; Klotzbach and Landsea 2015). Some past studies such as Danard and Murthy (1989) and Yu and Wang (2009) pointed out that there could be an increase in TC frequency and TC intensity over the NIO basin in a doubled CO_2 world. Here, it is noteworthy to mention that there is a reasonable degree of sophistication in the current generation climate models that are capable of reproducing not only the salient features of TCs but also the associated dynamical–physical processes behind their development (e.g., Vidale et al. 2010). One can readily expect that warmer and wetter climate may favor more TCs with higher intensities from the climate change experiments using climate models. Despite that there is a consistency with the observed globally declining frequency of TCs in response to global warming

from these experiments (Sugi et al. 2002; McDonald et al. 2005; Yoshimura et al. 2006; Knutson et al. 2010a, 2019b), the inferences were rather dubious for basin-wide TC changes and more for the NIO region in particular (see also Zhao and Held 2012). Long-period trend assessment (International Panel on Climate Change Fifth Assessment Report—IPCC AR5; IPCC 2014) in surface air temperatures from climate model simulations, participated in CMIP5, for the historical periods indicates that there is a detectable warming signal over the Indian Ocean from the beginning of twentieth century. But, the present-day climate model assessments for the NIO region apparently present larger ambiguity due to larger bias in the simulations concerned with NIO TC activity—while there is a realistic agreement for TC frequency and intensity changes against observations for the Atlantic and Pacific Ocean basins (Zhao et al. 2009; Bender et al. 2010; Knutson et al. 2014). Further, the reduction in TC frequency over most part of the Indian Ocean appears to come from equal contributions of rising CO_2 emissions and anomalous SST patterns—notably, the SST effect and reduction in vertical shear have more precedence to the rising numbers of TCs in the AS region (Sugi et al. 2014).

Through the use of reanalyzed archives, Ramesh Kumar and Sankar (2010) indicate that the declining frequency of TCs in the NIO region in the historical period has no clear bearing global warming signal in association with the rising SSTs, but the warming signal may have greater impact on observed changes in atmospheric parameters such as decreasing mid-tropospheric relative humidity, low-level vorticity, and vertical wind shear during cyclone seasons of the NIO region (see also Pattanaik 2005; Sebastian and Behera 2015). In corroboration with these investigations, Balaguru et al. (2014) further indicate that intensity of major TCs in the BOB region during the post-monsoon season tends to have a coupled response from increasing upper-ocean content consistent with the rising SSTs and enhanced convective instability in the atmosphere. Murakami et al. (2017) performed a suite of high-resolution coupled model experiments and also showed that increasing anthropogenic-induced warming has potentially increased the probability of extreme category storms in the AS region, while the role of natural variability is rather minimal for the unprecedented rise in TC activity in the AS region.

As compared to other ocean basins, future changes in TC activity in the NIO region have received less attention in particular. As for the projected changes in TC activity in NIO region based on the investigations from climate change experiments, there is a large inconsistency noted in the projections from various state-of-the-art climate models. Knutson et al. (2010a, b) documented that in comparison with present-day changes, there is a large variation between

—52% (Oouchi et al. 2006) and +79% (Sugi et al. 2009) in the future changes in TC frequency (see Table S1 of Knutson et al. 2010a, b), and between −13 and +17% (Oouchi et al. 2006) for the changes in TC intensity. The ambiguities largely appear to arise from factors in the model architecture and representation of physical processes such as model resolution, choice of convection parameterization, as well as model-simulated SST changes. In addition, detecting TCs in climate models is also apparently a form of modeling uncertainty (see also Murakami et al. 2012a, b). There is a low confidence in the climate projections of rising annual frequency of VSCS (category 4 and above; see Table 8.1) at the end of twenty-first century in the NIO region (2081–2100 relative to 1981–2005). However, in a 2° anthropogenic warming scenario, there is a medium-to-high confidence in the projected rise in TC precipitation rates and intensities for the NIO region (Knutson et al. 2019b).

Murakami et al. (2014) also indicated that future changes in TC activity in the NIO region have larger dependence on future changes in thermodynamic factors (e.g., low-level relative humidity, SST anomalies) than future changes in dynamic factors (e.g., vertical wind shear, low-level vorticity, mid-level vertical velocities). High-resolution climate simulations show a scenario of increasing TC frequency by about 46% in the AS region, while there is a reduction by 31% in the BOB region, and the reduction [rise] is notably during pre-monsoon [post-monsoon] cyclone season (Murakami et al. 2013). The TC frequency is projected to decline in the AS and BOB regions during pre-monsoon season where the frequency change is influenced by aforesaid dynamical factors, while an east–west contrast (i.e., a reduction in BOB and rise in AS) is noted during the post-monsoon season where the future changes in aforesaid thermodynamic parameters may be of greater importance. There is also an attribution of southerly surge enhancements over the western AS and southern BOB regions during the post-monsoon cyclone season due to active inter-tropical convergence zone situations to the rising frequency of SCS in these regions (Mohanty et al. 2012). The reader is also referred to Walsh et al. (2016) and Knutson et al. (2019a, b) for a comprehensive documentation of climate change influences on TCs from various oceanic basins.

The IPCC—AR5 report published in 2014 (IPCC 2014) enunciates that "the confidence on attribution of changes in TC activity to human influences still is low owing to inadequate observational evidence and physical understanding of the anthropogenic drivers of climate and TC activity, i.e., there is a low confidence in basin-scale projections of changes in TC intensity and frequency of all basins." Various recent studies indicate that the projected changes in global frequency of TCs will decrease while severe TCs (categories 4 and above; see Table 8.1) in anthropogenic warming world exhibit a general rise over the global ocean

basins, but the confidence from the information available till date is still debatable owing to the limited literature (see also Table 1 of Knutson et al. 2019a). Hitherto, scientific investigations only led to various contentions in reference to anthropogenic climate change and its connections to changes in TC activity pertinent to NIO region. However, Knutson et al. (2019a) describe that contribution of anthropogenic forcing signal to rising severe category TS in the AS region is rather not just coincidental, but there is a medium confidence based on available investigations till date. The latest assessment by Knutson et al. (2019b) provides a medium confidence to projected TC intensity rise and medium-to-high confidence to TC precipitation intensity in the NIO basin, while the confidence is yet low for the projected rise in the annual frequency of VSCS during the twenty-first century.

8.3 Localized Severe Storms

Localized severe weather outbreaks occurring on meso-γ scale (on the spatial scale of 2–20 km, a few hours temporally; Orlankski 1975) in association with high winds, hail, thunder, lightning, etc., are generally categorized into severe convective storms. Some examples include thunderstorms, hail storms, dust storms, etc. Various studies (Manohar and Kesarkar 2005; Kandalgaonkar et al. 2005; Tyagi 2007; Kulkarni et al. 2009; Singh et al. 2011) have investigated the severe storm activity for the Indian region. Cloudbursts are a special class which falls under the short-lived intensely precipitating convective storms (see Table 8.1). These are predominant over the mountainous regions of northern part of India (Das et al. 2006; Dimri et al. 2017; Deshpande et al. 2018).

Thunderstorms are localized convective storms, which are more frequent at low latitudes, where there is a greater expediency of convective overturning occurrences in association with heated low-level atmospheric layers coming in contact with warm ground or water. Synoptic and meteorological conditions generally favorable for the occurrences of thunderstorm include conditional and convective instability in the atmosphere, ample supply of moisture at low levels, strong wind shear, and a dynamical mechanism (lifting or destabilization by advective processes) to release the instability present in the atmosphere (e.g., Doswell 2001; Bhardwaj and Singh 2018). Thunderstorms occur all through the year in different parts of India; however, their frequency and intensity are found to be maximum from March to May owing to prevalence of unstable atmospheric conditions and high temperatures at lower levels (Tyagi 2007; Singh et al. 2011; Saha et al. 2014; Das 2015b). The thunderstorm activity generally remains low during the mid-ISM months (July and August) throughout the country, whereas it is pronounced during the onset and withdrawal phase months (June and September). The eastern and northeastern states of India (West Bengal, Bihar, Assam, Jharkhand, and Orissa) and Bangladesh experience a special type of violent thunderstorms known as "Nor'westers (also known as Kal-Baisakhi by the locals; Desai 1950)" during the pre-monsoon months (March–May). These storms are triggered by the intense convection activity over the Chhotanagpur Plateau (Jharkhand region) and are generally associated with sudden violent gusty winds from northwest, known as squalls, with wind speeds exceeding 28 m s^{-1}. These are usually accompanied with heavy rainfall (about 100 mm h^{-1}) and oftentimes hail of sizes 30 mm in diameter, causing flash floods, severe property and agricultural damages (e.g., Chakrabarty et al. 2007; Pradhan et al. 2012).

Thunderstorm is always characterized by cumulonimbus cloud type, in which individual cells produce copious rains, and sometimes hail, strong winds, and even occasional tornadoes in northeast India and Bangladesh from Nor'westers, e.g., a F3 tornado (on the Fujita–Pearson scale) over Rajkanika block of Kendrapara district of Orissa, India, on 31 March 2009 (Litta et al. 2012). Over the southern and northeastern parts of India, the pre-monsoon thundershowers are called by various names in conjunction with rain favorable to local plantations, viz Tea showers in Assam, Mango showers in Kerala, Konkan, and Goa, and Coffee showers in Kerala. Thunderstorms in India generally fall under two categories: strong rains and maximum wind speeds ranging from 8 to 20 m s^{-1} with loud blasts of thunder and frequent lightning flashes in the category of moderate thunderstorms, while continuous thunder and lightning with strong rains and wind speeds >20 m s^{-1} in the category of severe thunderstorms. The lightning activity from thunderstorms is also significantly noticeable despite lower values of convective available potential energy, over the northern, central India and Bangladesh as compared to other parts of Indian subcontinent (Murugavel et al. 2014). Based on the genesis and life cycle of thunderstorms, they are grouped into different types, viz single-cell thunderstorms (a single cloud cell moving independently without combining with any other cell with a life cycle of 30 min to 1 h, moves slowly with mean environmental wind, and generally the vertical wind shear is small), multiple-cell thunderstorms (consists of a series of evolving cells with a life cycle of 3–6 h, possible hail storms and damaging winds and isolated tornadoes and travel over a few hundreds of kilometers), and squall lines (a chain of thunderstorms connected together and moving like a single entity usually accompanied by high-speed winds). Regardless of the type of thunderstorms, they go through three stages: the cumulus stage, the mature stage, and the dissipation stage.

In addition, two types of high-impact dust storms known as "Andhi" co-occur with thunderstorms of northwest India during the later part of the pre-monsoon season following the development of a heat low over this region. These dust storms are essentially thunderstorms in the arid regions of northwest India, where the local instability mechanisms are similar to thunderstorms and occur in association with large cumulus or cumulonimbus clouds (Raipal and Deka 1980; Middleton 1986). The first type is pressure-gradient dust storm which is characterized by higher temperatures and relatively low moisture content in the lowest atmospheric layers which makes the thunderstorms to have much higher cloud bases above the ground (3–4 km). On occasions, a strong pressure gradient develops to the south of the heat low causing strong winds lifting the dust and sand that are amply available in this region. Another type of dust storm is convective dust storm which develops in association with strong negatively buoyant downdraft of cumulonimbus cloud. The pressure-gradient dust storms generally last for longer periods with suspended dust in air, whereas convective dust storms last for a few minutes to few hours. The annual mean frequencies of Andhi are higher over the Indian states of Rajasthan, Punjab, Haryana, and Delhi. India is also among the countries in the world having large frequency of hail storms, associated with severe thunderstorms in association with multiple updrafts and downdrafts. These are more commonly seen along the Himalayan foothills causing severe large-scale agricultural damages (De et al. 2005). The frequency of hailstorms is small in winter; however, it increases from February and peaks during the pre-monsoon months (March and April). Hailstorms are rare during the ISM season. Unprecedented hailstorms lasting for a week or more are observed on occasions over north peninsular India (Kulkarni et al. 2015).

Generally, cloudburst is a severe weather outbreak feature commonly seen along the southern Himalayan slopes of the Indian subcontinent during the ISM months (Dimri et al. 2017; Deshpande et al. 2018). These are associated with thunderstorms where strong updrafts from intense vortices on smaller scale tend to hold up a large amount of water and upon sudden cessation results in catastrophic rainfall in a short period of time (greater than 10 cm h^{-1}) concentrated over a limited geographical area (Bhan et al. 2004; Dimri et al. 2017; see references therein). Romatschke and Houze (2011) indicated that precipitation is predominantly associated with smaller spatial scale but highly convective systems along the western Himalayan region and bulk of the cloudburst episodes are observed between 1000 m and 2500 m elevations of the Greater Himalayas. Das et al. (2006) quote "Cloudbursts in India occur when monsoon clouds associated with low-pressure area travel northward from the Bay of Bengal across the Ganges Plains onto the Himalayas and 'burst' in heavy downpours." For example, there were seven cloudburst instances witnessed over the western Himalayan state of Uttarakhand during the ISM season of 2018 (https://sandrp.in/2018/07/21/uttrakhand-cloudburst-incidents-2018/).

A comprehensive documentation of 30 instances of cloudbursts and associated damages in the Himalayas from 1970 is shown in Dimri et al. (2017). Several researchers also highlight the role of large-scale forcing and anomalous atmospheric circulation conditions leading to flood situations in and along the western Himalayas (e.g., Houze et al. 2011; Vellore et al. 2016) with implications to cloudburst instances though not directly. The modeling community observes large limitations to capture the cloudburst events due to outstanding issues such as understanding of microscale interactions with rugged and variable orography, paucity of observations, and representation of physical processes at a much-localized scale setting. The essentiality of more modeling studies with multiple events and observations for better understanding these severe convective events over the Himalayan region is also clearly suggested. Indian scientists have additionally coined another new term "mini-cloudbursts (more than 5 cm in two consecutive hours)" (Deshpande et al. 2018) to define incidences of heavy rains over short period of time. Instances include heavy rainfall in Mumbai during the ISM season, e.g., 944 mm rainfall on July 27, 2005, and 304 mm rainfall on September 20, 2017, that caused massive urban flood. The new definition distinguishes the cloudbursts associated with high topography (e.g., Leh event in August 2010; Uttarakhand event in June 2013) and rainfall incidences over the plain region exceeding 10 cm h^{-1} that can result in urban flooding. The aforesaid Mumbai events fall under mini-cloudburst classification.

8.3.1 Historical Changes

The thunderstorm data available for the Indian region has been comprehensively documented in the recent studies of Tyagi (2007), Bharadwaj et al. (2017), Bharadwaj and Singh (2018). Based on the compiled dataset, Fig. 8.5 shows the state-wise distribution of thunderstorm events over the Indian region during the 1978–2012 period. Five states (West Bengal, Orissa, Bihar, Assam, and Jharkhand) were the worst hit of thunderstorms in terms of fatalities, injuries, and casualties. West Bengal experienced most intense thunderstorm events and highest casualties during this period. Maharashtra and Kerala experienced 147 and 79 events, respectively, during this period, the fatalities and casualties are comparatively less in these states. Delhi experienced only seven events with significant injuries and casualties (Bhardwaj et al. 2017). Figure 8.6 shows mean number of thunderstorm days for different seasons over the Indian region for two periods 1951–1980 and 1981–2010. On the annual scale, high thunderstorm activity days (100 days or

more) are seen over northern Assam, Meghalaya, and West Bengal. The annual frequency of thunderstorm days latitudinally increases from lower to higher latitudes; however, there is a sharp decline in their frequency noticed from 1980s, i.e., during the 1981–2010 period, relative to 1951–1980 time period, there was 34% decline of frequency of thunderstorm days (Bhardwaj and Singh 2018; Singh and Bhardwaj 2019). The observed changes in thunderstorm activity are found to be mostly dependent upon latitude and season, and they are consistent with the seasonal migration of the inter-tropical convergence zone (ITCZ) and the solar heating of the Indian landmass (Manohar et al. 1999; Manohar and Kesarkar 2005) along with strong regional influences from the topography. The thunderstorm activity also exhibits an increasing number from the western side of the subcontinent and moves northeastwards toward the Himalayan foothills, and more notably the highest and lowest number of thunderstorm days are observed over the mountainous terrain of Jammu and Kashmir and Ladakh region, respectively. Over the Gangetic plains, West Bengal and surrounding regions record between 80 and 100 days of thunderstorm activity annually while Kerala records the highest (80–100 days) thunderstorm frequency over the peninsular regions of India (Tyagi 2007).

The pre-monsoon increase in thunderstorm activity is primarily attributed to the topography, insolation, and advection of moisture under favorable wind conditions. The spatial distribution of thunderstorm occurrences during pre-monsoon season indicates a maximum occurrence over the north, northeast, and southern parts of India (Fig. 8.6). The highest frequency of thunderstorms is noticed over the country during monsoon season as a whole, and mostly over the northern and northeastern parts of India. During the post-monsoon season, highest number of thunderstorms occurs over Kerala and the neighboring state of Tamil Nadu. Lowest number of thunderstorms over India is generally observed during the winter season due to stable and dry atmospheric conditions prevailing over most parts of the Indian subcontinent. The declining thunderstorm activity over the Indian subcontinent is suggestively attributed to reductions in rainfall activity and in the moisture amount, due to a fall in the frequency of monsoon depressions, and enhanced intensities of natural variability sources such as ENSO, and PDO. Also, a recent study notes that there is a decline in the dust loading of the atmosphere, or a decrease in intensity of dust storms, during the period 2000–2017 due to increasing pre-monsoon rains over the northwestern states and the Indo-Gangetic Plains (Pandey et al. 2017). Das (2015a) suggests that frequency of cloudburst events in the western Himalayan region has been on the continuous rise due to faster evaporation rates from glacial lakes at high altitudes, as a consequence of global warming. It is also noted that most of the cloudburst events are reported from

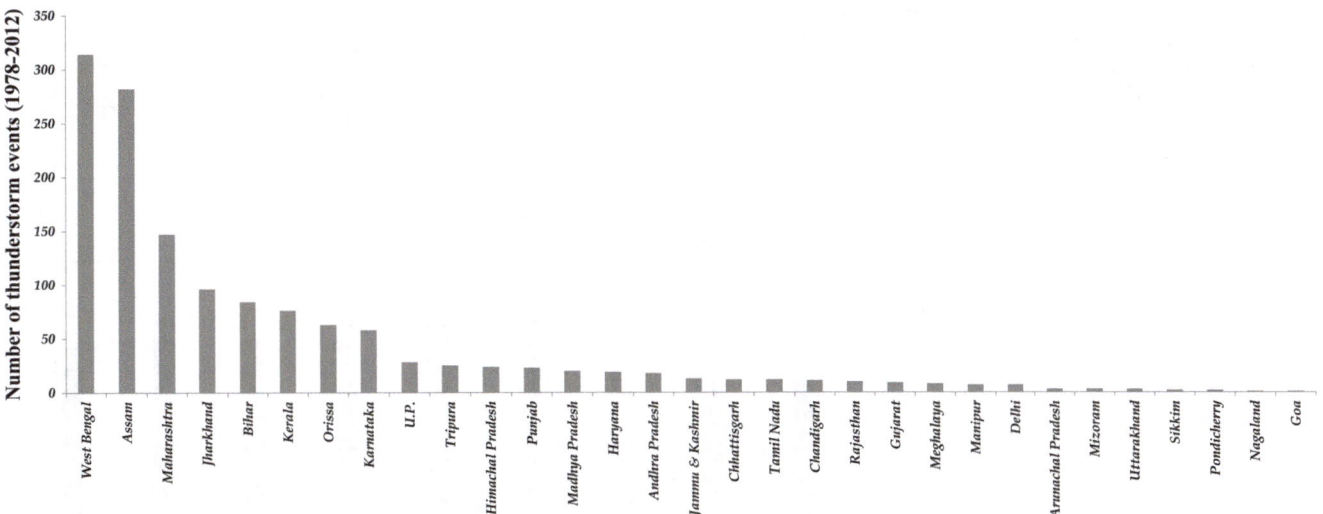

Fig. 8.5 State-wise distribution of thunderstorm events in India for the period 1978–2012. *Source* IMD, see also Bharadwaj et al. (2017)

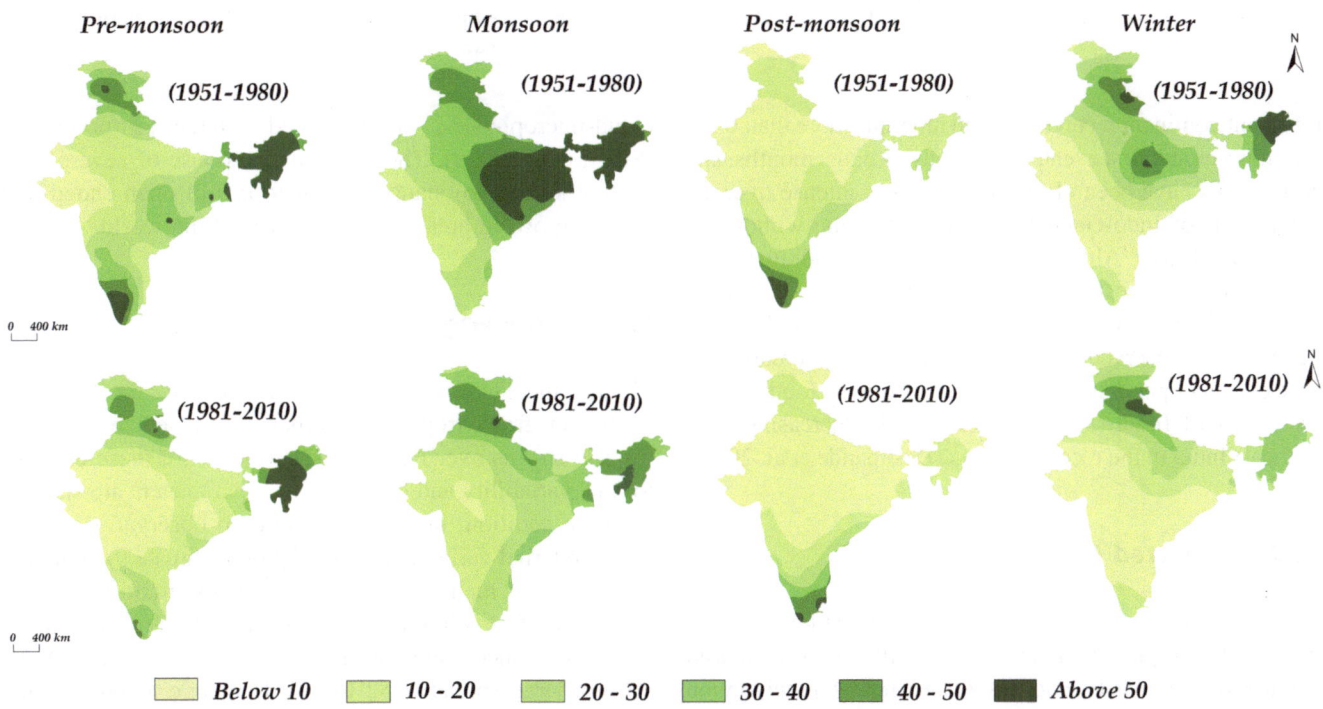

Fig. 8.6 Number of thunderstorm days in India: mean frequency during different Indian seasons for the period 1951–1980 and 1981–2010. *Source* IMD; Adopted from Bharadwaj and Singh (2018) under CC BY-NC 4

ISM months, and the cloudburst events are found to have a dynamical sequence of convective triggering followed by orographic locking mechanisms (Dimri et al. 2017).

Figure 8.7 shows distribution of cloudburst (rainfall rate exceeding 100 mm h^{-1}) and mini-cloudburst (rainfall in consecutive 2 h exceed 5 cm) events over the Indian subcontinent during the period 1969–2015. Based on the information from 126 station observations during this period, on an average, the Indian subcontinent experienced 130 mini-cloudburst events and 1 cloudburst event per year. During the ISM season, there are 2–3 mini-cloudburst events witnessed along the Himalayan foothills, west coast of India, Indo-Gangetic Plains and Saurashtra, and as high as five such events are noticed over the monsoon trough region of central India. The onset period of ISM typically witnesses more intense mini-cloudbursts (average rain intensities of 7 cm per event) prominent over the coastal regions of Gujarat and Odisha, and northeastern parts of India. Thus, a major part of

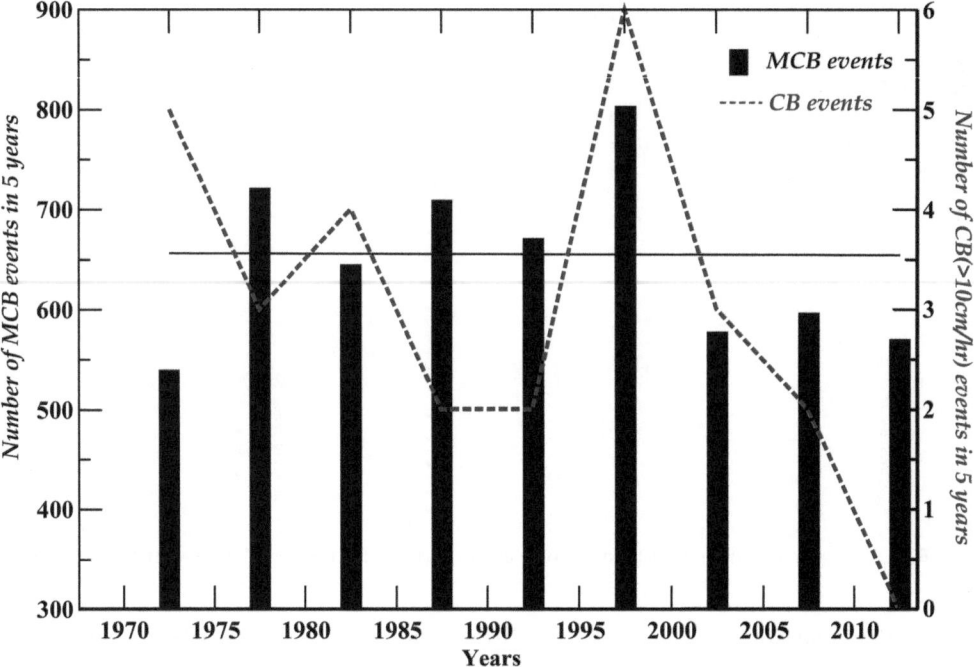

Fig. 8.7 Annual (pentad) frequency of cloudburst (CB) and mini-cloudburst (MCB) events observed over the Indian subcontinent during the period 1969–2015. *Data source* India Meteorological Department; see also Deshpande et al. (2018)

Indian subcontinent receives rainfall amounts more than 6 cm from mini-cloudburst events during July–August months, and these events have been observed to be very intense over the eastern part of Indo-Gangetic plains during the withdrawal phase of ISM. Although short-lived cloudburst and mini-cloudburst occurrences are generally projected to decline in frequency (not statistically significant at 5%) (see Fig. 8.7), it is observed that there is a significant increase of in these events (1 per decade) along the Himalayan foothills and west coast of India (5 per decade), while decreasing over northeast India in the recent decades (Deshpande et al. 2018).

8.3.2 Projected Changes

It is generally correlated that the temperature increase associated with global climate change will lead to increased thunderstorm intensity and associated heavy precipitation events. As for the changes in severe convective storms (thunderstorms, hail storms, cloudbursts) due to climate variability and anthropogenic modifications of atmospheric environment, still there is "low confidence" in observed trends because of historical data inhomogeneities and inadequacies in monitoring systems. Projections of aforementioned severe weather outbreaks in the future are very difficult to pronounce at this time due to the indispensable need of enhancing the meso-γ scale rainfall observational network and ultra-fine resolution model architecture with improved representations of physical processes (convective,

cloud-microphysics, aerosol-cloud interactive processes) (Singh et al. 2019). Therefore, the response of severe convective storms to changing climate is still open-ended and rapidly growing area of research around the world.

8.4 Knowledge Gaps

Though there is a rise in intensification rates in the transformation from tropical disturbances to severe category tropical storms over NIO basin, attribution of these changes to SST variability and environmental parameters are still not clear. In addition, inadequate long-term observational data, strong internal variability of the regional climate system, and limitations in realistically representing the multi-scale processes of TC evolution in climate models also add to the underlying uncertainty. Moreover, the climate simulations show a large spread (−52 to +79% relative to present-day changes) in the future projections of TC frequency in the NIO region (Knutson et al. 2010a, b). These factors demand for an improvement in the existing climate/Earth-system models and promising downscaling methods to reduce the uncertainties and to provide finer details of TC activity (see Knutson et al. 2015; see also Vishnu et al. 2019).

The attributions of trends in localized convective storms still have a low confidence owing to data inhomogeneities and insufficient monitoring systems, and the response of localized convective storms to changing climate still remains an open-ended research around the world.

8.5 Summary

A status on the current understanding of the changes in high-impact, in terms of socio-economic implications, stormy weather phenomena pertinent to the Indian subcontinent [i.e., severe category tropical cyclonic storms in the NIO region, thunderstorms and associated dust storms, short-span intense rain-producing cloudbursts] is documented in this chapter. Considerable progress has been generally realized in the understanding of changes in TC activity over the global ocean basins (see Walsh et al. 2016; Knutson et al. 2010a, 2019a, b), while a clear understanding of the reasoning behind the changes in NIO TC activity and extreme rain or convective storm occurrences over the Indian subcontinent is still rudimentary. Observed TC changes during the 1951–2018 (relative to pre-1950 period) period indicate that there is a rise in severe category TCs by 49% (relative to pre-1950 period) in the BOB region, and 52% in the AS region. There is also a marked rise of these storms in the NIO basin by 105% during the post-monsoon (October–December) season. There is a significant decline in the annual frequency of TCs in the NIO basin, i.e., −0.23 per decade for the entire NIO, and −0.26 per decade for the BOB. Observations also indicate a rising trend in VSCS (category 4 and above TC; see Table 8.1) in the NIO region during the 2000–2018 period which is apparently controlled by the post-monsoon VSCS trend (+0.86 per decade) from BOB. Another growing concern is the rising number and severe TCs in the AS region in the recent years—i.e., 6 out of 11 TCs formed in AS reached greater severity during the 2000–2018 period (see Table 8.2). Based on the investigations available till date for reasoning behind this rise, there is a consensus of medium confidence in attributing the observed rise in the AS post-monsoon TCs to human-induced SST warming (Murakami et al. 2017; Knutson et al. 2019a, b).

Localized convective storms such as thunderstorms over the Indian subcontinent indicate a declining frequency by 34% in the post-1980 period which is suggestively attributed to reductions in rainfall activity and in the moisture amount due to a fall in the frequency of monsoon depressions, and enhanced intensities of natural variability climate drivers. Although short-lived cloudburst and mini-cloudburst occurrences generally indicate a decline in frequency in the recent decades, it is observed that there is a significant increase in these events along the Himalayan foothills (1 per decade) and west coast of India (5 per decade).

References

Anthes RA (1982) Tropical cyclones: their evolution, structure and effects. Meteorological monograph No. 41. American Meteorological Society, 208 pp

Anthes RA et al (2006) Comments on hurricanes and global warming—potential linkages and consequences. Bull Am Met Soc 87:623–628

Balaguru K, Taraphdar S, Leung LR, Foltz GR (2014) Increase in the intensity of postmonsoon Bay of Bengal tropical cyclones. Geophys Res Lett 41:3594–3601

Balaji M, Chakraborty A, Mandal M (2018) Changes in tropical cyclone activity in north Indian Ocean during satellite era (1981–2014). Int J Climatol 38:2819–2837

Bender MA et al (2010) Modeled impact of anthropogenic warming of the frequency of intense Atlantic hurricanes. Science 327:454–458

Bengtsson L et al (2007) How may tropical cyclones change in a warmer climate? Tellus 59:539–561

Bhan SC, Paul S, Kharbanda KL (2004) Cloudbursts in Himachal Pradesh. Mausam 55:712–713

Bhardwaj P, Singh O (2018) Spatial and temporal analysis of thunderstorm and rainfall activity over India. Atmósfera 31:255–284

Bhardwaj P, Singh O, Kumar D (2017) Spatial and temporal variations in thunderstorm casualties over India. Singap J Trop Geogr 38:293–312

Bohra AK et al (2006) Heavy rainfall episode over Mumbai on 26 July 2005: assessment of NWP guidance. Curr Sci 90:1188–1194

Chakrabarty KK, Nath AK, Sengupta S (2007) Nor'wester over West Bengal and comfortability. Mausam 58:177–188

Chinchole PS, Mohapatra M (2017) Some characteristics of translational speed of cyclonic disturbances over North Indian ocean in recent years. In: Tropical cyclone activity over the North Indian Ocean. Springer, Cham, pp 165–179

Danard MTS, Murty TS (1989) Tropical cyclones in the Bay of Bengal and CO₂ warming. Nat Hazards 2:387–390

Das PK (2015a) Global warming, glacial lakes and cloud burst events in Garhwal-Kumaon Himalaya: A hypothetical analysis. Int J Env Sci 5:697

Das Y (2015b) Some aspects of thunderstorm over India during pre-monsoon season: a preliminary report. J Geosci Geomat 3:68–78

Das S, Ashrit R, Moncrieff MW (2006) Simulation of a Himalayan cloudburst event. J Earth Syst Sci 115:299–313

De US, Dube RK, Rao GP (2005) Extreme weather events over India in the last 100 years. J Ind Geophys Union 9:173–187

Deo AA, Ganer DW (2014) Tropical cyclone activity over the Indian Ocean in the warmer climate. In: Mohanty UC et al (eds) Monitoring and prediction of tropical cyclones in the indian ocean and climate change, pp 72–80. https://doi.org/10.1007/978-94-007-7720-0_7

Deshpande NR, Kothawale DR, Kumar V, Kulkarni JR (2018) Statistical characteristics of cloud burst and mini-cloud burst events during monsoon season in India. Int J Climatol 38:4172–4188

Dimri AP et al (2017) Cloudbursts in Indian Himalayas: a review. Earth-sci rev 168:1–23

Doswell CA (2001) Severe convective storms—an overview. Severe convective storms. American Meteorological Society, Boston, pp 1–26

Dvorak VF (1984) Tropical cyclone intensity analysis using satellite data. NOAA technical report NESDIS 11:1–47

Elsner JB, Kocher B (2000) Global tropical cyclone activity: a link to the North Atlantic Oscillation. Geophys Res Lett 27:129–132

Elsner JB, Kossin JP, Jagger TH (2008) The increasing intensity of the strongest tropical cyclones. Nature 455:92–95. https://doi.org/10.1038/nature07234

Emanuel K (2005) Increasing destructiveness of tropical cyclones over the past 30 years. Nature 436:686

Evan AT, Kossin JP, Ramanathan V (2011) Arabian Sea tropical cyclones intensified by emissions of black carbon and other aerosols. Nature 479:94

Evan AT, Kossin JP, Chung C, Ramanathan V (2012) Intensified Arabian Sea tropical storms. Nature 2012:E2–E3

Ghosh A, Lohar D, Das J (2008) Initiation of Nor'wester in relation to mid-upper and low-level water vapor patterns on METEOSAT-5 images. Atmos Res 87:116–135

Girishkumar MS, Ravichandran M (2012) The influences of ENSO on tropical cyclone activity in the Bay of Bengal during October–December. J Geophy Res 117:C02033. https://doi.org/10.1029/2011jc007417

Girishkumar MS, Prakash VT, Ravichandran M (2015) Influence of Pacific Decadal oscillation on the relationship between ENSO and tropical cyclone activity in the Bay of Bengal during October–December. Clim Dyn 44:3469–3479

Goswami BN et al (2006) Increasing trend of extreme rain events over India in a warming environment. Science 314:1442–1445

Gray WM (1968) Global view of the origin of tropical disturbances and storms. Mon Wea Rev 96:669–700

Guhathakurta P, Sreejith OP, Menon PA (2011) Impact of climate change on extreme rainfall events and flood risk in India. J Earth Syst Sci 120:359–373

Gupta S, Jain I, Johari P, Lal M (2018) Impact of climate change on tropical cyclones frequency and intensity on Indian Coasts. In: Rao PJ et al (eds) Proceedings of international conference on remote sensing for disaster management. Springer Series in Geomechanics and Geoengineering. https://doi.org/10.1007/978-3-319-77276-9_32

Haggag M, Yamashita T, Kim KO, Lee HS (2010) Simulation of the North Indian ocean tropical cyclones using the regional environment simulator: application to cyclone Nargis in 2008. In: Charabi Y (eds) Indian Ocean tropical cyclones and climate change, pp 73–82. Springer, Dordrecht

Houze RA, Rasmussen KL, Medina S, Brodzik SR, Romatschke U (2011) Anomalous atmospheric events leading to the summer 2010 floods in Pakistan. Bull Am Meteor Soc 92:291–298

Hoyos CD, Agudelo PA, Webster PJ, Curry JA (2006) De-convolution of the factors contributing to the increase in global hurricane intensity. Science 312:94–97

Hunt KM, Turner AG, Inness PM, Parker DE, Levine DERC (2016) On the structure and dynamics of Indian monsoon depressions. Mon Wea Rev 144:3391–3416

India Meteorological Department (IMD) (2003) Cyclone Manual. IMD, New Delhi

IPCC (2007) Climate change 2007: Synthesis report. Contribution of Working Groups I, II and III to the Fourth Assessment Report of the Intergovernmental Panel on Climate Change. Geneva, Switzerland, 72 pp

IPCC (2014) Climate change 2014: synthesis report. Contribution of Working Groups I, II and III to the Fifth Assessment Report of the Intergovernmental Panel on Climate Change. Geneva, Switzerland, 151 pp

Kandalgaonkar SS, Tinmaker MIR, Nath A, Kulkarni MK, Trimbake HK (2005) Study of thunderstorm and rainfall activity over the Indian region. Atmosfera 18:91–101

Kikuchi K, Wang B (2010) Formation of tropical cyclones in the northern Indian Ocean associated with two types of tropical intra-seasonal oscillation modes. J Meteorol Soc Jpn 88:475–496. https://doi.org/10.2151/jmsj.2010-313.p-0

Klotzbach PJ, Landsea CW (2015) Extremely intense hurricanes: revisiting Webster et al. (2005) after 10 years. J Clim 28:7621–7629

Knutson TR et al (2010a) Tropical cyclones and climate change. Nat Geosci 3:157–163. https://doi.org/10.1038/ngeo779

Knutson TR (2010b) Tropical cyclones and climate change: an Indian Ocean perspective. In: Charabi Y (ed) Indian Ocean tropical cyclones and climate change, pp 47–49. https://doi.org/10.1007/978-90-481-3109-9_7

Knutson TR et al (2014) Recent research at GFDL on surface temperature trends and simulations of tropical cyclone activity in the Indian Ocean region. In: Mohanty UC et al (eds), Monitoring and prediction of tropical cyclones in the Indian Ocean and climate change, pp 50–55

Knutson TR et al (2015) Global projections of intense tropical cyclone activity for the late twenty-first century from dynamical downscaling of CMIP5/RCP4. 5 scenarios. J Clim 28:7203–7224

Knutson T et al (2019a) Tropical cyclones and climate change assessment: part I. Detection and attribution. Bull Am Meteorol Soc. https://doi.org/10.1175/bams-d-18-0189.1

Knutson T et al (2019b) Tropical cyclones and climate change assessment: Part II. Projected response to anthropogenic warming. Bull Am Meteorol Soc. https://doi.org/10.1175/bams-d-18-0194.1

Knutson TK, Tuleya RE (2004) Impact of CO_2-induced warming on simulated hurricane intensity and precipitation: sensitivity to the choice of climate model and convective parameterization. J Clim 17:3477–3495

Knutson TR, Tuleya RE, Kurihara Y (1998) Simulated increase of hurricane intensities in a CO_2-warmed climate. Science 279:1018–1020

Knutson TR, Tuleya RE, Shen W, Ginis I (2001) Impact of CO_2-induced warming on hurricane intensities as simulated in a hurricane model with ocean coupling. J Climate 14:2458–2468

Knutson TR, Delworth T, Dixon K, Held I, Lu J, Ramaswamy V, Schwarzkopf M, Stenchikov G, Stouffer R (2006) Assessment of twentieth-century regional surface temperature trends using the GFDL CM2 coupled models. J Clim 19:1624–1651

Kossin JP (2018) A global slowdown of tropical-cyclone translation speed. Nature 558:104–107

Krishnan R et al (2016) Deciphering the desiccation trend of the South Asian monsoon hydroclimate in a warming world. Clim Dyn 47:1007–1027

Krishnan R et al (2019) Unraveling climate change in the Hindu Kush Himalaya: rapid warming in the mountains and increasing extremes. In: Wester P et al (eds) The Hindu Kush Himalaya assessment—mountains, climate change, sustainability and people, pp 57–98

Kulkarni JR et al (2015) Unprecedented hailstorms over north peninsular India during February–March 2014. J Geophy Res 120:2899–2912

Kulkarni MK, Tinmaker MIR, Manohar GK (2009) Characteristics of thunderstorm activity over India. Int J Meteorol 34(344): 341

Kumar SVJ, Ashtikar SS, Mohapatra M (2017) Life period of cyclonic disturbances over the NIO during recent years. In: Mohapatra M et al (eds) Tropical cyclone activity over the North Indian Ocean. Springer, Berlin 390 pp

Litta AJ et al (2012) Simulation of tornado over Orissa (India) on March 31, 2009, using WRF–NMM model. Nat Hazards 61:1219–1242

Lotus S (2015) Heavy rainfall over Jammu & Kashmir during 3–6 September, 2014 leading to flooding condition. Monsoon 2014: a report (ESSO/IMD/ SYNOPTIC MET/01(2015)/17). India Meteorological Department, National Climate Center, Pune, India

Mandal GS, Krishna P (2009) Global warming, climate change and cyclone related destructive winds—discussion of results from some selected studies with emphasis on the north Indian Ocean. Glob Environ Res 13:141–150

Manohar GK, Kesarkar AP (2005) Climatology of thunderstorm activity over the Indian region: III. Latitudinal and seasonal variation. Mausam 56:581–592

Manohar GK, Kandalgaonkar SS, Tinmaker MIR (1999) Thunderstorm activity over India and the Indian southwest monsoon. J Geophy Res 104:4169–4188

McDonald RE et al (2005) Tropical storms: representation and diagnosis in climate models and the impacts of climate change. Clim Dyn 25:19–36

Middleton NJ (1986) A geography of dust storms in South-west Asia. J Climatol 6:183–196

Mishra A (2014) Temperature rise and trend of cyclones over the eastern coast of India. J Earth Sci Clim Change 5–9. https://doi.org/10.4172/2157-7617.1000227

Mishra V, Shah HL (2018) Hydroclimatological perspective of the Kerala flood of 2018. J Geol Soc India 92:645–650

Mohanty UC, Pattanayak S, Osuri KK (2010) Changes in frequency and intensity of tropical cyclones over Indian seas in a warming environment. Disaster Dev 4:53–77

Mohanty UC, Osuri KK, Pattanayak S, Sinha P (2012) An observational perspective on tropical cyclone activity over Indian seas in a warming environment. Nat Hazards 63:1319–1335

Mohapatra M, Bandyopadhyay BK, Tyagi A (2012) Best track parameters of tropical cyclones over the North Indian Ocean: a review. Nat Hazards 63:1285–1317

Mohapatra M, Bandyopadhyay BK, Tyagi A (2014) Construction and quality of best tracks parameters for study of climate change impact on tropical cyclones over the north indian ocean during satellite era. Monitoring and prediction of tropical cyclones in the Indian ocean and climate change. Springer, Dordrecht, pp 3–17

Mohapatra M, Bandyopadhyay BK, Rathore LS (eds) (2017) Tropical cyclone activity over the North Indian Ocean. Springer, Berlin, 390 pp. https://doi.org/10.1007/978-3-319-40576-6

Mooley DA (1980) Severe cyclonic storms in the Bay of Bengal 1877–1977. Mon Wea Rev 108:1647–1655

Mooley DA, Mohile CM (1984) Cyclonic storms of the Arabian Sea, 1877–1980. Mausam 35:127–134

Mukherjee S, Aadhar S, Stone D, Mishra DV (2018) Increase in extreme precipitation events under anthropogenic warming in India. Weather Clim Extremes 20:45–53

Mukhopadhyay P, Singh HAK, Singh SS (2005) Two severe nor'westers in April 2003 over Kolkata, India, using Doppler radar observations and satellite imagery. Weather 60:343–353

Murakami H et al (2012b) Future changes in tropical cyclone activity projected by the new high-resolution MRI-AGCM. J Clim 25:3237–3260

Murakami H, Mizuta R, Shindo E (2012) Future changes in tropical cyclone activity projected by multi-physics and multi-SST ensemble experiments using the 60-km-mesh MRI-AGCM. Clim Dyn 39:2569–2584. https://doi.org/10.1007/s00382-011-1223-x

Murakami H, Sugi M, Kitoh A (2013) Future changes in tropical cyclone activity in the North Indian Ocean projected by high-resolution MRI-AGCMs. Clim Dyn 40:1949–1968

Murakami H, Sugi M, Kitoh A (2014) Future changes in tropical cyclone activity in the North Indian Ocean projected by the new high-resolution MRI-AGCM. In: Mohanty UC et al (eds) Monitoring and prediction of tropical cyclones in the Indian Ocean and climate change, pp 63–71. https://doi.org/10.1007/978-94-007-7720-0_6

Murakami H, Vecchi GA, Underwood S (2017) Increasing frequency of extremely severe cyclonic storms over the Arabian Sea. Nat Clim Change 7:885–889

Murugavel P, Pawar SD, Gopalakrishan V (2014) Climatology of lightning over Indian region and its relationship with convective available potential energy. Int J Climatol 34:3179–3187

Niyas NT, Srivastava AK, Hatwar HR (2009) Variability and trend in the cyclonic storms over north Indian Ocean. National Climate Centre, Office of the Additional Director General of Meteorology (Research), India Meteorological Department

Oouchi K et al (2006) Tropical cyclone climatology in a global-warming climate as simulated in a 20 km-mesh global atmospheric model: frequency and wind intensity analyses. J Meteorol Soc Jpn 84:259–276

Orlanski I (1975) A rational subdivision of scales for atmospheric processes. Bull Am Meteorol Soc 56:527–530

Pandey SK, Vinoj V, Landu K, Babu KSS (2017) Declining pre-monsoon dust loading over South Asia: signature of a changing regional climate. Sci Rep 7(1). https://doi.org/10.1038/s41598-017-16338-w

Pattanaik DR (2005) Variability of oceanic and atmospheric conditions during active and inactive periods of storms over the Indian region. Int J Climatol 25:1523–1530

Pielke RA et al (2005) Hurricanes and global warming. Bull Am Meteorol Soc 86:1571–1575

Pradhan D, De UK, Singh UV (2012) Development of nowcasting technique and evaluation of convective indices for thunderstorm prediction in Gangetic West Bengal (India) using Doppler Weather Radar and upper air data. Mausam 63:299–318

Prajeesh AG, Ashok K, Rao DVB (2013) Falling monsoon depression frequency: a Gray-Sikka conditions perspective. Nat Scientific Rep. https://doi.org/10.1038/srep02989

Priya P, Krishnan R, Mujumdar M, Houze RA (2017) Changing monsoon and midlatitude circulation interactions over the Western Himalayas and possible links to occurrences of extreme precipitation. Clim Dyn 49:2351–2364

Raghavan S, Rajesh S (2003) Trends in tropical cyclone impact—a study in Andhra Pradesh, India. Bull Am Meteorol Soc 84:635–644. https://doi.org/10.1175/BAMS-84-5-635

Raghavendra VK (1973) A statistical analysis of the number of tropical storms and depressions in the Bay of Bengal during 1890–1969. Ind J Meteorol Geophys 24:125–130

Raipal DK, Deka SN (1980) ANDHI, the convective dust storm of northwest India. Mausam 31:31–442

Rajeevan M, Srinivasan J, Niranjan Kumar K, Gnanaseelan C, Ali MM (2013) On the epochal variation of intensity of tropical cyclones in the Arabian Sea. Atmos Sci Lett 14:249–255

Ramesh Kumar MR, Sankar S (2010) Impact of global warming on cyclonic storms over north Indian Ocean. Indian J Mar Sci 39:516–520

Ramsay H (2017) The global climatology of tropical cyclones. Oxford research encyclopedia of natural hazard science, Oxford University Press, 34 pp

Ranalkar MR et al (2016) Incessant rainfall event of June 2013 in Uttarakhand, India: observational perspectives. In: High-impact weather events over the SAARC Region. Springer, Cham, pp 303–312

Rao DVB, Srinivas D, Satyanarayana GC (2019) Trends in the genesis and landfall locations of tropical cyclones over the Bay of Bengal in the current global warming era. J Earth Syst Sci 128. https://doi.org/10.1007/s12040-019-1227-1

Rasmussen KL, Houze RA (2012) A flash-flooding storm at the steep edge of high terrain: disaster in the Himalayas. Bull Amer Meteorol Soc 93:1713–1724

Romatschke U, Houze RA (2011) Characteristics of precipitating convective systems in the South Asian monsoon. J Hydrometeorol 12:3–26

Saha U, Maitra A, Midya SK, Das GK (2014) Association of thunderstorm frequency with rainfall occurrences over an Indian urban metropolis. Atmos Res 138:240–252

Sahoo BP, Bhaskaran K (2016) Assessment on historical cyclone tracks in the Bay of Bengal, east coast of India. Int J Climatol 36:95–109

Sebastian M, Behera MR (2015) Impact of SST on tropical cyclones in North Indian Ocean. Procedia Eng 116:1072–1077

Sikka DR (2006) Major advances in understanding and prediction of tropical cyclones over the north Indian Ocean: a perspective. Mausam 57:165–196

Singh OP (2007) Long-term trends in the frequency of severe cyclones of Bay of Bengal: observations and simulations. Mausam 58:59–66

Singh O, Bhardwaj P (2019) Spatial and temporal variations in the frequency of thunderstorm days over India. Weather 74:138–144

Singh OP, Khan TMA (1999) Changes in the frequencies of cyclonic storms and depressions over the Bay of Bengal and the Arabian Sea. SAARC Meteorological Research Centre Report 2, 121 pp

Singh OP, Rout RK (1999) Frequency of cyclonic disturbances over the North Indian Ocean during ENSO years. In: Meteorology beyond 2000: Proceedings of TROPMET-99, Chennai, India, pp 297–301

Singh OP, Khan TA, Rahman MS (2000) Changes in the frequency of tropical cyclones over the North Indian Ocean. Meteorol Atmos Phys 75:11–20

Singh OP, Khan TMA, Rahman S (2001) Has the frequency of intense tropical cyclones increased in the north Indian Ocean? Curr Sci 80:575–580

Singh C, Mohapatra M, Bandyopadhyay BK, Tyagi A (2011) Thunderstorm climatology over northeast and adjoining east India. Mausam 62:163–170

Singh K, Panda J, Osuri KK, Vissa NK (2016) Progress in tropical cyclone predictability and present status in the North Indian Ocean region. In: Lupo A (ed) in recent developments in tropical cyclone dynamics, prediction, and detection, pp 193–215. https://doi.org/10.5772/64333

Singh D, Ghosh S, Roxy MK, McDermid S (2019) Indian summer monsoon: extreme events, historical changes, and role of anthropogenic forcings. Wiley Interdiscip Rev Clim Change 10(2):e571

Sugi M, Noda A, Sato N (2002) Influence of the global warming on tropical cyclone climatology: an experiment with the JMA global model. J Meteorol Soc Jpn 80:249–272

Sugi M, Murakami H, Yoshimura J (2009) A reduction in global tropical cyclone frequency due to global warming. Sola 5:164–167

Sugi M, Murakami H, Yoshimura J (2014) Mechanism of the Indian Ocean tropical cyclone frequency changes due to global warming. In: Mohanty UC et al (eds) Monitoring and prediction of tropical cyclones in the Indian Ocean and climate change, pp 40–49. https://doi.org/10.1007/978-94-007-7720-0_4

Sumesh KGMR, Kumar MR (2013) Tropical cyclones over north Indian Ocean during El-nino modoki years. Nat Hazards 68:1057–1074

Thayyen RJ, Dimri AP, Kumar P, Agnihotri G (2013) Study of cloudburst and flash floods around Leh, India, during August 4–6, 2010. Nat Hazards 65:2175–2204

Tsuboi A, Takemi T (2014) The interannual relationship between MJO activity and tropical cyclone genesis in the Indian Ocean. Geosci Lett 1:9. https://doi.org/10.1186/2196-4092-1-9

Tyagi A (2007) Thunderstorm climatology over Indian region. Mausam 58:189–212

Tyagi A, Bandyopadhyay BK, Mohapatra M (2010) Monitoring and prediction of cyclonic disturbances over North Indian ocean by regional specialised meteorological centre, New Delhi (India): problems and prospective. Indian Ocean tropical cyclones and climate change. Springer, Dordrecht, pp 93–103

Tyagi A, Sikka DR, Goyal S, Bhowmick M (2012) A satellite based study of pre-monsoon thunderstorms (Nor'westers) over eastern India and their organization into mesoscale convective complexes. Mausam 63:29–54

Velden C et al (2006) The Dvorak tropical cyclone intensity estimation technique: a satellite-based method that has endured for over 30 years. Bull Am Meteorol Soc 87:1195–1210

Vellore RK et al (2014) On the anomalous precipitation enhancement over the Himalayan foothills during monsoon breaks. Clim Dyn 43:2009–2031

Vellore RK et al (2016) Monsoon-extratropical circulation interactions in Himalayan extreme rainfall. Clim Dyn 46:3517–3546

Vellore RK et al (2019) Sub-synoptic variability in the Himalayan extreme precipitation event during June 2013. Met Atmos Phy https://doi.org/10.1007/s00703-019-00713-5

Vidale PL, Roberts M, Hodges K, Strachan J, Demory ME, Slingo J (2010) Tropical cyclones in a hierarchy of climate models of increasing resolution. In: Charabi Y (ed) Indian Ocean tropical cyclones and climate change. Springer, Berlin, pp 9–14

Vishnu S, Sanjay J, Krishnan R (2019) Assessment of climatological TC activity over the North Indian Ocean in the CORDEX-South Asia regional climate models. Clim Dyn. https://doi.org/10.1007/s00382-019-04852-8

Walsh KJ et al (2016) Tropical cyclones and climate change. Wiley Interdiscip Rev Clim Change 7(1):65–89

Wang B, Xu S, Wu L (2012) Intensified Arabian Sea tropical storms. Nature 489:E1–E2. https://doi.org/10.1038/nature11470

Webster PJ, Holland GJ, Curry JA, Chang HR (2005) Changes in tropical cyclone number, duration, and intensity in a warming environment. Science 309:1844–1846

Xie S-P, Deser C, Vecchi GA, Ma J, Teng H, Wittenberg AT (2010) Global warming pattern formation: sea surface temperature and rainfall. J Clim 23:966–986

Yoshimura J, Masato S, Noda A (2006) Influence of greenhouse warming on tropical cyclone frequency. J Meteorol Soc Jpn 84:405–428

Yu J, Wang Y (2009) Response of tropical cyclone potential intensity over the north Indian Ocean to global warming. Geophys Res Lett 36. https://doi.org/10.1029/2008gl036742

Zhao M, Held IM (2012) TC-permitting GCM simulations of hurricane frequency response to sea surface temperature anomalies projected for the late twenty-first century. J Clim 25:2995–3009

Zhao M, Held IM, Lin S-J, Vecchi GA (2009) Simulations of global hurricane climatology, interannual variability, and response to global warming using a 50 km resolution GCM. J Clim 22:6653–6678

Coordinating Lead Authors

P. Swapna, Indian Institute of Tropical Meteorology (IITM-MoES), Pune, India,
e-mail: swapna@tropmet.res.in (corresponding author)
M. Ravichandran, National Centre for Polar and Ocean Research (NCPOR-MoES),
Vasco da Gama, Goa, India

Lead Authors

G. Nidheesh, Indian Institute of Tropical Meteorology (IITM-MoES), Pune, India
J. Jyoti, Indian Institute of Tropical Meteorology (IITM-MoES), Pune, India
N. Sandeep, Indian Institute of Tropical Meteorology (IITM-MoES), Pune, India
J. S. Deepa, Indian Institute of Tropical Meteorology (IITM-MoES), Pune, India

Review Editor

A. S. Unnikrishnan, formerly at CSIR-National Institute of Oceanography (NIO), Dona Paula, Goa, India

Corresponding Author

P. Swapna, Indian Institute of Tropical Meteorology (IITM-MoES), Pune, India,
e-mail: swapna@tropmet.res.in

R. Krishnan et al. (eds.), *Assessment of Climate Change over the Indian Region*,
https://doi.org/10.1007/978-981-15-4327-2_9

Key Messages

- *There is high confidence that the rate of global mean sea level (GMSL) rise has increased. Human-caused climate change has made a substantial contribution to the rise since 1900.*
- *The GMSL has risen by 1.7 (1.5 to 1.9) mm year^{-1} since 1901 and the rate of rise has accelerated to 3.3 mm year^{-1} since 1993.*
- *Sea-level rise in the Indian Ocean is non-uniform and the rate of north Indian Ocean rise is 1.06–1.75 mm year^{-1} from 1874 to 2004 and is 3.3 mm year^{-1} in the recent decades (1993–2015), which is comparable to the current rate of GMSL rise.*
- *Indian Ocean sea-level rise is dominated by the ocean thermal expansion, while the addition of water mass from terrestrial ice-melting is the major contributor to the GMSL rise.*
- *Interannual to decadal-scale variability in the Indian Ocean sea level is dominated by El Niño Southern Oscillation and Indian Ocean Dipole events.*
- *Relative to 1986–2005, GMSL is very likely to rise by ~ 26 cm by 2050 and ~ 53 cm by 2100 for a mid-range, mitigation scenario.*
- *Steric sea level along the Indian coast is likely to rise by about 20–30 cm at the end of the twenty-first century and the corresponding estimate for global mean steric sea-level rise is 18±5 cm (relative to 1986–2005), under RCP4.5 (for a mid-range emission scenario, excluding ice-melt contributions).*
- *Extreme sea-level events are projected to occur frequently over the tropical regions (high confidence) and along the Indian coast (medium confidence) associated with an increase in the mean sea level and climate extremes.*

9.1 Introduction

The global ocean plays a critical role in regulating the energy balance of the climate system. Over 90% of the anthropogenic excess heat goes into the oceans (Church et al. 2013a, b), remaining goes into melting both terrestrial and sea ice, and warming the atmosphere and land (Hansen et al. 2011; Church et al. 2011b; Trenberth et al. 2014). One of the consequences of warming of the global ocean and the melting of ice and glaciers is the rise in mean sea level. Sea-level rise can exert significant stress on highly populated coastal societies and low-lying island countries around the world. Indian Ocean region is heavily populated, comprises of many low-lying islands and coastal zones and is highly

rich in marine ecosystems. The regions in and around the Indian Ocean are home to roughly 2.6 billion people, which is 40% of the global population. One-third of the Indian population and the majority of the Asian population are located near coastal regions. Therefore, the rise in sea level can pose a growing challenge to population, economy, coastal infrastructures and marine ecosystems. Despite considerable progress during recent years, major gaps remain in our understanding of sea-level changes and their causes, particularly at regional scales.

Changes in mean sea level are the result of the complex interplay of a number of factors. Even though there is an unabated rise in observed global mean sea level, the spatial distribution of sea-level trends is not globally uniform (Church et al. 2013a, b). Regionally, sea-level variations can deviate considerably from the global mean. It is very likely that in the twenty-first century and beyond, the sea-level change will have a strong regional pattern, with some places experiencing significant deviations from the global mean sea-level rise (IPCC AR5). The detailed sea-level change along coastlines can therefore potentially be far more substantial than the global mean sea-level rise. The underlying causes of regional sea-level changes are associated with dynamic variations in the ocean circulation as part of climate modes of variability, changes in the wind pattern and with an isostatic adjustment of Earth's crust to past and ongoing changes in polar ice masses and continental water storage (Stammer et al. 2013). Assessment of vulnerability to rising sea levels requires consideration of physical causes, historical evidence and projections.

This chapter reviews the physical factors driving changes in global mean sea level (GMSL) as well as those causing additional regional variations in relative sea level (RSL). Geological and instrumental observations of historical sea-level changes in the global ocean and for the RSL in the Indian Ocean are presented here. The chapter then describes a range of scenarios for future levels and rates of sea-level change, for the Indian Ocean as well as for the global ocean. Finally, an assessment of the impact of changes in sea level on extreme water levels is discussed.

9.2 Physical Factors Contributing to Sea-Level Rise

Sea level is measured either with respect to the surface of the solid Earth, known as relative sea level (RSL) or a geocentric reference such as the reference ellipsoid, known as the geocentric sea level. RSL estimates have been obtained from tide gauges and geological records for the past few centuries. Geocentric sea level has been measured over the past two decades using satellite altimetry. The sea level, when averaged globally, provides global mean sea level (GMSL). The physical processes causing GMSL rise and regional changes in RSL are not identical, although they are related. The primary contributors to current GMSL rise are the thermal expansion of sea waters, land ice loss and freshwater mass exchange between oceans and land water reservoirs and groundwater storage change. The recent trends of these contributions are most likely resulted from the climate change induced by anthropogenic greenhouse gas emissions.

9.2.1 Ocean Warming

Analyses of in-situ ocean temperature data collected over the past 50 years by ships and recently by Argo profiling floats (Argo Data Management Team 2008; Roemmich et al. 2009) reveal that ocean has been warming and increases the upper ocean heat content (OHC). Hence, the sea level, due to the thermal expansion of sea water, has significantly increased since 1950 (e.g. Levitus et al. 2009; Ishii and Kimoto 2009; Domingues et al. 2008; Church et al. 2011a). A recent study by Cheng et al. (2017) has shown that the changes in OHC were relatively small before about 1980; since then, OHC has increased fairly steadily and, since 1990, has increasingly involved deeper layers of the ocean. In the climate system, the ocean acts as a 'buffer' for the atmospheric temperature by storing a large amount of heat from the atmosphere and transporting it to deeper depths via the ocean conveyor belt. On average, over the last 50 years, 93% of the excess heat accumulated in the climate system because of greenhouse gas emissions has been stored in the ocean owing its large heat capacity, the remaining 7% warm the atmosphere and continents, and melt sea and land ice (Levitus et al. 2012; von Schuckmann et al. 2016). Consequently, ocean warming explains about 30–40% of the observed sea-level rise of the last few decades (e.g. Church et al. 2011b).

9.2.2 Glaciers Melting

Apart from the global ocean thermal expansion, melting of the continental ice storage in a warming climate is turned out to be another factor for global mean sea-level rise. Being very sensitive to global warming, mountain glaciers and small ice caps have retreated worldwide during recent decades. The contribution of glacier ice melt to sea-level rise has been estimated based on the mass balance studies of a large number of glaciers (Meier et al. 2007; Kaser et al. 2006). In fact, studies have shown that glaciers have accounted for ∼21% of the global sea-level rise since 1993 (e.g. WCRP 2018).

9.2.3 Ice Sheets

The mass balance of the ice sheets was less known before the 1990s due to inadequate and incomplete observations. Different remote sensing techniques available since then have provided important results on the changing mass of Greenland and (west) Antarctica (e.g. Allison et al. 2009). These data indicate that both ice sheets are currently losing mass at an accelerated rate (e.g. Steffen et al. 2011). For the period 1993–2003, <15% of the rate of global mean sea-level rise was due to the melting of ice sheets (IPCC AR4). But their contribution has increased to ∼40% from 2003 to 2004. The ice sheets mass loss explains ∼25% of the rate of global sea-level rise during 2003–2010 (Cazenave and Remy 2011; Church et al. 2011a). A near-complete loss of Greenland ice sheet over a million years or more leading to the global mean sea-level rise of about 7 m can be caused by sustained global warming greater than a certain threshold above pre-industrial conditions (IPCC AR5). A schematic representation of different processes contributing to global and regional sea-level changes is shown in Fig. 9.1.

9.2.4 Regional Sea-Level Change

Sea-level rise pattern varies substantially from region to region. Geographical patterns of sea-level rise can result in different processes: changes in sea-water density due to changes in temperature and salinity (known as 'steric' sea-level changes) are the dominant process, especially in the tropical oceans. Steric sea-level changes are primarily associated with the atmosphere-ocean coupled dynamics driven mainly by surface winds and ocean circulation. Solid Earth's deformation and geoid changes in response to past and ongoing mass redistribution caused by land ice melt and land water storage changes (known as 'static' factors) also make regional changes in sea level (Stammer et al. 2013). It was shown that the dominant contribution to observed regional sea-level changes comes from steric effects caused by non-uniform thermal expansion and salinity variations (Church et al. 2013a, b; Stammer et al. 2013). Contributions from other effects, in particular, the static factors, are little in the present time but will become important in the future (Milne et al. 2009).

9.3 Mean Sea-Level Change

9.3.1 Global

At the time of the last interglacial period, about 125,000 years ago, sea level was likely 4–6 m higher than it was during the twentieth century, as polar average

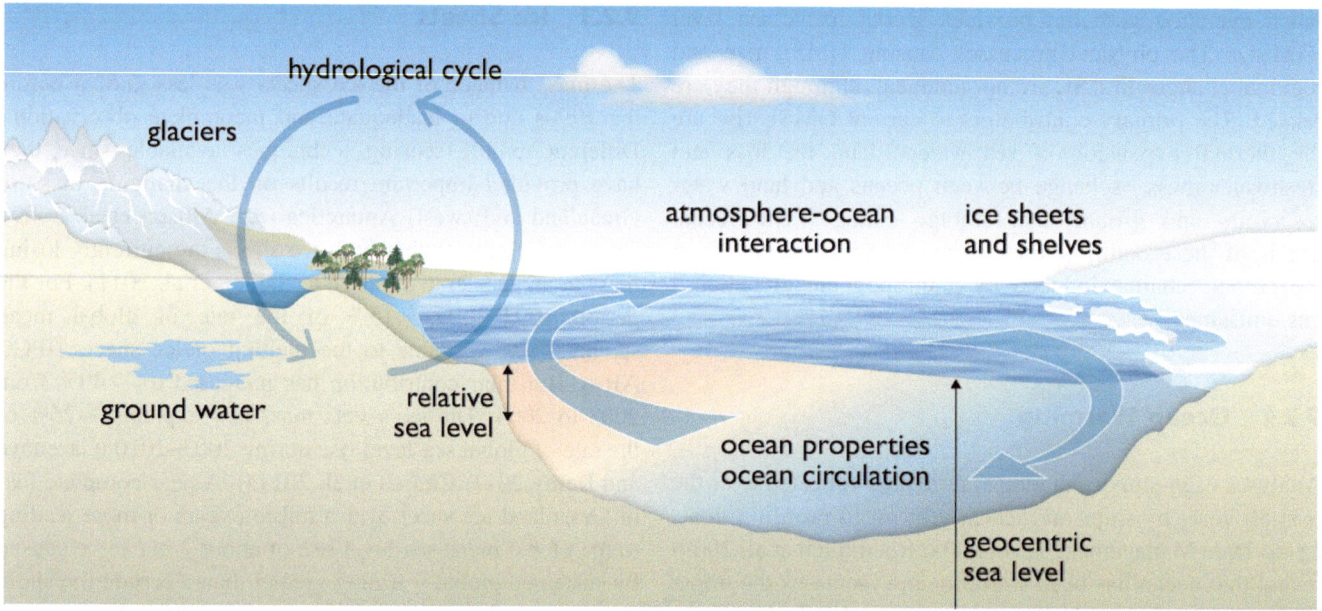

Fig. 9.1 Schematic representation of climate-sensitive processes and components that can influence the global mean sea level and regional sea level. Adapted from Fig. 13.1, IPCC AR5

temperatures were 3–5 °C higher than present values (Dutton and Lambeck 2012). It was shown that loss of ice from the Greenland ice sheet must have contributed to about 4 m of this higher sea level and there may also have been a contribution from the Antarctic ice sheet (Shepherd et al. 2018). Sea level was 120 m or more below present-day values at the last glacial maximum about 21,000 years ago (Peltier and Fairbanks 2006). Further, the Third Assessment Report (TAR) of the IPCC reported that during the disintegration of the northern hemisphere ice sheets at the end of the last glacial maximum, sea level rose at an average rate of 1 m century^{-1}, with peak rates of about 4 m century^{-1}. Since the end of the last deglaciation about 3000 years ago, sea level remained nearly constant (e.g. Lambeck et al. 2010; Kemp et al. 2011a, b).

Proxy and instrumental sea-level data indicate a transition in the late nineteenth century to the early twentieth century from relatively low mean rates of rising over the previous two millennia to higher rates of rise as shown in Fig. 9.2a. There have been many studies of twentieth-century sea-level rise based on analysis of past tide gauge data. For example, an estimate of global mean sea-level change over the last century based mainly on tide gauge observations is 1.5 ± 0.5 mm year^{-1} (Church et al. 2001). Since the beginning of the twentieth century, the sea-level rise was at an average rate of 1.7 (1.5–1.9) mm year^{-1} between 1901 and 2010 (Church and White 2011). This rise has accelerated recently, and the rate of global sea-level rise estimated from satellite altimetry during 1993–2010 is 3.3 (2.8–3.6) mm year^{-1}, significantly higher than the rate estimated for the

entire twentieth century. The spatial trend in sea-level anomalies from satellite data for the period 1993–2017 is shown in Fig. 9.2b. The most recent global mean sea-level trend during the period 1993–2017 amounts to 3.3 ± 0.5 mm year^{-1} at a 90% confidence level (WCRP 2018). This increase, however, has not happened at a constant rate and also sea-level rise is not globally uniform (e.g. Woodworth and Player 2003; Bindoff et al. 2007), i.e. sea-level variability, as well as trends, differs from region to region.

Accurate assessment of present-day global mean sea-level variations and its components (ocean thermal expansion, ice sheet mass loss, glaciers mass change, changes in land water storage, etc.) are highly essential. GMSL change as a function of time t is usually expressed by the sea-level budget equation:

$$GMSL(t) = GMSL(t)_{steric} + GMSL(t)_{ocean\,mass} \quad (1)$$

where $GMSL(t)_{steric}$ refers to the contributions of ocean thermal/haline expansion/contraction to sea-level change, and $GMSL(t)_{ocean\,mass}$ refers to the change in mass of the ocean caused mainly by the melting of ice sheets. The closure of the global sea-level budget was examined by WCRP (2018), comparing the observed global mean sea level with the sum of its components. Ocean thermal expansion, glaciers, Greenland and Antarctica contribute 42, 21, 15 and 8% to the global mean sea level over the 1993–present period (WCRP 2018). The time evolution of the global sea-level budget based on WCRP (2018) is shown in Fig. 9.2c. It can be seen from Fig. 9.2c that the sum of thermal expansion

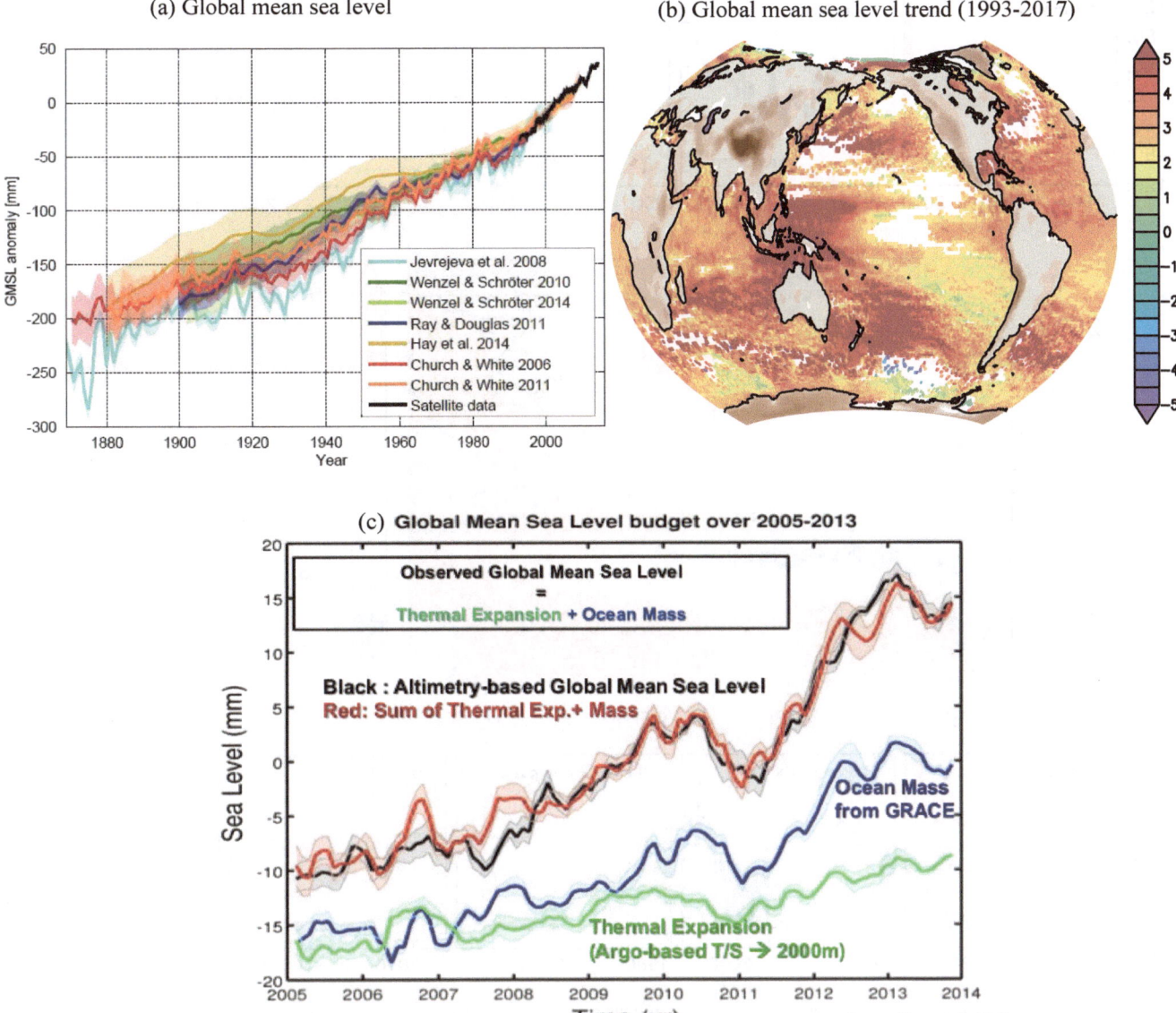

Fig. 9.2 a Time series of global mean sea-level anomaly (mm) from sea-level reconstructions and altimetry, **b** spatial map of trend of sea-level anomalies (mm year^{-1}) from AVISO (1993–2017) and **c** time series of components of sea-level budget (mm). Adapted from Dieng et al. (2015)

and mass changes nearly explains the observed sea-level rise in recent decades.

9.3.2 Regional Changes: Indian Ocean Sea-Level Rise

Regionally, sea-level variations can deviate considerably from the global mean due to various geophysical processes. These include changes in ocean circulations, which partially can be attributed to natural, internal modes of variability in the complex Earth's climate system and anthropogenic influence. The Indian Ocean sea-level trends since the 1960s

exhibit a basin-wide pattern, with sea-level falling in the south-west tropical basin and rising elsewhere (Han et al. 2010). Regionally, thermosteric changes are dominant; halosteric contributions can, however, be important in certain regions, like the Bay of Bengal and south-east Indian Ocean (e.g. Nidheesh et al. 2013; Llovel and Lee 2015). The instrumental record of sea-level change is mainly comprised of tide gauge measurements and satellite-based radar altimeter measurements.

Tide gauge records: understanding the changes in RSL is best determined based on continuous records from tide gauge measurements. The sea-level measurements by tide gauges along the west and east coast of India during the

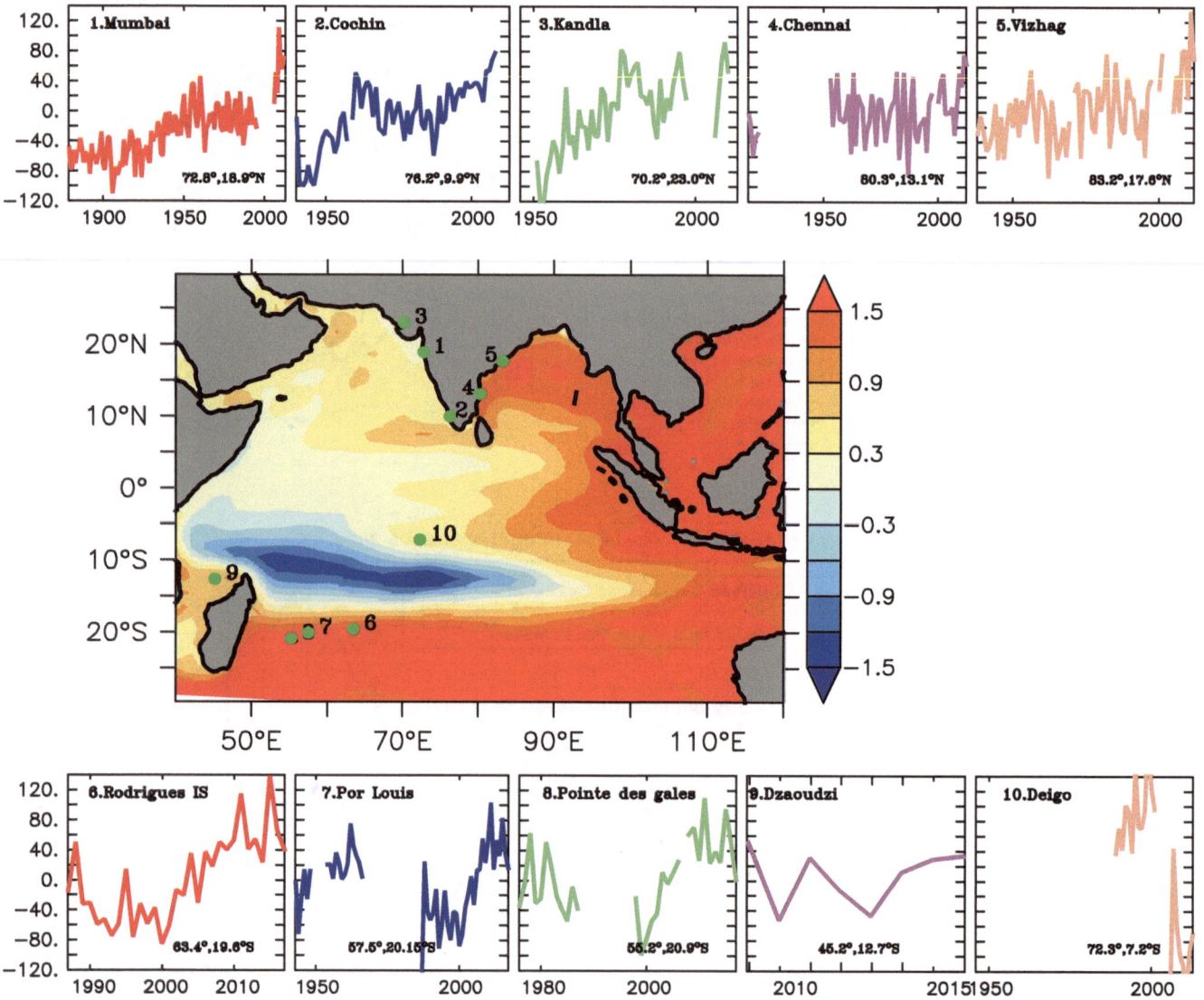

Fig. 9.3 Spatial map of sea-level trend (mm year^{-1}) in the Indian Ocean from ORAS4 reanalysis for the period 1958–2015 and time series of long-term tide gauge records along the Indian coast and open ocean. The tide gauge locations are marked by green circles. The anomalies are computed with the base period 1976–2005

twentieth century show evidence of significant sea-level rise (Fig. 9.3). Long-term sea-level trend estimates using tide gauge observations (having different time span) available along the coasts of India and the rim of the eastern Bay of Bengal show a rate of sea-level rise of about 1.06–1.75 mm year^{-1} in the Indian Ocean during 1874–2004 (Unnikrishnan et al. 2006; Unnikrishnan and Shankar 2007), similar to the global sea-level rise trend of 1.7 mm year^{-1} estimated for the period 1880–2009 (Church and White 2011). Tide gauge observations after corrections for vertical land movement from a glacial isostatic adjustment (GIA) model and for sea-level pressure changes show that sea level along the Indian Ocean coasts has increased since the 1960s, except for the fall at Zanzibar (Han et al. 2010).

Satellite Altimeter and Reanalysis With the beginning of satellite altimetry, it is possible not only to obtain a more robust estimate of global mean sea-level rise but also to examine the spatial patterns that enable assessment of regional sea-level variability. Sea level in the Indian Ocean has shown a distinct spatial pattern with a substantial increase in the north (Unnikrishnan et al. 2015; Swapna et al. 2017; Srinivasu et al. 2017; Thompson et al. 2016) and the south Indian Ocean as seen from Fig. 9.3. Mean sea-level rise in the Indian Ocean from satellite altimetry shows a rise of 3.28 mm year^{-1} during 1993 to 2017, which is higher compared to estimates from tide gauges over the historical period, but still close to the GMSL rise estimated over the same period (Unnikrishnan et al. 2015). The time evolution

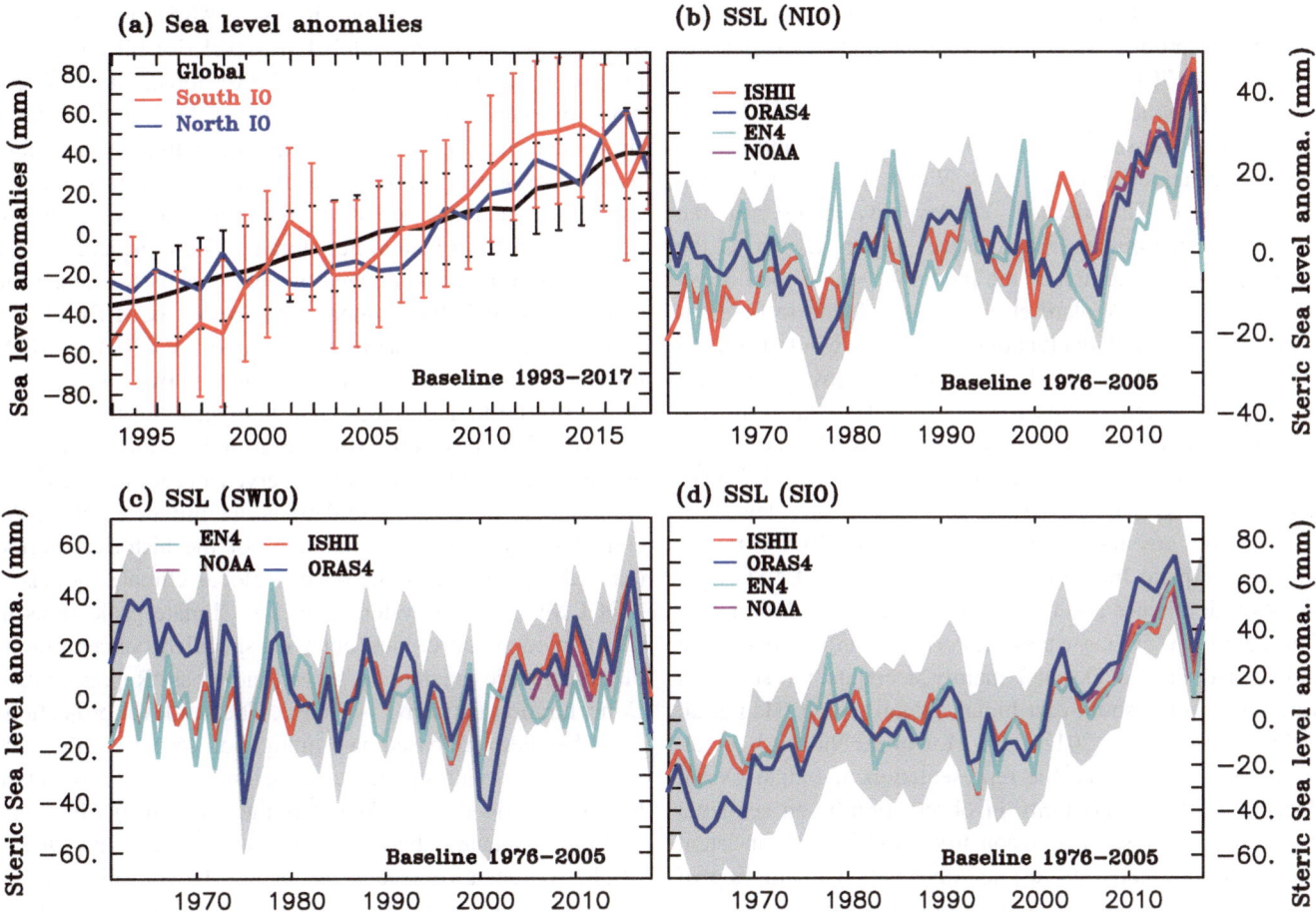

Fig. 9.4 Time series of annual mean sea-level anomalies (mm) from **a** AVISO averaged in the North Indian Ocean (blue curve; 50°E–110° E, 5°S–25°N), South Indian Ocean (red curve; 50°E–110°E, 20°S–30° S) and global mean (black curve); the error bars based on one standard deviation is shown for the global and south Indian Ocean mean sea level with respective colours; time series of steric sea level (mm) for upper the 2000 m from different reanalysis and observational datasets in the **b** North Indian ocean **c** South-west Indian Ocean and **d** South Indian Ocean. The anomalies are computed with the base period 1976–2005

of sea-level anomalies in the global ocean and the north Indian Ocean from satellite-derived Aviso data is shown in Fig. 9.4a. The north Indian Ocean sea level is rising at the same rate as that of the GMSL. In the north Indian Ocean (north of 5°S), sea level experienced a basin-wide rise from 2004 to 2013 (Srinivasu et al. 2017) with a rate of 6 mm year^{-1}. The time evolution of sea-level anomalies in the north (50°E–100°E; 5°S–25°N; NIO), south (50°E–100°E; 20°S–30°S; SIO) and in the open ocean upwelling dome (55°E–75°E; 5°S–15°S; SWIO) is shown in Fig. 9.4. We can see a sharp rise in sea level in the north and south, and a sea-level fall in the open ocean upwelling dome (SWIO). In the Indian Ocean, thermosteric sea level is the primary contributor for the spatial patterns of sea-level variability (Nidheesh et al. 2013), with halosteric sea level having apparent contributions in some regions particularly in the south-east tropical Indian Ocean and near the West Australian coast (Llovel and Lee 2015). Sea-level budget analysis performed for the Indian Ocean also indicates that more

than 70% of sea-level rise in the north Indian Ocean is contributed by the thermosteric component (Srinivasu et al. 2017; Swapna et al. 2017). The halosteric sea-level changes are dominant in the south-east Indian Ocean region with a halosteric sea-level rise of about 6.41 ± 0.62 mm year^{-1}, twice as large as that of the thermosteric sea-level rise (3.72 ± 1.04 mm year^{-1}, Llovel and Lee 2015).

Temporal Variability in Indian Ocean Sea Level

Indian Ocean sea level shows large regional variability in all temporal scales, interannual to decadal and multi-decadal scales. Superimposed on the trend, the Indian Ocean sea level shows large temporal variability (Han et al. 2014). On interannual scales, steric effects are reported to be the major contributing factors to the sea-level changes in the Indian Ocean, with contribution from El Niño Southern Oscillation (ENSO) and Indian Ocean Dipole (IOD) (Parekh et al. 2017; Palanisamy et al. 2014). Co-occurrence years (years when IOD and El Niño events co-occur) contribute significantly towards the sea-level variability in the Indian Ocean and account for around 30% of the total interannual variability (Deepa et al. 2018a, b).

Decadal sea-level variability is driven primarily by the variations in the surface wind forcing over the Indo-Pacific Ocean (Lee and McPhaden 2008; Nidheesh et al. 2013; Han et al. 2017). The decadal sea-level variability of the south-western tropical Indian Ocean region known as the thermocline ridge region of the Indian Ocean or open ocean upwelling region (Vialard et al. 2009) is found to be associated with decadal fluctuations of surface wind stress (Li and Han 2015; Deepa et al. 2018a, b). Decreasing sea-level trends were noted in this region from the 1960s and are driven by the changes in the surface winds associated with combined changes in the Indian Ocean Hadley and Walker cells, which is partly attributable to the rising levels of atmospheric greenhouse gases (Han et al. 2010). In the southern tropical Indian Ocean region, decadal ENSO contribution dominates and the Pacific influence via Indonesian Throughflow (ITF; known as the *oceanic bridge* between the Indian and Pacific Oceans) mainly accounts for sea-level variability in the south-east Indian Ocean region (Han et al. 2018; Deepa et al. 2018a, b). However, Nidheesh et al. (2017) have noted that the representation of Indian Ocean decadal sea-level variability in observation-based sea-level products (reanalyses and reconstructions) is not consistent across the products due to poor observational sampling of this basin as compared to other tropical oceans. Hence, the salient features of Indian Ocean decadal sea-level variability, briefly summarized above from various studies that used any single of those sea-level products or OGCMs, need to be considered with caution.

Multi-decadal sea-level variability in the Indian Ocean is dominated by the thermosteric changes forced by changes in the winds. For example, Swapna et al. (2017) have shown that the multi-decadal rise in north Indian Ocean sea level is caused by the weakened summer monsoon circulation, which reduces the southward ocean heat transport and results in an increased heat storage and thermosteric sea-level rise of about 3.3 mm year^{-1} during 1993–2015 in the north Indian Ocean (Swapna et al. 2017).

9.4 Future Mean Sea-Level Change

9.4.1 Global

Sea level has been rising over the past century, and the rate has accelerated in recent decades. The onset of modern sea-level rise coincided with increasing global temperature (e.g. Kemp et al. 2011a, b); sea-level rise over the coming centuries is perhaps the most damaging side of rising temperature. The Intergovernmental Panel on Climate Change's

(IPCC) Fifth Assessment Report (AR5) provided an assessment of projected global sea-level rise till the end of the twenty-first century (up to 2100) forced by different emission scenarios (Taylor et al. 2012). Projected sea-level rise under each scenario is the sum of individual contributions from steric changes and melting of glaciers and ice caps, the Greenland ice sheet, the Antarctic ice sheet and contribution from land water storage. These projections are derived from the co-ordinated modelling activities under the Coupled Model Intercomparison Project (CMIP) of the World Climate Research Programme (WCRP). CMIP5 provides projections of future climate on two time scales, near term (up to about 2035) and long term (up to 2100 and beyond). The near-term simulations (simulation over 10–30 years) are initialized with observed ocean state and sea ice, and the long-term simulations are initialized from the end of freely evolving simulations of the historical period (carried out by atmosphere-ocean global climate models—AOGCMs or Earth system models). Climate projections in CMIP models are carried out with specified concentrations of atmospheric greenhouse gases (known as 'Representative Concentration Pathways'—RCPs). The sea-level projections described here are based on a 'mitigation scenario' (RCP4.5, in which the anthropogenic emission leading to radiative forcing is limited to 4.5 Wm^{-2} in the year 2100) which is a mid-range scenario between a higher (RCP8.5) and lower (RCP2.5) scenarios (see Moss et al. 2010). Although the performance of CMIP5 models in simulating the sea level has substantially increased compared to previous versions, these models still do not account for the net ocean mass changes induced by melting ice sheets and glaciers (Flato et al. 2013). Consequently, it is not possible to evaluate the 'total' sea-level rise directly from CMIP sea-level simulations. However, the dynamic sea-level changes are given directly, and the methods by which the mass contributions are estimated (also the method of uncertainty calculation) are given in Church et al. 2013a, b (refer to their Supplementary material).

The observed global mean sea-level rise about 1–2 mm year^{-1} for 1900–2000 as inferred from tide gauges (see Fig. 9.5a) is within the range of hindcasts by CMIP models over the historical period (1870–2005; Church et al. 2013a, b), giving confidence in future projections from those models. Figure 9.5 provides the central estimates and likely ranges of projected evolution of GMSL for the twenty-first century for two emission scenarios (RCP8.5 corresponding to high emission and RCP2.6 corresponding to a very low emission scenario). Combining paleo data with historical tide gauge data confirms that the rate of sea-level rise has increased from a low rate of change during the pre-industrial period (of order tenths of mm year^{-1}) to rates of about 2 mm year^{-1} over the twentieth century, with a likely continuing acceleration during the twenty-first century

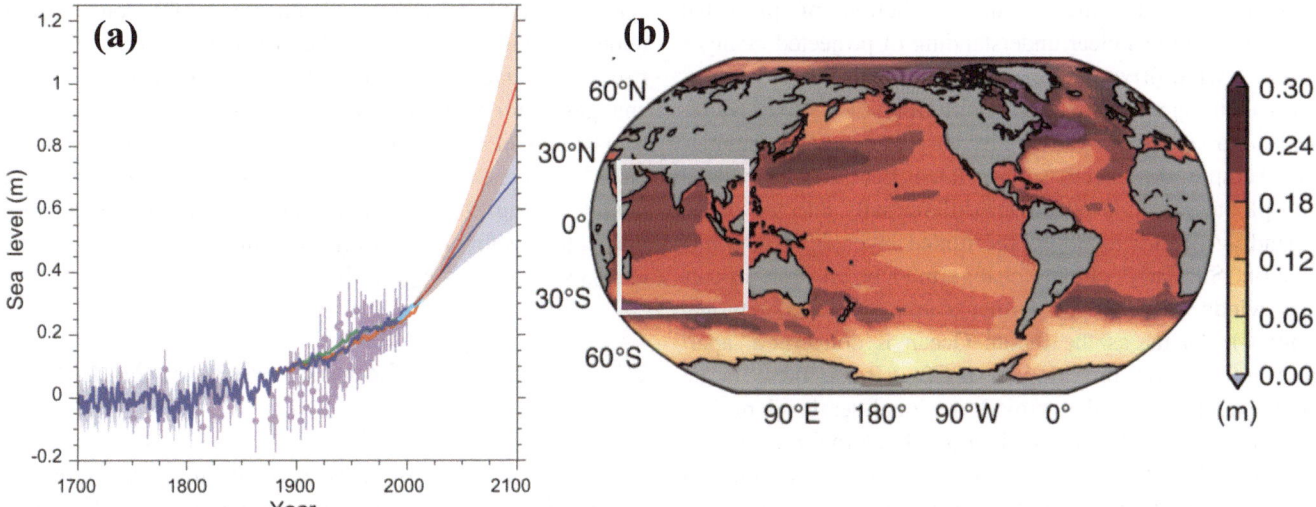

Fig. 9.5 **a** Global mean sea-level evolution derived from the compilation of palaeo sea-level data (purple), three different tide gauge reconstructions (Church and White 2011—orange, Jevrejeva et al. 2009 —blue, Ray and Douglas 2011—green), altimeter data (bright blue), and central estimates and likely ranges for future projections of global mean sea-level rise for RCP2.6 (very low emissions—blue) and RCP8.5 (very high emissions—red) scenarios, all relative to pre-industrial values. **b** Ensemble mean projection of the dynamic and steric sea-level changes for the period 2081–2100 relative to the reference period 1986–2005 from 21 CMIP5 models, using the RCP4.5 experiment. Note that, these regional sea-level projections do not include the effects of terrestrial ice-melting. The Indian Ocean is highlighted by a white rectangle. Adapted from Church et al. (2013a, b)

Table 9.1 Sea level estimates based on CMIP5 RCP scenarios (*Source* IPCC AR5; Chap. 13)

Global mean sea-level rise (relative to 1986–2005)

SSH	RCP2.6	RCP4.5	RCP6.0	RCP8.5
2081–2100 (m)	0.40 [0.26–0.55]	0.47 [0.32–0.63]	0.48 [0.33–0.63]	0.63 [0.45–0.82]
2046–2065 (m)	0.24 [0.17–0.32]	0.26 [0.19–0.33]	0.25 [0.18–0.32]	0.30 [0.22–0.38]
By the end of 2100 (m)	0.44 [0.28–0.61]	0.53 [0.36–0.71]	0.55 [0.38–0.73]	0.74 [0.52–0.98]
GMSL rise rate (mm year^{-1}) 2081–2100	4.4 [2.0–6.8]	6.10[3.5–8.8]	7.4 [4.7–10.3]	11.2 [7.5–15.7]

(Fig. 9.5a). For the high emission scenario, CMIP5 models predict a GMSL rise by 52-98 cm by the year 2100, which would threaten the survival of coastal cities and entire island nations all around the world. Even with a highly optimistic emission scenario (RCP2.5), this rise would be about 28–61 cm (Fig. 9.5a), with serious impacts on many coastal areas, including coastal erosion and a greatly increased risk of flooding. Global sea-level estimates are given in Table 9.1.

Projections incorporating Antarctic ice sheet dynamics indicate that sea levels may rise 70–100 cm under RCP4.5 and 100–180 cm under RCP8.5, though major uncertainty in sea level projections arise from the representation of ice-sheet dynamics in the models. Translating the sea-level projection into potential exposure of population, recent study by Kulp and Strauss (2019) reveals triple estimates of global vulnerability to sea-level rise and coastal flooding. However, their study is based on digital elevation model (DEM) utilizing neural networks for reducing errors in satellite-based DEM, and global coverage with widely distributed ground truth is highly essential for better understanding coastal inundations and population vulnerability.

9.4.2 Regional Sea-Level Projections for the Indian Ocean

While the global mean sea-level rise has strong societal implications, we will now see that regional sea-level changes can considerably deviate from the global mean. As shown in many studies (Zhang and Church 2012; Han et al. 2014; Hamlington et al. 2013), the observed sea-level rise over the altimeter period is indeed not uniform over the world oceans. While sea level rises at a faster rate in some oceanic regions, such as in the north Indian Ocean, sea level has shown a fall in the thermocline ridge region south of the equator in the Indian Ocean (Han et al. 2010). This contrasting spatial distribution of sea-level rise makes regional sea-level projections challenging as the attribution of projected changes requires a clear understanding of projected changes in ocean wind-driven circulation, steric changes and terrestrial ice storages (Air-sea momentum and heat fluxes variations are associated with spatially varying sea-level changes. For example, the excess surface latent heat flux associated with global warming penetrates differentially into the ocean depending on, for example, the mixed layer depth and circulation. Similarly, climate change induces changes in surface winds, which will further cause sea-level regional changes). As far as the Indian Ocean is concerned, apart from information provided in IPCC AR4/5, there is no independent studies that provide regional sea-level projections[1] and its attribution (i.e. the role of external forcing and natural variability as well as the respective contributions from global (regional) ocean steric and mass variations are still far from precise).

Figure 9.5b shows the spatial distribution of projected changes in regional sea level in the world oceans, derived from 21 CMIP models, for the RCP4.5 scenario (which is mid-range emission scenario (RCP4.5) between the high/low-end scenarios (RCPs 8.5 and 2.5). These sea-level projections reveal a clear regional pattern in sea-level changes, with complex ridge-and-trough patterns superimposed on a generally rising global mean sea level. For instance, in the Indian Ocean, the highest changes (trends) are seen in the north and western tropical Indian Ocean with a mean sea-level change of about 0.25–0.3 m at the end of the twenty-first century (Fig. 9.5b). It should be noted that north and western tropical IO is one of the few oceanic regions where maximum changes are predicted by CMIP climate models (maximum steric and dynamic sea-level rise). In a similar study, Carson et al. (2016) showed that these projected changes in mean sea level for the twenty-first century are considerably larger than the (projected) noise (natural variability) everywhere in the Indian Ocean, suggesting that the projected changes in the Indian Ocean mean sea level seen in Fig. 9.5b would have an anthropogenic origin. The contribution of thermal expansion to the GMSL rise by 2100 is estimated to be about 0.19 m for the RCP4.5 (Church et al. 2013a, b), and the slightly higher changes seen regionally in the western tropical Indian Ocean could be either related to additional changes from dynamic sea-level rise or could be resulting from an excess projected warming of this region. Note that, these projected changes (shown in Fig. 9.5b) do not include a contribution from ice-melting (mainly glaciers and ice sheets). Though, the dynamic adjustment of the world oceans to additional mass input from ice-melting is complex and occurs over decadal to century time scales (e.g. Stammer 2008), our current understandings indicate that the projected regional sea-level rise from glaciers and ice sheets is comparable to the changes from thermal expansion and circulation changes (see Church et al. 2013a, b). This is true for the Indian Ocean also, which means, combining the effects of steric changes (shown in Fig. 9.5) with mass addition may be higher than the values shown for projected Indian Ocean sea-level changes, shown in Fig. 9.5b.

9.5 Extreme Sea-Level Changes in the Indian Ocean

One of the main consequences of mean sea-level rise on human settlements is an increase in flood risk due to an increase in the intensity and frequency of extreme sea levels (ESL). Coastal areas become threatened when high tides coincide with extreme weather events and drive ESL (Wahl et al. 2017). Extreme weather (climate extremes) contributes to ESL through wind-waves and storm surges. Storm surge is an episodic increase in sea level driven by shoreward wind-driven water circulation and atmospheric pressure. Wind-waves are generated when wind energy is transferred to the ocean through surface friction and is transformed into wave energy fluxes. When waves reach the coast, they interact with the bathymetry and drive an additional increase in water levels through wave set-up and run-up. ESLs are exacerbated by tropical cyclones (TCs), which significantly intensify wind-waves and storm surge (Peduzzi et al. 2012).

ESL can be defined as a combination of surges/waves, tides and MSL, where MSL is the mean sea-level. Though there are several statistical techniques to evaluate ESL, there is no universally accepted standard or best approach for broadscale impact and adaptation analysis (Arns et al. 2013).

The IPCC AR5 (Church et al. 2013a, b; Rhein 2014) included a review of ESL reveal that the recent increase in observed extremes worldwide has been caused primarily by an increase in MSL, although the dominant modes of climate variability (particularly the El Niño Southern Oscillation (ENSO), Indian Ocean Dipole (IOD), North Atlantic Oscillation (NAO) and other modes) also have a measurable influence on extremes in many regions. Since the IPCC AR5, there have been a number of studies relating to ESL. Wahl (2014) reported rapid changes in the seasonal cycle of MSL along the Gulf Coast of the United States. The spatial variability of ESL is found to be considerably lower than the global mean trend (Vousdoukas et al. 2018). However, recent studies have shown that global warming will induce changes in storm surges and wind-waves, while cyclonic activity may also be affected (Hemer et al. 2013; Woodruff

---

[1]Future projections for the Indian Ocean are based on thermosteric component (thermal expansion only).

et al. 2013). These climate extremes along with sea-level rise will affect ESL and intensify coastal flood risk (Vousdoukas et al. 2018).

The recent evolution of extremely high waters along the severe cyclone-risk coasts of the Bay of Bengal (the east coast of India and Bangladesh) was assessed using long-term (24–34 years) hourly tide gauge data available from five stations by Antony et al. 2016. They have noticed the highest water levels above mean sea level with the greatest magnitude towards the northern part of the Bay, which decreases towards its south-west. Extreme high waters were also observed resulting from the combination of moderate, or even small, surges with large tides at these stations in most of the cases. In the Bay of Bengal, return period and return level estimations of extreme sea levels have been provided by Unnikrishnan et al. (2004) and Lee (2013) using hourly tide gauge data and Unnikrishnan et al. (2011) using storm surge models, driven by regional climate models. In the Indian Network for Climate Change Assessment Report-II, Chap. 4, Unnikrishnan et al. (2010) have shown that higher flood risks are also associated with storm surge along the southern part of the east coast of India, where tidal ranges are low. The study by Rao et al (2015) also suggested that in an extreme climate change scenario, there is a high risk of inundation over many regions of Andhra Pradesh, which is along the east coast of India. Extreme sea-level projections for Ganga-Brahmaputra-Meghna delta (Kay et al. 2015) show an increased likelihood of high water events through the twenty-first century. The First Biennial Update Report (BUR) to UNFCCC by the Government of Indian 2015 emphasized the need to promote sustainable development based on scientific principles taking into account the dangers of natural hazards in the coastal areas, and sea-level rise due to global warming.

Long and good quality tide gauge sea-level records are geographically biased towards the European and North American coasts, reflecting the historical limitations of the worldwide tide gauge dataset. Therefore, major ESL analyses cover the ocean's basin to the extent as possible, but there is a substantial lack of information especially in the Southern Hemisphere and the Indian Ocean (Marcos et al. 2015). Extreme sea levels, caused by storm surges and high tides, can have devastating societal impacts. It was shown that up to 310 million people residing in low elevation coastal zones are already directly or indirectly vulnerable to ESL (Hinkel et al. 2014). Since the extreme sea level is projected to increase with an increase in mean sea level and climate extremes, an extensive network of tide gauges with co-located GPS systems is needed along the Indian coastline, as our coastline is among the most vulnerable and densely populated regions of the globe.

The projected rise in mean sea level and ESL is a real caution for countries off the rim of north Indian Ocean, and for India, and to a large number of islands in the Indian Ocean, many of them have fragile infrastructures and population density is projected to become the largest in the world by 2030, with about 340 million people exposed to coastal hazards (Neumann et al. 2015). The sea-level rise-related coastal hazards include loss of land, salinization of freshwater supplies and an increased vulnerability to flooding. For instance, the Bay of Bengal already witnesses more than 80% of the total fatalities due to tropical cyclones, while only accounting for 5% of these storms globally (Paul 2009). The storm surges associated with the cyclone conflates with the climate change-induced sea-level rise (e.g. Han et al. 2010) to increase vulnerability. In general, the regional projections of the Indian Ocean mean sea level for the twenty-first century (see Fig. 9.5b) from climate models are high compared to other tropical oceans and the consequences could become double-fold considering the fact that the Indian Ocean basin hosts millions of people on its rim lands. In further research, an improved understanding of the factors controlling the Indian Ocean long-term sea-level rise and continuous monitoring of sea-level variability is essential for better assessing its socio-economic and environmental impacts in a changing climate.

9.6 Knowledge Gaps

Lack of long sea-level observations for the Indian Ocean is a major caveat to derive the reliable basin-scale pattern of sea-level rise and multi-decadal variability in this basin. For example, previous studies suggested that the multi-decadal oscillations in regional sea level call for a minimum of 50–60 years of sea-level data in order to establish a robust long-term trend (e.g. Douglas 1997; Chambers et al. 2012). There are only two tide gauges in the Indian Ocean that go back to the nineteenth century: Mumbai (west coast of India) and Fremantle (west coast of Australia). Figure 9.6 indeed shows that, except for a few gauge stations along the coastal India and west coast of Australia, there are no gauge records available in the interior ocean that spans over a minimum of 40 years. On the other hand, satellite altimetry provides high-resolution sea-level measurements over the entire basin since 1992, but the data are so short in terms of providing reliable estimates of regional sea-level rise trends given the presence of multi-decadal oscillations (e.g. Unnikrishnan et al. 2015; Swapna et al. 2017). In the same lines, the unavailability of long-term hydrographic profiles in the Indian Ocean limits our knowledge of long-term heat content and salinity variations in the basin and hence the steric sea level. In a recent study, Nidheesh et al. (2017) showed that representation of Indian Ocean decadal sea-level variations in observation-based sea-level datasets (reanalyses and reconstructions) is not robust and the inconsistencies are

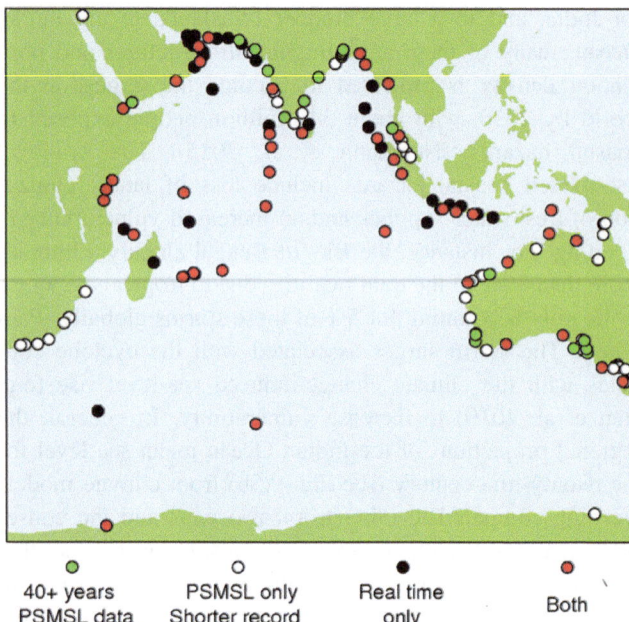

Fig. 9.6 Active tide gauge stations in the Indian Ocean. PSMSL stations are considered active if data are available for 2011 or later. Real-time stations are considered active if they have supplied data in 2017. Adapted from Beal et al. (2019)

largely attributed to the lack of past sea-level data in the interior ocean especially prior to 1980. The international venture of enhancing and sustaining the Indian Ocean Observing System (IndOOS, Beal et al. 2019) aims to fill those many knowledge gaps which arise mainly from lack of long-term observations in the Indian Ocean.

9.7 Summary

One of the major consequences of warming of the global ocean and the melting of ice and glaciers is the rise in mean sea level. There is high confidence that sea level has been rising in the global oceans as well as in the Indian Ocean. Over 90% of the anthropogenic excess heat goes into the oceans (Church et al. 2013a, b), remaining goes into melting both terrestrial and sea ice, and warming the atmosphere and land (Hansen et al. 2011; Church et al. 2011b; Trenberth et al. 2014). As a result, global mean sea level has risen by 1.7 (1.5–1.9) mm year^{-1} since 1901 and the rate of rise has accelerated to 3.3 mm year^{-1} since 1993. Sea-level rise in the Indian Ocean is non-uniform and the rate of north Indian Ocean rise is 1.06–1.75 mm year^{-1} from 1874 to 2004 and is 3.3 mm year^{-1} in the recent decades (1993–2015). Indian Ocean sea-level rise is distinct, dominated by the thermal expansion, while mass contribution is the major contributor to the GMSL rise. Steric sea level along the Indian coast is likely to rise by about 20 to 30 cm at the end of the twenty-first century and the corresponding estimate for global

mean steric sea-level rise is 18 ± 5 cm (relative to 1986–2005), under RCP4.5 (for a mid-range emission scenario, excluding ice-melt contributions). Considering the fact that coasts are home to approximately 28% of the global population, including 11% living on land less than 10 m above sea level, long-term sustained observations and continued modelling are critical for detecting, understanding and predicting ocean and cryosphere change, thus providing knowledge to inform risk assessments and adaptation planning (IPCC, Special Report on Ocean and Cryosphere in a Changing Climate).

References

Allison EH, Perry AL, Badjeck MC, Neil Adger W, Brown K, Conway D, Halls AS, Pilling GM, Reynolds JD, Andrew NL, Dulvy NK (2009) Vulnerability of national economies to the impacts of climate change on fisheries. Fish Fish 10:173–196

Antony C, Unnikrishnan AS, Woodworth PL (2016) Evolution of extreme high waters along the east coast of India and at the head of the Bay of Bengal. Global Planet Change 140:59–67. https://doi.org/10.1016/j.gloplacha.2016.03.008

Arns A, Wahl T, Haigh ID, Jensen J, Pattiaratchi C (2013) Estimating extreme water level probabilities: a comparison of the direct methods and recommendations for best practise. Coast Eng. https://doi.org/10.1016/j.coastaleng.2013.07.003

Beal LM, Vialard J, Roxy MK and lead authors (2019) IndOOS-2: a roadmap to sustained observations of the Indian Ocean for 2030. CLIVAR-4/2019. https://doi.org/10.36071/clivar.rp.4-1.2019

Bindoff NL, Willebrand J, Cazenave A, Gregory J, Gulev S, Hanawa K, Nojiri Y, Shum CK, Unnikrishnan A (2007) Observations: oceanic climate change and sea level. In: Climate change. Contribution of working group I to the fourth assessment report of the intergovernmental panel on climate change, pp 386–432

Carson M, Köhl A, Stammer D et al (2016) Coastal sea level changes, observed and projected during the 20th and 21st century. Clim Change 134:269–281. https://doi.org/10.1007/s10584-015-1520-1

Cazenave A, Remy F (2011) Sea level and climate: measurements and causes of changes. WIREs Clim Change 2:647–662

Cheng L, Trenberth KE, Fasullo J, Boyer T, Abraham J, Zhu J (2017) Improved estimates of ocean heat content from 1960 to 2015. Sci Adv 3:e1601545. https://doi.org/10.1126/sciadv.1601545

Church JA, White NJ (2011) Sea-level rise from the late 19th to the Early 21st Century. Surv Geophys 32(4–5):585–602. https://doi.org/10.1007/s10712-011-9119-1

Church JA, Gregory JM, Huybrechts P, Kuhn M, Lambeck K, Nhuan MT, Qin D, Woodworth PL (2001) Changes in sea level. In: Houghton JT, Ding Y, Griggs DJ, Noguer M, van der Linden P, Dai X, Maskell K, Johnson CI (eds) Climate change 2001: the scientific basis. Contribution of working group 1 to the third assessment report of the intergovernmental panel on climate change. Cambridge University Press, Cambridge, pp 639–694

Church JA, Gregory JM, White NJ, Platten SM, Mitrovica JX (2011a) Understanding and projecting sea-level change. Oceanography 24:130–143. https://doi.org/10.5670/oceanog.2011.33

Church JA, White NJ, Konikow LF, Domingues CM, Cogley JG, Rignot E, Gregory JM, Vanden Broeke MR, Monaghan AJ, Velicogna I (2011b) Revisiting the Earth's sea-level and energy budgets from 1961 to 2008. Geophys Res Lett 38:L18601. https://doi.org/10.1029/2011GL048794

Church JA, Monselesan D, Gregory JM, Marzeion B (2013a) Evaluating the ability of process based models to project sea-level change. Environ Res Lett 8:014051

Church JA, Clark PU, Cazenave A, Gregory JM, Jevrejeva S, Levermann A, Merrifield MA, Milne GA, Nerem RS, Nunn PD, Payne AJ, Pfeffer WT, Stammer D, Unnikrishnan AS (2013) Sea level change. In: Stocker TF, Qin D, Plattner G-K, Tignor M, Allen SK, Boschung J, Nauels A, Xia Y, Bex V, Midgley PM (eds) Climate change 2013: the physical science basis. Contribution of working group I to the fifth assessment report of the intergovernmental panel on climate change. Cambridge University Press, Cambridge, United Kingdom and New York, NY, USA

Chambers D, Merrifield MA, Nerem RS (2012) Is there a 60-year oscillation in global mean sea level? Geophys Res Lett. doi 10.1029/2012GL052885

Deepa JS, Gnanaseelan C, Kakatkar R, Parekh A, Chowdary JS (2018a) The interannual sea level variability in the Indian Ocean as simulated by an ocean general circulation model. Int J Climatol 38:1132–1144. https://doi.org/10.1002/joc.5228

Deepa JS, Gnanaseelan C, Mohapatra S, Chowdary JS, Karmakar A, Kakatkar R, Parekh A (2018b) The Tropical Indian Ocean decadal sea level response to the Pacific decadal oscillation forcing. Clim Dyn. https://doi.org/10.1007/s00382-018-4431-9

Dieng HB, Champollion N, Cazenave A, Wada Y, Schrama E, Meyssignac B (2015) Total land water storage change over 2003–2013 estimated from a global mass budget approach. Environ Res Lett 10(12):124010. https://doi.org/10.1088/1748-9326/10/12/124010

Domingues CM, Church JA, White NJ, Gleckler PJ, Wijffels SE, Barker PM, Dunn JR (2008) Improved estimates of upper-ocean warming and multi-decadal sea-level rise. Nature 453(7198):1090–1093. https://doi.org/10.1038/nature07080

Douglas B (1997) Global sea rise: a redetermination. Surv Geophys 18(2–3):279–292. https://doi.org/10.1023/a:1006544227856

Dutton A, Lambeck K (2012) Ice volume and sea level during the last interglacial science 337:216–219

First Biennial Update Report to the United Nations Framework Convention on Climate Change (2015). https://unfccc.int/resource/docs/natc/indbur1.pdf

Flato G, Marotzke J, Abiodun B et al (2013) Evaluation of climate models. In: Stocker TF, Qin D, Plattner GK et al (eds) Climate change 2013: the physical science basis. Contribution of working group I to the fifth assessment report of the intergovernmental panel on climate change. Cambridge University Press, Cambridge, pp 741–866

Han W et al (2010) Patterns of Indian Ocean sea-level change in a warming climate. Nat Geosci 3(8):546–550. https://doi.org/10.1038/ngeo901

Han W, Vialard J, McPhaden MJ et al (2014) Indian Ocean decadal variability: a review. Bull Amer Meteor Soc 95:1679–1703

Han W, Meehl GA, Stammer D, Hu A, Hamlington B, Kenigson J, Palanisamy H, Thompson P (2017) Spatial patterns of sea level variability associated with natural internal climate modes. Surv Geophys 38(1):217–250

Han W, Stammer D, Meehl G, Hu A, Sienz F, Zhang L (2018) Multi-decadal trend and decadal variability of the regional sea level over the Indian Ocean since the 1960s: roles of climate modes and external forcing. Climate 6(2):51. https://doi.org/10.3390/cli6020051

Hansen J, Sato M, Kharecha P, von Schuckmann K (2011) Earth's energy imbalance and implications. Atmos Chem Phys 11:13421–13449. https://doi.org/10.5194/acp-11-13421-2011

Hamlington BD, Leben RR, Strassburg MW, Nerem RS, Kim K-Y (2013) Contribution of the Pacific decadal oscillation to global mean sea level trends. Geophys Res Lett 40: 5171–5175

Hemer MA, Fan Y, Mori N, Semedo A, Wang XL (2013) Projected changes in wave climate from a multi-model ensemble. Nat Clim Change 3(5):471–476. https://doi.org/10.1038/nclimate1791

Hinkel J, Lincke D, Vafeidis AT, Perrette M, Nicholls RJ, Tol RSJ et al (2014) Coastal flood damage and adaptation costs under 21st century sea-level rise. Proc Natl Acad Sci. https://doi.org/10.1073/pnas.1222469111

Intergovernmental Panel on Climate Change (2007) Technical summary. In: Solomon S et al (eds) Climate change 2007: the physical science basis. Contribution of working group I to the fourth assessment report of the intergovernmental panel on climate change. Cambridge University Press, Cambridge, pp 19–91

Ishii M, Kimoto M (2009) Reevaluation of historical ocean heat content variations with an XBT depth bias correction. J Oceanogr 65:287–299. https://doi.org/10.1007/s10872-009-0027-7

Jevrejeva S, Grinsted A, Moore JC (2009) Anthropogenic forcing dominates sea level rise since 1850. Geophys Res Lett 36:L20706

Kaser G, Cogley JG, Dyurgerov MB, Meier MF, Ohmura A (2006) Mass balance of glaciers and ice caps: consensus estimates for 1961–2004. Geophys Res Lett 33:L19501. https://doi.org/10.1029/2006GL027511

Kay S, Caesar J, Wolf J, Bricheno L, Nicholls RJ, Saiful Islam AKM et al (2015) Modelling the increased frequency of extreme sea levels in the Ganges-Brahmaputra-Meghna delta due to sea level rise and other effects of climate change. Environ Sci Process Impacts. https://doi.org/10.1039/c4em00683f

Kemp AC, Horton BP, Donnelly JP, Mann ME, Vermeer M, Rahmstorf S (2011a) Climate related sea-level variations over the past two millennia. Proc Natl Acad Sci USA 108:11017–11022

Kemp PS, Worthington TA, Langford TE, Tree AR, Gaywood MJ (2011b) First published: 01 June (2011). https://doi.org/10.1111/j.1467-2979.2011.00421

Kulp SA, Strauss BH (2019) New elevation data triple estimates of global vulnerability to sea-level rise and coastal flooding. Nat Commun 10:4844. https://doi.org/10.1038/s41467-019-12808-z

Lee HS (2013) Estimation of extreme sea levels along the Bangladesh coast due to storm surge and sea level rise using EEMD and EVA. J Geophys Res Oceans 118:4273–4285. https://doi.org/10.1002/jgrc.20310

Lee T, McPhaden MJ (2008) Decadal phase change in large-scale sea level and winds in the Indo-Pacific region at the end of the 20th century. Geophys Res Lett 35:L01605. https://doi.org/10.1029/2007GL032419

Levitus S et al (2012) World ocean heat content and thermosteric sea level change (0–2000 m), 1955–2010. Geophys Res Lett 39: L10603. https://doi.org/10.1029/2012GL051106

Levitus S, Antonov JI, Boyer TP, Garcia HE, Locarnini RA, Mishonov AV, Garcia HE (2009) Global ocean heat content 1955–2008 in light of recently revealed instrumentation problems. Geophys Res Lett 36:L07608. https://doi.org/10.1029/2008GL037155

Li Y, Han W (2015) Decadal sea level variations in the Indian Ocean investigated with HYCOM: roles of climate modes, ocean internal variability, and stochastic wind forcing. J Clim 28:9143–9165. https://doi.org/10.1175/JCLI-D-15-0252.1

Lambeck K, Purcell A, Zhao J, Svensson N-O (2010) The Scandinavian Ice sheet: from MIS 4 to the end of the last glacial maximum. Boreas 39:410–435

Llovel W, Lee T (2015) Importance and origin of halosteric contribution to sea level change in the southeast Indian Ocean during 2005–2013, 1148–1157. https://doi.org/10.1002/2014GL062611

Marcos M, Calafat FM, Berihuete Á, Dangendorf S (2015) Long-term variations in global sea level extremes. J Geophys Res Oceans 120(12):8115–8134. https://doi.org/10.1002/2015JC011173

Meier MF et al (2007) Glaciers dominate eustatic sea-level rise in the 21st century. Science 317:1064. https://doi.org/10.1126/science.1143906

Milne GA, Gehrels WR, Hughes CW, Tamisiea ME (2009) Identifying the causes of sea-level change. Nat Geosci 2:471–478. https://doi.org/10.1038/ngeo544

Moss RH, Coauthors (2010) The next generation of scenarios for climate change research and assessment. Nature 463:747–756. doi 10.1038/nature08823

Neumann B, Vafeidis AT, Zimmermann J, Nicholls RJ (2015) Future coastal population growth and exposure to sea-level rise and coastal flooding—a global assessment. PLoS ONE 10(3). https://doi.org/10.1371/journal.pone.0118571

Nidheesh AG, Lengaigne M, Vialard J, Unnikrishnan AS, Dayan H (2013) Decadal and long-term sea level variability in the tropical Indo-Pacific Ocean. Clim Dyn 41(2):381–402. https://doi.org/10.1007/s00382-012-1463-4

Nidheesh AG, Lengaigne M, Vialard J, Izumo T, Unnikrishnan AS, Meyssignac B, Hamlington B, de Boyer Montégut C (2017) Robustness of observation-based decadal sea level variability in the Indo-Pacific Ocean. Geophys Res Lett 44:7391–7400. https://doi.org/10.1002/2017GL073955

Paul BK (2009) Why relatively fewer people died? The case of Bangladesh's cyclone sidr. Nat Hazards 50(2):289–304. https://doi.org/10.1007/s11069-008-9340-5

Palanisamy H, Cazenave A, Meyssignac B, Soudarin L, Wöppelmann G, Becker M (2014) Regional sea level variability, total relative sea level rise and its impacts on islands and coastal zones of Indian Ocean over the last sixty years. Global Planet Change 116:54–67. https://doi.org/10.1016/j.gloplacha.2014.02.001

Parekh A, Gnanaseelan C, Deepa JS, Karmakar A, Chowdary JS (2017) Sea level variability and trends in the North Indian Ocean. In: Rajeevan MN, Naik S (eds) Observed climate variability and change over the Indian Region. Springer Singapore, pp 181–192. doi https://doi.org/10.1007/978-981-10-2531-0

Peduzzi P, Chatenoux B, Dao H, De Bono A, Herold C, Kossin J et al (2012) Global trends in tropical cyclone risk. Nat Clim Change 2(4):289–294. https://doi.org/10.1038/nclimate1410

Peltier W, Fairbanks RG (2006) Global glacial ice volume and last glacial maximum duration from an extended Barbados Sea level record. Quatern Sci Rev 25:3322–3337. https://doi.org/10.1016/j.quascirev.2006.04.010

Rao AD, Poulose J, Upadhyay P, Mohanty S (2015) Local-scale assessment of tropical cyclone induced storm surge inundation over the coastal zones of India in probabilistic climate risk scenario. In: Ravela S, Sandu A (eds) Dynamic data-driven environmental systems science. DyDESS 2014. Lecture notes in computer science, vol 8964. Springer, Cham

Ray RD, Douglas BC (2011) Experiments in reconstructing twentieth-century sea levels. Prog Oceanogr 91:495–515

Rhein M (2014) Observations: ocean pages. Climate Change 2013—The Physical Science Basis, 255–316. https://doi.org/10.1017/CBO9781107415324.010

Roemmich D, and the Argo Steering Team (2009) Argo: the challenge of continuing 10 years of progress. Oceanography 22:46–55. DOI:10.5670/oceanog.2009.65

Shepherd A et al (2018) Mass balance of the Antarctic ice sheet from 1992 to 2017. Nature 558:219–222

Srinivasu U, Ravichandran USM, Han W, Rahman SSH (2017) Causes for the reversal of North Indian Ocean decadal sea level trend in recent two decades. Clim Dyn 49:3887. https://doi.org/10.1007/s00382-017-3551-y

Stammer D (2008) Response of the global ocean to Greenland and Antarctic ice melting. J Geophys Res Oceans 113:C06022

Stammer D et al (2013) Causes for contemporary regional sea level changes. Ann Rev Mar Sci 5:21–46

Steffen W et al (2011) The anthropocene: from global change to planetary Stewardship. Ambio 40(7):739–761. https://doi.org/10.1007/s13280-011-0185-x

Swapna P, Jyoti J, Krishnan R, Sandeep N, Griffies SM (2017) Multidecadal weakening of Indian summer monsoon circulation induces an increasing northern Indian Ocean sea level. Geophys Res Lett 44:560–10. https://doi.org/10.1002/2017GL074706

Taylor KE, Stouffer RJ, Meehl GA (2012) An overview of CMIP5 and the experiment design. Bull Am Meteor Soc 93:485–498

Thompson PR, Piecuch CG, Merrifield MA, McCreary JP, Firing E (2016) Forcing of recent decadal variability in the equatorial and North Indian Ocean. J Geophys Res Oceans 121:6762–6778. https://doi.org/10.1002/2016JC012132

Trenberth KE, Fasullo JT, Balmaseda MA (2014) Earth's energy imbalance. J Clim 27:3129–3144. https://doi.org/10.1175/jcli-d-13-00294.1

Unnikrishnan A, Kumar R, Sindhu B (2011) Tropical cyclones in the bay of Bengal and extreme sea level projections along the east coast of India in a future climate scenario. Curr Sci 101:327–331

Unnikrishnan A, Nidheesh G, Lengaigne M (2015) Sea-level-rise trends off the Indian coasts during the last two decades. Curr Sci 108:966–971

Unnikrishnan AS, Shankar D (2007) Are sea-level-rise trends along the coasts of the North Indian Ocean consistent with global estimates? Glob Planet Change 57(3–4):301–307. https://doi.org/10.1016/j.gloplacha.2006.11.029

Unnikrishnan A, Sundar SD, Blackman D, Kumar KR, Michael GS (2006) Sea level changes along the coast of India. Obs Proj 90(3)

Unnikrishnan AS, Manimurali M, Kumar R (2010) Observed sea level rise, extreme events and future projections. Indian Netw Clim Change Assess 4:47–66

Unnikrishnan AS, Sundar D, Blackman D (2004) Analysis of extreme sea level along the east coast of India. J Geophys Res–Oceans 109: C06023

Vousdoukas MI, Mentaschi L, Voukouvalas E, Verlaan M, Jevrejeva S, Jackson LP, Feyen L (2018) Global probabilistic projections of extreme sea levels show intensification of coastal flood hazard. Nat Commun 9(1):1–12. https://doi.org/10.1038/s41467-018-04692-w

Von Schuckmann K et al (2016) An imperative to monitor Earth's energy imbalance. Nat Clim Change 6:138–144. https://doi.org/10.1038/nclimate2876

Vialard J, Duvel J, McPhaden M, Bouruet Aubertot P, Ward B, Key E, Bourras D, Weller R, Minnett P, Weill A (2009) Cirene: air-sea interactions in the Seychelles-Chagos thermocline ridge region. Bull Am Meteor Soc 90:45–61. https://doi.org/10.1175/2008BAMS2499.1

Wahl T (2014) Along the US Gulf coast from the late 20th century, 1990, 1–8. https://doi.org/10.1002/2013GL058777.Received

Wahl T, Haigh ID, Nicholls RJ, Arns A, Dangendorf S, Hinkel J, Slangen ABA (2017) Understanding extreme sea levels for broad-scale coastal impact and adaptation analysis. Nat Commun 8(May):1–12. https://doi.org/10.1038/ncomms16075

WCRP Global Sea Level Budget Group: Global sea-level budget 1993–present (2018) Earth Syst. Sci. Data 10:1551–1590. https://doi.org/10.5194/essd-10-1551-2018

Woodruff JD, Irish JL, Camargo SJ (2013) Coastal flooding by tropical cyclones and sea-level rise. Nature 504(7478):44–52. https://doi.org/10.1038/nature12855

Woodworth PL, Player R (2003) The permanent service for mean sea level: an update to the 21st century. J Coastal Res 19:287–295

Zhang X, Church JA (2012) Sea level trends, interannual and decadal variability in the Pacific Ocean. Geophys Res Lett 39:L21701

Indian Ocean Warming

Coordinating Lead Authors

M. K. Roxy, Indian Institute of Tropical Meteorology (IITM-MoES), Pune, India
C. Gnanaseelan, Indian Institute of Tropical Meteorology (IITM-MoES), Pune, India,
e-mail: seelan@tropmet.res.in (corresponding author)

Lead Authors

Anant Parekh, Indian Institute of Tropical Meteorology (IITM-MoES), Pune, India
Jasti S. Chowdary, Indian Institute of Tropical Meteorology (IITM-MoES), Pune, India
Shikha Singh, Indian Institute of Tropical Meteorology (IITM-MoES), Pune, India
Aditi Modi, Indian Institute of Tropical Meteorology (IITM-MoES), Pune, India
Rashmi Kakatkar, Indian Institute of Tropical Meteorology (IITM-MoES), Pune, India
Sandeep Mohapatra, Indian Institute of Tropical Meteorology (IITM-MoES), Pune, India
Chirag Dhara, Indian Institute of Tropical Meteorology (IITM-MoES), Pune, India

Review Editors

S. C. Shenoi, Indian National Centre for Ocean Information Services (INCOIS-MoES), Hyderabad, India
M. Rajeevan, Ministry of Earth Sciences, Government of India, New Delhi, India

Corresponding Author

C. Gnanaseelan, Indian Institute of Tropical Meteorology (IITM-MoES), Pune, India,
e-mail: seelan@tropmet.res.in

R. Krishnan et al. (eds.), *Assessment of Climate Change over the Indian Region*,
https://doi.org/10.1007/978-981-15-4327-2_10

Key Messages

- The tropical Indian Ocean has experienced rapid basin-wide sea surface temperature (SST) warming, with an average rise of 1.0 °C (0.15 °C/decade) during 1951–2015, over which period the global average SST warmed about 0.7 °C (0.11 °C/decade) (*high confidence*). The SST warming is spatially non-uniform and about 90% of the warming is attributed to anthropogenic emissions.

- The basin-wide non-uniform SST warming trend in the tropical Indian Ocean is to continue in the future, under both medium and high emission scenarios (high confidence)

- The frequency of extreme positive Indian Ocean Dipole (IOD) events is projected to increase by almost a factor of three, with one-in-seventeen-year events in the twentieth century to one-in-six-year by the end of the twenty-first century (*low confidence*).

- The heat content of the upper 700 m of the Indian Ocean has exhibited an increasing trend during 1955–2015 (*high confidence*), with spatially non-uniform heating.

- SST warming has *very likely* contributed to the decreasing trend observed in oxygen (O$_2$) concentrations in the tropical Indian Ocean, and the declining trend in pH and marine phytoplankton over the western Indian Ocean. These trends are projected to continue with global warming.

10.1 Introduction

About one third of the global population lives around the Indian Ocean—many in low-lying coastal regions, small islands or low-to-middle income nations with low adaptive capacity—that are especially vulnerable to climate change impacts. Ocean-atmospheric conditions over the Indian Ocean regulate the regional weather-climate system over these regions. Hence, variability and changes in this basin are of great significance to the food, water and power security in India and neighbouring countries.

Warm sea surface temperatures (SSTs >28 °C), known as the Indian Ocean warm pool, occur over a large part of the tropical Indian Ocean (TIO, 40 °E:115 °E; 30 °S:30 °N), which is a part of the larger Indo-Pacific warm pool. It favours deep atmospheric convection (Graham and Barnett 1987) and energizes the global atmospheric circulation, particularly the Hadley circulation and the Walker circulation thereby modulating the major elements of global climate such as the Indian monsoon and the El Niño Southern Oscillation (ENSO). The strong monsoon winds during June–July–August–September force intense coastal and open-ocean upwelling in the Arabian Sea, and modulate evaporation and moisture transport towards India (Izumo et al. 2008). They also provide a globally significant source of atmospheric CO$_2$ (Valsala and Murtugudde 2015), and foster intense marine primary productivity (Roxy et al. 2016). Unlike the other tropical ocean basins, the Indian Ocean is landlocked in the north by the vast Asian landmass and hence the only trans-basin exchanges are from the West Pacific via the Indonesian Seas, known as the Indonesian Through flow (ITF), and from the south.

The rate of warming in the tropical Indian Ocean is the fastest among tropical oceans and accounts for about one quarter of the increase in global oceanic heat content over the last two decades (Beal et al. 2019) despite being the smallest of the tropical oceans (representing only 13% of the global ocean surface). The Indian Ocean is home to 30% of the world's coral reefs and 13% of global wild-catch fisheries. This marine ecosystem, including corals and phytoplankton, and fisheries are being impacted by a rise in heat waves in the ocean, known as marine heat waves (Collins et al. 2019). Moreover, an expansion of the Indian Ocean warm pool is changing the subseasonal weather variability such as the monsoon intraseasonal oscillation (Sabeerali et al. 2014) and the Madden-Julian oscillation (Roxy et al. 2019), which originate in the Indian Ocean. This has an impact on rainfall characteristics, particularly on extreme rain over the tropics, including India. Heat waves (Chap. 2), droughts and floods (Chap. 6), tropical cyclones (Chap. 8) and extreme sea-level changes (Chap. 9) are becoming more frequent and intense around the Indian Ocean as regional climate patterns respond to anthropogenic climate change (Collins et al. 2019). Hence, there is an urgent need to understand the status and future evolution of the Indian Ocean warming and its role in influencing the regional climate under a global warming environment.

10.2 Observed and Projected Changes in Indian Ocean SST

10.2.1 Observed Changes in SST

The oceans have absorbed approximately 93% of the additional heat due to anthropogenic global warming since the 1950s (Cheng et al. 2017) and have resulted in a significant increasing trend in the global average ocean SST, as evidenced by modern instrumental records (Deser et al. 2010). Among the oceans, Indian Ocean stands out as one of the most rapidly warming ocean basins (Gnanaseelan et al. 2017; Beal et al. 2019). The global average rise in SST

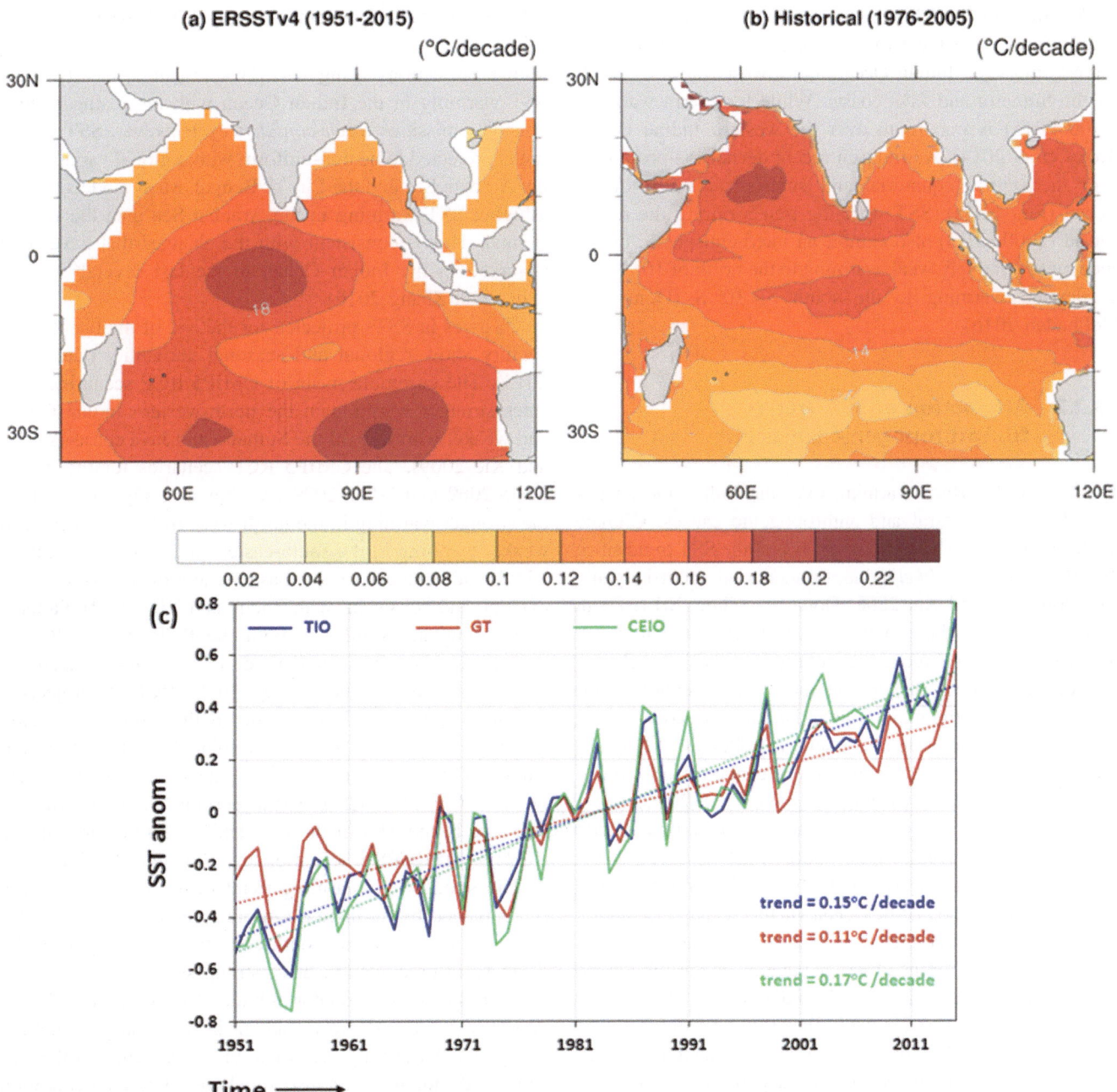

Fig. 10.1 SST trend (°C/decade) from: **a** observed Extended Reconstructed Sea Surface Temperature, version 4 (ERSST v4) during 1951–2015, **b** CMIP5 historical simulations for 1976–2005 and **c** Time series of SST anomalies (solid lines) and their linear trends (°C/decade; dotted line) over the tropical Indian Ocean (TIO, blue), central equatorial Indian Ocean (CEIO, 60 °E:90 °E; 10 °S:10 °N, green) and global tropics (GT, 0°:360°; 30 °S:30 °N, red)

during 1951–2015 is 0.7 °C (0.11 °C/decade), while the TIO SST has risen by about 1.0 °C on average (0.15 °C/decade) (*high confidence*) (Fig. 10.1). A 138-year (1870–2007) monthly observed SST time series averaged along the ship track extending from the Gulf of Aden through the Malacca Strait also reveals a warming of 1.4 °C during the entire period (Chowdary et al. 2012). Coherently, historical simulations by climate models participating in the Coupled Model Inter-comparison Project (CMIP5), with CO_2 forcing at observed rates, show warming trends of 0.1–0.18 °C per decade during 1976–2005 (Fig. 10.1b), with maximum warming trends over the northern Arabian Sea.

Warming in the TIO has been basin-wide but spatially non-uniform, with the largest increasing trends seen in the central equatorial Indian Ocean and lowest warming trends off the Sumatra and Java coasts. While long-term warming (1900–2015) is maximum over the western Indian Ocean (Roxy et al. 2014), warming in recent decades is prominent over the central equatorial and subtropical south Indian Ocean (Fig. 10.1a). SST warming trends during the recent period (2000–2013) also exhibit an interhemispheric difference, with relatively weak warming to the north of 10° S and accelerated warming to the south of 10° S (Dong and McPhaden 2016).

10.2.2 Attribution of SST Trends to Anthropogenic Emissions

The observed surface warming over the Indian Ocean has been linked to natural and anthropogenic causes. Climate model simulations show that over 90% of the SST trend since the 1950s is *very likely* due to increased anthropogenic emissions (Du and Xie 2008; Dong and Zhou 2014; Dong et al. 2014), while the remaining is due to internal variability (Dong et al. 2014). Among the anthropogenic causes, change in radiative forcing due to the increased greenhouse gas concentrations is the major factor (Du and Xie 2008). Over recent decades, increasing atmospheric pollutants, known as aerosols, have *likely* dampened the greenhouse gas forced warming of the Indian Ocean (Dong and Zhou 2014).

Changes in the tropical circulation, ocean-atmospheric interaction and dynamics play a role in the observed distribution of the warmer waters in the Indian Ocean. Redistribution of heat from the Pacific Ocean via the Walker circulation (Roxy et al. 2014) and the ITF (Dong and McPhaden 2016) is one of the reasons for the observed patterns of warming in the Indian Ocean. Other than the ITF, Indian Ocean may also be receiving a warming signal via the deep meridional overturning circulation, and from the Southern Ocean (Gille 2002). Contrary to expectation, there is a negative trend in the net heat flux despite the warming trend in the Indian Ocean SSTs (Rahul and Gnanaseelan 2013). This suggests that local ocean dynamics and ocean-atmosphere interaction also have a major role in the observed warming pattern in the Indian Ocean (Lau and Nath 2000; Du et al. 2009; Rahul and Gnanaseelan 2016; Pratik et al. 2019; Rao et al. 2012). The period 2000–2013 witnessed strong warming in the southern Indian Ocean south of 10° S inducing a north–south SST gradient. This interhemispheric gradient is forced primarily by an increased ITF, from the Pacific into the Indian Ocean, induced by stronger Pacific trade winds (Dong and McPhaden 2016; Lee et al 2015).

10.2.3 Future Projections of SST Warming

Future projections using CMIP5 simulations clearly show SST warming in the Indian Ocean with increasing anthropogenic emissions (*high confidence*). However, SST warming is projected to be non-uniform with regional variations in the Arabian Sea and the Bay of Bengal. Most models project a higher SST warming in the Arabian Sea than the Bay of Bengal. This is consistent with the changes that have been observed in the Indian Ocean in the last several decades (Zhao and Zhang 2016).

Changes in SSTs projected for the end of the twenty-first century show regional and seasonal variability (Cai et al. 2013). The ensemble mean of CMIP5 RCP scenarios indicates stronger warming in the north-western Indian Ocean and weaker warming off the Sumatra and Java coasts (Zheng and Xie 2009). The CMIP5 RCP scenarios for the period 2040–2069 and 2070–2099 are shown in Fig. 10.2, where the stronger warming in the north-western Indian Ocean and weaker warming in the south-eastern Indian Ocean south of 15° S with respect to the base period of 1976–2005 are evident, and consistent with Zheng and Xie (2009). Patterns of warming are similar in both the RCP4.5 and RCP8.5 scenarios, although the magnitude of warming is much larger in the latter scenario (Fig. 10.2). RCP4.5 projects a warming rate of 0.13 °C/decade in the TIO (Fig. 10.2e), similar to the current rate of warming (Fig. 10.1c). Meanwhile, RCP8.5 indicates an accelerated warming at the rate of 0.35 °C/decade. CMIP5 (the ensemble mean) projected SST rise in the TIO in the near and far future for both the RCP4.5 and RCP8.5 scenarios are tabulated in Table 10.1.

The strong SST warming trend in the north-west basin accompanied by an increase in precipitation and weak SST warming trend in the south-east basin accompanied by decrease in precipitation drive strong surface easterly wind anomalies along the equatorial Indian Ocean (Li et al. 2016). Along with the east–west gradient in SST and precipitation under global warming scenarios, the thermocline in the east equatorial Indian Ocean also shoals (a favourable condition for the formation of IOD) as a result of a weakened Indian Ocean Walker cell and easterly wind change along the equator (Zheng and Xie 2009). The shoaling of the thermocline then strengthens the thermocline feedback in this region. Such a pattern of SST, precipitation and thermocline feedback would result in more frequent occurrences of extreme positive IOD (pIOD) events in the future, from one event every 17.3 years over the twentieth century to one event every 6.3 years by the end of twenty-first century (*low confidence*) (Cai et al. 2014). This suggests increased risks of climate and weather extremes in regions impacted by extreme pIOD events.

Fig. 10.2 CMIP5 projected SST change (°C) in **a** RCP4.5 for 2040–2069, **b** RCP4.5 for 2070–2099, **c** RCP8.5 for 2040–2069 and **d** RCP8.5 for 2070–2099, with respect to the reference period 1976–2005. **e** Time series of projected SST anomalies with the trends (°C/decade) over the tropical Indian Ocean, in RCP4.5 (blue) and RCP8.5 (red) scenarios for 2040–2099

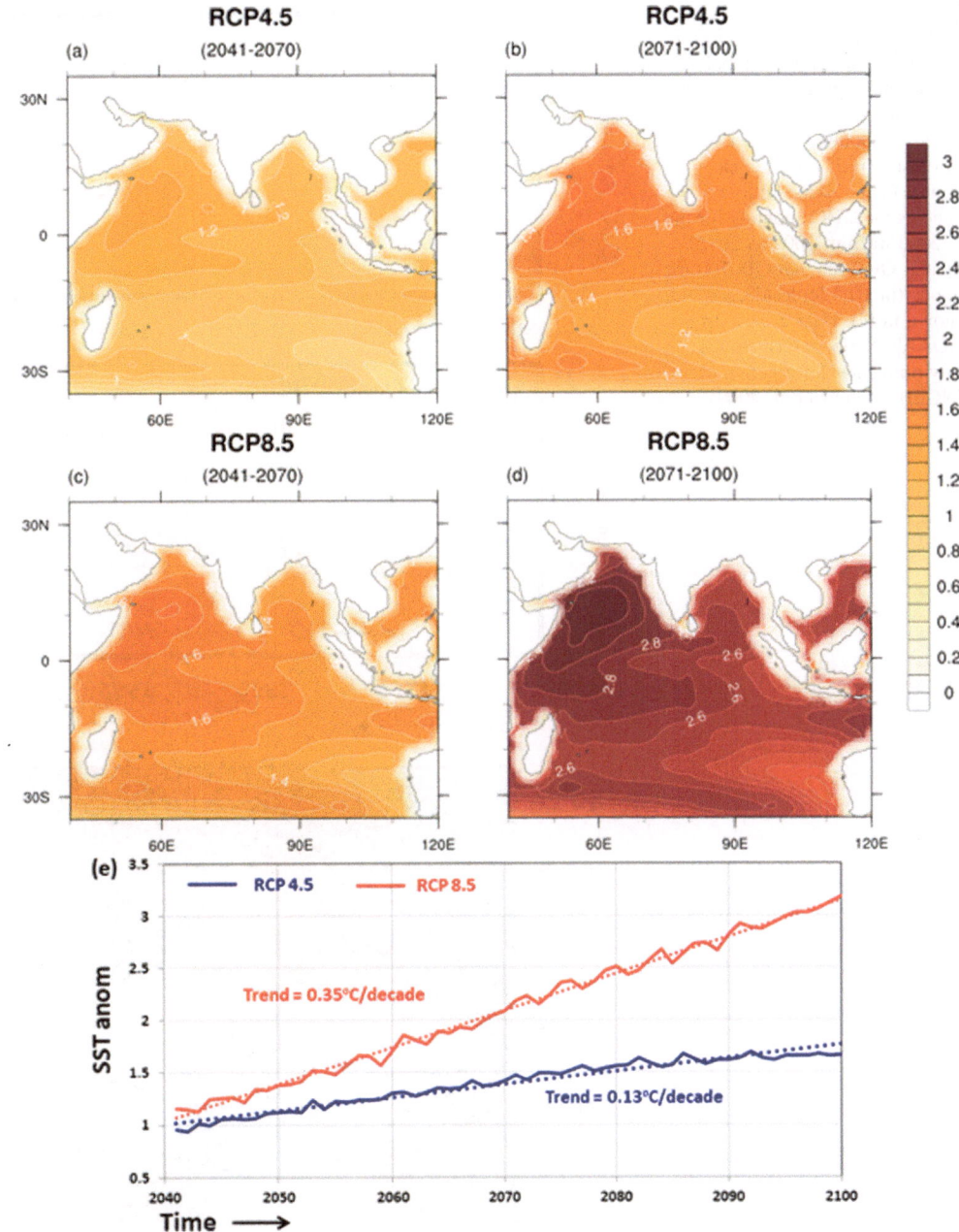

Table 10.1 CMIP5 (ensemble mean) projected SST rise (°C) averaged over the tropical Indian Ocean basin (40 °E:115 °E, 30 °S:30 °N) in the near future (2040–2069) and far future (2070–2099) in the RCP4.5 and RCP8.5 scenarios w.r.t. the reference period 1976–2005

Scenario	Near future	Far future
RCP4.5	1.2 °C (± 0.29 °C)	1.6 °C (± 0.40 °C)
RCP8.5	1.6 °C (± 0.55 °C)	2.7 °C (± 0.64 °C)

Fig. 10.3 Time series of heat content anomaly of the 0–700 m (in 10^{22} Joules) for **a** Global Ocean, **b** Tropical Indian Ocean (blue) and North Indian Ocean (yellow). Dotted lines denote the respective 1955–2015 linear trends. Green dashed curve denotes the OHC anomaly for the 0–2000 m from 2005 to 2015. Spatial OHC anomaly (0–700 m) trends (in 10^{22} Joules/decade) are shown in (**c**). Data used is from NOAA's National Centers for Environmental Information (NCEI) (Levitus et al. 2009)

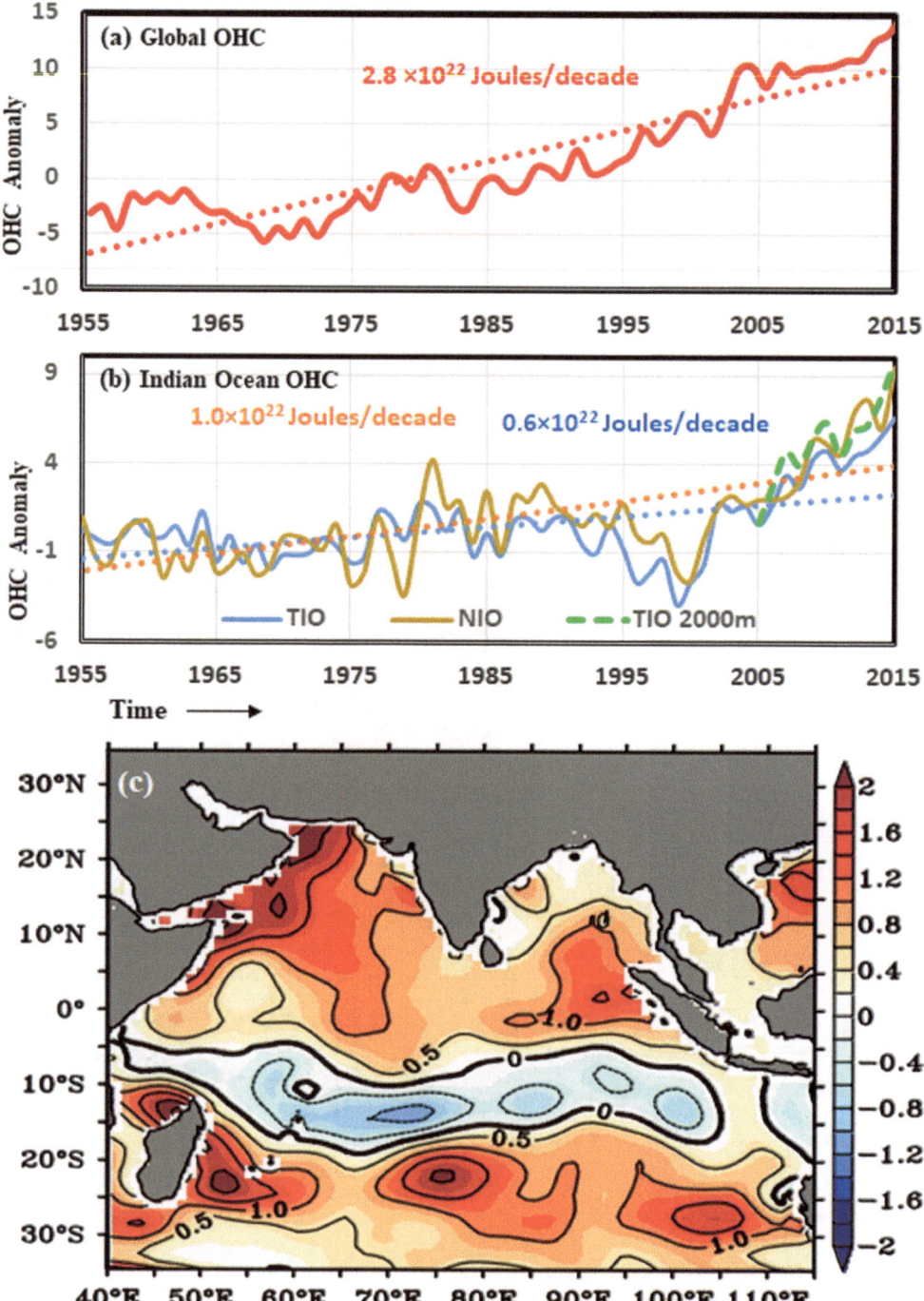

10.3 Changes in Ocean Heat Content

Contemporary global warming is driven by the additional heat trapped by greenhouse gases (GHGs) due to the Earth's energy imbalance between the energy absorbed and emitted. More than 90% of this additional heat is stored in the ocean, increasing ocean heat content (OHC), while the residual heat is manifested in the form of melting of both land and sea ice, and in warming of the atmosphere and land surface. This makes it critical to monitor changes in OHC in order to understand the rate and extent of global warming (Von Schuckmann et al. 2014). The ocean's interior is more sensitive to small external forcing than the global surface ocean because it is highly sensitive to heat exchange in the high-latitudes (e.g. Rosenthal et al. 2017). Monitoring regional variations in OHC are also important for understanding climate variability and change (Allison et al. 2019).

OHC changes also contribute substantially to sea-level rise, changes in ocean circulation, and energy transfer between ocean and atmosphere, making it a vital task to estimate and study historical OHC. Increases in OHC in the recent past have been attributed to the increase in greenhouse gases in earth's atmosphere (e.g. Levitus et al. 2001 and Barnett et al. 2001, 2005). The primary impacts of an increase in OHC are its effects on marine biodiversity and the melting of glaciers in the regions like Greenland and Antarctica. Other consequences of OHC rise include declining ocean oxygen (Schmidtko et al. 2017), bleaching and death of corals (Hughes et al. 2018), ice shelves directly through bottom heating, exacerbating marine heat waves (Oliver et al. 2018), and altered impacts of natural variability such as ENSO, IOD and Pacific Decadal Oscillation (PDO; e.g. Fasullo et al. 2018). Over long time periods the ocean's interior acts like a capacitor and builds up large heat anomalies. These persisting large-scale OHC anomalies are a source of ocean climate predictability from seasonal-to-decadal timescales (e.g. Smith and Murphy 2007).

10.3.1 Trends and Variability of Indian Ocean Heat Content

Upper OHC exhibits an increasing trend in the Indian Ocean since the 1950s (*high confidence*, Levitus et al. 2009; Xue et al. 2012; Han et al. 2014). The global OHC700 has risen at a rate of 2.8×10^{22} Joules per decade over the period 1955–2015 (Fig. 10.3a). Over the same period, the rate of rise in OHC700 in the tropical (30 °S to 30 °N and 40 °E to 115 °E) and north (5 °N to 30 °N and 40 °E to 100 °E) Indian Ocean basins were more muted at 0.62×10^{22} Joules per decade and 1.0×10^{22} Joules per decade, respectively (Fig. 10.3b). In other words, there was greater net heat content gain in the north Indian Ocean (NIO) than the TIO over 1955–2015 despite being a smaller basin in areal extent.

It is important to note that since the year 2000 both the TIO and NIO have experienced a steep rise in OHC (Cheng et al. 2017), which is absent in the global OHC signal (Fig. 10.3a, b). This abrupt increase of OHC in TIO has accounted for more than 70% of the global ocean heat gain in the upper 700 m during the same period (e.g. Lee et al. 2015), despite of representing only about 15% of the global ocean area. It is very likely that a significant portion of the heat due to greenhouse warming now resides in the upper 700 m of the Indian Ocean. Given the fact that the OHC700 in the Indian Ocean did not increase significantly during 1971–2000, the rapid increase during the recent period is striking.

In addition, higher variability in the TIO and NIO upper OHC compared to the global OHC is evident from Fig. 10.3. Changes in OHC700 in the NIO from 1955 to 2015 (Fig. 10.3c) show wide spatial variations with warming in the Arabian Sea, Bay of Bengal, equatorial Indian Ocean and in the region south of 20° S (Fig. 10.3c) (e.g. Nagamani et al. 2016; Anandh et al. 2018), but a zonally extended cooling from 20° S to 5° S. Trends in TIO OHC2000 (ocean heat content in the upper 2000 m) during 2005–2015 are similar to that of OHC700 (Fig. 10.3b).

There are significant differences in the recent OHC700 trends between the major ocean basins, particularly the Pacific and Indian Oceans. Despite of large natural variability in OHC700, increasing trends due to anthropogenic influence are evident (high confidence). The abrupt increase in the OHC700 during 2003–2012 was not due to surface heating, but due almost entirely to horizontal heat convergence in the form of an enhanced ITF (Lee et al. 2015). This Indian Ocean OHC increase corresponds to a concurrent Pacific Ocean OHC decrease in the 0–100 m, suggesting a transfer of heat from Pacific Ocean to Indian Ocean (Liu et al. 2016). The net surface heat flux into the Pacific Ocean increased greatly during 2003–2012, consistent with the La Niña-like condition across the Pacific Ocean. However, the anomalous surface heat uptake was completely masked by horizontal heat divergence in the tropical Pacific, which was mainly traced to the increased heat transport to the Indian Ocean. This has, in fact, inhibited the increase in tropical Pacific OHC700, which instead shows a slight decrease during this period.

Among the global oceans, the ensemble-mean trends for the 700–6000 m layer show the largest increasing trends in the North Atlantic sub-polar gyre, north Indian Ocean and Southern Ocean (e.g. Palmer et al. 2017). As there are fewer observations available in the Indian Ocean between 3000 and 5000 m compared to the other oceans, deep OHC estimates in the Indian Ocean, particularly in the north-western part, are relatively less accurate (Purkey and Johnson 2010). Although the Atlantic and Indian oceans contributed to the global OHC increase at 3500 m depth, the OHC increases in these two basins have been weak below 4000 m during the past two decades. The smaller contribution of the Indian Ocean to the global OHC increase might be due to the smaller bottom layer volume of the Indian Ocean than that of the Pacific Ocean (e.g. Kouketsu et al. 2011).

10.3.2 Multidecadal Variability of the Indian Ocean Heat Content

The observed upper OHC of the Indian Ocean reveals significant decadal variations (Han et al. 2014; Mohapatra et al. 2020). Model experiments suggest that the observed multidecadal trends in OHC are partially associated with anthropogenic forcing. The observed decadal variability in the basin-wide distribution of sea level and thermal structure in the Indian Ocean results primarily from forcing by Indian Ocean winds (Deepa et al. 2019; Srinivasu et al. 2017), with a significant contribution from the ITF transport to the interior of the south Indian Ocean after 1990 (Han et al. 2014). For the Pacific and Indian Oceans, decadal shifts are primarily observed in the upper 350 m, likely due to shallow subtropical circulation, leading to an abrupt increase of OHC in the Indian Ocean carried by the ITF from the Pacific Ocean over the last decade (Cheng et al. 2017). During the slowdown in Pacific SST warming, there was anomalous warming in the Indian Ocean and an accelerated OHC rise below 50 m, which is associated with a La Niña-like climate shift, and an enhanced heat transport of the ITF (Liu et al. 2016). Transmission of the multidecadal signal occurs via an oceanic pathway through the ITF and is manifested across the Indian Ocean centred along 12° S as westward-propagating Rossby waves modulating the thermocline and subsurface heat content variations (Rahul and Gnanaseelan, 2016; Deepa et al. 2019). Changes in Pacific wind forcing in recent decades and associated rapid increases in Indian Ocean subsurface heat content can thus affect the basin's leading mode of variability (Ummenhofer et al. 2017). Jin et al (2018) argued that the western Indian Ocean subsurface heat content is influenced by Interdecadal Pacific oscillation (IPO) through wind driven Ekman pumping via the atmospheric bridge, whereas the eastern Indian Ocean is largely affected through the oceanic pathway via ITF. The recent negative phase of IPO (1998–2012) enhanced the Pacific easterlies, which eventually led to export of anomalous heat from the Pacific towards the Indian Ocean (Gastineau et al 2018). The OHC300-based definition of PDO takes into account variations throughout the upper ocean and is better suited to capture various characteristics of PDO variability (Kumar and Wen 2016). Based on multiple observational datasets, ocean reanalysis products and an ocean model simulation, Li et al. (2018) reported the presence of prominent multidecadal variations in the Indian Ocean OHC400. They suggested that the upper Indian Ocean first experienced a heat content increase at the rate of $5.9 \pm 2.5 \times 10^{21}$ J decade^{-1} during 1965–79, followed by a decrease at the rate of $-5.2 \pm 2.5 \times 10^{21}$ J decade^{-1} during 1980–96, and subsequently an enhanced rate of increase of $13.6 \pm 1.1 \times 10^{21}$ J decade^{-1} from 2000 to 2014. This suggests that the Indian Ocean OHC underwent tremendous decadal variations which might continue into the future. CMIP5 simulations have limited skill in capturing decadal variability in upper OHC during the past 45 years (Collins et al. 2013; Cheng et al. 2015).

10.3.3 Future Projections of Indian Ocean Heat Content

Monitoring and understanding OHC change and the role of circulation in shaping the patterns of increase remain key to predicting global and regional climate change, and sea-level rise (Zanna et al. 2019). Under the low-to-medium (RCP4.5) emissions scenario, half of the energy taken up by the ocean by the end of the twenty-first century will be in the uppermost 700 m, and 85% will be in the uppermost 2000 m (*low confidence*). Future changes in wind and air-sea fluxes, and ocean transport, likely to have serious implications for regional sea-level rise and coastal flood risk. Additionally, the spatial patterns of OHC change under global warming contribute to the regional sea-level projections (e.g. Slangen et al 2014). There is a large spread among CMIP5 models in projections of future changes in OHC, suggesting an urgent need for further refinement (Cheng et al 2016; Allison et al. 2019).

10.4 Impacts of Indian Ocean Warming

10.4.1 Consequences of Indian Ocean Warming on Regional Climate

Rapid warming of the Indian Ocean during 1950–2015, along with substantial changes in land use and anthropogenic aerosols have altered the Indian summer monsoon (Singh et al. 2019, Chaps. 3 and 6). During 1950–2015, there has been a significant decline in the summer monsoon rainfall over central India and parts of north India due to a reduction in the tropospheric thermal contrast that is associated with the rapid warming of the Indian Ocean (Mishra et al. 2012; Saha et al. 2014; Roxy et al. 2016). At the same time, rapid warming in the Arabian Sea has resulted in a rise in widespread extreme rains over Western Ghats and central India, since warming induces increased fluctuations in the monsoon winds, with ensuing episodes of enhanced moisture transport from the Arabian Sea towards the Indian subcontinent (Roxy et al. 2017). Indian Ocean warming is also found to reduce rainfall over India during the onset phase and increase it during the withdrawal phase (Chakravorty et al. 2016).

In terms of tropical cyclones, the Bay of Bengal region witnesses more than 80% of the global fatalities associated with tropical cyclones, while only accounting for 5% of these storms globally (Beal et al. 2019). Since tropical cyclones primarily draw their energy from evaporation at the ocean surface, SST and OHC strongly constrain cyclone intensity (Rajeevan et al. 2013). Global warming appears to have increased the intensity of tropical cyclones during the post-monsoon period in the Bay of Bengal (Chap. 8) and the pre-monsoon period in the Arabian Sea. For example, Cyclone Nilofar in 2014 was the first severe tropical cyclone to be recorded in the Arabian Sea in the post-monsoon season. Though the cyclone did not make landfall, it produced heavy rainfall along the western coast of India. Future projections suggest a *likely* increase in the number of extremely severe tropical cyclones in response to Indian Ocean warming, particularly in the Arabian Sea, while changes in frequency remain uncertain (Chap. 8).

Indian Ocean SSTs have a role in regulating the surface air temperatures over the Indian subcontinent (Chowdary et al. 2014). Associated with the basin-wide warming and frequent El Niños, the frequency and duration of heat waves have increased over the Indian subcontinent (Rohini et al. 2016). Rising ocean temperatures have also resulted in instances of marine heat waves in the Indian Ocean. Marine heat waves are similar to heat waves over the land, with periods of extremely high ocean temperatures that persist for days to months (Collins et al. 2019). Recent marine heat waves, including the one in 2016 that co-occurred with an extreme El Niño event of 2015–16, resulted in mass bleaching of coral reefs and adversely impacted aquaculture industries along the Indian Ocean rim countries (Collins et al. 2019). Satellite observations reveal that the intensity of marine heat waves have increased and that they have *very likely* doubled in frequency over 1982–2016. Climate projections indicate that at the high emissions RCP 8.5 GHG scenario, a one-in-100-day marine heat wave event (with pre-industrial CO_2 levels) is *very likely* to become a one-in-four-day event by 2031–2050 and a one-in-two-day event by 2081–2100 (Collins et al. 2019).

The impact of Indian Ocean warming is reflected in the sea-level changes, as thermal variations have dominated these changes in recent decades (Chap. 9). The largest sea-level changes were observed along the northern and eastern coasts of the Bay of Bengal (Chap. 9). Tide gauge data corroborates the increasing sea level among the coastal regions of the Indian Ocean, such as Mumbai, Kochi, Visakhapatnam on the Indian coast and Durban, Fremantle, Port Hedland on the Australian coast. The consistent increase in sea level is attributed to the thermal expansion of sea water, due to a basin-wide surface warming in the Indian Ocean. Besides, the observed changes in the wind circulation in the Indian Ocean has modulated the ocean heat transport, and distributed the heat across the basin, resulting in a large thermosteric response in the sea level (Chap. 9).

Beyond the regional climate, Indian Ocean warming has global and remote impacts also. The Indian Ocean has contributed to more than 21% of the global oceanic heat uptake over the last two decades and contributed strongly to the temporary slowdown in global warming during 1998–2013 (Lee et al. 2015; Cheng et al. 2017). Climate model experiments indicate that the rapid warming of the Indian Ocean is strengthening the Atlantic meridional overturning circulation (Hu and Fedorov 2019; Cherchi 2019). The basin-wide warming could modulate the Pacific climate, affect the North Atlantic oscillation and enhance the positive Southern Annular Mode, and may cause West Sahel and Mediterranean droughts (Beal et al. 2019). Atmospheric blocking triggered by tropical convection in the Indian and Pacific oceans can cause persistent anticyclonic circulation that not only leads to severe drought but also generates marine heat waves in the adjacent ocean. Warming in the Indian Ocean and associated deep convection is found to trigger droughts in South America and marine heat waves in the adjoining South Atlantic (Rodrigues et al. 2019). A strong negative IOD event in 2016 strongly impacted East African rainfall, with some regions recording below 50% of normal rainfall, leading to devastating drought, food insecurity and unsafe drinking water for over 15 million people in Somalia, Ethiopia and Kenya (Collins et al. 2019).

10.4.2 Consequences of Indian Ocean Warming on Ocean Biogeochemistry

The biogeochemical properties of the Indian Ocean are distinct from other ocean basins for mainly two reasons—the land boundary in the north due to Indian subcontinent and the large amplitude of seasonally reversing monsoon cycle. Climate driven physical fluctuations are expected to impact the marine ecosystem substantially by modifying the biotic and abiotic environments, which can lead to severe repercussions for the oceanic primary production. The warming of SSTs in the western Indian Ocean leads to increased stratification in the basin. Roxy et al. (2016) show a declining trend in marine phytoplankton in the western Indian Ocean (*high confidence*), by 30% in the observations during 1998–2013 and 20% in the CMIP5 simulations during 1950–2015 (*medium confidence*). This significant decline in phytoplankton is attributed to enhanced stratification of the oceanic water column as a result of rapid surface warming, thereby suppressing the mixing of nutrients from subsurface layers into the surface. Downward trends in primary production over the Indian Ocean can be detrimental to the marine food web and the fishing industry, especially the economically valuable tuna industry (Lee et al. 2005). However, careful gathering of in situ observations in the open ocean is needed to build a substantial understanding of the bio-physical interactions in the Indian Ocean.

Approximately 30% of the historical anthropogenic CO_2 emissions have been absorbed by the oceans since the pre-industrial era (e.g. Canadell et al. 2007). The increasing oceanic uptake of CO_2 has changed seawater chemistry and resulted in ocean acidification, with profound impacts on biological ecosystems in the upper ocean. Long-term increasing trends in ocean acidification, consistent with the increase in atmospheric CO_2, are evident over the past several decades (Dore et al. 2009).

Fig. 10.4 Projected changes in multiple stressor intensity in 2090–2099 relative to 1990–1999 under the RCP8.5 scenario. Red indicates where sea surface warming exceeds +3.5 °C, hatched yellow indicates where subsurface (200–600 m) oxygen concentrations decrease by more than 20 µmol/m^3 and hatched blue indicates where vertically integrated annual NPP decreases by more than 100 gC/m^2. In addition, hatched orange indicates present-day simulated low-oxygen (<50 mmol/m^3) in the subsurface waters Figure adapted from Bopp et al. (2013)

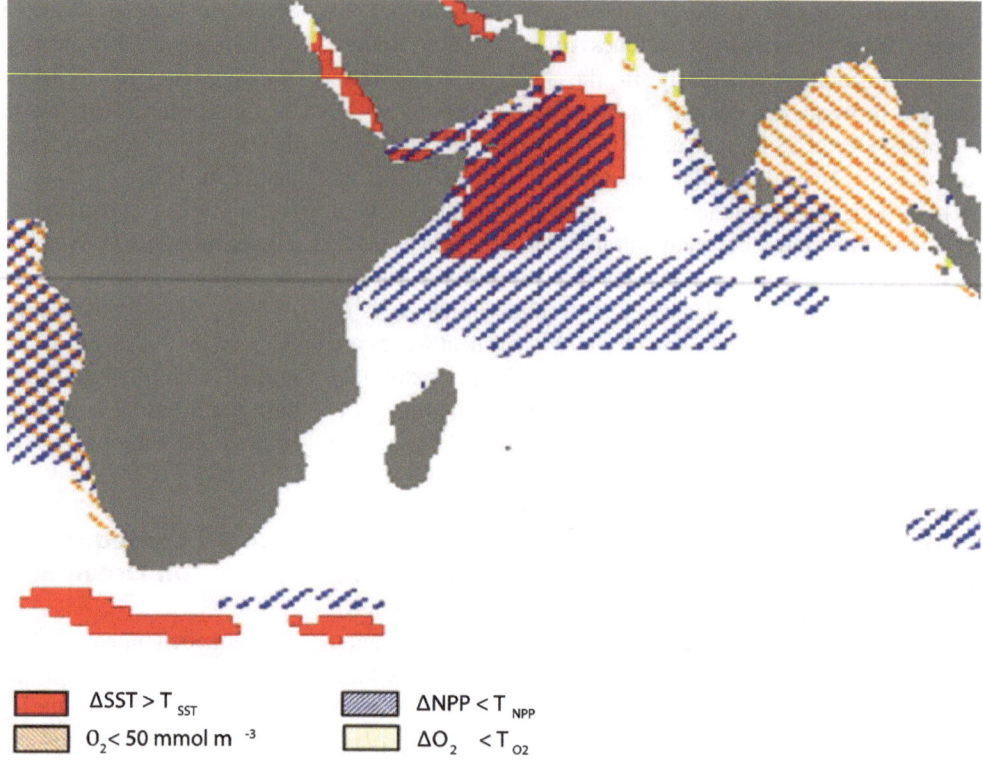

Recent work by Sreeush et al. (2019a) has found that the Indian Ocean is acidifying due to the accumulation of anthropogenic CO_2 from the atmosphere, and this storage of anthropogenic carbon in the Indian Ocean is comparable (after normalizing for the size of the basins) with the other major oceans (Sabine et al. 2004). Surface ocean pH over the Indian Ocean has declined by about 0.1 unit (current mean is 8.1) relative to pre-industrial levels and is larger over the western Indian Ocean (e.g. Sreeush et al. 2019a). This increase in ocean acidity may be responsible for the functional collapse of reef building corals. The western Arabian Sea has undergone more rapid acidification than the rest of the TIO basin due to strong upwelling in this region drawing up anthropogenic CO_2 embedded in the deeper ocean. Moreover, SST warming also accelerates acidification due to the endothermic nature of CO_2 dissolution in water. An ocean biogeochemical model-based simulations from 1960 to 2009 show that western Arabian Sea has acidified by 108% due to dissolved inorganic carbon, −36% due to buffering due to alkalinity, 16% due to SST warming, 6% due to salinity changes and remaining due to changes in other minor ions (Sreeush et al. 2019a). Considering that the western Arabian Sea is a highly productive zone of the Indian Ocean (Roxy et al. 2016), the role of SST warming in exacerbating acidification needs to be monitored carefully.

Dissolved O_2 is a major determinant of the abundance and distribution of the marine habitat. The Indian Ocean

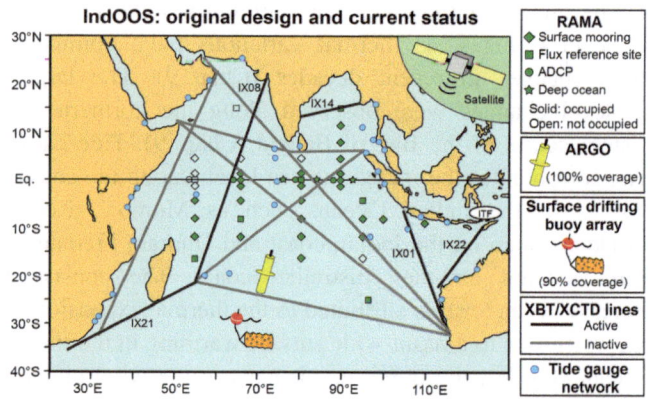

Fig. 10.5 Status of the Indian Ocean Observing System (IndOOS) in 2018. The sustained observing system in the Indian Ocean comprises of the Argo, RAMA, XBT/XCTD, surface drifting buoy and tide gauge networks. It is supported by satellite observations and the GO-SHIP program. The empty symbols indicate RAMA sites that were not implemented due to logistical constraints Source: Beal et al. 2019

contains one of the oceans' most pronounced oxygen minimum zones (OMZs) encompassing more than 50% of the area containing OMZs (e.g. Helly and Levin 2004). The tropical open ocean O_2 concentration has decreased at a rate of 0.1–0.3 µ mol kg^{-1} year^{-1} during the past five decades (Stramma et al. 2008). Long-term measurements over the TIO also show a pronounced decrease (at a rate of 20–30 mol m^{-2} per decade) in O_2 concentration (Koslow et al.

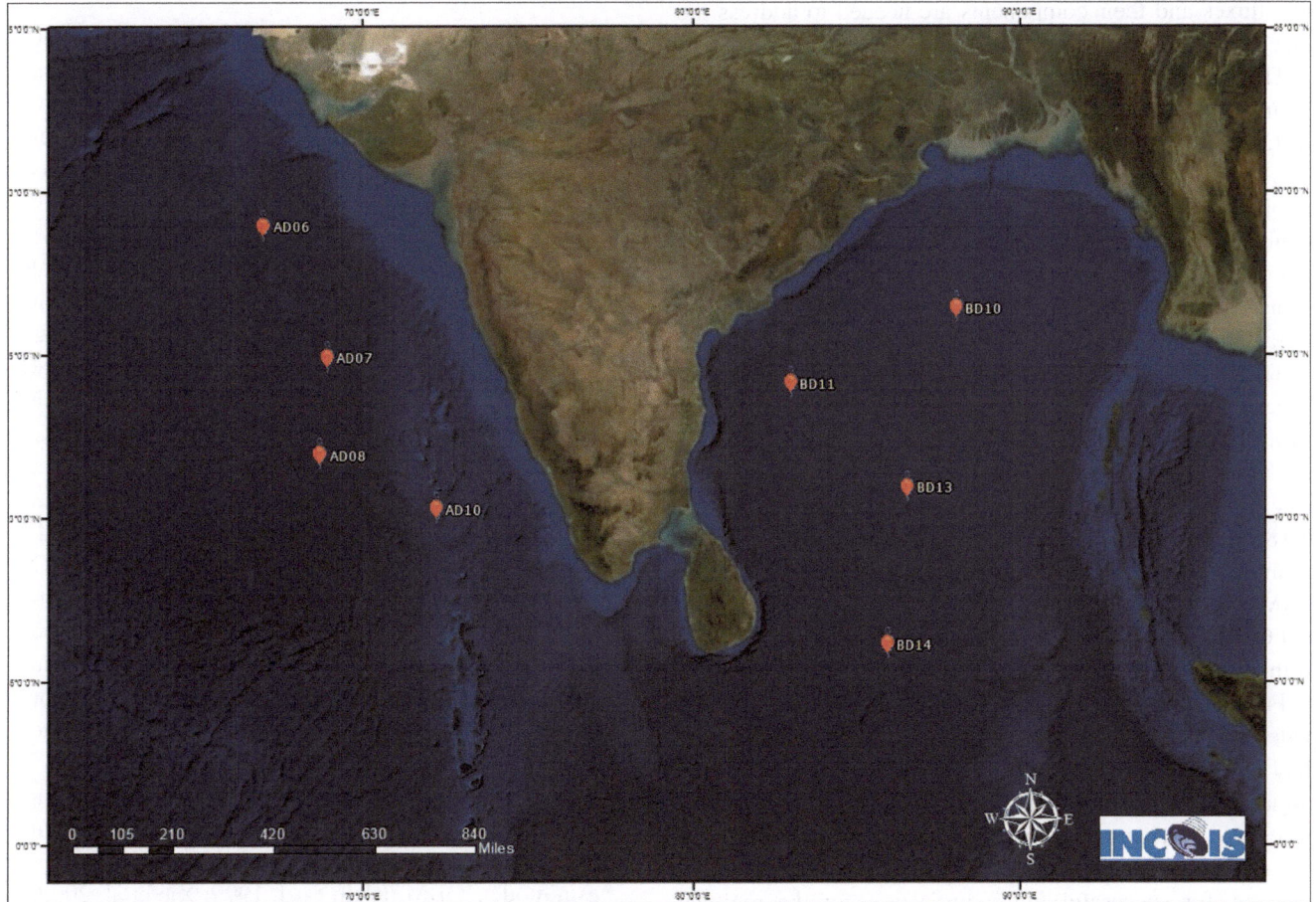

Fig. 10.6 Status of the OMNI moorings as on Oct 2019, installed by INCOIS/NIOT in the north Indian Ocean

2011). This oxygen decline is a result of warming-induced decline in oxygen solubility and reduced ventilation of deep ocean due to enhanced ocean surface stratification (Keeling et al. 2010).

Future projections based on the CMIP5 models indicate further warming of the Indian Ocean and a resultant decline in marine primary productivity (Fig. 10.4, Bopp et al. 2013). CMIP5 models also project a continuing decline in open-ocean surface pH from the current mean of 8.1–7.8 by the end of the twenty-first century, as a response to the projected rise in CO_2 emissions (Bopp et al. 2013). Most models project decreasing oxygen concentrations in the ocean by 1–7% from present-day concentrations by 2100, with global warming (Keeling et al. 2010; Bopp et al. 2013).

Interactions of rising temperatures, ocean acidification and oxygen minimum zones narrow thermal ranges and enhance sensitivity to temperature extremes in marine organisms ranging from corals to fish. These negative effects of increase in warming are projected to be most significant in developing nations in tropical regions (high confidence, IPCC AR5 2014). However, understanding the combined effect of both low O_2 and low pH on marine ecosystems in the Indian Ocean demands more active research in the Indian Ocean.

10.5 Knowledge Gaps

Though we understand the extent and magnitude of the warming trend, we still do not have a clear understanding on the amount of heat entering and exiting the Indian Ocean through the various ocean-atmospheric pathways. One of the reasons is the inability to accurately perform a heat budget analysis, due to lack of long-term high frequency data over regions like the western Indian Ocean and the ITF (Fig. 10.5). This need to be resolved with a combination of Research Moored Array for African-Asian-Australian Monsoon Analysis and Prediction (RAMA) moorings and Argo buoys (and gliders where both do not function) which measure ocean data at various timescales. Also, though Argo floats can generally provide measurements up to a depth of 2000 m, only 60% of the floats do so. Hence, the uncertainty in warming trends in the deeper ocean could be large, and need to be addressed. Sustained measurements of air-sea heat fluxes and their components are needed to address the uncertainties in SST trends (Beal et al. 2019).

The lack of sufficient observation from Indian Ocean for the upper ocean carbon cycle research is a lacuna at present. Surface ocean parameters such as ocean partial pressure of CO_2 (pCO_2) and nutrients such as PO_4 and NO_3 are worthy of observing because they can cascade through the ocean solubility and biological pumps to constrain the variables and parameters in the upper ocean carbon cycle yielding robust estimates of upper ocean carbon budget of the Indian Ocean (Sreeush et al. 2019b).

There is a mooring network in the northern Indian Ocean, Ocean Moored Buoy Network for northern Indian Ocean (OMNI), led by India that also records high resolution ocean and near-surface meteorological data (Fig. 10.6). The Ministry of Earth Sciences (MoES) of India announced in June 2018 that the data from OMNI outside the Indian EEZ would be made freely available at data standards similar to that of RAMA. A coordinated moored network combining RAMA and OMNI is a priority for successful monitoring of changes in the Indian Ocean and also for skilful forecasting.

Further, CMIP5 models fail to reproduce the observed pattern of SST warming in the Indian Ocean, and as a result they fail to represent the local and remote impact on the climate system in the models (Saha et al. 2014). Hence, it is necessary to accurately monitor changes over the Indian Ocean, and also improve the models to simulate these changes, for successful future projections of the regional climate.

Box 10.1 Modes of climate variability in the Indo-Pacific

El Niño Southern Oscillation (ENSO): ENSO is a coupled ocean-atmosphere mode of interannual variability with a periodicity of about 4–7 years. El Niño is the positive phase of ENSO, characterized by anomalous surface warming of eastern and central equatorial Pacific, lasting for several months. The negative phase with cooler (than normal) SST in the eastern equatorial Pacific is called La Niña. There are several indices to quantify the strength, nature and duration of El Niño based on the anomalous SST over different regions of eastern and central equatorial Pacific. The atmospheric counterpart to this interannual warming/cooling is known as southern oscillation, which is quantified as the anomalous sea-level pressure difference between Darwin and Tahiti. The southern oscillation index quantifies the intensity of Walker circulation and the ocean atmosphere coupling associated with El Niño and La Niña. El Niño is found to weaken the Indian summer monsoon and warm the

Indian Ocean basin, whereas La Niña strengthens the monsoon and cools the Indian Ocean basin. Different flavours of El Niño such as east Pacific (cold tongue) El Niño, central Pacific (warm pool) El Niño (or Modoki, Ashok et al. 2007) and their regional impacts are significant over the Indian landmass and Indian Ocean.

Pacific decadal oscillation (PDO): PDO is a decadal mode of variability in the north Pacific (north of 20° N), with warm SST anomalies in the eastern and north Pacific and cold SST anomalies in the central north Pacific during the positive phase, and vice versa in the negative phase. In the equatorial Pacific, PDO imprints similar spatial structure as that of ENSO but with a longer time scale. When PDO and ENSO are in the same phase, the impact of ENSO is amplified. Interdecadal Pacific oscillation (IPO) is similar to PDO but has a wider spatial structure covering both southern hemisphere and northern hemisphere, with the pattern of warming and cooling in the north Pacific similar to that of PDO. The typical cycle of a PDO or IPO is about 15–30 years, but in some period, one phase itself may last more than 20–30 years.

Indian Ocean basin mode (IOBM) and Indian Ocean Dipole (IOD): TIO is characterized by several modes of climate variability such as the Indian Ocean basin mode (IOBM, Klein et al. 1999; Xie et al. 2002; Chowdary and Gnanaseelan 2007), Indian Ocean Dipole (IOD, Saji et al. 1999; Webster et al. 1999), and subsurface mode (Sayantani and Gnanaseelan 2015). IOBM is mainly caused by ENSO forcing and the associated changes in the net heat flux. IOD is an ocean atmosphere coupled climate mode of variability (east west) in the tropical Indian Ocean, defined as the difference in SST anomalies of western (50 °E to 70 ° E;10 °S to 10 °N) and south-eastern (90 °E to 110 °E; 10 °S to equator) equatorial Indian Ocean (Saji et al 1999). In contrast to the SST variability, subsurface temperature (at thermocline) displays a north–south mode of variability. Generally, a positive IOD favours enhanced summer monsoon rainfall over the Indian subcontinent while negative IOD favours less rainfall. More details on the ENSO-IOD-monsoon interactions are given in Chapter 3, Box 3.2.

10.6 Summary

The ocean-atmospheric conditions in the Indian Ocean region strongly modulate the subcontinental climate. This chapter has assessed changes in the Indian Ocean in the recent past, as well as changes anticipated over the twenty-first century based on simulations by state-of-the-art global climate models.

The SST of the TIO has warmed by about 1 °C over the period 1951–2015, which is much higher than the global average SST rise of about 0.7 °C over the same period. Most of this temperature rise is attributed to anthropogenic emissions. Heat content in the upper Indian Ocean (OHC700) has also exhibited an increasing trend since the 1950s, with a notably abrupt rise after the year 2000. Observed declining trends in phytoplankton and oxygen concentrations in the TIO, attributed to SST warming, and the increasing acidification of the Indian Ocean due to excess CO_2 uptake, have also likely impacted marine ecosystems in the western Indian Ocean.

Climate models project a rise in surface temperatures of the TIO by 1.2–1.6 °C and 1.6–2.7 °C in the near and far futures across GHG emissions scenarios (RCP4.5 and RCP8.5) relative to the reference period 1976–2005. Trends in other observed changes are also projected to continue with global warming.

Lack of sufficient observations in the western Indian Ocean and the Indonesian Through flow has hampered the understanding of changes in the heat budget of the Indian Ocean. Increasing the skill of Indian Ocean forecasts requires coordinated efforts in monitoring changes in the Indian Ocean at different time scales and at different depths. Efforts in these directions are already underway.

The central role of the Indian Ocean in modulating the regional climate implies that changes in this basin have serious implications for both the densely populated coastal regions around this basin, and for marine ecosystems. Hence, efforts towards detailed and continuous monitoring of ongoing changes and improving climate models are essential for developing effective adaptation and mitigation strategies to reduce risk due to climate change.

References

Allison LC, Roberts CD, Palmer MD, Hermanson L, Killick RE, Rayner NA, Smith DM, Andrews MB (2019) Towards quantifying uncertainty in ocean heat content changes using synthetic profiles. Environ Res Lett 14(8):084037

Ashok K, Behera SK, Rao SA, Weng H, Yamagata T (2007) El Niño Modoki and its possible teleconnection. J Geophys Res Oceans 112: C11007. https://doi.org/10.1029/2006JC003798

Anandh TS, Das BK, Kumar B, Kuttippurath J, Chakraborty A (2018) Analyses of the oceanic heat content during 1980–2014 and satellite-era cyclones over Bay of Bengal. Int J Climatol 38:5619–5632. https://doi.org/10.1002/joc.5767

Barnett TP, Pierce DW, Schnur R (2001) Detection of anthropogenic climate change in the world's oceans. Science 292:270–274. https://doi.org/10.1126/science.1058304

Barnett TP, Pierce DW, AchutaRao KM, Gleckler PJ, Santer BD, Gregory JM, Washington WM (2005) Penetration of human-induced warming into the world's oceans. Science 309:284–287. https://doi.org/10.1126/science.1112418

Beal LM, Vialard J, Roxy MK et al (2019) IndOOS-2: a roadmap to sustained observations of the Indian Ocean for 2020–2030. CLIVAR-4/2019. https://doi.org/10.36071/clivar.rp.4-1.2019

Bopp L, Resplandy L, Orr JC, Doney SC, Dunne JP, Gehlen M, Vichi M (2013) Multiple stressors of ocean ecosystems in the 21st century: projections with CMIP5 models. Biogeosciences 10 (10):6225–6245. https://doi.org/10.5194/bg-10-6225-2013

Cai W, Santoso A, Wang G et al (2014) Increased frequency of extreme Indian Ocean Dipole events due to greenhouse warming. Nature 510:254–258. https://doi.org/10.1038/nature13327

Cai W, Zheng X-T, Weller E et al (2013) Projected response of the Indian Ocean Dipole to greenhouse warming. Nat Geosci 6:999–1007. https://doi.org/10.1038/ngeo2009

Canadell JG et al (2007) Contributions to accelerating atmospheric CO_2 growth from economic activity, carbon intensity, and efficiency of natural sinks. Proc Natl Acad Sci USA 104:18866–18870

Chakravorty S, Gnanaseelan C, Pillai PA (2016) Combined influence of remote and local SST forcing on Indian Summer Monsoon Rainfall variability. Clim Dyn 47(9–10):2817–2831

Cheng LJ, Zhu J, Abraham J (2015) Global upper ocean heat content estimation: recent progress and the remaining challenges. Atmos Oceanic Sci Lett 8:333–338. https://doi.org/10.3878/AOSL20150031

Cheng L, Trenberth KE, Palmer MD, Zhu J, Abraham JP (2016) Observed and simulated full-depth ocean heat-content changes for 1970–2005. Ocean Sci 12(4):925–935. https://doi.org/10.5194/os-12-925-2016

Cheng L, Trenberth K, Fasullo J, Boyer T, Abraham J, Zhu J (2017) Improved estimates of ocean heat content from 1960 to 2015. Sci Adv 3:e1601545. https://doi.org/10.1126/sciadv.1601545

Cherchi A (2019) Connecting AMOC changes. Nat Clim Chang 9:729–730

Chowdary JS, Gnanaseelan C (2007) Basin-wide warming of the Indian Ocean during El Nino and Indian Ocean dipole years. Int J Climatol 27:1421–1438

Chowdary JS, John N, Gnanaseelan C (2014) Interannual variability of surface air-temperature over India: impact of ENSO and Indian Ocean Sea surface temperature. Int J Climatol 34(2):416–429

Chowdary JS, Xie SP, Tokinaga H, Okumura YM, Kubota H, Johnson N, Zheng XT (2012) Interdecadal variations in ENSO teleconnection to the Indo–western Pacific for 1870–2007. J Clim 25(5):1722–1744. https://doi.org/10.1175/JCLI-D-11-00070.1

Collins M, Knutti R, Arblaster J, Dufresne JL, Fichefet T, Friedlingstein P, Gao X, Gutowski WJ, Johns T, Krinner G, Shongwe M, Tebaldi C, Weaver AJ, Wehner M (2013) Long-term climate change: projections, commitments and irreversibility. In: Stocker TF, Qin D, Plattner GK, Tignor M, Allen SK, Boschung J, Nauels A, Xia Y, Bex V, Midgley PM (eds) Climate change 2013: the physical science basis. Contribution of working group I to the fifth assessment report of the intergovernmental panel on climate change. Cambridge University Press, Cambridge

Collins M, Sutherland M, Bouwer L, Cheong SM, Frölicher T, Jacot Des Combes H, Roxy MK, Losada I, McInnes K, Ratter B, Rivera-Arriga E, Susanto RD, Swingedouw D, Tibig L (2019) Extremes, abrupt changes and managing risks. In: Portner et al (eds) IPCC special report on oceans and cryosphere in a changing climate. Cambridge University Press, Cambridge

Deepa JS, Gnanaseelan C, Mohapatra S, Chowdary JS, Karmakar A, Kakatkar R, Parekh A (2019) The tropical Indian Ocean decadal sea level response to the Pacific decadal oscillation forcing. Clim Dyn 52:5045–5058. https://doi.org/10.1007/s00382-018-4431-9

Deser C, Phillips AS, Alexander MA (2010) Twentieth century tropical sea surface temperature trends revisited. Geophys Res Lett 37: L10701. https://doi.org/10.1029/2010GL043321

Dong L, McPhaden MJ (2016) Interhemispheric SST gradient trends in the Indian Ocean prior to and during the Recent global warming hiatus. J Clim 29:9077–9095. https://doi.org/10.1175/JCLI-D-16-0130.1

Dong L, Zhou T, Wu B (2014) Indian Ocean warming during 1958–2004 simulated by a climate system model and its mechanism. Clim Dyn 42:203–217. https://doi.org/10.1007/s00382-013-1722-z

Dong L, Zhou T (2014) The Indian Ocean sea surface temperature warming simulated by CMIP5 models during the twentieth century: competing forcing roles of GHGs and anthropogenic aerosols. J Clim 27(9):3348–3362

Dore JE, Lukas R, Sadler DW, Church MJ, Karl DM (2009) Physical and biogeochemical modulation of ocean acidification in the central North Pacific. Proc Natl Acad Sci USA 106:12235–12240

Du Y, Xie S-P (2008) Role of atmospheric adjustments in the tropical Indian Ocean warming during the 20th century in climate models. Geophys Res Lett 35:L08712. https://doi.org/10.1029/2008GL033631

Du Y, Xie S-P, Huang G, Hu K (2009) Role of air-sea interaction in the long persistence of El Niño-induced north Indian Ocean warming. J Clim 22(8):2023–2038. https://doi.org/10.1175/2008JCLI2590.1

Fasullo JT, Otto-Bliesner BL, Stevenson S (2018) ENSO's changing influence on temperature, precipitation, and wildfire in a warming climate. Geophys Res Lett 45:9216–9225. https://doi.org/10.1029/2018GL079022

Gastineau G, Friedman AR, Khodri M, Vialard J (2018) Global ocean heat content redistribution during 1998–2012 Interdecadal Pacific Oscillation negative phase. Clim Dyn 53:1187–1208. https://doi.org/10.1007/s00382-018-4424-8

Gnanaseelan C, Roxy MK, Deshpande A (2017) Variability and trends of sea surface temperature and circulation in the Indian Ocean. In: Rajeevan MN, Nayak S (eds) Observed climate variability and change over the Indian Region, vol 10. Springer, Singapore, pp 165-179. Doi:10.1007/978-981-10-2531-0

Graham NE, Barnett TP (1987) Sea surface temperature, surface wind divergence, and convection over tropical oceans. Science 238 (4827):657–659

Gille ST (2002) Warming of the Southern Ocean since the 1950s. Science 295(5558):1275–1277

Han W, Vialard J, McPhaden MJ, Lee T, Masumoto Y, Feng M, De Ruijter WP (2014) Indian Ocean decadal variability: a review. Bull Amer Meteor Soc 95(11):1679–1703. https://doi.org/10.1175/BAMS-D-13-00028.1

Helly JJ, Levin LA (2004) Global distribution of naturally occurring marine hypoxia on continental margins. Deep-Sea Res I 51:1159–1168

Hu S, Fedorov AV (2019) Indian Ocean warming can strengthen the Atlantic meridional overturning circulation. Nat Clim Chang 9:747–751

Hughes TP et al (2018) Global warming transforms coral reef assemblages. Nature 556:492–496. https://doi.org/10.1038/s41586-018-0041-2

IPCC (2014) Climate change 2014: synthesis report. Contribution of working groups I, II and III to the fifth assessment report of the intergovernmental panel on climate change, vol 151. IPCC, Geneva, Switzerland. (Core Writing Team, Pachauri RK, Meyer LA (eds))

Izumo T, Montégut CB, Luo JJ, Behera SK, Masson S, Yamagata T (2008) The role of the western Arabian Sea upwelling in Indian monsoon rainfall variability. J Clim 21(21):5603–5623

Jin X, Kwon Y, Ummenhoffer CC, Seo H, Schwarzkopf FU, Biastoch A, Boning CW, Wright JS (2018) Influence of Pacific climate variability on decadal subsurface ocean heat content in the

Indian Ocean. J Clim 31:4157–4171. https://doi.org/10.1175/JCLI-D-17-0654.1

Keeling RF, Arne Körtzinger A, Gruber N (2010) Ocean deoxygenation in a warming world. Annu Rev Mar Sci 2:199–229

Klein SA, Soden BJ, Lau NC (1999) Remote sea surface temperature variations during ENSO: evidence for a tropical atmospheric bridge. J Clim 12:917–932. https://doi.org/10.1175/1520-0442(1999)012,0917:RSSTVD.2.0.CO;2

Koslow JA, Goericke R, Lara-Lopez A, Watson W (2011) Impact of declining intermediate-water oxygen on deepwater fishes in the California Current. Mar Ecol-Prog Ser 436:207–218

Kouketsu S, Kawano T, Masuda S, Sugiura N, Sasaki Y, Toyoda T, Igarashi H, Kawai Y, Katsumata K, Uchida H, Fukasawa M (2011) Deep ocean heat content changes estimated from observation and reanalysis product and their influence on sea level change. J Geophys Res 116:C03012. https://doi.org/10.1029/2010JC006464

Kumar A, Wen CH (2016) An oceanic heat content-based definition for the Pacific decadal oscillation. Mon Weather Rev 144(10):3977–3984. https://doi.org/10.1175/mwr-d-16-0080.1

Lau NC, Nath MJ (2000) Impact of ENSO on the variability of the Asian-Australian monsoons as simulated in GCM experiments. J Clim 13(24):4287–4309. https://doi.org/10.1175/1520-0442(2000)013%3c4287:IOEOTV%3e2.0.CO;2

Lee P-F, Chen I-C, Tzeng W-N (2005) Spatial and temporal distribution patterns of bigeye tuna (Thunnus obesus) in the Indian Ocean. Zool Stud Taipei 44(2):260

Lee SK, Park W, Baringer M, Gordon AL, Huber B, Liu Y (2015) Pacific origin of the abrupt increase in Indian Ocean heat content during the warming hiatus. Nat Geosci 8:445–449. https://doi.org/10.1038/NGEO2438

Levitus S, Antonov JI, Boyer TP, Locarnini RA, Garcia HE, Mishonov AV (2009) Global ocean heat content 1955–2008 in light of recently revealed instrumentation problems. Geophys Res Lett 36:L07608. https://doi.org/10.1029/2008GL037155

Levitus S, Antonov JI, Wang J, Delworth TL, Dixon KW, Broccoli AJ (2001) Anthropogenic warming of earth's climate system. Science 292(5515):267–270. https://doi.org/10.1126/science.1058154

Li G, Xie S-P, Du Y (2016) A robust but spurious pattern of climate change in model projections over the tropical Indian Ocean. J Clim 29:5589–5608. https://doi.org/10.1175/JCLI-D-15-0565.1

Liu W, Xie SP, Lu J (2016) Tracking ocean heat uptake during the surface warming hiatus. Nat Commun 7:10926. https://doi.org/10.1038/ncomms10926

Li Y, Han W, Hu A, Meehl GA, Wang F (2018) Multidecadal changes of the upper Indian Ocean heat content during 1965–2016. J Clim 31:7863–7884. https://doi.org/10.1175/JCLI-D-18-0116.1

Mishra V, Smoliak BV, Lettenmaier DP, Wallace JM (2012) A prominent pattern of year-to-year variability in Indian Summer Monsoon Rainfall. Proc Natl Acad Sci 109(19):7213–7217

Mohapatra S, Gnanaseelan C, Deepa JS (2020) Multidecadal to decadal variability in the equatorial Indian Ocean subsurface temperature and the forcing mechanisms. Clim Dyn 54:3475–3487. https://doi.org/10.1007/s00382-020-05185-7

Nagamani PV, Ali MM, Goni GJ, Bhaskar TU, McCreary JP, Weller RA, Rajeevan M, Krishna VG, Pezzullo JC (2016) Heat content of the Arabian Sea Mini Warm Pool is increasing. Atmos Sci Lett 17(1):39–42. https://doi.org/10.1002/asl.596

Oliver ECJ et al (2018) Longer and more frequent marine heatwaves over the past century. Nat Commun 9:1324. https://doi.org/10.1038/s41467-018-03732-9

Palmer MD, Roberts CD, Balmaseda M, Chang YS, Chepurin G, Ferry N, Fujii Y, Good SA, Guinehut S, Haines K, Hernandez F (2017) Ocean heat content variability and change in an ensemble of ocean reanalyses. Clim Dyn 49(3):909–930. https://doi.org/10.1007/s00382-015-2801-0

Pratik K, Parekh A, Karmakar A, Chowdary JS, Gnanaseelan C (2019) Recent changes in the summer monsoon circulation and their impact on dynamics and thermodynamics of the Arabian Sea. Theor Appl Climatol 136:321–331. https://doi.org/10.1007/s00704-018-2493-6

Purkey SG, and Johnson GC (2010) Warming of global abyssal and deep Southern Ocean waters between the 1990s and 2000s: Contributions to global heat and sea level rise budgets. J Climate 23 (23):6336–6351

Rahul S, Gnanaseelan C (2013) Net heat flux over the Indian Ocean: Trends, driving mechanisms, and uncertainties. IEEE Geosci Remote Sens Lett 10(4):776–780. https://doi.org/10.1109/LGRS.2012.2223194

Rahul S, Gnanaseelan C (2016) Can large scale surface circulation changes modulate the sea surface warming pattern in the Tropical Indian Ocean? Clim Dyn 46(11):3617–3632. https://doi.org/10.1007/s00382-015-2790-z

Rosenthal Y, Kalansky J, Morley A, Linsley BK (2017) A paleo-perspective on ocean heat content: lessons from the Holocene and Common Era. Quat Sci Rev 155:1–2

Rajeevan M, Srinivasan J, Niranjan Kumar K, Gnanaseelan C, Ali MM (2013) On the epochal variation of intensity of tropical cyclones in the Arabian Sea. Atmos Sci Lett 14(4):249–255

Rao SA, Dhakate AR, Saha SK, Mahapatra S, Chaudhari HS, Pokhrel S, Sahu SK (2012) Why is Indian Ocean warming consistently? Clim Change 110:709–719. https://doi.org/10.1007/s10584-011-0121-x

Rodrigues RR, Taschetto AS, Gupta AS, Foltz GR (2019) Common cause for severe droughts in South America and marine heatwaves in the South Atlantic. Nat Geosci 12(8):620–626

Rohini P, Rajeevan M, Srivastava AK (2016) On the variability and increasing trends of heat waves over India. Scientific reports 6:26153

Roxy MK, Ghosh S, Pathak A, Athulya R, Mujumdar M, Murtugudde R, Terray P, Rajeevan M (2017) A threefold rise in widespread extreme rain events over central India. Nat Commun 8:78

Roxy MK, Dasgupta P, McPhaden MJ, Suematsu T, Zhang C, Kim D (2019) Twofold expansion of the Indo-Pacific warm pool warps the MJO lifecycle. Nature 575:647–651. https://doi.org/10.1038/s41586-019-1764-4

Roxy MK, Modi A, Murtugudde R, Valsala V, Panickal S, Prasanna Kumar S, Ravichandran M, Vichi M, Lévy M (2016) A reduction in marine primary productivity driven by rapid warming over the tropical Indian Ocean. Geophys Res Lett 43(2):826–833. https://doi.org/10.1002/2015GL066979

Roxy MK, Ritika K, Terray P, Masson S (2014) The curious case of Indian Ocean warming. J Clim 27(22):8501–8509. https://doi.org/10.1175/JCLI-D-14-00471.1

Sabeerali CT, Rao SA, George G, Rao DN, Mahapatra S, Kulkarni A, Murtugudde R (2014) Modulation of monsoon intraseasonal oscillations in the recent warming period. J Geophys Res Atmos 119(9):5185–5203

Sabine CL, Freely RA, Gruber N, Key RM, Lee K, Bullister JL, Wanninkhof R, Wong CS, Wallace DWR, Tilbrook B, Millero FJ, Peng T-H, Kozyr A, Ono T, Rios AF (2004) The oceanic sink for anthropogenic CO_2. Science 305:367–371

Saha A, Ghosh S, Sahana AS, Rao EP (2014) Failure of CMIP5 climate models in simulating post-1950 decreasing trend of Indian monsoon. Geophys Res Lett 41(20):7323–7330

Saji NH, Goswami BN, Vinayachandran PN, Yamagata T (1999) A dipole mode in the tropical Indian Ocean. Nature 401:360–363

Sayantani O, Gnanaseelan C (2015) Tropical Indian Ocean subsurface temperature variability and the forcing mechanisms. Clim Dyn 44:2447–2462. https://doi.org/10.1007/s00382-014-2379-y

Schmidtko S, Stramma L, Visbeck M (2017) Decline in global oceanic oxygen content during the past five decades. Nature 542:335–339. https://doi.org/10.1038/nature21399

Singh D, Ghosh S, Roxy MK, McDermid S (2019) Indian summer monsoon: extreme events, historical changes, and role of anthropogenic forcings. Wiley Interdiscip Rev Clim Chang 10(2):e571

Slangen ABA, Carson M, Katsman CA, van de Wal RSW, Koehl A, Vermeersen LLA, Stammer D (2014) Projecting twenty-first century regional sea-level changes. Clim Change 124:317–332. https://doi.org/10.1007/s10584-014-1080-9

Smith D, Murphy JM (2007) An objective ocean temperature and salinity analysis using covariances from a global climate model. J Geophys Res Oceans 112(C2)

Sreeush MG, Rajendran S, Valsala V, Pentakota S, Prasad KV, Murtugudde R (2019a) Variability, trend and controlling factors of Ocean acidification over Western Arabian Sea upwelling region. Marine Chem. https://doi.org/10.1016/j.marchem.2018.12.002. (2018)

Sreeush MG, Valsala V, Santanu H, Pentakota S, Prasad KVSR, Naidu CV, Murtugudde R (2019b) Biological production in the Indian Ocean upwelling zones—part 2: data based estimates of variable compensation depth for ocean carbon models via cyclo-stationary Bayesian inversion, deep sea research part II: topical studies in oceanography, 2019, ISSN 0967-6645. https://doi.org/10.1016/j.dsr2.2019.07.007

Srinivasu U, Ravichandran M, Han W, Sivareddy S, Rahman H, Li Y, Nayak S (2017) Causes for the reversal of north Indian Ocean decadal sea level trend in recent two decades. Clim Dyn 49:3887–3904. https://doi.org/10.1007/s00382-017-3551-y

Stramma L, Johnson GC, Sprintall J, Mohrholz V (2008) Science 320 (5876):655–658. https://doi.org/10.1126/science.1153847

Ummenhofer CC, Biastoch A, Böning CW (2017) Multidecadal Indian Ocean variability linked to the Pacific and implications for preconditioning Indian Ocean dipole events. J Clim 30(5):1739–1751. https://doi.org/10.1175/JCLI-D-16-0200.1

Valsala V, Murtugudde R (2015) Mesoscale and intraseasonal air–sea CO_2 exchanges in the western Arabian Sea during boreal summer. Deep Sea Res Part I 103:101–113

Von Schuckmann K, Sallée JB, Chambers D, Le Traon PY, Cabanes C, Gaillard F, Speich S, Hamon M (2014) Monitoring ocean heat content from the current generation of global ocean observing systems. Ocean Sci 10:547–557. https://doi.org/10.5194/os-10-547-2014

Webster PJ, Moore AM, Loschnigg JP, Leben RR (1999) Coupled ocean-atmosphere dynamics in the Indian Ocean during 1997–1998. Nature 401:356–360

Xie S-P, Annamalai H, Schott FA, McCreary JP Jr (2002) Structure and mechanisms of South Indian Ocean climate variability. J Clim 15:864–878

Xue Y et al (2012) A comparative analysis of upper-ocean heat content variability from an ensemble of operational ocean reanalyses. J Clim 25:6905–6929. https://doi.org/10.1175/JCLI-D-11-00542.1

Zanna L, Khatiwala S, Gregory JM, Ison J, Heimbach P (2019) Global reconstruction of historical ocean heat storage and transport. Proc Natl Acad Sci 116(4):1126–1131

Zhao Y, Zhang H (2016) Impacts of SST warming in tropical Indian Ocean on CMIP5 model-projected summer rainfall changes over Central Asia. Clim Dyn 46:3223–3238. https://doi.org/10.1007/s00382-015-2765-0

Zheng XT, Xie S (2009) Indian Ocean dipole response to global warming: analysis of ocean–atmospheric feedbacks in a coupled model. J Clim 23:1240–1253. https://doi.org/10.1175/2009JCLI3326.1

Climate Change Over the Himalayas

Coordinating Lead Authors

T. P. Sabin, Indian Institute of Tropical Meteorology (IITM-MoES), Pune, India,
e-mail: sabin@tropmet.res.in (corresponding author)
R. Krishnan, Indian Institute of Tropical Meteorology (IITM-MoES), Pune, India

Lead Authors

Ramesh Vellore, Indian Institute of Tropical Meteorology (IITM-MoES), Pune, India
P. Priya, Indian Institute of Tropical Meteorology (IITM-MoES), Pune, India
H. P. Borgaonkar, Indian Institute of Tropical Meteorology (IITM-MoES), Pune, India
Bhupendra B. Singh, Indian Institute of Tropical Meteorology (IITM-MoES), Pune, India
Aswin Sagar, Indian Institute of Tropical Meteorology (IITM-MoES), Pune, India

Review Editor

Anil Kulkarni, Indian Institute of Science, Bangalore, India

Contributing Authors

Thamban Meloth, National Centre for Polar and Ocean Research (NCPOR-MoES), Goa, India
Paramanand Sharma, National Centre for Polar and Ocean Research (NCPOR-MoES), Goa, India
M. Ravichandran, National Centre for Polar and Ocean Research (NCPOR-MoES), Goa, India
Mahesh Ramadoss, Indian Institute of Tropical Meteorology (IITM-MoES), Pune, India

Corresponding Author

T. P. Sabin, Indian Institute of Tropical Meteorology (IITM-MoES), Pune, India,
e-mail: sabin@tropmet.res.in

© The Author(s) 2020
R. Krishnan et al. (eds.), *Assessment of Climate Change over the Indian Region*,
https://doi.org/10.1007/978-981-15-4327-2_11

Key Messages

- The Himalayas and the Tibetan Plateau have experienced substantial warming during the twentieth century. The warming trend has been particularly pronounced over the Hindu Kush Himalaya (HKH) which is the largest area of permanent ice cover outside the North and South Poles.

- The annual mean surface-air-temperature in the HKH increased at a rate of about 0.1 °C per decade during 1901–2014, with a faster rate of warming of about 0.2 °C per decade during 1951–2014, which is attributable to anthropogenic climate change (*High confidence*). Additionally, high elevations (> 4000 m) of the Tibetan Plateau have experienced stronger warming, as high as 0.5 ° C per decade, which is commonly referred to as elevation-dependent warming (EDW).

- Several areas in the HKH have exhibited declining trends in snowfall and retreating glaciers during the recent decades. Parts of the high-elevation Karakoram Himalayas have, in contrast, experienced increased wintertime precipitation in association with enhanced amplitude variations of synoptic western disturbances (*Medium confidence*).

- Future climate projections under various CMIP5 scenarios suggest warming of the HKH region in the range of 2.6–4.6 °C by the end of the twenty-first century. While future projections indicate significant decrease of snowfall in several regions of the HKH, high-elevation locations (> 4000 m) in the Karakoram Himalayas are projected to experience an increase in annual precipitation during the twenty-first century.

11.1 Introduction

The name "Himalaya" means "the abode of snow" in *Sanskrit*. The continental drift theory suggests that the Himalayas were formed about 50 million years ago when the Indian plate collided with the Eurasian plate (Kious and Tilling 1996). The large spatial extent of the Himalayas (Fig. 11.1) spans across eight countries of the Asian continent and is the source of ten major river systems (Sharma et al. 2019, HIMAP) providing water for drinking, irrigation and power for over 1.3 billion people in Asia—which is nearly 20% of the world's population (e.g. Bookhagen and Burbank 2006; Rashul 2014). The Himalayan mountain range is the world's tallest and is notably the home to 10 of the 14 world's highest peaks, while the Karakoram and the Hindu Kush are generally viewed as separate ranges in the literature (Godin et al. 1999). The area that encompasses the Hindu Kush Himalaya (HKH) mountain range and the

Tibetan Plateau (TP) is popularly known as the "*Third Pole*" as it contains the largest reserve of freshwater outside the north and south poles. The meltwater generated from the Himalayan glaciers supplies the rivers and streams of the region, including the Indus, Ganges and Brahmaputra river systems of India. These rivers collectively provide about 50% of the country's total utilisable surface water resources (Srivastava and Misra 2012). Scientific evidence also shows that most glaciers in the HKH region are subjected to loss in volume and mass under the propensity of rising temperatures due to climate change (Kulkarni and Karyakarte 2014; ICIMOD 2007, 2011; Armstrong 2011; Wester et al. 2019; IPCC SROC 2019). Yet, a clear understanding and quantification of its consequences in these mountain ranges remain challenging.

The climate of the HKH is characterised by tropical/subtropical climatic conditions from the foothill region of the mountains to permanent ice and snow-covered peaks at higher altitudes (Pant et al. 2018). Flora and fauna of the Himalayas vary with climate, rainfall, altitude and soils. The amount of annual rainfall increases from west to east along the southern front of the range. Further, the Himalaya is delineated by different climatic sub-zones due to diverse geographical variability, which is closely linked to topographical distribution of the region (Bookhagen and Burbank 2006). The annual cycle of temperature and precipitation differs substantially in these different zones. Seasonal variations in the mean climate of HKH are closely tied to the seasonal cycle of the regional atmospheric processes (Box 11.1).

The valleys experience mean summer temperatures between 15 and 25 °C and much colder in winter. Regions with elevations above 4500 m experience severe winter, with temperatures far below freezing point and precipitation in the form of snow, e.g. the Karakoram range of the Himalayas experiences an average maximum temperature of about 20 °C during the summer, and average minimum temperature goes below −3 °C in February (Hasson et al. 2014; Kapnick et al. 2014). The north-western peaks of Himalayas typically experience dry conditions, with surface temperatures ranging between 3 and 35 °C in summer and −20 and −35 °C in winter together with heavy snowfall.

The climate of the HKH has been experiencing significant temperature changes since twentieth century where the warming trend during the first (second) half of the twentieth century was about 0.10 °C (0.16 °C) per decade, which later doubled to 0.32 °C per decade from the beginning of the twenty-first century (Yan and Liu 2014). The warming rate is reported to be more substantial in winter as compared to other seasons in most parts of the HKH region (Bhutiyani et al. 2007; Shrestha et al., 2010). Studies by Dimri and Dash (2012), Negi et al. (2018), etc., also confirm that most of the

Fig. 11.1 Hindu Kush Himalayan (HKH) region and the three sub-regions (rectangular black box) of interest: the northwest Himalaya and Karakoram (HKH1), central Himalaya (HKH2) and southeast Himalaya and Tibetan Plateau (HKH3)

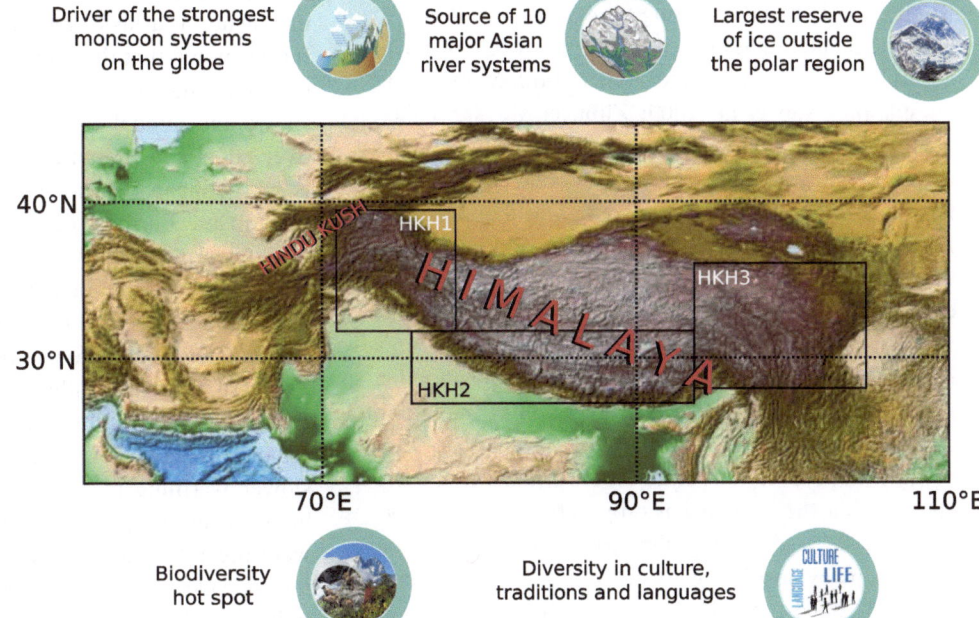

western Himalayan (WH) stations recorded a significant warming trend especially from 1975 onwards. This is also supported by tree-ring chronologies of the region which indicated rapid growth of tree rings in the recent decades especially at higher altitudes (Borgaonkar et al. 2009). This chapter provides an assessment of the current state of knowledge of the weather and climate aspects of the HKH region. Thematic assessments of socio-economic and other sectorial impacts are covered in the *"Hindu Kush Himalaya Assessment report"* (Wester et al. 2019) and the *"IPCC Special report on Ocean and Cryosphere in a Changing Climate, Chapter 2: High Mountain Areas"* (IPCC SROC 2019).

Box 11.1 Weather and climate of Himalaya

The HKH1 sub-region generally receives more precipitation during the winter months, while more than 80% of annual precipitation in the central and eastern sub-region (HKH 2 and 3) of the Himalayas is received during the summer monsoon season. The three major Himalayan river basins, viz. Indus, the Ganga and the Brahmaputra, receive an annual rainfall of about 435, 1094 and 2143 mm, respectively (Shrestha et al. 2015). Weather dynamics is intricate in the Himalayan region arising due to extensive interactions of tropical and extra-tropical weather systems. Vellore et al. (2014) reported that heavy rainfall over the WH region during summer is often associated with the interaction between westward-moving monsoon low-pressure systems and subtropical westerly winds.

Precipitation during the winter months contributes nearly half of the annual precipitation over the Karakoram and the Hindu Kush Mountain ranges (Kapnick et al. 2014; Cannon et al. 2015; Hunt et al. 2018; Krishnan et al. 2019a). The accumulation of snow occurs during the winter months which is the primary water reserve for the subsequent dry periods (Lang and Barros 2004; Rees and Collins 2006; Immerzeel et al. 2010; Bolch et al. 2012; Hasson et al. 2013). The western side of the Karakoram Himalaya is prone to large amounts of snowfall in winter from frequent passage of midlatitude synoptic-scale systems known as the western disturbances (Dimri et al. 2015; Madhura et al. 2015; Cannon et al. 2015; Krishnan et al. 2019a). Tropical and extra-tropical climate drivers such as the El Niño-Southern Oscillation (ENSO; Diaz and Markgraf 2000), North Atlantic Oscillation (NAO; Branstator 2002), Madden–Julian Oscillation (MJO; Madden and Julian 1971), Arctic Oscillation (AO; Thompson and Wallace 1998; Wallace 2000) and the Indian Ocean Dipole mode (IOD; Saji et al. 1999) generally appear to exert significant influence in regulating the weather and climate of the HKH region (see Barlow et al. 2005; Bhutiyani et al. 2009; Cannon et al. 2017). More details on these aforesaid teleconnection patterns can be seen in Panagiotopoulos et al. (2002) and Sheridan et al. (2012). Archer and Fowler (2004) also note that there is an in-phase relation between NAO and precipitation variability over the

Karakoram Himalayas during winter months. It is also widely accepted that there is a significant role of HKH and TP in maintaining the Asian summer monsoon circulation (Nan et al. 2009; Zhou et al. 2009), i.e., heating of the TP in summer raises air temperatures thereby enhancing the pressure gradient which drives the South Asian summer monsoon, and therefore, the HKH and TP act as a major heat sink (source) during the winter (summer) (Yanai and Li 1994). The presence of the HKH topographical barrier restricts the upper-level subtropical westerly winds to regions poleward of 30° N during the boreal summer months, thereby allowing warm and moist summer monsoon circulation to extend northward into the Indian subcontinent. Observations and model simulations clearly suggest that the Himalayan orography has an important role in maintaining the South Asian monsoon circulation by insulating warm, moist air over continental India from the cold and dry extra-tropics (Chakraborty et al. 2006; Krishnan et al. 2009; Krishnamurti et al. 2010; Boos and Kuang 2010; Turner and Annamalai 2012; Sabin et al. 2013).

11.2 Observed Trends in Mean Surface Temperature and Precipitation

Significant rise in surface temperatures is noted throughout the HKH region during the past six decades (see Kulkarni et al. 2013; Rajbhandari et al. 2016). The warming was reported progressively over the western and eastern Himalayan river basins, and the long-term trend of minimum temperatures is noted slightly higher than the trend seen in maximum temperatures (Rajbhandari et al. 2015). Using century-long historical time series, Ren et al. (2017) showed that there were epochs of rise and fall in temperature trend over the HKH region, i.e., mean temperature exhibited a moderate rising trend from 1901 to the early 1940s, while a falling trend is seen between 1940 and 1970 followed by a rapid warming. The spatial pattern of trends of annual mean temperatures over the HKH region during the 1901–2014 period shows that the warming rates were more than 0.3 °C (decade)$^{-1}$ in the TP region and about 0.2 °C (decade)$^{-1}$ over the eastern side of the HKH range (Ren et al. 2017). It can also be noted that most of the grids consistently showed a positive trend in the annual warming signal; however, the warming rates are significantly different. Various studies attributed this observed warming signal to the increase in anthropogenic greenhouse gas concentrations (IPCC 2007, 2013; You et al. 2017).

Pepin et al. (2015) show that mountain temperatures are increasing at a faster rate than the global average. We have further noted that the Himalayas are also warming at a faster rate than that of the nearby Indian land mass. Figure 11.2 shows the annual mean temperature time series averaged over HKH and Indian land mass from 1951 to 2018, which is indicating a warmer Himalaya comparing to the Indian land mass. Further, the observations also reveal that the recent warming rates are not seen to be uniform over the HKH region where the annual average warming rates change with altitudes which are commonly referred as the elevation dependency of climate warming (EDW) (e.g. Liu et al. 2009; Ren et al. 2017; Shrestha et al. 1999; Thompson et al. 2003). Figure 11.3 shows the trend per decade for different altitude sectors of HKH. Low-elevation sites (<500 m) show a slower warming rate (<0.2 °C per decade) as compared to high elevations (>2000 m) of the eastern TP where higher warming rate (0.61 °C per decade) is seen during the past few decades (Liu et al. 2009; Ren et al. 2017). Northern India and the Sichuan Basin of China showed the weakest warming trend with annual warming rates less than 0.10 °C (decade)$^{-1}$, as well as the Karakoram range during the northern summer (Forsythe et al. 2017).

The spatial distribution of the trends of annual precipitation based on APHRODITE dataset during 1951–2015 is shown in Fig. 11.4. The observed map of trend in the annual mean precipitation depicts substantial spatial heterogeneity over the HKH region (Fig. 11.4). While there is a subtle rise in the precipitation trend over the Karakoram and Western Himalayas and the eastern part of Himalayas, a declining trend can be noticed over many areas. A wet trend is evident over northwest China, including the TP (Ren et al. 2015; You et al. 2015). The triangular markings in Fig. 11.4 correspond to precipitation trends per decade based on CMA-GMLP dataset for a longer period for 1901–2013. Both the datasets show consistency in precipitation enhancement over WH and decline over central and north Indian plains. The reduction in annual precipitation over northern India is also consistent with the reported declining trend of the Indian summer monsoon precipitation during the post-1950 (Krishnan et al. 2013, 2016).

Recent observational studies also suggest that there is an increasing trend in the number of wet days over the WH during the past few decades (Klein Tank et al. 2006; Choi et al. 2009). It has been reported that increases in wintertime heavy precipitation over the Karakoram region of northwest Himalayas and falling precipitation trend over Central Himalayas (CH) are linked to an increasing trend in the synoptic-scale activity of the western disturbances (WDs), while the CH region experiences a falling local precipitation trend (Cannon et al. 2015; Madhura et al. 2015; Krishnan

HIMALAYA VS. INDIA

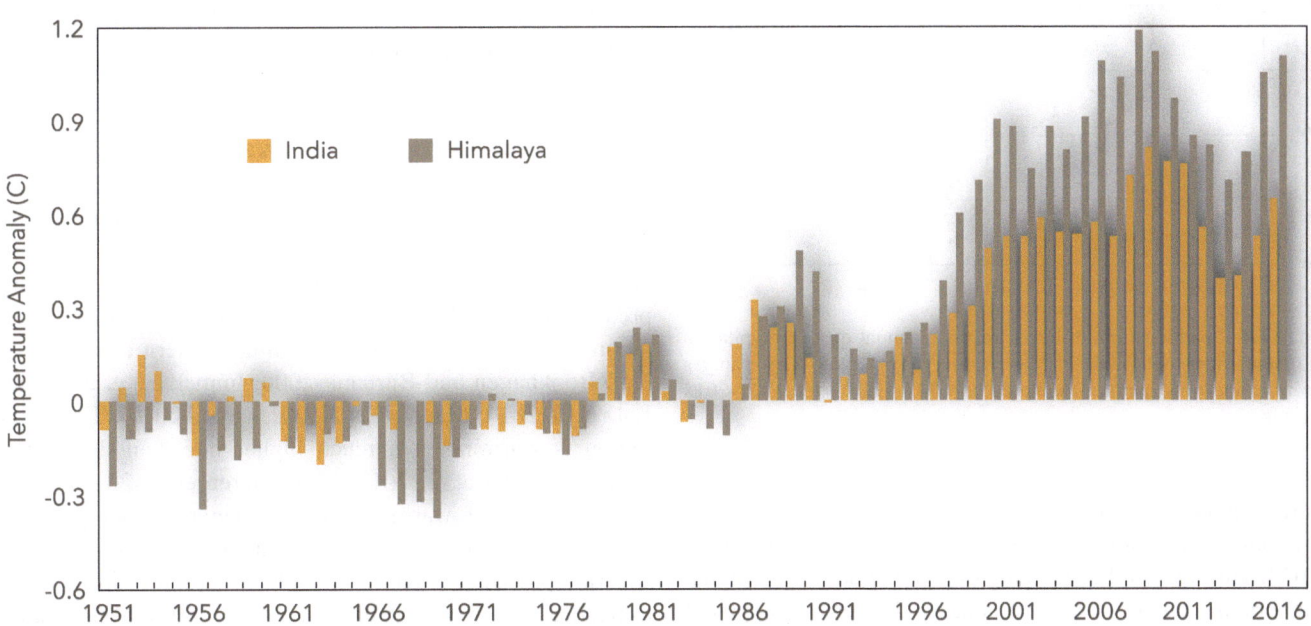

Fig. 11.2 Annual mean temperature time series (5-year running mean) averaged over HKH (grey) and Indian land mass (yellow) from 1951 to 2018

Fig. 11.3 Elevation-dependent warming over and around HKH. Shown is the trend/decade of surface air temperature for respective altitude ranges

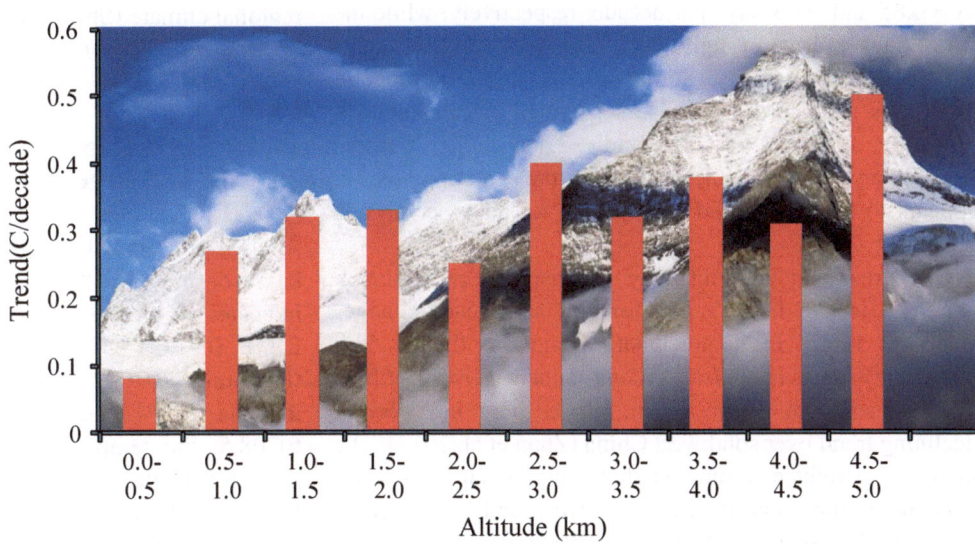

et al. 2019a). A few studies also attribute the increase in the mass of some glaciers in the Karakoram and Western Himalaya to changes in the WD activity (Forsythe et al. 2017). In contrast to other mountain peaks across the globe, which have largely experienced decreasing snowfall amounts and glacial extent during the recent decades, the Karakoram Himalayas appear to have slightly gained glacial mass in the early twenty-first century (Gardelle et al. 2012; Hewitt 2005; Kapnick et al. 2014).

11.3 Observed Changes in Temperature and Precipitation Extremes

Significant rise in extreme warm events and a substantial fall in extreme cold events observed over the HKH during the last few decades (e.g. Sun et al. 2017). In particular, the magnitude of trends in warm events is more significant than those of the cold events. A significant rise in the number of

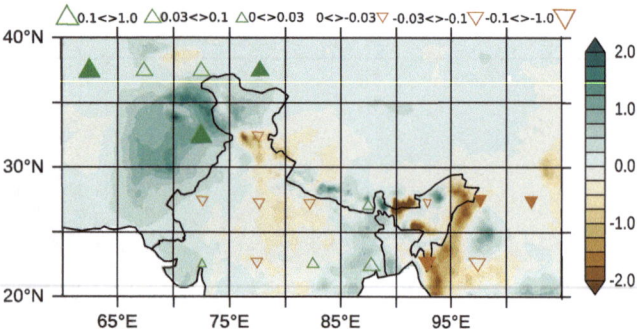

Fig. 11.4 Spatial pattern of linear trends in annual mean precipitation anomalies from APHRODITE data from 1951 to 2015. The triangles are from the trend per decade based on CMA-GMLP for 1901–2013 periods

warm nights (Tn90p) and a decrease in the number of frost days are seen from early-1990s (Fig. 2h–g of Sun et al. 2017). Throughout the HKH region, the annual mean diurnal temperature range (DTR) anomalies showed an apparent decline before 1980s, while DTR exhibits a rising trend after the mid-1980s. The spatial distribution of linear trends for extreme temperature indices indicate that extremely cold days and nights in the Tibetan Plateau (TP) region decreased by −0.85 and −2.3 days per decade, respectively, while the warm days and warm nights increased by 1.2 and 2.5 days per decade, respectively (Sun et al. 2017). The number of frost days and ice days is also seen to be decreasing significantly at a rate of −4.3 and −2.4 days per decade, respectively. Overall, the length of the growing season appears to have increased at a rate of 4.5 days per decade (Liu et al. 2006).

The frequency and intensity of observed precipitation extremes in the HKH region exhibit significant changes since the 1960s. Light precipitation amounts (below 50th percentile) show a significant rising tendency over northern TP and southern Tarim Basin, while there has been a declining trend over southwest China (Zhan et al. 2017). The intensity of light precipitation shows significant reductions over the northern part of the Hindu Kush and Central India (cf. Figs. 2 and 3 of Zhan et al. 2017). In addition, the frequency and intensity of heavy precipitation show significant increasing trends mostly over the TP, with opposite trends over southwest China, and South-Central Asia. Linear trends of regional average maximum 5-day consecutive precipitation (RX5DAY) index show a clear increasing trend over the HKH region by 2.3% per decade during 1961–2012. Consecutive wet days significantly increased over the Indian side of Himalayan and Karakoram ranges and moderate rise over most of the other locales of the TP. The spatial distribution pattern of consecutive dry-day trend is nearly opposite to that of consecutive wet days. In summary,

several areas in the TP indicate a rising tendency in intense precipitation, whereas the change is heterogeneous over other areas of the HKH region.

11.4 Future Projections Over HKH

Precipitation from the summer monsoon rainfall is an important source of water for the river basins in the eastern and central HKH. River basins originating in the WH are predominantly fed by snow and glacial melt with precipitation largely coming from wintertime western disturbances (e.g. Bookhagen and Burbank 2006; Immerzeel et al. 2009; Lutz et al. 2014; Madhura et al. 2015; SAC 2016). The HKH region is warming at a much higher rate than the global mean (Shrestha et al. 2015; Van Vuuren et al. 2011). With continued global warming, future changes in temperature and precipitation are expected to alter the sensitive cryospheric processes over the HKH region substantially (Shrestha and Aryal 2011; Xu et al. 2008). Accelerated warming over the ice-covered mountain peaks and valleys exert profound impacts on the climate-dependent sectors like agriculture and water resources of the HKH region, thereby warranting a robust and reliable future outlook of the regional climate (Shrestha et al. 2015; Krishnan et al. 2019b; Sharma et al. 2019).

11.4.1 Projected Changes in Mean Temperature and Precipitation

Analysis of annual mean surface temperature projections based on the CMIP5 multi-models (Table 3.2b of Chap. 3) indicates an increase of temperature in the HKH region by 2.2 ± 0.9 °C (3.3 ± 1.4°C) for the near future; 2040–2069 (far future; 2070–2099) of the twenty-first century, following the RCP4.5 scenario. Under the extreme scenario RCP8.5, the temperature increase in the HKH region is projected to be 2.8 ± 1.2 °C (4.8 ± 1.7 °C) for the near future (far future) of the twenty-first century (Fig. 11.5a). Wintertime (DJF) temperatures are projected to increase by 2.4 ± 1 °C (3.5 ± 1.4 °C) for the near future (far future) of the twenty-first century, following RCP4.5. The corresponding wintertime temperatures for the two epochs under the RCP8.5 scenario are projected to increase by 3.1 ± 1.4 °C (5.1 ± 1.8 °C), respectively (Fig. 11.5b).

Significant warming is projected over the HKH region in the near and far future, with prominent temperature increase projected over the Tibetan Plateau with magnitudes exceeding 5 °C under the RCP8.5 scenario by the end of the twenty-first century (Xu et al. 2014; Wu et al. 2017). The projected warming also differs by more than 1 °C between

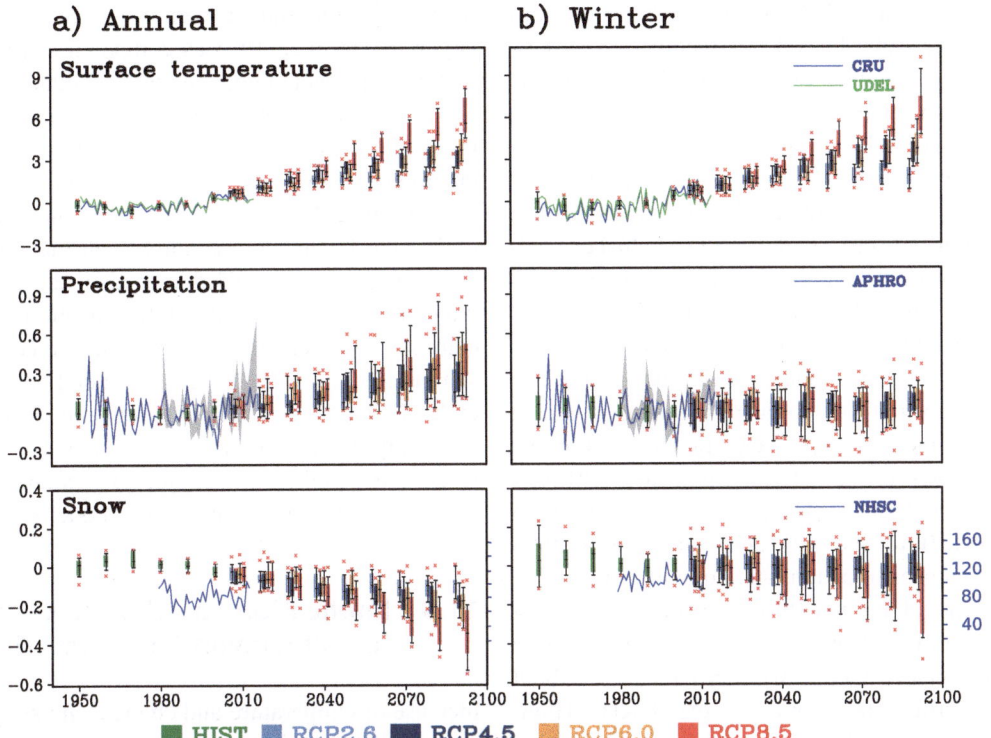

Fig. 11.5 Box whiskers of variations in surface air temperature (°C), precipitation (mm/day) and snow (mm/day) over the HKH from the CMIP5 projections for the different future scenarios (Annual mean (**a**) and Winter season (**b**)). Variations in ten-year means, with respect to the reference period (1976–2005), are presented as box whiskers plots. Observed temperature (CRU) and precipitation (APHRODITE) are shown during 1951–2015. Precipitation from GPCP and CMAP are also shown during 1981–2015. Grey shade is the ensemble spread during 1981–2015. The snow cover observation is based on the NSIDC dataset from 1979 to 2012 (NHSC: Northern Hemisphere State of Cryosphere, right y-axis). Note that variations in observed snow cover extent are expressed in x 10000 km². The CMIP5 projected variations in snowfall are expressed in mm/day (left y-axis)

Table 11.1 Projected changes in mean annual precipitation (%) over HKH region with respect to 1976–2005 period

	Scenario	Spatial range of annual precipitation change (%)		
		CMIP5	CORDEX	NEX
Near future	RCP4.5	2–8	5–15	5–20
	RCP8.5	4–15	10–20	5–25
Far future	RCP4.5	5–15	10–20	15–25
	RCP8.5	10–20	15–30	20–40

the eastern and western HKH, with relatively higher warming in winter based on the diagnosis from high-resolution CORDEX model outputs (Sanjay et al. 2017). The downscaled multi-RCMs analysis from CORDEX projection shows a significant warming over the HKH as a whole and its hilly sub-regions, with a projected change of 5.4 °C during winter and 4.9 °C during the summer season by the end of the twenty-first century under the high-end emissions (RCP8.5) scenario (Sanjay et al. 2017). Projected changes in annual mean temperature over the HKH region is also calculated using the Reliability Ensemble Approach (REA; Box 2.4) based on the CORDEX models. The near future (2040–2069) increase in annual

mean temperature over the HKH is projected to be 2.26 °C ± 0.45 °C (3.2 °C ± 0.58 °C) in the RCP4.5 (RCP8.5) scenarios, respectively. The corresponding projections for the two scenarios in the far future (2070–2099) are noted to be 2.76 °C ± 0.61 °C (5.23 °C ± 0.91 °C), respectively. The changes are relative to the baseline period (1976–2005).

Precipitation response to climate change is subject to greater uncertainties, as compared to temperature changes, particularly over the complex topographical terrains of HKH. To better understand projected changes in precipitation, we additionally used the multi-model ensemble (MME) from the high-resolution (0.5° resolution) dynamically downscaled CORDEX-SA models (Table 3.2a of Chap. 3; 16 members) and statistically downscaled NEX-GDDP (0.25° resolution) (marked in Table 3.2b of Chap. 3; 19 members) datasets. Future projections of precipitation over the HKH, from the three approaches along with their spatial range in ensemble mean, are summarised in Table 11.1.

The projected precipitation changes are similar until the 2050s for both the RCP4.5 and RCP8.5 scenarios, with a higher increase after 2050 in RCP8.5. A box-whisker analysis is provided to demonstrate the projected changes based on CMIP5 models (Fig. 11.5). While a significant rise is projected for annual mean rainfall, a moderate increase is projected for the winter precipitation over the HKH. The increase in annual mean precipitation could be in part due to overall summertime increase in projected precipitation in the CMIP models, which is also consistent with CORDEX simulation (Sanjay et al. 2017). The projected changes are in general, less than 12% for the near future under the RCP4.5 scenario. The differences in pattern and amount of projected precipitation by end of the twenty-first century, following RCP4.5, show similar changes as those of near future changes projected under RCP8.5, with an increase of about 16% over the north-eastern areas of the HKH. High-resolution CORDEX and NEX simulations show value additions in capturing the precipitation variability, as compared to the coarse-resolution CMIP5 models (Kapnick et al. 2014; Singh et al. 2017; Sanjay et al. 2017).

11.4.2 Projected Changes in Temperature and Precipitation Extremes

Future changes in temperature and precipitation extremes over the HKH based on the indices, such as (a) Maximum of daily maximum temperature (TXx), (b) Minimum of daily minimum temperature (TNn), (c) Annual total precipitation when the daily amount exceeds the 95th percentile of wet-day precipitation (R95p), (d) Maximum consecutive 5-day precipitation (RX5day) based on the RCP4.5 and RCP8.5 scenarios, relative to 1976–2005, are presented in Fig. 11.6.

The projected changes of both TXx and TNn over the HKH indicate a tendency for extreme warm days and extreme cold nights to become warmer in the future, with a significant increase in TNn compared to TXx. The maxima of TXx is projected to increase by 2.8 °C (3.4 °C) under RCP4.5 (RCP8.5) in the near future, 4.0 °C (5.2 °C) by the end of twenty-first century, respectively (Fig. 11.6: middle panel). The north-western part of the HKH region is projected to experience substantial increases of TXx compared to other areas, while pronounced warming in TNn is projected over the Eastern Himalaya and TP. In particular, changes in TNn are projected to increase by as much as 5.5 °C in the Eastern Himalayas and TP under the RCP8.5

scenario by the end of twenty-first century, with relatively larger spread in RCP8.5 as compared to RCP4.5.

Future projected changes in precipitation extremes show significant increases of R95p in both RCP4.5 and RCP8.5, indicating the enhanced likelihood of occurrence of extreme precipitation over the HKH. In particular, a substantial increase in R95p is projected over the central Himalayas during the twenty-first century. The maximum consecutive 5-day precipitation (RX5day) also shows a general rise indicative of the future intensification of precipitation extremes. The changes in extreme indices over HKH are summarised in Table 11.2. In general, the MME medians under RCP8.5 are larger as compared to those of RCP4.5, especially for the temperature extremes. For changes in mean precipitation, the projected median change over the HKH is positive with large inter-model spread.

11.5 Implications of Climate Change for Himalayan Snow and Glacier Mass

Increase of temperature and changes in precipitation patterns over the HKH region is a major concern for the health of the Himalayan snow cover and glaciers. This region has experienced significant melting of snow and retreat of glaciers during the past five decades (Kulkarni and Karyakarte 2014). While global climate change significantly affects the environment over the high mountain regions of Asia, its impact on the Himalayan cryosphere is a major threat to the regional water resources (ICIMOD 2007, 2011; Armstrong 2011). In addition to global warming, the absorbing aerosols at high elevations can also enhance the warming rate and indirectly amplify the melting of snowpacks and glaciers (Ramanathan and Carmichael 2008). Significant decrease of wintertime snow over the HKH region in the recent decades is evident from the MODIS satellite snow products which affect the river flow regimes and water resources availability (Maskey et al. 2011). Satellite observations of snow cover area (NSIDC) over the HKH show large variability during the historical period (1980–2018), with moderate decline since 2000 (Fig. 11.5: bottom panel) which is consistent with MODIS analysis. Analysis of CMIP5 projections indicates decrease of annual average snow over the HKH throughout the twenty-first century, with large inter-model spread (Fig. 11.5: bottom panel).

Heavy precipitation over the Western Himalayas (WH), during the winter and early spring, is strongly linked to the activity of western disturbances (WD) (Krishnan et al. 2019a, b). High-resolution climate model projections suggest that increasing amplitude variations of the WD in warming world can favour enhancement of wintertime precipitation over the Karakoram and WH (Fig. 11.7) and provide a plausible explanation for stable snow/glacier mass in the Karakoram

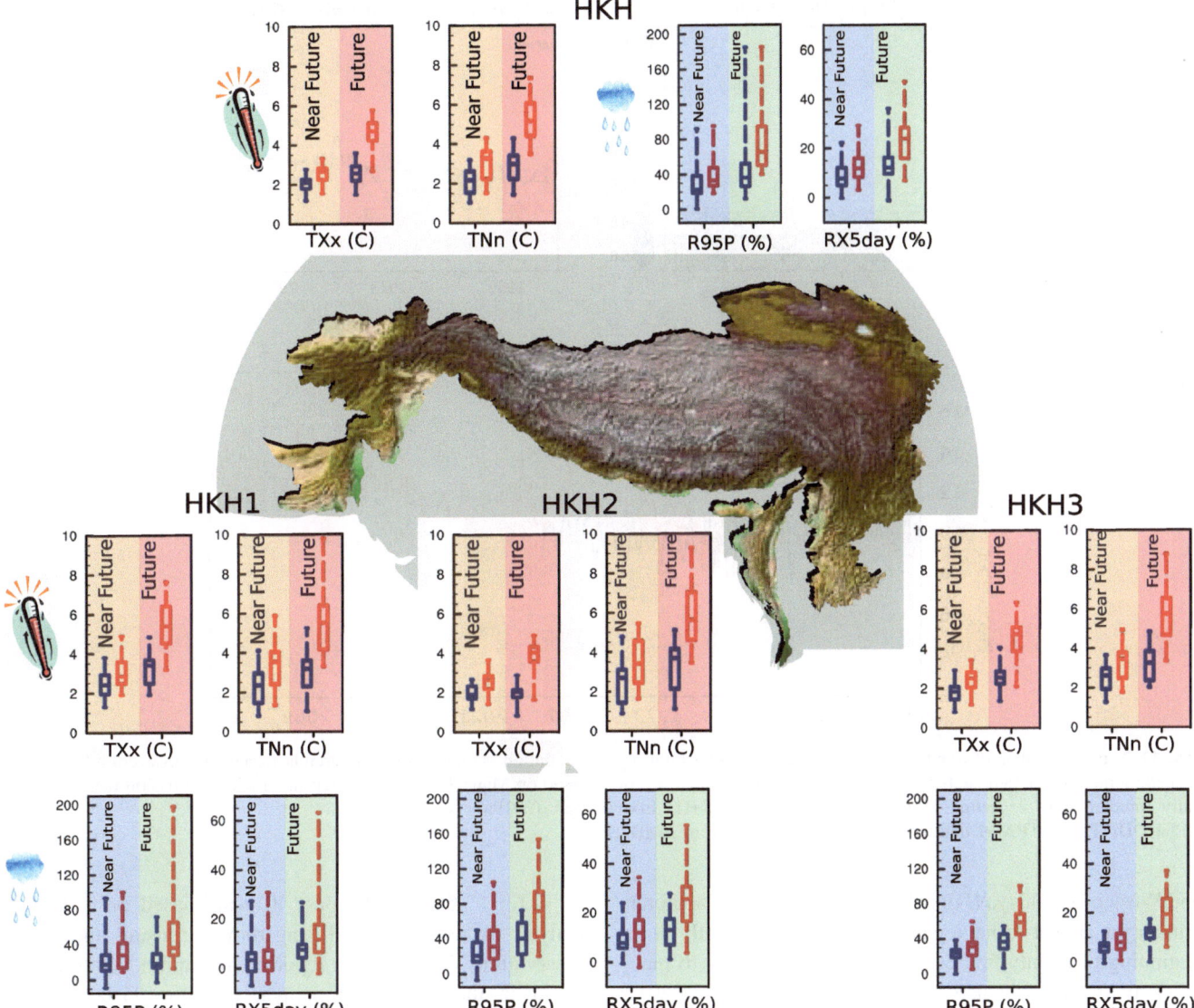

Fig. 11.6 Projected changes in temperature in °C (yellow and pink) and precipitation in % (blue and green) extremes over HKH and its three sub-regions (as shown in Fig. 11.1) from CMIP5 models. Changes with respect to present-day mean are shown as box whiskers from RCP4.5 (blue colour) and RCP8.5 (red colour). The ranges between the 25th and 75th quantiles are indicated by boxes, the MME medians are indicated by the horizontal lines within boxes, and the extreme ranges of models are indicated by whiskers

Table 11.2 Projected changes in extreme indices over HKH represented by the MME median for near and far future relative to 1976–2005, under the RCP4.5 and RCP8.5 scenarios from CMIP5 projections

Indices	Near future		Far future	
	RCP4.5	RCP8.5	RCP4.5	RCP8.5
TXx (°C)	2.2	2.8	2.7	4.8
TNn (°C)	2.3	3.3	3.0	5.1
R95p (%)	22	32	33	62
RX5day (%)	8	12	13	24

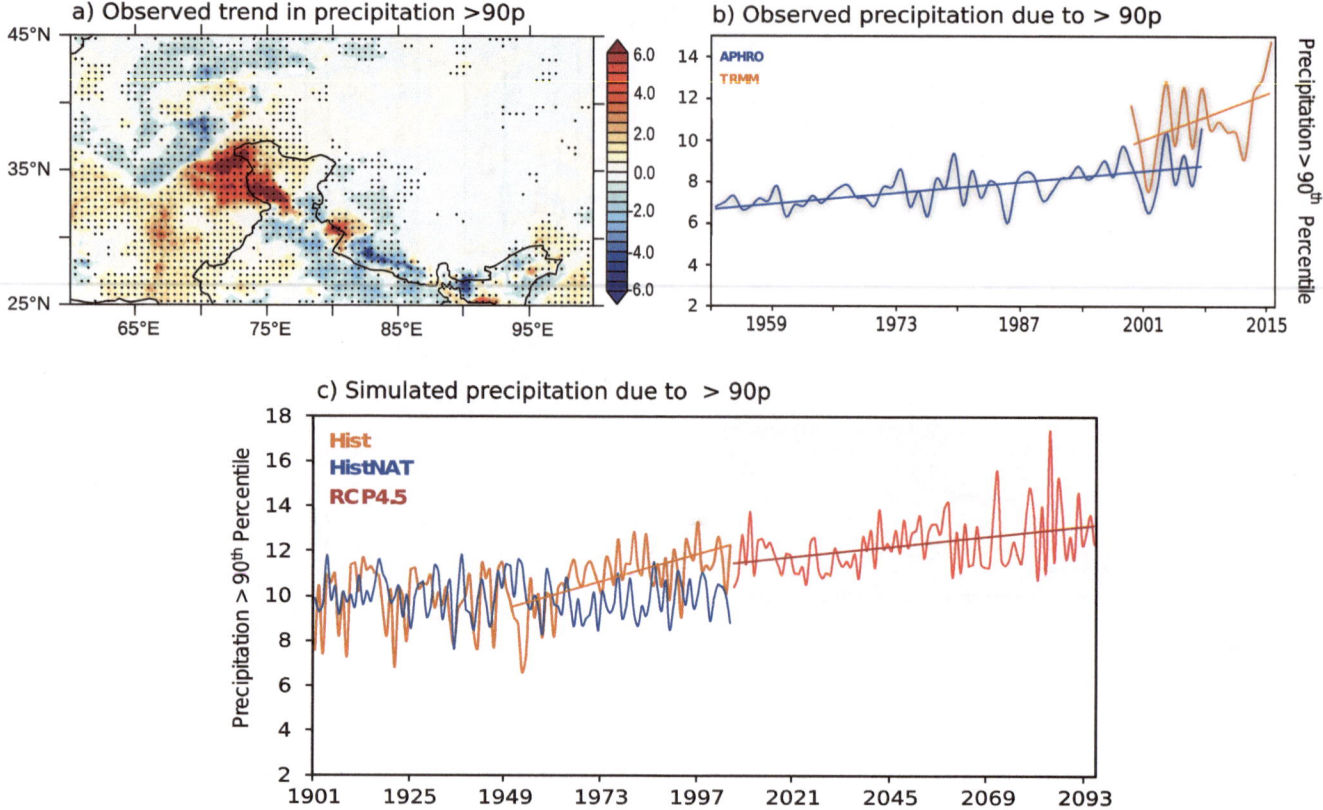

Fig. 11.7 a Spatial map of trend in daily precipitation (APHRODITE) exceeding the 90th percentile. **b** Time series of interannual variations of daily precipitation exceeding 90th percentile over HKH1 from APHRODITE and TRMM datasets. **c** Is same as **b** but from LMDZ simulation. Dotted areas are trends exceeding 95% confidence level based on Mann-Kendall test. Reprinted with permission from Krishnan et al. (2019a)

and Western Himalaya (Forsythe et al. 2017). The Karakoram Himalayas appear to have gained glacial mass slightly at the beginning of twenty-first century, which is contrary to many other mountain peaks across the globe (Gardelle et al. 2012; Hewitt 2005; Krishnan et al. 2019a). [More details regarding WD are provided in Chap. 7.]

More than 30,000 km^2 of the Himalayan region is covered by glaciers which provide about 8.6 million m^3 of water every year (NASA-LCLUC). Climate change over the past several decades has influenced the Himalayan mountain glaciers significantly (Rai and Gurung 2005); however, detailed information is available only for a few glaciers. Glaciers in the Chinese Himalayas are reported to have significantly retreated since the 1950s, with accelerated retreat after 1990s (Rai and Gurung 2005). Almost all glaciers in the region of the Mount Everest have experienced a retreat from the late 1990s. Rai and Gurung (2005) reported that Rongbuk glacier, which is an important source of freshwater into Tibet, was retreating at the rate of 20 m/a. The Gangotri glacier in India, which provides a significant source of meltwater for the Ganges, has retreated about 30 m every year during the recent decade (Rai and Gurung 2005). Kulkarni and Pratibha (2018) reported a loss in glacier extent

by 12.6 ± 7.5% for the past 40 years based on a detailed analysis of 83 glaciers utilising remote sensing and in situ observations. They further reported maximum loss of glacier extent in the Eastern Himalaya, near the Tista and Mt. Everest region, followed by Bhutan and Western Himalayas, with minimum loss over the Karakoram Himalayas (Brahmbhatt et al. 2015).

Glacier inventory estimated from satellite images, topographical maps and aerial photographs provides information about 9040 glaciers covering an area of 18,528 km^2 in the HKH region (Sangewar and Shukla 2009; Sangewar and Kulkarni 2010). An illustration of glacier retreat between 1960 and 2000 from Kulkarni and Karyakarte (2014) is shown in Fig. 11.8. They further showed that the retreat of the Himalayan glaciers ranges from a few metres to almost 61 m $year^{-1}$, depending on the terrain and meteorological parameters (Kulkarni and Karyakarte 2014). Mapping of nearly 20,060 out of ~40,000 km^2 of glaciated area, distributed in all major climatic zones of the Himalayas, suggests a loss in glacier area by about 13% during the last four decades (Kulkarni et al. 2017). Using a high-resolution cryospheric–hydrological model, Lutz et al. (2014) quantified the run-off of hydrological regimes of major Himalayan

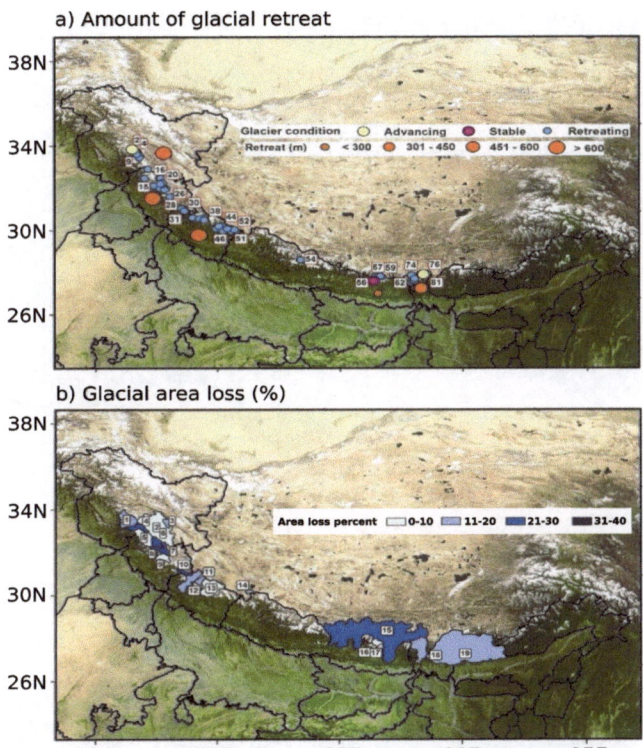

a) Amount of glacial retreat

b) Glacial area loss (%)

Fig. 11.8 a Amount of glacial retreat between 1960 and 2000. **b** Glacial area loss in different regions of the Himalaya from 1960 to 2000. The number represents names of glaciers/basins/regions as given in Tables 1 and 2 of Kulkarni and Karyakarte (2014). From Kulkarni and Karyakarte (2014)

river basins (Indus, Ganges, Brahmaputra, Salween and Mekong rivers) for the near future following the moderate RCP4.5 scenario. Despite differences in run-off composition, they noted an increase in the projected run-off by 2050 caused primarily by an increase in precipitation in the upper Ganges, Brahmaputra, Salween and Mekong Basins and from the accelerated glacier melt in the upper Indus Basin (Fig. 1 in Lutz et al. 2014).

Integrated glaciological studies in the Chandra River basin, located in the monsoon-arid transition zone of the Upper Indus, led by the National Centre for Polar and Ocean Research (NCPOR) have provided valuable information for understanding glacier retreat and its variability in the Western Himalayas. As part of this activity, a detailed glacier mass/energy/water balance of five glaciers, viz. Sutri Dhaka, Batal, Bara Shigri, Samundra Tapu and Gepang, has been studied since 2014. While these glaciers are in similar geographical disposition, they exhibit comparatively different characteristics such as debris cover, aspect and size. The glaciers with high debris cover and varying thickness (e.g. Batal Glacier) revealed low surface melting under debris cover, as compared to clean ice and thin debris-covered ice (e.g. Sutri Dhaka) (Sharma et al. 2016). The thickness (2–100 cm) of debris has attenuated melting rates up to 70% of

total melting, and debris cover of <2 cm thickness has accelerated melting up to 10% of the total melting (Patel et al. 2016). Further, the role of air temperature was evident with higher melting rate ($\sim 80\%$ of total yearly melt) during the short summers (Pratap et al. 2019). Moisture source for precipitation over the study region is dominantly (>70%) derived from the Mediterranean regions by western disturbances (WDs) during winter and early spring, with minor (<20%) contributions from the Indian Summer Monsoon (ISM) during the summer monsoon season (June–September). A three-component hydrograph separation based on oxygen isotope fingerprinting and field-based ablation measurements for one of the glacier basin (Sutri Dhaka) revealed that glacier ice melting is the dominant (65–80%) contributor to the river water, followed by snow melt (20–35%) (Singh et al. 2019). Spatial mass balance gradient varied with specific glacier's location and topography. Results of six years (2014–19) of in situ mass balance observation by NCPOR in this basin show a dominantly negative mass balance (-0.45 ± 0.09 to -1.37 ± 0.27 m water equivalent per year), with the glacier snout retreating at a rate of 13–33 m per year (NCPOR, unpublished data).

It is noteworthy to mention that regions in the Karakoram Himalayas have experienced relatively stable glacier behaviour in recent decades, as opposed to glacier shrinkage observed in many other places (Hewitt 2005; Gardelle et al. 2012; Kapnick et al. 2014; Kääb et al. 2015; Forsythe et al. 2017). Climate model simulations indicate that changes in non-monsoonal wintertime frozen precipitation over the Karakoram Himalayas appear to possibly shield this region from significant glacier thickness losses under warming climate (Kapnick et al. 2014; Kääb et al. 2015; Krishnan et al. 2019a). Robust assessments of future projections of precipitation and snowfall over the Western and Karakoram Himalayas need further research given the inherent complexities of the HKH region and large uncertainties in model projections over this region (Forsythe et al. 2017; Ridley et al. 2013).

11.6 Knowledge Gaps

The accelerated anthropogenic warming over this ice-covered mountain peaks and valleys of the HKH have profound impacts such as loss in glacier mass and snow cover which can directly affect agriculture food production. Enhanced glacier mass loss can cause increased streamflow and flooding of the Himalayan river basins, and further affect downstream agricultural activity. Current generation climate models and downscaling methodologies have limitations in capturing the observed hydroclimatic variations of the Himalayan river basins (Hasson et al. 2014, 2018). Increases in snowmelt can also result from deposition of

Fig. 11.9 HIMANSH: high-altitude observatory of NCPOR, MoES, during a typical Himalayan winter

light-absorbing aerosols (e.g. black carbon) over snow-covered regions, which causes warming by reducing surface albedo (Bond et al. 2013; Lau et al. 2018; Lau and Kim 2010). However, the climatic effects of black carbon on the Himalayan glaciers are not adequately understood, in part due to the large spatio-temporal variability of black carbon in the region (Kopacz et al. 2011, Chap. 5).

Long-term monitoring of weather and climate in the complex and rugged terrain of the Himalayan cryosphere is essential to fill information gaps in the region and to better represent the regional cyrospheric processes in climate models. Towards this end, the National Centre for Polar and Ocean Research (NCPOR) of the Ministry of Earth Sciences (MoES) has set up a high-altitude station named "HIMANSH" in Spiti at an altitude of 13,500 feet (Fig. 11.9).

11.7 Summary

In summary, human-induced climate change has led to accelerated warming of the Himalayas and the Tibetan Plateau at a rate of 0.2 °C per decade during 1951–2014. High-elevation areas (altitude > 4 km), in particular, underwent amplified warming at a rate of about 0.5 °C per decade. Many areas in the HKH, except the high-elevation Karakoram Himalayas, experienced significant decline in wintertime snowfall and glacier retreat in recent decades. Future warming in the HKH region, which is projected to be in the range of 2.6–4.6 °C by the end of the twenty-first century, will further exacerbate the snowfall and glacier decline leading to profound hydrological and agricultural impacts in the region.

References

Archer DR, Fowler H (2004) Spatial and temporal variations in precipitation in the Upper Indus Basin, global teleconnections and hydrological implications. Hydrol Earth Syst Sci 8(1):47–61

Armstrong RL (2011) The glaciers of the Hindu Kush-Himalayan region: a summary of the science regarding glacier melt/retreat in the Himalayan, Hindu Kush, Karakoram, Pamir, and Tien Shan mountain ranges. ICIMOD, Kathmandu

Barlow M, Matthew W, Bradfield L, Heidi C (2005) Modulation of daily precipitation over southwest Asia by the Madden-Julian Oscillation. Mon Weather Rev 133(12):3579–3594

Bhutiyani MR, Vishwas SK, Pawar NJ (2007) Long-term trends in maximum, minimum and mean annual air temperatures across the northwestern Himalaya during the twentieth century. Clim Change 85:59–177

Bhutiyani MR, Kale VS, Pawar NJ (2009) Climate change and the precipitation variations in the northwestern Himalaya: 1866–2006. Int J Climatol 30(4):535–548

Bolch T et al (2012) The state and fate of Himalayan glaciers. Science 336(6079):310–314. https://doi.org/10.1126/science.1215828

Bond TC et al (2013) Bounding the role of black carbon in the climate system: a scientific assessment. J Geophys Res Atmos. https://doi.org/10.1002/jgrd.50171

Bookhagen B, Burbank DW (2006) Topography, relief, and TRMM-derived rainfall variations along the Himalaya. Geophys Res Lett 33(8)

Boos W, Kuang Z (2010) Dominant control of the South Asian monsoon by orographic insulation versus plateau heating. Nature 463:218–222. https://doi.org/10.1038/nature08707

Borgaonkar HP, Ram S, Sikder AB (2009) Assessment of tree-ring analysis of high-elevation *Cedrus deodara* D. Don from western Himalaya (India) in relation to climate and glacier fluctuations. Dendrochronologia 27(1):59–69

Brahmbhatt R, Bahuguna I, Rathore B, Singh S, Rajawat A, Shah R, Kargel J (2015) Satellite monitoring of glaciers in the Karakoram from 1977 to 2013: an overall almost stable population of dynamics glaciers. Cryosphere Discuss 9:1555–1592. https://doi.org/10.5194/tcd-9-1555-2015

Branstator G (2002) Circumglobal teleconnections, the jet stream waveguide, and the North Atlantic Oscillation. J Climate 15:1893–1910. https://doi.org/10.1175/1520-0442(2002)015%3c1893:CTTJSW%3e2.0.CO;2

Cannon F, Carvalho LMV, Jones C, Bookhagen B (2015) Multi-annual variations in winter westerly disturbance activity affecting the Himalaya. Clim Dyn 44:441–455. https://doi.org/10.1007/s00382-014-2248-8

Cannon F, Carvalho LMV, Jones C, Hoell A, Norris J, Kiladis GN et al (2017) The influence of tropical forcing on extreme winter precipitation in the western Himalaya. Clim Dyn 48(3):1213–1232

Chakraborty A, Nanjundiah RS, Srinivasan J (2006) Theoretical aspects of the onset of Indian summer monsoon from perturbed orography simulations in a GCM. Ann Geophys 24:2075–2089

Choi G et al (2009) Changes in means and extreme events of temperature and precipitation in the Asia-Pacific Network region, 1955–2007. Int J Climatol 29(13):1906–1925

Diaz HF, Markgraf V (2000) El Niño and the Southern Oscillation: multiscale variability and global and regional impacts. Cambridge University Press, Cambridge

Dimri A, Dash SK (2012) Wintertime climatic trends in the Western Himalayas. Clim Change 111:775–800

Dimri AP, Niyogi D, Barros AP, Ridley J, Mohanty UC, Yasunari T et al (2015) Western disturbances: a review. Rev Geophys 53(2):225–246

Forsythe N, Fowler HJ, Li X-F, Blenkinsop S, Pritchard D (2017) Karakoram temperature and glacial melt driven by regional atmospheric circulation variability. Nat Clim Change 7(9). https://doi.org/10.1038/nclimate3361

Gardelle J, Berthier E, Arnaud Y (2012) Slight mass gain of Karakoram glaciers in the early twenty-first century. Nat Geosci 5(5):322–325. https://doi.org/10.1038/NGEO1450

Godin L et al (1999) High strain zone in the hanging wall of the Annapurna detachment, central Nepal Himalaya. In: MacFarlane A, Sorkhabi RB, Quade J (eds) Himalaya and Tibet: mountain roots to mountain tops, 328. GSA, p 201

Hasson S, Lucarini V, Pascale S (2013) Hydrological cycle over South and Southeast Asian river basins as simulated by PCMDI/CMIP3 experiments. Earth Syst Dyn 4(2):199–217. https://doi.org/10.5194/esd-4-199-2013

Hasson S, Lucarini V, Khan MR, Petitta M, Bolch T, Gioli G (2014) Early 21st century snow cover state over the western river basins of the Indus river system. Hydrol Earth Syst Sci 18(10):4077–4100

Hasson SU, Böhner J, Chishtie F (2018) Low fidelity of CORDEX and their driving experiments indicates future climatic uncertainty over Himalayan watersheds of Indus basin. Clim Dyn. https://doi.org/10.1007/s00382-018-4160-0

Hewitt K (2005) The Karakoram anomaly? Glacier expansion and the 'elevation effect,' Karakoram Himalaya. Mt Res Dev 25(4):332–340

Hunt KMR, Turner AG, Shaffrey LC (2018) The evolution, seasonality and impacts of western disturbances. Q J R Meteorol Soc. https://doi.org/10.1002/qj.3200

ICIMOD (2007) The melting Himalayas: regional challenges and local impacts of climates change on mountain ecosystems and livelihoods. Technical paper X. International Centre for Integrated Mountain Development, Kathmandu, 33 pp

ICIMOD (2011) The status of glaciers in the Hindu Kush-Himalayan region. International Centre for Integrated Mountain Development, Kathmandu

Immerzeel WW, Droogers P, Jong SMD, Bierkens MFP (2009) Large-scale monitoring of snow cover and runoff simulation in Himalayan river basins using remote sensing. Remote Sens Environ 113(1):40–49

Immerzeel WW, Beek LPH, Bierkens MFP (2010) Climate change will affect the Asian water towers. Science 328. https://doi.org/10.1126/science.1183188

IPCC (2007) In: Solomon S, Qin D, Manning M, Chen Z, Marquis M, Averyt KB et al (eds) Climate change 2007: the physical science basis. Contribution of working group I to the fourth assessment report of the Intergovernmental Panel on Climate Change. Cambridge University Press, Cambridge, New York, p 996

IPCC (2013) Climate change. The physical science basis. Work group contribution to the IPCC fifth assessment report (AR5). Intergovernmental Panel on Climate Change, Stockholm, Sweden

IPCC SR (2019) IPCC SR ocean and cryosphere in a changing climate, Chap 2. In: Hock R et al (eds) High mountain areas

Kääb A, Treichler D, Nuth C, Berthier E (2015) Brief communication: contending estimates of 2003–2008 glacier mass balance over the Pamir–Karakoram–Himalaya. Cryosphere 9(2):557–564. https://doi.org/10.5194/tc-9-557-2015

Kapnick SB, Delworth TL, Ashfaq M, Malyshev S, Milly PCD (2014) Snowfall less sensitive to warming in Karakoram than in Himalayas due to a unique seasonal cycle. Nat Geosci 7:834–840

Kious WJ, Tilling RI (1996) This dynamic earth: the story of plate tectonics. USGS report. https://doi.org/10.3133/7000097

Klein Tank AMG, Peterson TC, Quadir DA et al (2006) Changes in daily temperature and precipitation extremes in central and South Asia. J Geophys Res 11116105. https://doi.org/10.1029/2005JD006316

Kopacz M, Mauzerall DL, Wang J, Leibensperger EM, Henze DK, Singh K (2011) Origin and radiative forcing of black carbon transported to the Himalayas and Tibetan Plateau. Atmos Chem Phys 11:2837–2852. https://doi.org/10.5194/acp-11-2837-2011

Krishnamurti TN, Thomas A, Simon A, Kumar V (2010) Desert air incursions, an overlooked aspect, for the dry spells of the Indian summer monsoon. J Atmos Sci 67:3423–3441

Krishnan R, Kumar V, Sugi M, Yoshimura J (2009) Internal feedbacks from monsoon-midlatitude interactions during droughts in the Indian summer monsoon. J Atmos Sci 66:553–578

Krishnan R, Sabin TP, Ayantika DC, Kitoh A, Sugi M, Murakami H et al (2013) Will the South Asian monsoon overturning circulation stabilize any further? Clim Dyn 40(1–2):187–211. https://doi.org/10.1007/s00382-012-1317-0

Krishnan R, Sabin TP, Vellore R, Mujumdar M, Sanjay J, Goswami BN et al (2016) Deciphering the desiccation trend of the South Asian monsoon hydroclimate in a warming world. Clim Dyn 47(3–4):1007–1027. https://doi.org/10.1007/s00382-015-2886-5

Krishnan R, Sabin TP, Madhura RK, Vellore R, Mujumdar M, Sanjay J, Nayak S, Rajeevan M (2019a) Non-monsoonal precipitation response over the Western Himalayas to climate change. Clim Dyn. https://doi.org/10.1007/s00382-018-4357-2

Krishnan R et al (2019b) Unravelling climate change in the Hindu Kush Himalaya: rapid warming in the mountains and increasing extremes. In: Wester P, Mishra A, Mukherji A, Shrestha A (eds) The Hindu Kush Himalaya assessment. Springer, Cham

Kulkarni AV, Karyakarte Y (2014) Observed changes in Himalayan glaciers. Curr Sci 106(2):237–244

Kulkarni AV, Pratibha S (2018) Assessment of glacier fluctuations in the Himalaya. In: Goel P, Ravindra R, Chattopadhyay S (eds) Science and geopolitics of the White World. Springer, Cham

Kulkarni A, Patwardhan S, Kumar K, Ashok K, Krishnan R (2013) Projected climate change in the Hindu Kush-Himalayan region by using the high-resolution regional climate model PRECIS. Mt Res Dev 33(2):142–151

Kulkarni AV, Nayak S, Pratibha S (2017) Variability of glaciers and snow cover. In: Rajeevan M, Nayak S (eds) Observed climate variability and change over the Indian region. Springer geology. Springer, Singapore

Lang TJ, Barros AP (2004) Winter storms in the Central Himalayas. J Meteorol Soc Japan 82:829–844

Lau WKM, Kim KM (2010) Fingerprinting the impacts of aerosols on long-term trends of the Indian summer monsoon regional rainfall. Geophys Res Lett 37(16):L16705. https://doi.org/10.1029/2010GL043255

Lau WKM, Sang J, Kim MK, Kim KM, Koster RD, Yasunari TJ (2018) Impacts of aerosol snow darkening effects on hydro-climate over Eurasia during boreal spring and summer. J Geophys Res Atmos 123:8441–8461. https://doi.org/10.1029/2018JD028557

Liu X, Yin ZY, Shao X, Qin N (2006) Temporal trends and variability of daily maximum and minimum, extreme temperature events, and growing season length over the eastern and central Tibetan Plateau during 1961–2003. J Geophys Res Atmos 111(D19):4617–4632

Liu X, Cheng Z, Yan L et al (2009) Elevation dependency of recent and future minimum surface air temperature trends in the Tibetan Plateau and its surroundings. Glob Planet Change 68:164–174

Lutz AF, Immerzeel WW, Shrestha AB, Bierkens MFP (2014) Consistent increase in High Asia's runo due to increasing glacier melt and precipitation. Nat Clim Change. https://doi.org/10.1038/NCLIMATE2237

Madden R, Julian P (1971) Detection of a 40–50-day oscillation in the zonal wind in the tropical Pacific. J Atmos Sci 28:702–708

Madhura RK, Krishnan R, Revadekar JV, Mujumdar M, Goswami BN (2015) Changes in western disturbances over the Western Himalayas in a warming environment. Clim Dyn 44:1157–1168. https://doi.org/10.1007/s00382-014-2166-9

Maskey S, Uhlenbrook S, Ojha S (2011) An analysis of snow cover changes in the Himalayan region using MODIS snow products and in-situ temperature data. Clim Change 108:391–400

Nan S, Zhao P, Yang S, Chen J (2009) Springtime tropospheric temperature over the Tibetan Plateau and evolution of the tropical Pacific SST. J Geophys Res Atmos 114(10). https://doi.org/10.1029/2008jd011559

Negi HS, Neha K, Shekhar MS, Ganju A (2018) Recent wintertime climatic variability over the North West Himalayan cryosphere. Curr Sci 114(4):25

Panagiotopoulos F, Shahgedanova M, Stephenson DB (2002) A review of Northern Hemisphere winter time teleconnection patterns. J Phys IV France 12(10):27. https://doi.org/10.1051/jp4:20020450

Pant GB, Pradeep Kumar P, Revadekar JV, Singh N (2018) Climate change in the Himalayas. Springer. https://doi.org/10.1007/978-3-319-61654-4

Patel LK, Sharma P, Thamban M, Singh A, Ravindra R (2016) Debris control on glacier thinning—a case study of the Batal glacier, Chandra basin, Western Himalaya. Arab J Geosci 9(4):309. https://doi.org/10.1007/s12517-016-2362-5

Pepin N et al (2015) Elevation-dependent warming in mountain regions of the world. Nat Clim Change 5:424–430

Pratap B, Sharma P, Patel L, Singh AT, Gaddam VK, Oulkar S, Thamban M (2019) Reconciling high glacier surface melting in summer with air temperature in the semi-arid zone of Western Himalaya. Water 11:1561. https://doi.org/10.3390/w11081561

Rai SC, Gurung CP (2005) An overview of glaciers, glacier retreat and subsequent impacts in Nepal, India and China. WWF Program. http://assets.panda.org/downloads/himalayaglaciersreport2005.pdf

Rajbhandari R, Shrestha AB, Kulkarni A et al (2015) Projected changes in climate over the Indus river basin using a high resolution regional climate model (PRECIS). Clim Dyn 44:339. https://doi.org/10.1007/s00382-014-2183-8

Rajbhandari R, Shrestha AB, Nepal S, Wahid S (2016) Projection of future climate over the Koshi River basin based on CMIP5 GCMs. Atmos Clim Sci 6:190–204. https://doi.org/10.4236/acs.2016.62017

Ramanathan V, Carmichael G (2008) Global and regional climate changes due to black carbon. Nat Geosci 1(4):221

Rashul G (2014) Food, water, and energy security in South Asia: a nexus perspective from the Hindu Kush Himalayan region. Environ Sci Policy 39:35. https://doi.org/10.1016/j.envsci.2014.01.010

Rees HG, Collins DN (2006) Regional differences in response of flow in glacier-fed Himalayan rivers to climatic warming. Hydrol Process 20(10):2157–2169

Ren YY, Parker D, Ren GY, Dunn R (2015) Tempo-spatial characteristics of sub-daily temperature trends in mainland China. Clim Dyn 46(9–10):2737–2748. https://doi.org/10.1007/s00382-015-2726-7

Ren YY, Ren GY, Sun XB, Shrestha AB, You QL, Zhan YJ et al (2017) Observed changes in surface air temperature and precipitation in the Hindu Kush Himalayan region during 1901–2014. Adv Clim Change Res 8(3). https://doi.org/10.1016/j.accre.2017.08.001

Ridley J, Wiltshire A, Mathison C (2013) More frequent occurrence of westerly disturbances in Karakoram up to 2100. Sci Total Environ 468–469:S31–S35

Sabin TP et al (2013) High resolution simulation of the South Asian monsoon using a variable resolution global climate model. Clim Dyn 41:173–194

SAC (2016) Monitoring snow and glaciers of Himalayan region. Space Applications Centre, ISRO, Ahmedabad, p 413. ISBN 978-93-82760-24-5

Saji NH, Goswami BN, Vinayachandran PN et al (1999) A dipole mode in the tropical Indian Ocean. Nature 401(23):360–363

Sangewar CV, Kulkarni A (2010) Observational studies of the recent past, report of the study group on Himalayan glaciers. Prepared for Principal Scientific Advisor Government of India

Sangewar CV, Shukla SP (2009) Inventory of Himalayan glaciers (an updated edition). Special publication no. 34. Geological Survey of India

Sanjay J, Krishnan R, Shrestha AB, Rajbhandari R, Ren GY (2017) Downscaled climate change projections for the Hindu Kush Himalayan region using CORDEX South Asia regional climate models. Adv Clim Change Res 8(3):185–198. https://doi.org/10.1016/j.accre.2017.08.003

Sharma E et al (2019) Introduction to the Hindu Kush Himalaya assessment. In: Wester P, Mishra A, Mukherji A, Shrestha AB (eds) The Hindu Kush Himalaya assessment—mountains, climate change, sustainability and people. Springer-Nature, Switzerland

Sharma P, Patel LK, Ravindra R, Singh A, Mahalinganathan K, Thamban M (2016) Role of debris cover to control specific ablation of adjoining Batal and Sutri Dhaka glaciers in Chandra Basin (Himachal Pradesh) during peak ablation season. J Earth Syst Sci 125:459. https://doi.org/10.1007/s12040-016-0681-2

Sheridan SC, Lee CC, Allen MJ, Kalkstein LS (2012) Future heat vulnerability in California, part I: projecting future weather types and heat events. Clim Change 115:291–309

Shrestha AB, Aryal R (2011) Climate change in Nepal and its impact on Himalayan glaciers. Reg Environ Change 11:S65–S77

Shrestha AB, Wake CP, Mayewski PA, Dibb JE (1999) Maximum temperature trends in the Himalaya and its vicinity: an analysis based on temperature records from Nepal for the period 1971–94. J Clim 12:2775–2786

Shrestha AB, Devkota LP, John D et al (2010) Climate change in the Eastern Himalayas: observed trends and model projections. Climate change impact and vulnerability in the EH-technical report 1

Shrestha AB, Agrawal NK, Alfthan B, Bajracharya SR, Maréchal J, Van Oort B (2015) The Himalayan climate and water atlas: impact of climate change on water resources in five of Asia's major river basins. GRID-Arendal and CICERO, ICIMOD

Singh S, Ghosh S, Sahana AS, Vittal H, Karmakar S (2017) Do dynamic regional models add value to the global model projections of Indian monsoon? Clim Dyn 48(3–4):1375–1397. https://doi.org/10.1007/s00382-016-3147-y

Singh AT, Rahaman W, Sharma P, Laluraj CM, Patel LK, Pratap B, Gaddam VK, Thamban M (2019) Moisture sources for precipitation and hydrograph components of the Sutri Dhaka Glacier Basin, Western Himalayas. Water 11:2242. https://doi.org/10.3390/w11112242

Srivastava P, Misra DK (2012) Optically stimulated luminescence chronology of terrace sediments of Siang River, Higher NE Himalaya: comparison of quartz and feldspar chronometers. J Geol Soc India 79:252–258

Sun XB, Ren GY, Shrestha AB, Ren YY, You QL, Zhan YJ et al (2017) Changes in extreme temperature events over the Hindu Kush Himalaya during 1961–2015. Adv Clim Change Res 8(3):157–165. https://doi.org/10.1016/j.accre.2017.07.001

Thompson DWJ, Wallace JM (1998) The Arctic Oscillation signature in the wintertime geopotential height and temperature fields. Geophys Res Lett 25:1297–1300

Thompson LG, Mosley-Thompson E, Davis ME, Lin PN, Henderson K, Mashiotta TA (2003) Tropical glacier and ice core evidence of climate change on annual to millennial time scales. Clim Change 59 (1–2):137–155

Turner A, Annamalai H (2012) Climate change and the South Asian summer monsoon. Nat Clim Change 2:587–595. https://doi.org/10.1038/nclimate1495

Van Vuuren DP, Edmonds J, Kainuma M, Riahi K, Thomson A, Hibbard K et al (2011) The representative concentration pathways: an overview. Clim Change 109(1):5–31

Vellore R, Krishnan R, Pendharkar J, Choudhury AD, Sabin TP (2014) On anomalous precipitation enhancement over the Himalayan foothills during monsoon breaks. Clim Dyn 43(7):2009–2031. https://doi.org/10.1007/s00382-013-2024-1

Wallace JM (2000) North Atlantic oscillatiodannular mode: two paradigms—one phenomenon. Q J R Meteorol Soc 126:791–805. https://doi.org/10.1002/qj.49712656402

Wester P, Mishra A, Mukherji A, Shrestha AB (eds) (2019) The Hindu Kush Himalaya assessment—mountains, climate change, sustainability and people. Springer-Nature, Switzerland

Wu J, Xu Y, Gao X-J (2017) Projected changes in mean and extreme climates over Hindu Kush Himalayan region by 21 CMIP5 models. Adv Clim Change Res 8(3):176–184. https://doi.org/10.1016/j.accre.2017.03.001

Xu J, Shrestha AB, Vaidya R, Eriksson M, Nepal S, Sandstrom K (2008) The changing Himalayas. Impact of climate change on water resources and livelihoods in the greater Himalayas. ICIMOD, Kathmandu

Xu WH, Li QX, Yang S, Yan X (2014) Overview of global monthly surface temperature data in the past century and preliminary integration. Adv Clim Change Res 5(3):111–117. https://doi.org/10.1016/j.accre.2014.11.003

Yan LB, Liu XD (2014) Has climatic warming over the Tibetan Plateau paused or continued in recent years? J Earth Ocean Atmos Sci 1:13–28

Yanai MH, Li C (1994) Mechanism of heating and the boundary layer over the Tibetan Plateau. Mon Weather Rev 122:305–323

You QL, Min J, Zhang W, Pepin N, Kang S (2015) Comparison of multiple datasets with gridded precipitation observations over the Tibetan Plateau. Clim Dyn 45:791–806

You QL, Ren GY, Zhang YQ, Ren YY, Sun XB, Zhan YJ et al (2017) An overview of studies of observed climate change in the Hindu Kush Himalayan (HKH) region. Adv Clim Change Res 8(3):141–147. https://doi.org/10.1016/j.accre.2017.04.001

Zhan YJ, Ren GY, Shrestha AB, Rajbhandari R, Ren YY, Sanjay J (2017) Change in extreme precipitation events over the Hindu Kush Himalayan region during 1961–2012. Adv Clim Change Res 8(3). https://doi.org/10.1016/j.accre.2017.08.002

Zhou X, Zhao P, Chen J, Chen L, Li W (2009) Impacts of thermodynamic processes over the Tibetan Plateau on the Northern Hemispheric climate. Sci China Ser D Earth Sci 52(11):1679–1693

Possible Climate Change Impacts and Policy-Relevant Messages

Coordinating Lead Authors

Chirag Dhara, Indian Institute of Tropical Meteorology (IITM-MoES), Pune, India,
e-mail: chirag.dhara@tropmet.res.in (corresponding author)
R. Krishnan, Indian Institute of Tropical Meteorology (IITM-MoES), Pune, India

Review Editor

Dev Niyogi, Purdue University, West Lafayette, IN and University of Texas at Austin, Austin, TX, USA

Corresponding Author

Chirag Dhara, Indian Institute of Tropical Meteorology (IITM-MoES), Pune, India,
e-mail: chirag.dhara@tropmet.res.in

© The Author(s) 2020
R. Krishnan et al. (eds.), *Assessment of Climate Change over the Indian Region*,
https://doi.org/10.1007/978-981-15-4327-2_12

The Paris Agreement 2015, framed during the twenty-first Conference of Parties (COP21), called on all nations to collectively limit global warming to 2°C below pre-industrial levels and to pursue efforts to limit the increase to 1.5°C. In this context, the IPCC published a Special Report in October 2018 on the relative impacts of global warming levels of 1.5 and 2°C. According to the report, the impacts of a 2°C rise are not only severe but also considerably larger than for 1.5°C on a global scale.

Despite the dire projections from the IPCC Special Report, whether the 1.5°C target is practically achievable is contested. The pace at which human activities have already changed the climate since the pre-industrial era is unprecedented in the history of modern civilization, and the world is presently on course to far exceed the Paris targets. At current greenhouse gas emissions trajectories, global average temperature may rise 3–5°C, perhaps higher if tipping points[1] are triggered, above pre-industrial temperatures by the end of the twenty-first century. However, the extent and rate of climate change as well as its impacts will vary across the planet.

Chapters 1–11 have assessed the observed and future changes in climate over the Indian region. However, the impacts of these changes were not discussed since they lie outside the scope of this report.

This concluding chapter is an essay briefly discussing the possible impacts of climate change on India and the policy relevance of this report. Focused impact assessment studies based on the findings in the previous chapters can be the foundation for adaptation and mitigation strategies.

12.1 Potential Impacts Over India

The rapid changes in India's climate projected by climate models will place increasing stress on the country's natural ecosystems, agricultural output, and fresh water resources, while also causing escalating damage to infrastructure. These portend serious consequences for the country's biodiversity, food, water and energy security, and public health. In the absence of rapid, informed and far-reaching mitigation and adaptation measures, the impacts of climate change are likely to pose profound challenges to sustaining the country's rapid economic growth, and achieving the sustainable development goals (SDGs) adopted by UN Member States in 2015.

The impact of climate change on the availability of fresh water is a critical area of concern for India. Continued global warming through the twenty-first century, together with the high probability of future reductions in anthropogenic

aerosol emissions from North America, Europe, and Asia, will likely intensify summer monsoon precipitation and its variability over India. The growing propensity for droughts and floods because of changing rainfall patterns caused by climate change would be detrimental to surface and groundwater recharge, posing threats to the country's water security.

Likewise, the country's food security may be placed under progressively greater pressure due to rising temperatures, heat extremes, floods, droughts and increasing year-to-year rainfall variability that can disrupt rain-fed agricultural food production and adversely impact crop yield.

Studies indicate that climate change may seriously compromise human health in the absence of risk mitigation, adaptation, or acclimatization, particularly among children and the elderly. Higher temperatures, extreme weather events, and higher climate variability have been associated with an elevated risk of heat strokes, cardiovascular and neurological diseases, and stress-related disorders. Heat stress in urban areas is often compounded by the heat island effect. Warmer, higher moisture conditions, on average, are also more favourable for the spread of vector-borne diseases such as malaria and dengue fever. In addition, a decrease in the availability or affordability of food and potable water caused by climate change may lead to reduced nutritional intake, particularly among economically weaker sections.

Rising temperatures are also likely to increase energy demand for space cooling, which if met by thermal power would constitute a positive feedback to global warming by increasing GHG emissions. In addition, thermal power plants require substantial amounts of water for cooling to generate electricity. Power plants sited inland draw freshwater largely from dam reservoirs, rivers and canals. A rise in water withdrawal by power plants would directly compete with water withdrawal for agriculture and domestic consumption, particularly in water stressed areas. On the other hand, power plants sited around the coast that use sea water for cooling are vulnerable to damage from sea-level rise, cyclones, and storm surge. In short, climate change could impact the reliability of the country's energy infrastructure and supply.

India's long coastline, where some of its largest cities are located, is among the most densely populated regions of the planet, making it exceedingly vulnerable to the impacts of sea-level rise. Potential coastal risks include loss of land due to increased erosion, damage to coastal projects and infrastructure such as buildings, roads, monuments, and power plants, salinization of freshwater supplies and a heightened vulnerability to flooding. Higher sea levels and receding coastlines escalate the destructive potential of storm surge associated with cyclonic storms. These impacts of sea-level rise may be additionally compounded by land subsidence

---

[1]Tipping points refer to critical thresholds in a system that, when exceeded, can lead to an abrupt change in the state of the system, often with an understanding that the change is irreversible.

occurring in parts of the country due to factors such as the declining water table depth.

Several regions in India are global biodiversity hotspots with numerous endemic species of plants and animals. With the climate changing more rapidly than the usual pace of evolutionary adaptability of many species, they may face increasing threats on account of these changes. Species specially adapted to narrow environmental conditions are likely to be affected the most.

The risks posed by climate change can be considerably magnified when a cascade of climate-related hazards overlap or follow one another. For instance, a region may experience an abnormally long or intense summer heat wave followed by intense monsoon floods that alternate with lengthening dry spells. Low-lying coastal zones, especially on India's east coast, may witness rising sea levels damaging property and increasing groundwater salinity. A rise in cyclone intensities will likely result in increasing inundation from the accompanying storm surges that turn proximate agricultural lands and lakes saline, and imperil wildlife. Such sequences of events will become increasingly frequent if anthropogenic climate change continues unimpeded.

The aggregate result of these impacts according to the recent IPCC Special Report on the difference in impacts between 1.5°C versus 2°C warming is that tropical countries such as India are projected to experience the largest impacts on economic growth because of climate change.

12.2 Policy-Relevant Messages

Due to the multiplicity of threats posed by climate change, it is crucial to make vulnerability assessment central to long-term planning for developing adaptation and mitigation strategies.

Climate change impacts will differ widely across the country depending on local context such as geography (e.g. coastal, inland or mountains) and climate (e.g. arid or wet) among others. Inclusion of detailed, regional-scale climate change risk assessments would help develop region and sector-specific mitigation and adaptation measures to reduce vulnerability to climate change.

India's climate change adaptation and mitigation response is achievable by a greater emphasis on widening observational networks, sustained monitoring, expanding research on regional changes in climate and their impacts, development of integrated, multiscale models that are capable of aiding predictions, scenario synthesis and providing information required for vulnerability assessment, and by continued investment in education and outreach programs including national and international collaborative partnerships. For instance, networks of tide gauges with GPS along the Indian coastline would help monitor local changes in sea level, climate models would help project future changes, and outreach would help inform the requisite adaptation measures in coastal communities and cities. Outreach and communication of climate change risk to district and village-level communities would facilitate water-harvesting and farming decisions needed to adapt to a changing climate. Additionally, dedicated educational programmes from school to university would vastly improve awareness about climate change and its implications for humans and natural ecosystems. Such programs could encourage young minds to contribute through individual and collective efforts that are crucial for climate action.

It is necessary to develop useful to usable (U2U) research and application agenda that can translate research to on-ground, effective decision tools for adapting to climatic change. Cities being uniquely impacted by heat stress and localized flooding, there is a pressing need for research and strategies that are directed towards improving resilience in Indian cities. Specific additional examples such as passive reduction of indoor temperatures, water conservation and rainwater harvesting, groundwater regulation, reversing land degradation, reduction in food and water wastage, waste segregation and recycling, low impact urban development, expansion of urban green spaces and urban farming, pollution control, increasing the area under irrigation and improving the efficiency of agricultural water use, forest conservation and proactive afforestation, construction of coastal embankments and mangrove restoration, improvement in disaster response, phasing out fossil fuels and transition to renewables, electrification, expansion of walking, bicycling and public transport infrastructure, and carbon taxation, among others, have been successfully implemented in parts of the country and the world to reduce risk from climate change.

Equity and social justice are critical to building climate resilience since the most vulnerable people such as the poor, the disabled, outdoor labourers and farmers will bear the brunt of climate change impacts.

12.2.1 Multiple Benefits of Climate Change Action

Just as climate change impacts cascade into complex multi-hazards, several of these policy measures are likely to deliver multiple benefits. For instance, a strategic transition to renewables would reduce both GHG emissions as well as water consumption required to cool thermal power plants. Low impact development and green building infrastructure can reduce both urban heating and air pollution. A reduction in air pollution would greatly benefit human and environmental health, improve the efficiency of solar energy generation, and even potentially aid in increasing the quantum of monsoon rainfall over India (see Chaps. 1 and 3). An increase in rainfall together with measures for water harvesting would aid the restoration of groundwater levels. Restoration of groundwater levels would not only improve water security and resilience to droughts but also help check land subsidence, consequently reducing the impacts of sea-level rise and storm surge.

Ambitious afforestation efforts likewise offer myriad benefits. Aside from mitigating climate change through carbon sequestration, trees also enhance resilience to flash floods and landslides by improving soil retention, improve resilience to droughts by increasing percolation of surface water into the soil, improve resilience of coastal infrastructure and habitation by reducing coastline erosion due to storm surges and sea-level rise, reduce vulnerability to extreme heat by reducing ambient temperatures, and support native wildlife and biodiversity. In short, forests and urban green spaces will deliver substantial economic benefits to the country by mitigating a wide range of the expected impacts of climate change in India and is the safest, most reliable means of realizing several of India's sustainable development goals.

12.3 Summary

In conclusion, changes in climate need to be considered as a part of a complex system. The merits and trade-offs of different policy measures need to be assessed so as to extend well beyond climate change mitigation and adaptation. This warrants a continued and sustained push for research, cooperative inter-disciplinary work and a close coordination between researchers and policy-makers to convert knowledge into action.

This report has, using observations as well as model studies, assessed the extent and rate of climate change in India and how it may evolve in future. The impacts of these changes on infrastructure, environmental and public health can hamper efforts towards attaining India's developmental goals and impede the country's economic growth. Addressing these challenges will require developing periodically updated regional-scale assessments of climate change and its impacts, evaluating the efficacy of existing policies, learning from sustainable practices across India and around the world, and responding dynamically in line with scientific advancements.

Correction to: Assessment of Climate Change over the Indian Region

R. Krishnan, J. Sanjay, Chellappan Gnanaseelan, Milind Mujumdar, Ashwini Kulkarni, and Supriyo Chakraborty

---

Correction to:
R. Krishnan et al. (eds.), *Assessment of Climate Change over the Indian Region,*
https://doi.org/10.1007/978-981-15-4327-2

In the original version of the book, belated corrections as listed below are incorporated:

Chapter 1:

The caption of figure 1.5 has been changed as follows:

Spatial patterns of change in the June–to–September seasonal precipitation (mm day $^{-1}$) over the globe in the left-hand column, and over India in the right-hand column. In the top row are plotted the observed changes for the period (1951–2014) relative to (1900–1930) over the globe based on the CRU dataset, and over India based on the IMD dataset. Plots in the middle row are from the IITM-ESM simulations for the historical period, and the plots in the last row are from the IITM-ESM projections following the SSP5-8.5 scenario. The IITM-ESM simulated changes in the historical period (first and middle rows) are plotted as difference for the period (1951–2014) relative to (1850–1900). Changes under the SSP5-8.5 scenario (last row) are plotted as difference between the far-future (2070–2099) relative to (1850–1900).

Chapter 2:

On page 40, line 5 the word "business-as-usual" has been changed to "twenty-first century under this high emission scenario".

Chapter 4:

On page 88, the last sentence "The business as usual scenario will continue to increase atmospheric CO_2 and CH_4 loading for next several decades." has been changed to "Without rapid mitigation policies, atmospheric CO_2 and CH_4 loading will continue to increase for the next several decades."

The erratum chapters and book have been updated with the changes.

The updated version of these chapters can be found at
https://doi.org/10.1007/978-981-15-4327-2_1,
https://doi.org/10.1007/978-981-15-4327-2_2,
https://doi.org/10.1007/978-981-15-4327-2_4